T0155863

The Handbook of
Polyhydroxyalkanoates

The Handbook of Polyhydroxyalkanoates

Kinetics, Bioengineering, and Industrial Aspects

Edited by
Martin Koller

CRC Press

Taylor & Francis Group

Boca Raton London New York

CRC Press is an imprint of the
Taylor & Francis Group, an **informa** business

First edition published 2020
by CRC Press
6000 Broken Sound Parkway NW, Suite 300, Boca Raton, FL 33487-2742

and by CRC Press
4 Park Square, Milton Park, Abingdon, Oxon OX14 4RN

First issued in paperback 2023

Publisher's Note
The publisher has gone to great lengths to ensure the quality of this reprint but points out that some imperfections in the original copies may be apparent.

ISBN 13: 978-0-367-27562-4 (hbk)
ISBN 13: 978-0-367-54114-9 (pbk)
ISBN 13: 978-0-429-29663-5 (ebk)

DOI: 10.1201/9780429296635

Typeset in Times
by Deanta Global Publishing Services, Chennai, India

Dedicated to the fond memory of our father Josef Koller (1949–2019), who passed away during the creation of this book.

Contents

PART IV   Mixed Microbial Culture Approaches

PART V   Industrial Aspects

Foreword

Public concern about environmental issues has recently brought plastics under quite heavy pressure. We have to admit that to a certain extent it is deserved, especially bearing in mind the creepy plastic islands of menacing size in the Atlantic Ocean. On the other hand, certainly to claim that "plastic waste is the most dangerous issue jeopardizing human existence on Earth in the near future", as some environmental activists declare, seems to be not only inappropriate, but distracts attention from other, probably even more serious, environmental threats. Anyway, the public attitude towards plastics has become increasingly negative.

The attitude of the public towards biodegradable plastics (BDPs) is rather tortuous; on the one hand, they are plastics and therefore environmentally conspicuous as a considerable number of people think; on the other hand, just a few years ago BDPs have been repeatedly suggested by the media as viable environmental alternatives to "harmful", even "toxic" petroplastics and this belief is held by a substantial portion of the population. However, perhaps more important is the changing concern of some leading scientists active in BDP research, preferring the biobased origin of a particular plastic over biodegradability considering the environmental impact from the point of view of sustainability. Biodegradability is certainly a positive property, but the decisive factor is whether the plastic is made from renewable resources. All these topics are nowadays discussed and considered and we are at a time, first, in selecting what are the most important factors regarding the environment and, second, if an acceptable decision is made, to persuade the decisionmakers and the majority of the public to take corresponding steps towards supporting environmentally optimal solutions.

My good friend Prof. Ivan Tomka, retired member of ETH Zurich and an acknowledged expert on all aspects of thermoplastic starch, presented the following idea in one of his lectures: scientists should show what is possible, industry would select what is feasible (especially considering available technology and economic aspects of the production), while consumers are in charge to decide what is needed. Consumer opinion is influenced more and more by sometimes really aggressive and not always reliable advertisements. The selection of industry is to a certain extent affected by the primary goal of entrepreneurs, aiming at maximizing profit. Therefore, in many cases the selection offered by research is considered primarily from an economical point of view instead of choosing the most environmentally friendly solution.

Even more dangerous from the application point of view is extensive concern about innovations. A couple of years ago I was involved in a project financed by a Slovak agency aimed at analyzing and predicting the environmental impacts of megatrends. My task was to analyze and comment on the megatrend "Accelerating Rate of Technical Development". I presented a number of examples where innovations were applied with insufficient checking of possible negative effects of a particular product or technology, especially regarding the long-term impact on human health, global environment, or other key factors for maintaining life on Earth. The main reason for abandoning a more thorough investigation of possible negative influences was, in the majority of examples, the pressure on a company to start production as soon as possible.

A few years ago we prepared a project on new technologies for BDP applications, and one of the proposed targets was development of plastics which would biodegrade in marine conditions. The research on this topic is ongoing in several acknowledged research teams all over the world. However, two years ago I talked to Ramani Narayan (Michigan State University), who is one of the most experienced and acknowledged scientists investigating various topics of BDPs, and he commented: "oceans are certainly not the place for plastic waste disposal". In my view, he is right. 30 years ago, when the rate of plastics production was accelerating, nobody expected the appearance of menacing plastic islands in oceans. The same (although of much smaller impact) occurred with finding plastic microparticles in drinking water only a few years ago. Such events will certainly make the public hostile to plastics, which in the end may negatively affect the whole plastic production industry. So, if plastics that are biodegradable in sea water will really be developed, before starting high-volume production, all aspects of degradation should be thoroughly checked to avoid the possibility that we will get rid of plastic islands in the oceans, but the water will be poisoned with something unexpected.

One example of a development connected directly with BDPs seems to be the application of oxobiodegradable poly(ethylene) for packaging. This event also made quite a controversial image for plastics and I remember a number of my friends blaming me for working on these plastic bags, which were used for a short time, albeit without comprehensive testing, in one of the global supermarkets operating in Slovakia. The bosses of the supermarket were very keen to show their environmental concern but in the end the result was catastrophic. I also had a hard time to explain that these are not real biodegradable plastics and that I am investigating quite different materials with excellent properties. I am still afraid that some of my friends even now still do not believe that I was not involved in, or perhaps even directly responsible, for this failure.

Considering dozens of similar cases, from this point of view, the responsibility of scientists to reveal "what is possible" should be understood not only as "what positive is possible", but the question should be extended also to "what negative consequences may be expected". The most important is to be aware of the responsibility aimed not only at the effort to reveal, develop, and design something scientifically important or industrially applicable, but also the investigation of possible threats, especially those resulting from long-term applications. So that the nice epigram saying "science is the most efficient tool to deal with the problems caused by scientific research performed in the past" would serve just for the amusement of the audience and not as the description of alarming reality in the near future. Therefore, the chapter on life cycle assessment (LCA) written by Prof. Khurram Shahzad and colleagues in this volume of the Handbook of Polyhydroxyalkanoates edited by Dr. Martin Koller is directly connected to the previous paragraph. The correct, more frequent, and more detailed investigation, but also announcing the result to the general public and, even more important, to decisionmakers may be a good step to reinstate trust in plastics, but also to teach the public the appropriate way to deal with them.

Of course, this book deals with a number of other actual topics of utmost importance. All of them describe current trends from various aspects; some were never worked out to such a comprehensive extent at all before. It is worth mentioning that various processes of preparation of polyhydroxyalkanoates (PHA), and especially poly(3-hydroxybutyrate) (PHB) homopolyester, economic, and industrial aspects of PHB production are competently discussed, while another chapter even describes technical details of various bioreactors including feeding regimes in PHA production, so that the companies or individuals considering starting production may get really reliable, up-to-date information on the topic including a number of related aspects. Another interesting part, found in Volume 3, is devoted to postsynthetic processing, where a number of ideas can be introduced for modification of the PHA-based materials to obtain optimal properties for desired applications.

Another part which I appreciate very much consists of four chapters in Volume 3 devoted to applications. In these chapters, inspiring hints can be found on packaging, medical applications, as well as other innovative areas, where PHA may find space for a faster increase in production, contributing to a more sustainable environment.

In any case, I consider all three volumes of this Handbook to be extremely useful for everybody intending to acquire detailed knowledge on new advances in biodegradable, PHA-based materials.

Ivan Chodak
Slovak Academy of Science

Editor

Martin Koller was awarded his Ph.D. degree by Graz University of Technology, Austria, for his thesis on polyhydroxyalkanoate (PHA) production from dairy surplus streams, which was embedded into the EU-FP5 project WHEYPOL (Dairy industry waste as source for sustainable polymeric material production), supervised by Gerhart Braunegg, one of the most eminent PHA pioneers. As senior researcher, he worked on bio-mediated PHA production, encompassing the development of continuous and discontinuous fermentation processes, and novel downstream processing techniques for sustainable PHA recovery. His research focused on cost-efficient PHA production from surplus materials by eubacteria and haloarchaea and, to a minor extent, the development of PHA for biomedical use.

He currently holds about 80 Web-of-Science listed articles, often in high ranked scientific journals, has authored twelve chapters in scientific books, edited three scientific books and five special issues on PHA for diverse scientific journals, has given plenty of invited and plenary lectures at scientific conferences, and supports the editorial teams of several distinguished journals.

Moreover, Martin Koller coordinated the EU-FP7 project ANIMPOL (Biotechnological conversion of carbon-containing wastes for eco-efficient production of high added value products), which, in close cooperation between academia and industry, investigated the conversion of the animal processing industry's waste streams toward structurally diversified PHA and follow-up products. In addition to PHA exploration, he was also active in microalgal research and in biotechnological production of various marketable compounds from renewables by yeasts, chlorophyte, bacteria, archaea, fungi, or lactobacilli.

At the moment, Martin Koller is active as research manager, lecturer, and external supervisor for PHA-related projects.

Contributors

Tomas Alexandersson
Primozone Production AB
Löddeköpinge, Sweden

Nadeem Ali
Center of Excellence in Environmental
    Studies (CEES)
King Abdul Aziz University
Jeddah, Saudi Arabia

Simon Anterrieu
Veolia Water Technologies
Marseille, France

Simon Bengtsson
Promiko AB
Lomma, Sweden

Gerhart Braunegg
ARENA – Association for Resource
    Efficient and Sustainable
    Technologies
Graz, Austria

Guo-Qiang (George) Chen
Tsinghua University
Beijing, PR China

Ivan Chodak
Polymer Institute
Slovak Academy of Sciences
Bratislava, Slovakia

José Geraldo da Cruz Pradella
Braskem S.A., Campinas
São Paulo, Brazil

Alexandra Deeke
Waterschap De Dommel
Boxtel, The Netherlands

Geeta Gahlawat
Defence Institute of High Altitude
    Research
Defence Research and Development
    Organization (DRDO)
Chandigarh, India

Liya Ge
Residues and Resource Reclamation
    Centre
Nanyang Technological University
Singapore, Singapore

José Gregório Cabrera Gomez
Laboratory of Bioproducts
Department of Microbiology
Institute of Biomedical Sciences
University of São Paulo
São Paulo, Brazil

Emma Gustafsson
Orkla Foods
Malmö, Sweden

Markus Hjort
Concawe
Brussels, Belgium.

Predrag Horvat
Department of Biochemical
    Engineering
Faculty of Food Technology and
    Biotechnology
University of Zagreb
Zagreb, Croatia

Iqbal Muhammad Ibrahim Ismail
Center of Excellence in Environmental
    Studies (CEES)
King Abdul Aziz University
Jeddah, Saudi Arabia

Peter Johansson
Perstorp AB
Perstorp, Sweden

Lamija Karabegovic
Veolia Water Technologies
Lund, Sweden

Anton Karlsson
Sweco Environment AB
Gothenburg, Sweden

Guneet Kaur
Department of Biology
Hong Kong Baptist University
Kowloon Tong, Hong Kong
and
Department of Civil Engineering
Lassonde School of Engineering
York University
Toronto, Ontario, Canada

Martin Koller
University of Graz
NAWI Graz
Office of Research Management and
    Service
Institute of Chemistry
Graz, Austria
and
ARENA – Association for Resource
    Efficient and Sustainable
    Technologies
University of Graz
Graz, Austria

Maria R. Kosseva
University of Nottingham Ningbo
    China
PR China

Leon Korving
Waterschapp Brabantse Delta
Breda, The Netherlands

Adriana Kovalcik
Faculty of Chemistry
Brno University of Technology
Brno, Czech Republic

Po-Heng (Henry) Lee
Department of Civil and Environmental
    Engineering
Imperial College London
London, UK

Paulo Costa Lemos
LAQV, REQUIMTE
Chemistry Department
Faculty of Science and
    Technology
University Nova of Lisbon
Lisbon, Portugal

Laura Lorini
Department of Chemistry
Sapienza University of Rome
Rome, Italy

Per Magnusson
Veolia Water Technologies
Lund, Sweden

Mauro Majone
Department of Chemistry
Sapienza University of Rome
Rome, Italy

Ivana Marova
Faculty of Chemistry
Brno University of Technology
Brno, Czech Republic

Fernando Morgan-Sagastume
Veolia Water Technologies
Lund, Sweden

Michael Narodoslawsky
Institute for Process and Particle
  Engineering
Graz University of Technology
Graz, Austria.

Ivana Novackova
Faculty of Chemistry
Brno University of Technology
Brno, Czech Republic

Stanislav Obruca
Faculty of Chemistry
Brno University of Technology
Brno, Czech Republic

Edmar Ramos Oliveira-Filho
Laboratory of Bioproducts, Department
  of Microbiology, Institute of
  Biomedical Sciences
University of São Paulo
São Paulo, Brazil

Chaozhi Pan
School of Civil and Environmental
  Engineering
Nanyang Technological University
Singapore, Singapore

Joana Pereira
CICECO-Aveiro Institute of Materials
Chemistry Department
University of Aveiro
Aveiro, Portugal

Iva Pernicova
Faculty of Chemistry
Brno University of Technology
Brno, Czech Republic

Rosane Aparecida Moniz Piccoli
Laboratory of Industrial Biotechnology
Bionanomanufacturing Center
Institute for Technological Research of
  São Paulo State
São Paulo, Brazil

Muhammad Imtiaz Rashid
Center of Excellence in Environmental
  Studies (CEES)
King Abdul Aziz University
Jeddah, Saudi Arabia

Mohammad Reda Kabli
Department of Industrial Engineering
Faculty of Engineering
King Abdul Aziz University
Jeddah, Saudi Arabia

Edy Rusbandi
Spruson & Ferguson Pte. Ltd
Singapore
and
University of Nottingham
  Ningbo China
PR China

Petr Sedlacek
Faculty of Chemistry
Brno University of Technology
Brno, Czech Republic

Luísa Seuanes Serafim
CICECO-Aveiro Institute of Materials
Chemistry Department
University of Aveiro
Aveiro, Portugal

Khurram Shahzad
Center of Excellence in Environmental
  Studies (CEES)
King Abdul Aziz University
Jeddah, Saudi Arabia

Luc Sijstermans
Slibverwerking Noord-Brabant
Moerdijk, The Netherlands

Luiziana Ferreira da Silva
Laboratory of Bioproducts
Department of Microbiology
Institute of Biomedical Sciences
University of São Paulo
São Paulo, Brazil

Sujata Sinha
Department of Biochemical
  Engineering and Biotechnology
Indian Institute of Technology Delhi
  Hauz Khas
New Delhi, India

Eva Slaninova
Faculty of Chemistry
Brno University of Technology
Brno, Czech Republic

Ahmed Saleh Ahmed Summan
Center of Excellence in Environmental
  Studies (CEES)
King Abdul Aziz University
Jeddah, Saudi Arabia

Marilda Keico Taciro
Laboratory of Bioproducts
Department of Microbiology
Institute of Biomedical Sciences
University of São Paulo
São Paulo, Brazil

Giin-Yu Amy Tan
Department of Civil Engineering
The University of Hong Kong
Hong Kong

Martin Tietema
Foamplant BV
Groningen, The Netherlands

Cora Uijterlinde
STOWA
Amersfoort, The Netherlands

Francesco Valentino
Department of Chemistry
Sapienza University of Rome
Rome, Italy

Yede van der Kooij
Wetterskip Friesland
Leeuwarden, The Netherlands

Marianna Villano
Department of Chemistry
Sapienza University of Rome
Rome, Italy

Alan Werker
Promiko AB
Lomma, Sweden
and
School of Chemical Engineering
University of Queensland
Brisbane, Australia

Fuqing Wu
Tsinghua University
PR China

Etteke Wypkema
Waterschapp Brabantse Delta
Breda, The Netherlands

The Handbook of Polyhydroxyalkanoates, Volume 2

Introduction by the Editor

This second volume of *The Handbook of Polyhydroxyalkanoates* is technology-oriented, involving thermodynamic considerations and mathematical modeling of PHA biosynthesis, bioengineering aspects regarding bioreactor design and downstream processing for PHA recovery from microbial biomass, chapters on microbial mixed culture processes, which couple PHA biosynthesis to waste treatment, a strong industry-focused section with chapters on the economics of PHA production, industrial-scale PHA production from sucrose, next generation industrial biotechnology approaches for PHA production based on novel robust production strains, and holistic techno-economic and sustainability considerations on PHA manufacturing.

The 13 chapters of the second volume can be grouped into five different sections.

THERMODYNAMIC CONSIDERATIONS IN MATHEMATICAL MODELING OF PHA BIOSYNTHESIS AND ENVIRONMENTAL STRESS CONDITIONS

Microbial PHA production can occur both aerobically and anaerobically. Even though PHA production pathways are described, and optimization of PHA biosynthesis from a wide spectrum of substrates has been achieved, arbitrary understanding of only one of these two central aspects is typically available for a given process. This causes that the required holistic mechanistic understanding of PHA biosynthesis is missing. Therefore, a specialized chapter, provided by Chaozhi Pan and colleagues from the research team of Giin-Yu Amy Tan, provides an insight into microbial thermodynamics so as to comprehensively reveal the central domain governing in PHA formation both aerobically and anaerobically. In a nutshell, the chapter explains under which conditions PHA production is thermodynamically feasible. The interplay of energetics for PHA production, e.g., reducing power, carbon source, cell growth, and catabolic energy capture, is presented. Specific information about substrate effects is discussed, and, finally, a clear picture of aerobic PHA production is drawn and extended to the possibility of PHA production under anaerobic conditions, which indeed has a still severely underestimated and neglected potential, which should be tapped in future.

The next chapter provides a systematic overview of different mathematical modeling approaches, from low-structured and formal kinetic models to modern tools like metabolic models, cybernetic models, neural networks, and hybrid-type models. Importantly, such models are nowadays indispensable for every new bioprocess in development, thus also for PHA biosynthesis. It is shown by Predrag Horvat, Gerhart Braunegg, and myself how mathematical modeling identifies metabolic bottlenecks in strain-substrate combinations, how productivity in PHA production can be improved *in silico*, how intracellular processes can be better understood, how modeling can support improving feeding strategies and bioengineering, and how mathematical modeling saves the number of laboratory experiments normally needed during the development process.

ENVIRONMENTAL AND STRESS FACTORS

Another section of this volume is dedicated to stress factors provoking PHA biosynthesis; this section elucidates how PHA accumulation helps microbial cells to overcome these stress factors. As shown only recently, PHA display many more functions than simply acting as carbon and energy reserves. How PHA formation protects cells against unfavorable conditions like UV-radiation, desiccation, high concentration of heavy metals and xenobiotics, temperature shock, osmotic up- and down shock, or redox imbalances is summarized in a comprehensive chapter, written by the team of Stanislav Obruca and Ivana Marova.

In this context, a separate contribution by myself, Stanislav Obruca, and Gerhart Braunegg is specifically dedicated to halophilic microbial PHA producers; such organisms thrive in salt-rich environments and can often be used for PHA production under semi-sterile or even unsterile conditions, which makes the PHA production process economically more competitive, and constitutes a prime example of "Next Generation Industrial Biotechnology" (NGIB) processes. This chapter covers both PHA biosynthesis by haloarchaea and by halophilic eubacterial strains. Special emphasis is dedicated to genetic engineering tools to enhance product formation by halophiles, and to strategies dedicated to trigger the microstructure of PHA on a molecular basis. Moreover, the chapter presents the co-production of PHA and other valued bioproducts, such as pigments, halocins, extremozymes, or extracellular polysaccharides by halophiles, and details techno-economic assessments of individual processes.

BIOENGINEERING

Bioreactor design and the feeding regime during PHA biosynthesis are pivots in process optimization. By the respective chapter by Geeta Gahlawat, Sujata Sinha, and Guneet Kaur, the reader will get a clear impression about matching bioreactor types (stirred tank reactors, bioreactor cascades, tubular plug flow reactors, bubble columns, airlift bioreactors, etc.), and modes of nutrient supply (batch, fed-batch, continuous processing). Special emphasis in this section is dedicated to challenges during scale-up of PHA production processes (mixing, aeration, transport phenomena, etc.), and on development of non-sterile processes and contamination-resistant strains for maximum productivity, and process robustness.

The second engineering-oriented chapter covers the field of downstream process-ing to recover PHA as intracellular products from microbial biomass. Downstream processing for PHA recovery from biomass plays a decisive role in the PHA manu-facturing process with respect to cost performance, material quality, and eco-bal-ance. In this chapter by Maria R. Kosseva and Edy Rusbandi, the reader will learn how several factors affect the selection of the adequate PHA recovery method: the microbial production strain, type and composition of PHA, PHA load in biomass, required product purity for a defined application, availability of chemicals for PHA recovery, and impact of the recovery method on the physical properties of PHA. Established (solvent-based, enzymatic, and mechanical methods) and novel strate-gies (e.g., use of ionic liquid, supercritical solvents, hypo-osmotic shock, digestion of non-PHA biomass by animals, etc.) for PHA recovery from microbial biomass, always focusing on reducing solvent and energy input, are compared, and assess-ments of feasibility for large-scale application are comprehensively discussed.

MIXED MICROBIAL CULTURE APPROACHES

The use of mixed microbial cultures is increasingly being considered as one of the most promising future strategies to make PHA production more efficient by cou-pling it to waste water treatment. This field is definitely multi-faceted, and different engineering, microbiological, and downstream processing methods are currently in the status of development, both at laboratory scale and in long running pilot plant studies. Therefore, three independent chapters cover this scientific field, each deal-ing with particular aspects of PHA production by mixed microbial cultures. These chapters, considered as a whole, generate unprecedented synergism in presenting a holistic picture of the current state of PHA research by mixed cultures.

The first of these chapters, written by Luísa Seuanes Serafim, Joana Pereira, and Paulo Costa Lemos, constitutes an overarching compilation of the principles and current state of the art in using mixed microbial cultures for PHA biosynthesis, highlighting the pros and cons of this technology. Aspects like microbial culture dynamics and reproducibility of PHA quality from mixed cultures are specifically addressed. In the second chapter by Francesco Valentino, Marianna Villano, Laura Lorini, and Mauro Majone, special emphasis is dedicated to successful pilot-scale studies using mixed cultures for PHA production to remediate urban waste streams, while the third contribution, by Alan Werker's team, evaluates and interprets PHA copolyester properties produced from a full scale municipal activated sludge bio-mass used for PHA production by mixed cultures in a pilot plant, which was oper-ated for ten months, and delivered 52 batches of PHA-rich biomass.

INDUSTRIAL ASPECTS

The industrial point of view must not be missed in a Handbook like this; in this sense, four chapters provide the industry-related point of view about current and future trends in PHA production and processing. After a comprehensive review of the economic and industrial aspects of PHA production by José Geraldo da Cruz Pradella, who also focuses on the historic aspects of PHA commercialization and factors decisive for PHA's market success, special emphasis is dedicated by Fuqing Wu and Guo-Qiang Chen to the emerging concept of "next generation industrial biotechnology" (NGIB) for high-throughput and cost-efficient PHA production on an industrial scale, which involves the use of genetically optimized, stable, and robust production strains able to produce PHA under minimized energy demands from inexpensive substrates in simple tanks at high productivity, and tailored composition and properties. Hence, this chapter is dedicated to the presentation of a concept on the "ideal" PHA production process of the future, which is already in an advanced stage of development, and is based on a proper combination of production strain, genetic engineering, bioprocess engineering, feeding regime, and downstream processing.

This is followed by a highly specialized contribution provided by the authorship headed by Luiziana Ferreira da Silva, dealing with the integrated (semi)industrial process of PHA production from materials of the sugar industry in Brazil, which currently still constitutes the prime example for large-scale PHA production. This chapter illustrates the viability of cost-efficient PHA production by integration of biopolyester production into existing industrial production lines, where the feedstocks directly accrue, and how diverse surplus materials of sugar production can be utilized in a smart way to make PHA production economically feasible. Moreover, this chapter also highlights the close cooperation between academia and industry to develop PHA production processes, and explains how both biofuel and biopolymer production can exist in parallel within the same production lines according to the concept of circular economy.

According to the final chapter of this volume, written by the team of Khurram Shahzad, the number of reliable, well-grounded life cycle studies for PHA-based materials is only now starting to grow. Most of all, this chapter makes clear that all steps of the entire PHA production chain need to be considered in order to make these materials sustainable both in economic and environmental terms. Therefore, by focusing on the process of PHA production from byproducts of the animal-processing industry, the reader becomes familiar with how tools like life cycle assessment (LCA), cleaner production studies, techno-economic assessments, or the sustainable process index (SPI) can be applied to identify those hot topics, which determine the economic feasibility and environmental viability of a new PHA production process in development. This, of course, is the condition without which industrial-scale PHA production will not be possible. Moreover, it is shown how this has to go in parallel with the understanding of the current market situation and customer acceptance of new plastic materials.

In the end, I deeply hope that this second volume will meet the expectations of the scientific community, offer concrete answers to existing R&D questions, and, above all, will encourage further research activities in the intriguing realm of PHA biopolyesters!

Martin Koller
Graz, Austria

Part I

Kinetics and Mathematical Modeling

1 An Introduction to the Thermodynamics Calculation of PHA Production in Microbes

Chaozhi Pan, Liya Ge, Po-Heng (Henry) Lee and Giin-Yu Amy Tan

CONTENTS

1.1 INTRODUCTION

The large variance in food availability is not uncommon in the natural environment and microbes have evolved strategies to minimize the impact of food scarcity and starvation. The universal methods include sporulation with low metabolic rate [1], and intracellular storage of energy or nutrient source [2–4]. In the latter case, polyhydroxyalkanoate (PHA), which is composed of 3-hydroxyalkanoic acids, is a representative example [5]. Based on the carbon chain length, PHA is categorized into short-chain-length PHA (*scl*-PHA) with three to five carbon atoms per monomer, medium-chain-length PHA (*mcl*-PHA) with six to 14 carbon atoms, and long-chain-length PHA (*lcl*-PHA) with monomers of 15 carbon atoms or more [6]. PHA

3

synthesis is generally triggered by nutrient limitation with excessive carbon source, during which active microbial growth is halted [7]. However, some microbes such as *Bacillus* sp. also produce PHA alongside biomass growth [8]. During food scarcity, intracellular PHA would be degraded into its monomeric units [9] and catabolized via β-oxidation into carbon dioxide (CO_2) to provide energy. The capability to synthesize PHA provides the microbe with a competitive advantage in combating starvation. Simultaneously, the microbial PHA metabolism also means that PHA is readily biodegradable.

As a polymeric material, PHA shows comparable properties in elasticity and thermostability to synthetic chemical polymers [10, 11]. The insatiable consumption and inappropriate disposal of petroleum-based plastic products has raised the concerns of environmental contamination. Particularly, the recent discovery of the ubiquitous presence of microplastic particles in the aquatic sphere [12] further highlights the long-term ecological toxic effect and strengthens the calls for sustainable green packaging materials. Thus, the biodegradable property of PHA makes it a possible alternative for current packaging materials. Furthermore, PHA's biocompatibility with human tissue further expands its potential applications as biomedical and pharmaceutical materials [13]. Moreover, some researchers also propose that the chemical modification of PHA precursors would further create novel functional polymer materials [14]. All these applications became the driving forces for PHA industrial production.

Currently, reported PHA-producing strains are heterogenous species, and most are aerobic with distinct PHA yields. Some strains of *Pseudomonas putida*, *Cupriavidus necator* and *Azohydromonas lata* were reported to store PHA up to 90% of dry biomass, while other strains have varying PHA content of 40–70% in dry biomass [6, 15]. As the produced PHA is intracellularly stored as lipid granules, biomass growth must occur in order to accommodate PHA granules. Both PHA synthesis and biomass synthesis act as sinks for carbon sources. Simultaneously, a carbon source could be converted to reducing power, nicotinamide adenine dinucleotide plus hydrogen (NADH), and further to adenosine triphosphate (ATP) [16]. Thus, the carbon source in PHA-producing microbes could be used for biomass synthesis, PHA production and energy production. These three metabolisms are intertwined in a competition for carbon, redox balance, and energy balance. At present, there is lack of understanding on the dynamics between these three metabolisms.

Additionally, a limited number of studies on anaerobic PHA production is reported [17]. Recent evidence indicates that anaerobes produce PHA as an electron sink [18]. This raises the question on the feasibility of anaerobic microbial PHA production. Since oxygen is not available to act as an electron acceptor under anaerobic conditions, the excessive reducing power, produced during carbon metabolism, would require an electron sink to ensure the continuation of metabolic reactions. Methanogens could dispose of reducing power through methane formation to maintain the reaction [19]. However, for non-methane producing anaerobes, they must seek alternative forms of electron sink to dispose of the reducing power, e.g., PHA production. Anaerobic condition is regarded as an energy-constraint environment due to the lack of electron acceptors. The minor redox potential gap between electron donor and acceptor makes the substrate-level phosphorylation the main energy-producing mechanism [20]. If PHA synthesis is energy-intensive, it is less likely to be favored under anaerobic conditions.

This chapter aims to introduce the concept of thermodynamics to PHA production along with cell synthesis and energy production. Specific cases with common substrates under aerobic and anaerobic conditions are examined. This provides new perspectives for PHA production and enhances understanding of rarely reported scenarios such as anaerobic PHA production.

1.2   INTRODUCTION TO THERMODYNAMICS AND ITS APPLICATION TO PHA SYNTHESIS

Thermodynamics is categorized into equilibrium thermodynamics and non-equilibrium thermodynamics. Enthalpy (heat adsorbed under constant pressure) and entropy (the degree of order in a system) are also two key concepts of thermodynamics. Classic thermodynamics emphasizes equilibrium in a closed system, while a biological system is an open and dynamic system, which converts high-enthalpy and low-entropy substrates into low-enthalpy and high-entropy metabolites [21]. Since non-equilibrium thermodynamics also is concerned with the mass flow, it is beyond the scope of this chapter and will not be discussed here. In a standard reaction, the Gibbs energy change (ΔG_r^0) of a chemical reaction describes the effects of enthalpy (ΔH_r^0) and entropy (ΔS^0) changes:

$$\Delta G_R^0 = \Delta H_R^0 - T\Delta S^0$$

where T represents temperature (K) and "^{0}" represents standard conditions: 1 atm, 25°C, 1 mol/L active concentration of a given substance.

Thermodynamics emphasizes the energy gap between the reactants and products. Kinetic rate is not considered in thermodynamics of biological processes [22]. Thus, thermodynamics is helpful for predicting the energetic feasibility of a reaction but cannot be used to predict whether the reaction proceeds at a significant rate. For a chemical reaction, a stoichiometric reaction to achieve mass and redox balance needs to be defined. The temperature and pressure under which the reaction occurs need to be defined as well, hence the Gibbs energy change could be alternatively expressed as the sum of standard formation Gibbs energy (ΔG_f^0) relative to the standard state. Under standard conditions, the ΔG_f^0 values of several basic elements such as carbon and hydrogen are assigned 0 kJ/mol, while the ΔG_f^0 values of other compounds are calculated as a value relative to these basic elements [23]. A comprehensive summary of ΔG_f^0 values for various biochemical species can be found in existing reviews and textbooks [21, 24, 25]. For a detailed thermodynamics illustration and application, the reader could refer to the textbook written by Atkins and De Paula [23].

For reaction: $aA + bB = cC + dD$, reactants (A and B) generate products (C and D) with stoichiometric coefficients a, b, c and d respectively.

Based on the defined reaction, $\Delta G_R^0 = \sum_{i=1}^{n} Y_i G_f^i - \sum_{j=1}^{m} Y_j G_f^j$, is the energy gap between end products and the initial reactants, Y_i is the stoichiometric coefficient of species i; G_f^i: the standard formation Gibbs energy of species i relative to basic elements.

$$\Delta G = cG_f^c + dG_f^d - aG_f^a - bG_f^b = \Delta G^{0`} + RT\ln\left(\frac{C_C^c C_D^d}{C_A^a C_B^b}\right)$$

$$= -RT\ln\left(K_{eq}\right) + RT\ln\left(\frac{C_C^c C_D^d}{C_A^a C_B^b}\right)$$

R: gas constant, 8.3145 J/(mol·K); K_{eq}: reaction constant at equilibrium; T: reaction temperature; C_i: concentration of species i, mol/L.

Through the transformation, the $\Delta G^{0`}$ could be linked with reaction equilibrium constant as shown above.

Specific to PHA synthesis, there is a phase transition from aqueous PHA monomer, 3-hydroxyalkanoate to solid PHA polymer, and the corresponding Gibbs free energy change for the phase transition is currently unknown. Thus, in this chapter, the fed carbon source as the reactant and 3-hydroxyalkanyl-CoA as the product in PHA synthesis, is defined. The reaction stoichiometry is normalized to 1 mol of carbon source. For biomass synthesis, the biochemical formula $CH_{1.8}O_{0.5}N_{0.2}$ is used to represent simplified biomass [22] and ammonium nitrogen is used as the nitrogen source. Under the biological standard condition, the ionic species should be represented by the dominant species dependent on the pKa and pH. In aerobic reactions, the complete oxidation of carbon source with oxygen to carbon dioxide and water is used for energy production reaction. In contrast, in anaerobic conditions, the energy production reaction would vary according to specific scenarios. Figure 1.1 provides an overview of the carbon and PHA metabolism pathways that are covered in this chapter. The ΔG_f^0 and $\Delta G^{0`}$ values for common compounds and reactions involved in carbon and PHA metabolisms, including those discussed in this chapter, are summarized in Table 1.1 and Table 1.2, respectively.

1.3   PHA SYNTHESIS UNDER AEROBIC CONDITIONS

Although PHA could be synthesized along with cell growth under nitrogen rich conditions [8, 28], high PHA yields are usually triggered by the limitation of nutrients such as nitrogen, phosphorus and sulfur [7]. Amino acids consisting of nitrogen are rarely reported as the sole carbon source for PHA synthesis. Meanwhile, medium to long chain fatty acids could be used as carbon sources for PHA, but limited solubility in the aqueous phase and severe foaming issues restrain their application dosage and potential scale-up [29]. Thus, highly soluble and commonly used short chain fatty acids and glucose are chosen as substrates for thermodynamics analysis of PHA biosynthesis. Acetate, butyrate and glucose are used as examples for poly(3-hydroxybutyrate) (PHB) synthesis, while valerate is considered for poly(3-hydroxyvalerate) (PHV) and poly(3-hydroxybutyrate-*co*-3-hydroxyvalerate) (PHBV) synthesis.

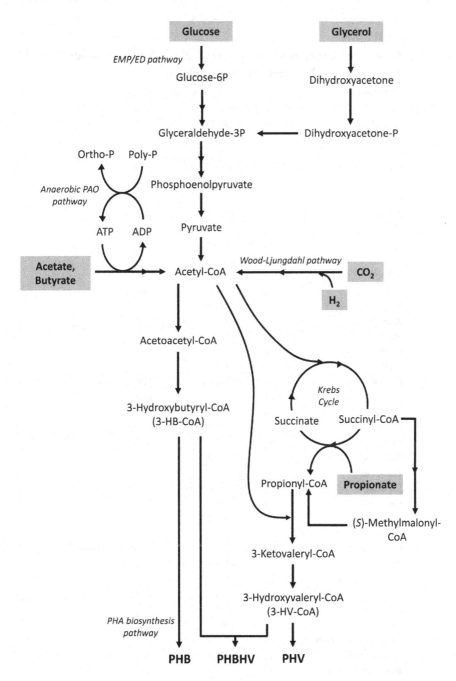

FIGURE 1.1 An overview of aerobic and anaerobic carbon and PHA metabolism pathways involving various substrates (shaded in gray). Adapted and modified from [6, 18, 26, 27]. (P, phosphate; EMP, Embden-Meyerhof-Parnas; ED, Entner-Doudoroff.)

TABLE 1.1

Gibbs Free Energies of Formation (at 25°C) for Compounds Involved in Carbon and PHA Metabolisms

Compound	State	$\Delta Gf^{0'}$ (kJ/mol)	Reference
H^+	aq	0	[45]
H^+ (pH 7)	aq	−39.87	[45]
H_2	g	0	[45]
H_2O	l	−237.178	[45]
HS^-	aq	12.05	[45]
OH^-	aq	−157.293	[45]
O_2	g	0	[45]
CO_2	g	−394.359	[45]
CO_2	aq	−386.20	[21]
Fe^{2+}	aq	−78.87	[45]
Fe^{3+}	aq	−4.6	[45]
HPO_4^{2-}	aq	−249.4	[21]
NH_4^+	aq	−79.37	[45]
SO_4^{2-}	aq	−744.63	[45]
ADP	aq	−230.5	[21]
ATP	aq	−252.7	[21]
GDP	aq	−431.8	[21]
GTP	aq	−453.5	[21]
Formate$^-$	aq	−351.04	[45]
Acetate$^-$	aq	−369.41	[45]
Propionate$^-$	aq	−361.08	[45]
Butyrate$^-$	aq	−352.63	[45]
Pyruvate$^-$	aq	−474.63	[45]
Succinate^{2-}	aq	−690.23	[21]
Fumarate^{2-}	aq	−604.21	[45]
(S)-Malate^{2-}	aq	−845.08	[45]
Oxaloacetate^{2-}	aq	−797.18	[45]
α-D-Glucose	aq	−917.22	[45]
Glycerol	aq	−488.52	[23]
Coenzyme A	aq	−270	Derived from [46]
Acetyl-CoA	aq	−138.5[a]	[21]
Propionyl-CoA	aq	−131.4[a]	[21]
Butyryl-CoA	aq	−124.3[a]	[21]
Acetoacetyl-CoA	aq	−245.2[a]	[21]
Crotonyl-CoA	aq	−46[a]	[21]
Methylmalonyl-CoA	aq	−459.4[a]	[21]
Succinyl-CoA	aq	−458.6[a]	[21]
Keothexanoyl-CoA	aq	−231[a]	[21]
3-Hydroxybutyryl-CoA	aq	−284.9[a]	[21]
3-Hydroxyhexanoyl-CoA	aq	−270.7[a]	[21]
3-Hydroxyvaleryl-CoA	aq	−277.8[a]	Derived from [21]

(Continued)

TABLE 1.1 (CONTINUED)

Gibbs Free Energies of Formation (at 25°C) for Compounds Involved in Carbon and PHA Metabolisms

Compound	State	$\Delta Gf^{0'}$ (kJ/mol)	Reference
Pyrophosphate (PPi)	aq	−240.2	[21]
Acetyl phosphate	aq	−387.9	Derived from [21]
Acetyl phosphate	aq	−392.93	[46]
Bacterial biomass $CH_{1.8}O_{0.5}N_{0.2}$	s	−67	[22]

[a] Value is relative to coenzyme A

TABLE 1.2

Gibbs Free Energies for Intracellular Redox and Common Reactions Involved in Carbon and PHA Metabolisms

Reaction	$\Delta G^{0'}$ (kJ/mol)	Reference
$ATP + H_2O \rightarrow ADP + HPO_4^{2-} + H^+$	−32	[46]
$MK + H_2 \rightarrow MKH_2$	−65.5	Derived from [45]
$NADH + H^+ \rightarrow NAD^+ + H_2$	18	[47]
$2Fd_{red}^{2-} + 2H^+ \rightarrow 2Fd_{ox} + H_2$	3	[47]
$SO_4^{2-} + H^+ + 4H_2 \rightarrow HS^- + 4H_2O$	−191.77	[45]

PPi, pyrophosphate; Fd, ferredoxin; MK, menaquinone

1.3.1 ACETATE CONVERSION TO PHB

Acetate is a dominant and abundant carbon source produced during the anaerobic digestion of polysaccharides, proteins and long chain fatty acids [30]. It is regarded as a cheap, abundant and readily available carbon source for PHB production. Acetyl-CoA, which is an intracellular intermediate of acetate, is a key metabolite in metabolism (anabolisms and catabolisms). As shown in Equation 1.1, the conversion of acetate to 3-hydroxybutyryl-CoA (3HB-CoA) is not energetically favorable, and this process is NADH consuming (78.667 kJ/mol). In contrast, acetate oxidation to CO_2 and H_2O would release a large amount of energy (Equation 1.2: −797.101 kJ/mol), which is ten times the energy needed in 3HB-CoA synthesis. Moreover, the biomass synthesis from acetate under aerobic conditions is also energetically favorable (Equation 1.3: −27.33 kJ/mol). Although the −27.33 kJ/mol in biomass is not comparable to cell respiration, the actual ΔG will be close to zero only when the reaction product increases by 1000 times ($RT \cdot \ln(1000) = 17.123$ kJ/mol) and along with biomass synthesis, there is sufficient energy to support PHB synthesis. When nitrogen is limited and carbon sources are still abundant, biomass synthesis would cease and more oxygen would be directed to cell respiration, thus driving 3HB-CoA synthesis. Both cell respiration and biomass synthesis would provide energy to drive

PHB synthesis. Cell respiration would likely constitute the dominant form of energy source. Therefore, PHB synthesis would play the role of carbon, reducing power and energy sink. In conditions of carbon source depletion, the degradation of PHB would release energy to support cell maintenance and activity.

$$CH_3COO^- + 0.5HSCoA + 0.5NADH + 0.5H^+$$

$$\rightarrow 0.5CH_3CHOHCH_2CO\text{-}SCoA + OH^- + 0.5NAD^+ \qquad (1.1)$$

$$\Delta G^{0'} = +78.667 \text{ kJ/mol}$$

$$CH_3COO^- + 2O_2 \rightarrow 2CO_2 + H_2O + OH^- \qquad \Delta G^{0'} = -797.101 \text{ kJ/mol} \qquad (1.2)$$

$$CH_3COO^- + 0.33\ NH_4^+ + 0.25\ O_2$$

$$\rightarrow 1.67\ CH_{1.8}O_{0.5}N_{0.2} + 0.33\ CO_2 + 0.33\ H_2O + 0.67OH^- \qquad (1.3)$$

$$\Delta G^{0'} = -27.33 \text{ kJ/mol}$$

1.3.2 BUTYRATE CONVERSION TO PHB

Butyrate could be a product of β-oxidation of long-chain fatty acids with an even carbon number or the condensation of two acetyl-CoA units with reduction. It is another dominant volatile fatty acid (VFA) present in anaerobic digestion effluent [31]. Butyrate is usually associated with PHB production. Similar to the acetate conversion to 3HB-CoA, the conversion of butyrate to 3HB-CoA is endergonic (Equation 1.4) and the energy consumption is comparable based on the 3HB-CoA produced, but unlike for acetate, this conversion is NADH generating. A larger energy generation is obtained for aerobic butyrate respiration compared to aerobic acetate respiration (Equation 1.5). Since the carbon redox status is more reduced than biomass (−1 vs −0.2 per carbon chemical valence on average), more energy is released in the biomass synthesis process (Equation 1.6). Thus, both biomass formation and cell respiration would drive the PHB synthesis with cell respiration likely providing the dominant energy source.

$$CH_3CH_2CH_2COO^- + HSCoA + NAD^+ + H_2O$$

$$\rightarrow CH_3CHOHCH_2CO - SCoA + OH^- + NADH + H^+ \qquad (1.4)$$

$$\Delta G^{0'} = +146.4 \text{ kJ/mol}$$

$$CH_3CH_2CH_2COO^- + 5O_2 \rightarrow 5CO_2 + 3H_2O + OH^-$$
$$\qquad\qquad\qquad\qquad (1.5)$$
$$\Delta G^{0'} = -2060.28 \text{ kJ/mol}$$

$$CH_3CH_2CH_2COO^- + 0.5NH_4^+ + 2.3750_2$$

$$\rightarrow 2.5CH_{1.8}O_{0.5}N_{0.2} + 1.5CO_2 + 2H_2O + 0.5OH^- \quad (1.6)$$

$$\Delta G^{0'} = -890.44 \text{ kJ/mol}$$

1.3.3 PROPIONATE CONVERSION TO 3HV

PHB is the most common unit in PHA produced by microbes from various carbon feedstocks. To generate an odd-number carbon unit such as 3-hydroxyvalerate (3HV) incorporated into PHA to enhance its material properties, a carbon source with odd-number carbon is added [32]. Here, propionate is used as the carbon source for PHV and PHBV synthesis. Propionate is usually found in the anaerobic digestion effluent, but a high accumulation of propionate is regarded as an indicator of poor anaerobic digestion due to its endergonic conversion to acetate and hydrogen accumulation [33]. However, from a resource recovery perspective, propionate build-up can be considered advantageous when coupled to PHA synthesis. Similar to 3HB-CoA produced from acetate and butyrate, 3-hydroxyvaleryl CoA (3HV-CoA) produced from propionate is endergonic as well (Equation 1.7). The reaction is NADH generating. Since propionate is not as reduced as butyrate, aerobic biomass synthesis releases less energy than that from butyrate. But the exergonic energy is still much higher than that released from acetate. Cell respiration is probably the main driving force for 3HV-CoA production.

$$CH_3CH_2COO^- + 0.5HSCoA + NAD^+ + H_2O$$

$$\rightarrow 0.5CH_3CH_2CHOHCH_2CO - SCoA + 0.5CO_2 + OH^- + NADH + H \quad (1.7)$$

$$\Delta G^{0'} = +91.78 \text{ kJ/mol}$$

$$CH_3CH_2COO^- + 3.5O_2 \rightarrow 3CO_2 + 2H_2O + OH^-$$
$$\Delta G^{0'} = -1427.91 \text{ kJ/mol} \quad (1.8)$$

$$CH_3CH_2COO^- + 0.5NH_4^+ + 0.8750_2$$

$$\rightarrow 2.5CH_{1.8}O_{0.5}N_{0.2} + 0.5CO_2 + H_2O + 0.5OH^- \quad (1.9)$$

$$\Delta G^{0'} = -274.85 \text{ kJ/mol}$$

1.3.4 GLUCOSE CONVERSION TO PHA

Glucose is one of the most naturally abundant monosaccharides and a bulk building block of many complex carbohydrate compounds. Due to its ubiquitous presence and biological importance, the substrate spectrum of most microbes includes glucose. Herein, glucose is considered as the representative sugar for 3HB-CoA synthesis. In contrast to the previous examples involving carboxylic acids (*cf.* Sections 1.3.1–1.3.3), glucose conversion to 3HB-CoA is energetically favorable (Equation

1.10). The high energy produced is probably from the glucose conversion to acetyl-CoA because the production of 3HB-CoA is shared with the conversion of acetate to 3HB-CoA. For 1 mol of 3HB-CoA produced from 1 mol glucose, 3 mol of NADH is produced as well. In cell respiration, the complete aerobic oxidation of glucose is highly exergonic, releasing up to 2821.98 kJ/mol of energy. Due to the minor redox status gap between glucose and biomass (0 vs −0.2 per carbon), most of the carbon in glucose will be directed to biomass in cell synthesis, while a small proportion of carbon would be converted to CO_2. Since the conversion of glucose to 3HB-CoA is energetically favorable, PHB synthesis does not require external energy. The synthesis of PHB from glucose would serve as a carbon sink when glucose is abundant.

$$C_6H_{12}O_6 + HSCoA + 3NAD^+$$

$$\rightarrow CH_3CHOHCH_2CO - SCoA + 2CO_2 + 3NADH + 3H^+ \qquad (1.10)$$

$$\Delta G^{0'} = -193.72 \text{ kJ/mol}$$

$$C_6H_{12}O_6 + 6O_2 \rightarrow 6CO_2 + 6H_2O \quad \Delta G^{0'} = -2821.98 \text{ kJ/mol} \qquad (1.11)$$

$$C_6H_{12}O_6 + NH_4^+ + 0.75O_2 \rightarrow 5CH_{1.8}O_{0.5}N_{0.2} + CO_2 + 3H_2O + H^+$$

$$(1.12)$$

$$\Delta G^{0'} = -474.94 \text{ kJ/mol}$$

1.4   PHA SYNTHESIS UNDER ANAEROBIC CONDITIONS

In anaerobic conditions, the absence of oxygen induces the disproportionation of fed substrate with most of the energy stored in the reduced products [25]. Low energy yield is a feature of anaerobic bioprocesses, and this drives the anaerobic microbe to become more efficient and effective in energy capture. Based on the thermodynamic analyses performed for aerobic PHA synthesis (Section 1.3), it was observed that the synthesis of PHA monomers from carboxylic acids requires energy to drive the reaction forward. Hence, the energy constraint nature of anaerobic processes could explain why PHA synthesis is less frequently reported. In this Section, several scenarios are presented to explore the possibility of anaerobic PHA synthesis from the thermodynamics perspective.

1.4.1   PHOSPHATE ACCUMULATING ORGANISMS (PAO) UTILIZE ACETATE FOR PHA SYNTHESIS

In a sequencing batch reactor (SBR), the interval of aeration and settling without aeration creates aerobic and regional anaerobic conditions. Accompanied with the operation, significant phosphate release and absorption was found with VFA reduction in the wastewater [34]. Although no model microbe responsible for this observation has yet been isolated, it is believed that the hydrolysis of polyphosphate would provide energy for the uptake of carbon source and storage as PHA under anaerobic conditions. The stored PHA would be used as an energy source for the passive transport of phosphate to form intracellular polyphosphate granules in the subsequent aerobic stage [35]. Therefore, the relative competitiveness of a phosphate accumulating organism (PAO) would be sustained in the cyclical aerobic and anaerobic conditions.

In anaerobic conditions, energy generation through the oxidation of acetate with oxygen is not possible. Meanwhile, the biomass synthesis from acetate is endergonic (Equation 1.15). 3HB-CoA production from acetate is identical to the aerobic process, requiring energy to drive the reaction (Equation 1.13). Since phosphate bonds are highly energetic bonds, the endergonic acetate metabolisms could be supported by the exergonic hydrolysis of phosphate bonds. For the hydrolysis of one phosphate from polyphosphate, the released energy is 61.29 kJ/mol (Equation 1.14). Compared to aerobic respiration, the released energy from the hydrolysis is smaller and it falls in the same order of energy required for 3HB-CoA synthesis. The coupling of polyphosphate hydrolysis with 3HB-CoA formation is thermodynamically feasible. However, it should be noted that biomass synthesis would compete for energy as well. Moreover, in wastewater, phosphate is typically present at a lower concentration compared to carbon source. As such, the stored polyphosphate is limited and the hydrolysis of polyphosphate to yield energy cannot continue indefinitely in anaerobic conditions [36]. Nevertheless, the cyclic aerobic and settling operation in SBR would regenerate the polyphosphate. The wastewater treatment plant operated in SBR provides a favorable niche for PAO. The production of PHA by PAO in anaerobic conditions with energy from the hydrolysis of polyphosphate is feasible but the quantity and concentration should be limited. Liu *et al.* reported on a 330 mg/L phosphate release accompanied with an equivalent amount of PHA synthesis in an SBR sludge, which accounted for 5.5% (w/w) of biomass [37].

$$CH_3COO^- + 0.5HSCoA + 0.5NADH + 0.5H^+$$

$$\rightarrow 0.5CH_3CHOHCH_2CO-SCoA + OH^- + 0.5NAD^+ \qquad (1.13)$$

$$\Delta \boldsymbol{G}^{0'} = +78.67 \text{ kJ/mol}$$

$$HP_2O_7^{3-} + H_2O \rightarrow 2HPO_4^{2-} + H^+ \quad \Delta \boldsymbol{G}^{0'} = -61.29 \text{ kJ/mol} \qquad (1.14)$$

$$CH_3COO^- + 0.33NH_4^+ + 0.5\ NAD^+ + 0.167H_2O$$

$$\rightarrow 1.67CH_{1.8}O_{0.5}N_{0.2} + 0.33CO_2 + 0.5NADH + 0.5H^+ + 0.67OH^- \quad (1.15)$$

$$\Delta \boldsymbol{G}^{0'} = +81.55 \text{ kJ/mol}$$

1.4.2 Syntrophic Microbes Degrading Fatty Acids to Synthesize PHA

Under standard biological conditions, both 3HB-CoA and biomass production from acetate and hydrogen are not energetically favorable (Equation 1.16 and 1.17). Due to a lack of energy source to drive 3HB-CoA, this scenario may be not feasible. In anaerobic conditions, acetyl-CoA can be produced from β-oxidation of long-chain fatty acids accompanied with hydrogen production. However, the accumulation of excessive hydrogen results in unfavorable anaerobic degradation of fatty acids and thus potential PHA synthesis from acetyl-CoA. To resolve the excessive hydrogen formation, H_2-producing microbes establish a syntrophic relationship with H_2-consuming microbes. MacInerney *et al.* reviewed the PHA production by strict anaerobes, *Syntrophomonas wolfei*, in crotonate degradation or co-culture with H_2-consuming methanogens [17]. In the former case, the unsaturated carbon bond in crotonate would act as an electron acceptor to resolve the issue of disposing reducing power by substrate disproportion. In the latter case, the syntrophic relation with H_2-consuming methanogens would help maintain lower hydrogen pressure. However, the reported PHA concentration was around the 10 mg/L level [38].

$$CH_3COO^- + 0.5HSCoA + 0.5H_2 \rightarrow 0.5CH_3CHOHCH_2CO - SCoA + OH^-$$

$$\Delta G^{0'} = +69.67 \text{ kJ/mol}$$

(1.16)

$$CH_3COO^- + 0.2H_2 + 0.4NH_4^+ \rightarrow 2CH_{1.8}O_{0.5}N_{0.2} + 0.4H_2O + 0.6OH^-$$

$$\Delta G^{0'} = +77.91 \text{ kJ/mol}$$

(1.17)

1.4.3 ACETOGENS UTILIZING CO_2 AND H_2 FOR PHA SYNTHESIS

Since PHB is the most common PHA detected in microbes, the production of PHB through the condensation and reduction of two acetyl-CoA units is among one of the most common PHA synthesis pathways. In addition to acetyl-CoA generation from the organic carbon source, acetyl-CoA produced from CO_2 at the cost of reducing power, via the Wood-Ljungdahl pathway possessed by acetogens, is also a feasible reaction [20, 39, 40]. In natural systems, acetate produced by acetogens is an important source for methanogens to produce methane. In engineered systems, syngas, which is a mixture of CO_2, CO and H_2, produced during the hydrothermal process or the pyrolysis of biomass in the absence of oxygen [40], could be a potential source for PHA production. Recently, several bacterial species which could utilize syngas to produce PHA in anaerobic conditions have also been identified [41]. In standard biological conditions, the 3-hydroxybutyryl-CoA (3HB-CoA) produced from CO_2 and H_2 is energetically favorable (Equation 1.18). Meanwhile, the biomass synthesis from CO_2 and H_2 also releases energy (Equation 1.19). Since both 3HB-CoA and biomass production could release energy, the issue of energy source is resolved, and the synthesized biomass could accommodate intracellular PHA accumulation.

$$CO_2 + 2.25H_2 + 0.25HSCoA$$

$$\rightarrow 0.25CH_3CHOHCH_2CO - SCoA + 1.5H_2O$$

(1.18)

$$\Delta G^{0'} = -40.98 \text{ kJ/mol}$$

$$CO_2 + 2.1H_2 + 0.2NH_4^+ \rightarrow CH_{1.8}O_{0.5}N_{0.2} + 1.5H_2O + 0.2H^+$$

$$\Delta G^{0'} = -28.85 \text{ kJ/mol}$$

(1.19)

1.4.4 ANAEROBIC RESPIRATION OF ACETATE TO SYNTHESIZE PHA

While an anaerobic/anoxic environment is devoid of oxygen, other oxidants, such as nitrate and sulfate, may be present, replacing oxygen as electron acceptors [25]. The anaerobic respiration of acetate with inorganic oxidant would help to yield energy for PHA synthesis. Both the 3HB-CoA and biomass production from acetate are not thermodynamically favorable, requiring an energy-producing reaction to drive 3-HB-SCoA synthesis. Potential oxidants including sulfate, nitrate and ferric ion (Fe^{3+}) are examined. Using sulfate as an electron acceptor in anaerobic respiration was not found to be energetically favorable (Equation 1.22). In contrast to sulfate, anaerobic respirations with nitrate and Fe^{3+} yield a negative Gibbs energy (Equation 1.23 and 1.24). This indicates that nitrate and Fe^{3+} are capable of supporting PHA production. However, Fe^{3+} and ferrous ion (Fe^{2+}) tend to form precipitate or complex with ligands such as hydroxide, phosphate or sulfide, limiting their bioavailability at pH 7 [42]. PHA formation associated with nitrate anoxic respiration is probably widely present in niche environments such as agricultural soil with nitrate fertilizer application.

$$CH_3COO^- + 0.5HSCoA + 0.5NADH + 0.5H^+$$

$$\rightarrow 0.5CH_3CHOHCH_2CO-SCoA + OH^- + 0.5NAD^+ \qquad (1.20)$$

$$\Delta G^{0'} = +78.67 \text{ kJ/mol}$$

$$CH_3COO^- + 0.4NH_4^+ + 0.2NADH + 0.2H^+$$

$$\rightarrow 2\ CH_{1.8}O_{0.5}N_{0.2} + 0.4H_2O + 0.6OH^- + 0.2NAD^+ \qquad (1.21)$$

$$\Delta G^{0'} = +81.51 \text{ kJ/mol}$$

$$CH_3COO^- + SO_4^{2-} \rightarrow HS^- + 2CO_2 + 2OH^- \quad \Delta G^{0'} = +39.47 \text{ kJ/mol} \qquad (1.22)$$

$$CH_3COO^- + NO^{3-} + 3H^+ \rightarrow NH_4^+ + 2CO_2 + H_2O$$
$$\Delta G^{0'} = -504.91 \text{ kJ/mol} \qquad (1.23)$$

$$CH_3COO^- + 8\ Fe^{3+} + 2\ H_2O \rightarrow 2\ CO_2 + 8Fe^{2+} + 7H^+$$
$$\Delta G^{0'} = -184.22 \text{ kJ/mol} \qquad (1.24)$$

1.4.5 ANAEROBIC FERMENTATION OF GLYCEROL TO SYNTHESIZE PHA

Glycerol is a component in plant and animal lipids, with a carbon redox status that is more reduced than that of glucose [43]. It is also produced as a low-value byproduct during biodiesel production through lipid with methanol in alkalic conditions [44]. As shown by Equation 1.25 and 1.26, both biomass synthesis and 3HB-CoA production from glycerol in anaerobic conditions are exergonic. The synthesis of 3HB-CoA is energetically self-sufficient and does not require any coupling to another energy-producing reaction. This indicates that PHB formation in anaerobic glycerol fermentation is feasible. Indeed, the presence of PHB has been previously detected in glycerol-fermenting anaerobic sludge [18].

$$C_3H_8O_3 + 0.5HSCoA + 2.5NAD^+$$

$$\rightarrow 0.5CH_3CHOHCH_2CO - SCoA + CO_2 + 2.5\ NADH + 2.5H^+ \quad (1.25)$$

$$\Delta G^{0'} = -84.95 \text{ kJ/mol}$$

$$C_3H_8O_3 + 0.5NH_4^+ + 1.75NAD^+$$

$$\rightarrow 2.5CH_{1.8}O_{0.5}N_{0.2} + 0.5CO_2 + 1.75NADH + 2.25H^+ + 0.75H_2O \quad (1.26)$$

$$\Delta G^{0'} = -62.37 \text{ kJ/mol}$$

1.5   CONCLUSIONS AND OUTLOOK

This chapter applies thermodynamics to assess potential microbial PHA synthesis scenarios under aerobic and anaerobic conditions. According to the analysis, in aerobic conditions, fatty acid conversion to PHA is endergonic and it needs the energy from biomass synthesis and cell respiration to drive PHA formation. For glucose however, the synthesis of 3HB-CoA is energetically favorable, and this highlights the impact of different types of carbon sources in 3HB-CoA synthesis. In anaerobic conditions, the absence of oxygen eliminates the possibility of high energy generation through carbon source oxidation. The conversion of acetate to 3HB-CoA is not energetically favorable, but it could be actualized by coupling with the hydrolysis of polyphosphate or anaerobic respiration with nitrate and ferric ions. However, anaerobic respiration with sulfate yields positive Gibbs free energy, suggesting that sulfate reduction cannot support PHA under standard biological conditions. For acetogenesis and glycerol fermentation, the synthesis of 3HB-CoA is exergonic, indicating the possibility of detecting PHA in these processes. On the other hand, syntrophic microbes with acetate and hydrogen production may not form PHA in standard biological conditions. Overall, aerobic PHA formation occurs more readily due to the highly oxidizing environment but should be coupled with an appropriate choice of carbon source for certain types of PHA. Conversely, anaerobic PHA formation is constrained by energy availability but is still found to be thermodynamically favorable in specialized microbes including PAO and acetogens, and conditions such as glycerol fermentation. Aside from the scenarios highlighted in this chapter, there remain many possibilities of carbon and PHA metabolisms, which could be similarly analyzed to predict their thermodynamic feasibility and explain PHA formation/consumption phenomena. Hence, thermodynamics analysis could be an important tool that the scientific community and industry could leverage on during a priori bioprocess design and set-up to maximize PHA yield.

REFERENCES

1. Hecker M, Völker U. *General Stress Response of Bacillus Subtilis and Other Bacteria. Advances in Microbial Physiology.* Vol. 44. London; New York: Academic Press, 2001; pp. 35–91.
2. Madison LL, Huisman GW. Metabolic engineering of poly(3-hydroxyalkanoates): From DNA to plastic. *Microbiol Mol Biol Rev* 1999; 63(1): 21–53.
3. Kulaev I, Kulakovskaya T. Polyphosphate and phosphate pump. *Ann Rev Microbiol* 2000; 54(1): 709–34.
4. Koller M, Gasser I, Schmid F, *et al.* Linking ecology with economy: Insights into poly-hydroxyalkanoate-producing microorganisms. *Eng Life Sci* 2011; 11(3): 222–37.
5. Kadouri D, Jurkevitch E, Okon Y, *et al.* Ecological and agricultural significance of bacterial polyhydroxyalkanoates. *Crit Rev Microbiol* 2005; 31(2): 55–67.
6. Tan G-YA, Chen C-L, Li L, *et al.* Start a research on biopolymer polyhydroxyalkanoate (PHA): A review. *Polymers* 2014; 6(3): 706–54.
7. Salehizadeh H, Van Loosdrecht MCM. Production of polyhydroxyalkanoates by mixed culture: Recent trends and biotechnological importance. *Biotechnol Adv* 2004; 22(3): 261–79.
8. Chen G-Q, König K-H, Lafferty RM. Occurrence of poly-D(–)-3-hydroxyalkanoates in the genus Bacillus. *FEMS Microbiol Lett* 1991; 84(2): 173–6.
9. Feng L, Wang Y, Inagawa Y, *et al.* Enzymatic degradation behavior of comonomer compositionally fractionated bacterial poly(3-hydroxybutyrate-*co*-3-hydroxyvalerate)s by poly(3-hydroxyalkanoate) depolymerases isolated from *Ralstonia pickettii* T1 and Acidovorax sp. TP4. *Polym Degrad Stab* 2004; 84(1): 95–104.
10. Verlinden RAJ, Hill DJ, Kenward MA, *et al.* Bacterial synthesis of biodegradable poly-hydroxyalkanoates. *J Appl Microbiol* 2007; 102(6): 1437–49.
11. Rai R, Keshavarz T, Roether JA, *et al.* Medium chain length polyhydroxyalkanoates, promising new biomedical materials for the future. *Mater Sci Eng: R: Rep* 2011; 72(3): 29–47.
12. Cole M, Lindeque P, Halsband C, *et al.* Microplastics as contaminants in the marine environment: A review. *Mar Pollut Bull* 2011; 62(12): 2588–97.
13. Zinn M, Witholt B, Egli T. Occurrence, synthesis and medical application of bacterial polyhydroxyalkanoate. *Adv Drug Del Rev* 2001; 53(1): 5–21.
14. Zinn M, Hany R. Tailored material properties of polyhydroxyalkanoates through bio-synthesis and chemical modification. *Adv Eng Mater* 2005; 7(5): 408–11.
15. Poblete-Castro I, Escapa IF, Jäger C, *et al.* The metabolic response of *P. putida* KT2442 producing high levels of polyhydroxyalkanoate under single- and multiple-nutrient-limited growth: Highlights from a multi-level omics approach. *Microb Cell Fact* 2012; 11(1): 34.
16. Madigan MT. *Brock Biology of Microorganisms.* 13 ed. San Francisco, CA: Benjamin Cummings, 2012.
17. McInerney MJ, Amos DA, Kealy KS, *et al.* Synthesis and function of polyhydroxyal-kanoates in anaerobic syntrophic bacteria. *FEMS Microbiol Lett* 1992; 103(2): 195–205.
18. Leng L, Nobu MK, Narihiro T, *et al.* Shaping microbial consortia in coupling glycerol fermentation and carboxylate chain elongation for co-production of 1,3-propanediol and caproate: Pathways and mechanisms. *Water Res* 2019; 148: 281–91.
19. Costa KC, Leigh JA. Metabolic versatility in methanogens. *Curr Opin Biotechnol* 2014; 29: 70–5.
20. Schuchmann K, Müller V. Energetics and application of heterotrophy in acetogenic bacteria. *Appl Environ Microbiol* 2016; 82(14): 4056–69.
21. Stephanopoulos G, Aristidou AA, Nielsen J. *Metabolic Engineering: Principles and Methodologies.* San Diego, USA: Academic Press, 1998.

22. Kleerebezem R, Van Loosdrecht MCM. A Generalized method for thermodynamic state analysis of environmental systems. *Crit Rev Environ Sci Technol* 2010; 40(1): 1–54.

23. Atkins PW, De Paula J. *Atkins' Physical Chemistry.* 9th ed. Oxford; New York: Oxford University Press, 2010.

24. Thauer RK, Jungermann K, Decker K. Energy conservation in chemotropic anaerobic bacteria. *Bacteriological Rev* 1977; 41(1): 100–80.

25. Rittmann BE, McCarty PL. *Environmental Biotechnology: Principles and Applications.* Boston: McGraw-Hill, 2001.

26. Bunce JT, Ndam E, Ofiteru ID, *et al.* A review of phosphorus removal technologies and their applicability to small-scale domestic wastewater treatment systems. *Front Environ Sci* 2018; 6(8).

27. Flamholz A, Noor E, Bar-Even A, *et al.* Glycolytic strategy as a tradeoff between energy yield and protein cost. *Proc Natl Acad Sci* 2013; 110(24): 10039.

28. Pan C, Tan G-YA, Ge L, *et al.* Microbial removal of carboxylic acids from 1,3-propanediol in glycerol anaerobic digestion effluent by PHAs-producing consortium. *Biochem Eng J* 2016; 112: 269–76.

29. Chen G-Q, Zhang J, Wang Y. Chapter 16 - White biotechnology for biopolymers: Hydroxyalkanoates and polyhydroxyalkanoates: Production and applications. In: Pandey A, Höfer R, Taherzadeh M, Nampoothiri KM, Larroche C, Eds. *Industrial Biorefineries & White Biotechnology.* Amsterdam: Elsevier, 2015; pp. 555–74.

30. Perry L. McCarty, Daniel P. Smith. Anaerobic wastewater treatment. *Environ Sci Technol* 1986; 20(12): 1200–1206.

31. Feng L, Chen Y, Zheng X. Enhancement of waste activated sludge protein conversion and volatile fatty acids accumulation during waste activated sludge anaerobic fermentation by carbohydrate substrate addition: The effect of pH. *Environ Sci Technol* 2009; 43(12): 4373–80.

32. Yang Y-H, Brigham C, Budde C, *et al.* Optimization of growth media components for polyhydroxyalkanoate (PHA) production from organic acids by *Ralstonia eutropha. Appl Microbiol Biotechnol* 2010; 87(6): 2037–45.

33. Felchner-Zwirello M, Winter J, Gallert C. Interspecies distances between propionic acid degraders and methanogens in syntrophic consortia for optimal hydrogen transfer. *Appl Microbiol Biotechnol* 2013; 97(20): 9193–205.

34. Pijuan M, Saunders AM, Guisasola A, *et al.* Enhanced biological phosphorus removal in a sequencing batch reactor using propionate as the sole carbon source. *Biotechnol Bioeng* 2004; 85(1): 56–67.

35. Mino T, van Loosdrecht MCM, Heijnen JJ. Microbiology and biochemistry of the enhanced biological phosphate removal process. *Water Res* 1998; 32(11): 3193–207.

36. Seufferheld MJ, Alvarez HM, Farias ME. Role of polyphosphates in microbial adaptation to extreme environments. *Appl Environ Microbiol* 2008; 74(19): 5867–74.

37. Liu WT, Nielsen AT, Wu JH, *et al. In situ* identification of polyphosphate- and polyhydroxyalkanoate-accumulating traits for microbial populations in a biological phosphorus removal process. *Environ Microbiol* 2001; 3(2): 110–22.

38. Amos DA, McInerney MJ. Poly-β-hydroxyalkanoate in *Syntrophomonas wolfei. Arch Microbiol* 1989; 152(2): 172–7.

39. Ljungdahl LG. *Biochemistry and Physiology of Anaerobic Bacteria.* New York: Springer, 2003.

40. Latif H, Zeidan AA, Nielsen AT, *et al.* Trash to treasure: Production of biofuels and commodity chemicals via syngas fermenting microorganisms. *Curr Opin Biotechnol* 2014; 27(0): 79–87.

41. Dürre P, Eikmanns BJ. C1-carbon sources for chemical and fuel production by microbial gas fermentation. *Curr Opin Biotechnol* 2015; 35(0): 63–72.

42. Stumm W, Morgan JJ. *Aquatic Chemistry: Chemical Equilibria and Rates in Natural Waters*. 3rd ed. New York: Wiley, 1996.

43. Clomburg JM, Gonzalez R. Anaerobic fermentation of glycerol: A platform for renewable fuels and chemicals. *Trends Biotechnol* 2013; 31(1): 20–8.

44. Ma FR, Hanna MA. Biodiesel production: A review. *Bioresour Technol* 1999; 70(1): 1–15.

45. Thauer RK, Jungermann K, Decker K. Energy conservation in chemotrophic anaerobic bacteria. *Bacteriological Rev* 1977; 41(1): 100–80.

46. González-Cabaleiro R, Lema JM, Rodríguez J, *et al.* Linking thermodynamics and kinetics to assess pathway reversibility in anaerobic bioprocesses. *Energy Environ Sci* 2013; 6(12): 3780–9.

47. Stams AJM, Plugge CM. Electron transfer in syntrophic communities of anaerobic bacteria and archaea. *Nat Rev Microbiol* 2009; 7(8): 568–77.

2 Mathematical Modeling for Advanced PHA Biosynthesis

Predrag Horvat, Martin Koller,
and Gerhart Braunegg

CONTENTS

2.1 INTRODUCTION

Mathematical modeling is a tool that allows the expression of biological facts through systems of mathematical equations. The mentioned systems are usually presented in differential or in integral form. The system is to be solved by well-known mathematical procedures (analytical, graphical, numerical, and computational). It is the question of why scientists have developed such tools, and why these tools were applied in microbiological, chemical, biochemical, and biotechnological fields. Among others, one (perhaps the main) reason is the attempt to substitute cumbersome experimental work with simulated results. Two different problems often appear in such approaches: the problem of interpolation and the problem of extrapolation, both connected with the validity of mathematical models. The development of mathematical models for biotechnological purposes can be described in a few consecutive steps:

- First, the experiments are performed, encompassing a certain range of conditions;
- The data from experiments are then subjected to stoichiometric and kinetic analysis;
- Based on kinetic analysis, the mathematical model is structured;
- The model is finally validated by comparison with experimental results.

The problem of interpolation is subdued on the calculation of results by mathematical models, respecting the trustable validity range of given models. That means, if the result is within the range of experimental results (conditions) that were used for establishing the model, then interpolation occurs. Further, if the simulated results lie out of the range of the experimental results (conditions) that were used for developing the model, the problem of extrapolation is "on the table". Interpolation is much more frequently used and significantly less sensitive than extrapolation. The accuracy of interpolated results is more trustable than that of extrapolated ones. Model theory classifies models into several classes, usually cited together with their antipodes: verbal/non-verbal, descriptive/explanatory, black-/grey-/white-box, unstructured/structured, non-mathematical/mathematical, deterministic/stochastic, discrete/continuous, and distributed (non-segregated)/segregated. Different combinations of these model classes can be established. Usually, a mathematical model established to handle a certain biological case belongs to more than one class. The consequence is that very different model types can be found in the literature describing multifarious biotechnological cases, encompassing cultivation techniques, mass transfer, growth, and metabolic network reaction kinetics. In this context, polyhydroxyalkanoates (PHA) biopolyester biosynthesis has grabbed the attention of the scientific community for a long time. The possibility that biosynthetic PHA could replace conventional plastic materials, which are typically produced from fossil fuels, constitutes a driving force for this increasing attention. When we look back in the past, an astonishing quantity of mathematical models dealing with different aspects of PHA synthesis was published in the scientific literature. That is why the intention of this chapter is to present a comprehensive review of mathematical models that were applied for describing and comparing the achieved results and emphasizing the benefits and shortcomings of different modeling approaches. Special attention is dedicated to those models contributing to process improvements, which encompasses yields, downstream processing, microbial strain selection, or genetic engineering. Hence, mathematical models used for the optimization of microbial metabolism, product formation, cultivation techniques, and metabolic networks are the target of this chapter. Special attention is dedicated to the cellular metabolism using analysis of metabolic pathways, gene expression, development of signaling procedures, and regulation strategies investigation. The knowledge of intracellular metabolic networks and bioinformatics techniques is currently indispensable for the optimization of microbial industrial-scale production processes. Therefore, this chapter encompasses kinetics of PHA biosynthesis both on the "micro" and "macro" level (enzyme kinetics, kinetics of microbial growth, and PHA synthesis), mathematical modeling of PHA biosynthesis (unstructured and low-structured formal kinetic mathematical models, monoseptic- and mixed culture dynamic models, metabolic models, cybernetic models in PHA biosynthesis, modeling of PHA biosynthesis by neural networks, and hybrid models), and metabolic pathway and metabolic flux analysis methods in modeling of PHA biosynthesis (encompassing elementary flux modes in microbial PHA biopolyester synthesis).

2.2 KINETICS OF PHA BIOSYNTHESIS

Considering PHA biosynthesis from the kinetic point of view, two different approaches can be used. The first one is the micro-kinetic approach that refers to the kinetic properties of intracellular enzymes involved in the formation of PHA granules. From a wider perspective, this aspect also covers the enzymatic depolymerization of PHA that occurs when these storage materials serve as a reserve carbon source during starvation periods. From a technological point of view, the macro-kinetic consideration is more attractive. It involves finding relations between nutrient concentrations (i.e. usually C-, N- and P-sources) and specific rates of microbial growth and depletion of nutrients as well as the product formation, respectively.

2.2.1 ENZYME KINETICS

The observations of enzyme kinetics described in this chapter are predominantly related to PHA-synthetizing prokaryotes, encompassing eubacteria and archaea. Here it is necessary to mention that there are a few hundred species reported to accumulate PHA, which makes it somewhat difficult to discuss enzyme kinetics in PHA biosynthesis in general. Moreover, there are certain differences in enzymes involved in short-chain-length (*scl-*) and medium-chain-length (*mcl-*) PHA synthesis, respectively. In addition, multiple isoenzymes are usually present in prokaryotes, yeasts, plants, and mammalian cells, and the PHA biosynthesis is not an exception here. Therefore, the behaviors of enzymes presented below should be considered as more or less typical examples. In the intracellular process of PHA synthesis, three enzymes are predominantly involved. Those are:

- E.C. 2.3.1.16: Acetyl-CoA C-acyltransferase, well known as 3-ketoacyl-CoA thiolase (formerly known as β-ketothiolase)

- E.C. 1.1.1.36: Acetoacetyl-CoA reductase, (alternatively E.C. 1.1.1.35: 3-hydroxyacyl-CoA dehydrogenase or β-keto-reductase)
- E.C. 2.3.1.B2 (B4, B5) PHB synthase (polymerase). /preliminary BRENDA (Braunschweig Enzyme Database)-supplied EC numbers

PHA degradation in prokaryotic cells is catalyzed by the action of PHA depolymerase:
- E.C. 3.1.1.75: Poly(3-hydroxybutyrate) depolymerase; known as PHB depolymerase, Poly((R)-hydroxyalkanoic acid) depolymerase, Poly(3HB) depolymerase, poly(scl-HA) depolymerase, and poly(HA) depolymerase.

2.2.1.1   E.C. 2.3.1.16: Acetyl-CoA C-Acyltransferase

Although earlier investigations can be found in the literature (e.g. by Oeding and Schlegel) [1], the typical PHA synthesis enzymes isolated from *Alcaligenes eutrophus* NCIB 11599 (today *Cupriavidus necator*) by Haywood et al. [2] will be described here. In this work, two different constitutive isoenzymes were reported, both performed a reversible reaction with a ping-pong mechanism:

$$acyl\text{-}CoA + acetyl\text{-}CoA \leftrightarrow CoA + 3\text{-}oxoacyl\text{-}CoA$$

The kinetic and molecular properties of isolated enzymes are presented in Table 2.1. Free enzyme A (CoA) inhibits the condensation reaction by both isoenzymes, which indicates the effective control of the activity in PHA biosynthesis during cell growth. During nutritionally balanced growth, when the intracellular concentration of CoA

TABLE 2.1
The Properties of Acetyl-CoA C-Acyltransferase Isolated from *Alcaligenes eutrophus* NCIB 11599 (Today *Cupriavidus necator*) [2]

Property	Enzyme A	Enzyme B	Units
Molecular mass (by gel filtration)	170,000	160,000	u(Da)
(by centrifugation)	171,000	175,000	u(Da)
Subunits (SDS_PAGE)	44,000	46,000	u(Da)
Activity (thiolysis reaction)	115	11.4	U/mg
K_m (acetoacetyl-CoA)	44	394	μM
K_m (CoA)	16	93	μM
K_m (acetyl-CoA)	1.1	0.23	mM
Inhibitor of condensation: CoA	Competitive, $K_i = 16$	Positive cooperativity or complex interaction	μM
pI	5.0	6,4	
Substrates	Acetoacetyl-CoA and weak reaction with 3-ketopentanoyl-CoA	Acetoacetyl-CoA and C_4-C_{10} 3-ketoacyl-CoAs	

is high (due to the action of citrate synthase), this compound acts as the inhibitor of acetyl-CoA C-acyltransferase and prevents the accumulation of PHA.

2.2.1.2 E.C. 1.1.1.36: Acetoacetyl-CoA Reductase

This enzyme catalyzes the second reaction step in PHA biosynthesis:

$$3\text{-oxoacyl-CoA} + \text{NADPH} \leftrightarrow (R)\text{-3-hydroxyacyl-CoA} + \text{NADP}^+$$

Two constitutive acetoacetyl-CoA (AcAc-CoA) reductases were purified from *Alcaligenes eutrophus* (today *Cupriavidus necator*) by Haywood et al. [3]. Based on incorporation of [C^{14}]-acetyl-CoA into poly(3-hydroxybutyrate) [PHB; a.k.a. P(3HB)] it was concluded that only reduced nicotinamide adenine dinucleotide phosphate (NADPH)-AcAc-CoA reductase is involved in PHB biosynthesis. The specificity of two reductases was different: AcAc-CoA NADH reacting reductase was active with all (C_4–C_{10}) D(-) and L(+)-3-hydroxyacyl-CoA substrates. On the contrary, the NADPH reacting AcAc-CoA reductase was active only with (C_4–C_6) D(-)-3-hydroxyacyl-CoAs. Interestingly, the products of AcAc-CoA reduction were also different. For the NADH-linked enzyme, the result of the reaction was L(+)-3-hydroxybutyryl-CoA, while the NADPH-linked enzyme produced D(-)-3-hydroxybutyryl-CoA. The NADH-linked enzyme had identical M_r 30,000 subunits and mass of $M_r = 150,000$, while the NADPH-linked enzyme seems to be a tetramer (identical subunits with M_r of 23,000) with a mass M_r of 84,000. K_m values of 22 µM and 5 µM for AcAc-CoA were estimated for the NADH- and NADPH-linked enzymes, respectively. Michaelis constants of 13 µM for NADH and 19 µM for NADPH were determined for the related enzymes.

A short overview of some microbial NADP-AcAc-CoA-reductases is summarized in Table 2.2.

TABLE 2.2

An Overview of Properties of Microbial NADP-AcAc-CoA-Reductases According to Belova et al. [4]

Microorganism	M_r (u)	Subunit M_r (u)	K_m (mM) NADPH	K_m (mM) AcAc-CoA	K_m (mM) NADP	K_m (mM) D-hydroxybutyryl-CoA
Methylobacterium extorquens 15	140,000	32,000	41	11	–	–
Methylobacterium rhodesianum MB 126	250,000	42,000	18	15	60	30
Alcaligenes eutrophus	84,000	23,000	19	5	31	33
Azotobacter vinelandii	77,000	–	20*	11#	–	–
Zoogloea ramigerea	92,000	25,500	21	8,3	–	–

* at 32 mM AcAc-CoA,
# at 100 mM NADPH

Respecting the high number of microbial species that produce PHA of various compositions, different polyester synthases were isolated and characterized. Their biocatalytic function is that of an acyltransferase, hence, it transfers acyl-groups. In the special case of PHB homopolyester biosynthesis, the following reaction can be defined:

$$(R)\text{-3-hydroxybutanoyl-CoA} + \left[(R)\text{-3- hydroxybutanoate}\right]_{(n)}$$

$$\rightarrow \left[(R)\text{-3- hydroxybutanoate}\right]_{(n+1)} + CoA$$

In addition, for other PHA, a more general formula can be applied:

$$(R)\text{-3-hydroxyalkyl-CoA} + \left[3(R)\text{-hydroxyalkanoate}\right]_{(n)}$$

$$\rightarrow \left[(R)\text{-3-hydroxyalkanoate}\right]_{(n+1)} + CoA$$

Here it is necessary to state that the action of these polyester synthases is tightly bounded with the carbonosome structure. So, such a structure with associated reactions can be considered as a heterogeneous system. Actually, this micro-location in the cells consists, among others, of the short chain primer, the PHA synthase (PhaC), their activator (PhaM), phasins, acetyl-CoA acetyltransferase, acyl-CoA synthetase, polyester depolymerase and a regulatory protein, PhaR. The central role of acetyl-CoA acetyltransferase and acyl-CoA synthetase enzymes is probably the reduction of the local concentration of free CoA because of its feedback inhibition of the main reaction.

Based on the sequential analysis, molecular mass, subunit composition, primary structures, and the substrate specificities, four classes of PHA synthases were proposed and reviewed [5] (Table 2.3).

According to reference [5], there are exceptions that are not in agreement with systematization in Table 2.3. These are the synthases from *Thiocapsa pfennigii* (85% identity to class III; two different subunits highly similar to the PhaC), the synthase

TABLE 2.3

Four Classes of Polyester Synthases According to Rehm [5]

Class	Preferential substrates	M_w (u)	Subunits	Microorganism
I	$C_3\text{-}C_5$ 3-hydroxyacyl-CoA 4-hydroxyacyl-CoA 5-hydroxyacyl-CoA	60,000–73,000	PhaC	*Ralstonia eutropha*
II	$C_6\text{-}C_{14}$ 3-hydroxyacyl-CoA	60,000–65,000	PhaC	*Pseudomonas aeruginosa*
III	$C_3\text{-}C_5$ 3-hydroxyacyl-CoA $C_6\text{-}C_8$ 3-hydroxyacyl-CoA 4-hydroxyacyl-CoA 5-hydroxyacyl-CoA	40,000 40,000	PhaC PhaE	*Allocromatium vinosum*
IV	$C_3\text{-}C_5$ 3-hydroxyacyl-CoA	40,000 22,000	PhaC PhaR	*Bacillus megaterium*

from *Aeromonas punctata* (45% similarity to class I; one type of subunit), from *Pseudomonas* sp. 61-3 (80% identity; two subunits PhaC1 and PhaC2; strong similarity to class II.), and PHA synthases from extremely halophilic archaea.

Here it is necessary to mention that the PHA granule structure is a complex system. The "scaffold model" is the perhaps most accurate model for PHA synthesis within bacterial cells. According to this model, initial points of PHA granules synthesis are placed, more or less, along the length axis of the cell. Within the cell, the PHA synthase of nascent PHB granules is attached to a scaffold (according to Jendrossek and Pfeiffer, as well as to Mezolla et al.) [6, 7]. PhaM reacts very specifically with PhaC1 protein and with phasin (PhaP5), but an interaction with DNA is present, too. That is probably why PHB granules may be found bounded (attached) to the cell nucleoid. Diversity and distribution of PhaC in bacteria, crystal structure, catalytic mechanism, mutation, and amino acid substitution studies are reviewed in reference [7] by Mezolla et al.

2.2.1.3 E.C. 3.1.1.75: Poly(3-hydroxybutyrate) Depolymerase

This enzyme is crucial for degradation of PHA polyesters when it becomes necessary that the cells mobilize internal carbon reserves. It catalyzes the subsequent reaction:

$$\left[-(R)\text{-3-hydroxybutanoate-}\right]_{(n)} + H_2O \Leftrightarrow \left[-(R)\text{-3-hydroxybutanoate-}\right]_{(n-1\,to\,-5)}$$

$$+ \left[-(R)\text{-3-hydroxybutanoate-}\right]_{(1-5)}$$

This reaction also occurs with esters of other short-chain-length (C(1)–C(5)) hydroxyalkanoic acids (HA). Here, it is necessary to state that one must clearly distinguish intracellular PHA depolymerases from extracellular PHA depolymerases. The first has an important function in PHB metabolism inside the cell. In contrast, the released PHB from dead PHB-accumulating bacteria is hydrolyzed by extracellular PHB depolymerases present in many microorganisms. E.C.3.1.1.75 is described for species from very different genera (Anonymous 1) [8]: *Acidovorax, Agrobacterium, Alcaligenes, Arthrobacter, Aspergillus, Azohydromonas, Azospirillum, Azotobacter, Bacillus, Comamonas, Cupriavidus, Delftia, Leptothrix, Mycobacterium, Paucimonas, Penicillium, Pseudomonas, Purpureocillium, Rhodobacter, Rhodospirillum, Schlegela, Sinorhizobium, Sphingomonas, Streptomyces, Talaromyces, Thermobifidia,* and *Thermus.* The K_m values reported for polyester depolymerases from different organisms were in the range 60, 10, and 1 mM (see Table 2.4). In this chapter, the properties of *Cupriavidus necator* polyester depolymerase will be discussed as a typical example.

There are some striking facts about the above reaction. According to earlier results based on PHB depolymerase activity in vitro (related to partially crystalline PHB with damaged protein outer layer) [9–11], the product of polyester cleavage is a free monomer acid. Recent findings revealed by Jendrossek and Pfeiffer [6] and Uchino et al. [12] indicate that the above reaction performed in vivo (related to native PHB granule covered by protein layer) is a thiolysis reaction with the CoA-bounded acyl as the final product. These results partially confirmed the earlier findings about

TABLE 2.4

Substrate Specificity of PhaZd and Other Intracellular PHB Depolymerases in *W. eutropha* H16 (Today *C. necator*) According to [9]

Substrate	Activity [μmol/(min mg)]			
	PhaZd	PhaZa1	PhaZb	PhaZc
Amorphous PHB granules	110	0.71	3	3
Native PHB granules	0.43	0	0	0
Semicrystalline PHB granules	0	0	0	0
3HB oligomers				
Dimer	1.2	0	58	31
Trimer	2.6	0	59	200
Tetramer	0	1.1	54	240
Pentamer	n.d.	1.2	50	310
p-Nitrophenyl acetate (*R*)	12	0.093	0	0.022
p-Nitrophenyl butyrate (*R*)	10	0.074	1.1	0.24
p-Nitrophenyl palmitate (*R*)	0	0	0	0
Olive oil	0	0	1.8	0

the frequently cited "cyclic nature of poly(3-hydroxyalkanoate) metabolism in *Alcaligenes eutrophus*", as proposed by Jendrossek; Doi et al.; Taidi et al. [13–15].

Furthermore, multiple PHB depolymerases and two 3HB oligomer hydrolase genes were reported to be included in PHB metabolism in *R. eutropha* (today *C. necator*) [12]. An overview of PHA proteins detected in purified PHB granules of *R. eutropha* H16 and proteins with prominent function in PHA metabolism in different bacteria is given by Jendrossek and Pfeiffer in reference [6]. Interestingly, only *PhaZa1*-encoded depolymerase is reported to be attached to PHB granules of *R. eutropha* (today *C. necator*) and involved in the PHB mobilization.

Considering data reported in literature published by Jendrossek and Pfeiffer [6], Mezolla *et al.* [7], Abe et al. [9], and Uchino et al. [12], it is unavoidable to mention that three different types of PHB in living cells have to be distinguished. They differ functionally and in chain length:

- storage PHB of high molecular mass ($>10^3$ monomer units); it appears as granules ("carbonosomes") in microorganisms,
- oligo-PHB with chain length of $\approx$100 to 200 3HB units,
- conjugated PHB which is low in numbers of monomer units ($<\approx$30) and is covalently linked to proteins.

Types (b) and (c) are molecules generally present in a broad spectrum of living cells, and display important biological functions [16, 17]. It seems to be that some extracellular microbial polyester depolymerases have the biological function to degrade such PHB chains when cells perished. On the contrary, intracellular polyester depolymerase activity (on native PHB granules) is probably tightly bonded with the carbonosome (PHB granule) structure, with the presence of regulatory proteins and phasins on the surface as well as with the presence of key metabolites (CoA, acetyl-CoA, 3HB-CoA, NAD/NADH), which affects the storage type PHB synthesis and degradation.

2.2.2 Microbial Growth and Kinetics of PHA Synthesis

It is well known from the early phases of microbial growth investigations that, considering cell number, unlimited nutritionally balanced growth is of exponential nature. After the concentration of growth-essential nutrients is reduced to a certain level, the growth rate slows down and progressively declines from the exponential pattern. Further nutrient depletion leads to an equilibrium between cell division and the death rate (described in the literature as the "stationary phase" of growth). After all nutrients are exhausted, the number of living cells decreases because the cell death rate reaches its maximum and no more growth (cell division) takes place. Again, autolyzed cells can be a source of nutrients for the surviving cells; in such cases, a small number of remaining live cells start to express so-called "cryptic growth". Details about microbial growth kinetics can be found elsewhere (e.g., in work published by Moser [18, 19]). From the biotechnological point of view, it is of interest to know the dynamic of useful metabolites (products) synthesis. Respecting the vast number of microbial species in nature, its different metabolic "constructions" and the enormous variety of diverse nutritional sources, very different relations between growth rates and product synthesis rates have been observed for different strain-substrate-product combinations. Gaden [20] has defined three general types of product formation patterns (considering different growth phases and their relationship with growth rates) that can also be applied for describing PHA biosynthesis:

(1) Growth-associated product formation (the rate of product synthesis is chronologically coordinated with the growth rate). Such growth-associated product formation is directly linked to the energy metabolism of carbohydrates/C-sources; such growth-associated products constitute products of the primary metabolism.
(2) Indirect products of carbohydrate (substrate) catabolism (the time course of synthesis is only partially overlapped with growth curve; "partially growth-associated product formation").
(3) Product formation is apparently unrelated to the rate of carbohydrate oxidation and to the growth rate ("not growth-associated product formation"); typical products of the secondary metabolism.

Gaden's study was based on excreted products synthesized from carbohydrates used as C-sources. Later on, these findings were confirmed for different other C-sources used by organisms for energy production and for synthesis of cell components. In this context, PHA are very different products if compared with those studied by Gaden: they are intracellular, insoluble in the cytosol and they serve as the C-storage materials. Despite that, Gaden's kinetic types of product synthesis can be, under some conditions, also applied to these processes. Accordingly, among the great number of described cultivations of PHA synthetizing microorganisms, three different classes of microbial PHA producers can be proposed:

(i) Class I – the strains with extremely strict separation of time courses related to biomass growth and PHA production phase. Onset of the production phase is usually provoked by N-source or P-source growth limitation combined with obligate excess of exogenous C-source (typical organisms: *Pseudomonas* sp. 2F, *Methylomonas extorquens*) [21].

(ii) Class II – those strains that accumulate certain, rather small, quantities of PHA during the main growth phase (i.e. under balanced nutritional conditions), but which display accelerated PHA biosynthesis and accumulation during the non-growth phase (usually provoked by simultaneous excess of C-source and limitation of N-, P- or other nutrient sources). Typical organism: *Cupriavidus necator* [22].

(iii) Class III – those strains able to hold high PHA production rates during whole growth phase (nutritional limitation is not necessary to have high PHA production rates) [23–26]. Typical organisms: *Azohydromonas lata* DSM 1122, or *Pseudomonas putida* GPo1 ATTC 29347.

The cited kinetic characteristics of PHA-producing bioprocesses are the basis for the development of appropriate cultivation techniques intended to improve the production of bio-based PHA. Definitively, it is practically impossible to find any cultivation technique that was not applied and described for PHA biosynthesis in the literature: batch, fed-batch, repeated batch, repeated fed-batch as well as continuous processes in one-, two-, and multi-stage systems were published. They have been exhaustively reviewed for both pure and mixed microbial cultures [22, 27].

Keeping in mind the great metabolic and systematic diversity of microbial species that produce PHA, it is not surprising that a variety of different C-sources were applied in cultivations. These encompass mainly traditional C-sources like carbohydrates; however, the search for inexpensive C-sources is steadily emerging, and non-traditional materials from agriculture, forestry, food, and industrial production, usually treated as a waste in these sectors (i.e. whey, molasses, wood hydrolysates, alcohols, fatty acids, hydrocarbons, meat and bone meal, hydrolysates, etc.) are used more and more as raw materials to cultivate PHA-accumulating strains. As a consequence, a great plurality of kinetic behaviors of PHA-producing microorganisms growing on such C-sources was described. Some of these substrates were inhibitory and toxic in certain concentration ranges. Hence, kinetic analysis and mathematical modeling were applied with the aim of enhancing the rate of PHA synthesis and PHA productivity by different production strains on different substrates. An overview of these works is given in the next section of this chapter.

2.3 MATHEMATICAL MODELING OF PHA BIOSYNTHESIS

According to the theory of mathematical modeling, models can be classified into several types. Often, each subtype is opposed to its antipode. So, in the literature, verbal/non-verbal, descriptive/explanatory, black-/grey-/white-box, unstructured/ structured, non-mathematical/mathematical, deterministic/stochastic, discrete/ continuous, and distributed (non-segregated)/segregated model types have been reported. Furthermore, the combinations of the above-cited model classes lead to new divisions: kinetic, dynamic, logistic, and cybernetic models. The real situation in daily praxis teaches us that one individual model, developed for a certain biological situation, can be deployed to more than one class-type. Kinetic models contain two general sub-types: unstructured and structured. In addition, unstructured kinetic models are reported as formal kinetic and mechanistic models, and structured kinetic models are classified in several subclasses: morphological, compartmental, metabolic, and chemically related. Structured, mathematical, continuous, and segregated sub-types are more or less considered as the dynamic model types. Hundreds of microbial PHA producers combined with various different types of substrates and with all possible cultivation strategies have led to the application of very different models. In the current literature, it is impossible to find any model type that has not yet been applied to describe PHA biosynthesis. Such a multitude of mathematical modeling material (extensive, comprehensive, and sophisticated in its nature) is not easy to systematically summarize in accordance with scientific principles. Despite that, the reader can refer to some reviews dealing with PHA biosynthesis modeling [28–30]. In this chapter, attention is given to recent advances in this field, especially to applications of mathematical models for the upgrading of biotechnological processes, for improvement of product yields, strain development, and investigation of metabolic networks.

Modern mathematical modeling of bioprocesses follows a certain procedure consisting of a few steps. In certain auspicious fortunate circumstances, some steps can be avoided, but, in general, all of them should be performed. According to Moser [18, 19] and Kossen and Oosterhuis [31], the steps necessary for developing successful mathematical modeling are:

(a) Perform kinetic analysis (specific growth-, substrate consumption-, product synthesis rates, yields, and stoichiometric coefficients).
(b) Recognize the kinetic behaviors (e.g. inhibitions, activations, limitations, etc.).
(c) Determine the related kinetic constants (or use those reported in the literature).
(d) Set up the appropriate mathematical model in differential or integral form respecting the aim and the scope of the model (interdependent system of mass balance and kinetic equations).
(e) Transfer the model on a computer equipped with a self-made or commercial program for solving of equations system.
(f) Simulate the time course of the bioprocess in silico on the computer and, if necessary, make the model corrections.
(g) Use an appropriate optimization procedure to optimize all kinetic- and model parameters by comparison of experimental and simulated data (error minimization).
(h) Test the ability of the model for interpolation and extrapolation.

When validated, a mathematical model can be used for different purposes: broth optimization, maximization of productivity, energy saving, time courses optimization, feeding strategies, development and comparison of strains, investigation of metabolic networks, etc.

2.3.1 Unstructured and Low-Structured Formal Kinetic Mathematical Models

This type of mathematical model belongs to the "modest" ones in terms of mathematical sophistication. They are the simplest among plenty of models applied for describing PHA biosynthesis. The first step in the development of models is usually the kinetic analysis of a bioprocess. Its goal is to determine the time courses (changes) in biochemical rates of biomass growth (dX/dt), substrate(s) consumption (dS/dt) and product(s) formation (dP/dt), respectively. After that, the related specific rates are calculated:

specific growth rate μ = (dX/dt)*(1/X);
specific substrate consumption rate q_s = (dS/dt)*(1/X); and
specific product formation rate q_p = (dP/dt) *(1/X).

In parallel, yield coefficients and stoichiometric coefficients (biomass to substrate and product to substrate) must be determined based on experimental data. The next step in mathematical modeling is the establishment of kinetic relations (dependence of rates on biomass (X), growth-determining substrate (S) and product (P, i.e. PHA) concentrations, respectively). In this type of mathematical modeling, the internal biochemical cell metabolism is not considered; therefore, such models are labeled as "formal kinetic" and as "black box" type models. In the early days of PHA biosynthesis modeling, the formal-kinetic, black box model type was exclusively performed. Applying the second Monod equation to mathematically connect the substrate (S) concentration (the argument) and the specific growth rate (μ) (the function), this approach was similar to models used for synthesis of primary metabolites. Solutions like those just mentioned were successful for Class III (iii) (growth-associated) PHA producers, but they were not applicable for the other two classes, where microbial growth and product synthesis are more or less temporarily separated. For these early works in this field, the reader should refer to references revealed by [32–36]. The main problems at that time can be addressed to the applied cultivation media ingredients (or, better to say, to the related concentrations). Microbial cultivation media in these cultivations were C-source(s) acting as growth-determining nutrient(s), while the N-, P-, O-sources were available in surplus. Soon after that, it was noticed that such cultivation strategies are not the best way to achieve high PHA biosynthesis yields (except for the above-mentioned kinetic class III producers). Additionally, it was proven that the PHA mass fraction of the whole biomass (X) reaches a high value if compared with the residual, biologically active part (X_r; up to 0.9 g/g of X); that gave the reason for launching a new modeling approach based on the application of "biomass structuring" by Sonnleitner et al. [32] and Heinzle and Lafferty [33], who have established the principle of "two main biomass compartments" in the modeling of PHA biosynthesis:

Compartment (1) was named "residual biomass" (Xr), which constitutes the biocatalytically active fraction of cells (encompassing nucleic acids, glycolipids, glycoproteins, phospholipids, cytosol, cytosolic, as well as membrane-bound proteins, i.e. active enzymes and structural proteins).

Compartment (2) refers to the PHA fraction, considered as the intracellular biocatalytically inactive (inert) biomass part. This compartment consists of the polymeric storage material.

This compartmental segregation is applicable and still valid, but it becomes more and more obvious that it needs to be updated; this is mainly in light of the carbonosome and scaffold hypothesis, which postulates the presence of phasins and catalytic enzymes on granule surfaces, which contradicts the paradigm of PHA granules being just inert inclusion bodies.

This biomass structuring has been of considerable influence on mathematical modeling. Whole biomass (X; usually experimentally determined by dry matter gravimetric method and expressed as volumetric concentration in [g/L]) was defined as the sum of masses: residual biomass (Xr) plus storage material (X_{PHA}). All specific rates, i.e. those related to microbial growth, substrate consumption and product formation were defined by residual biomass (Xr). Product concentrations (i.e. intracellular storage polyesters) were presented as a mass fraction [g_{PHA}/g_X] in biomass (X) and, if necessary, recalculated to its volumetric concentration.

For some producers from Class I and II, it was experimentally proven that the nitrogen (N-) and phosphate (P-) sources can be used as limiting substrates if a surplus of C-substrates is present in the cultivation broth. This way, the cultivation can be divided into two parts:

• Nutritionally balanced growth, and
• Phase of enhanced (or triggered) PHA production.

Based on the above mentioned, N- and/or P-source limitation was reported as the inducing factor that triggers (or enhances) PHA synthesis in the second stage of cultivation. In order to encompass both parts of cultivation for the mathematical modeling of Class I and II producers, some authors assumed that the rate of PHA formation is related to both: the growth rate (during the first phase) and the residual biomass (Xr) concentration (during the second phase of cultivation). The Luedeking-Piret equation [37] applied in these cases has been very powerful in this context, giving good correlation between model values and experimental results. However, N- and/or P-source growth-limitations challenged the adaptation of the Monod relationship to describe specific growth rate (μ); therefore, double- or even triple- substrate versions of the Monod relationship (previously reported by [38]) have been introduced instead of the single substrate type. When using natural media of higher complexity (or byproducts from industrial production usually constituted of several C- and/or N-sources), this relationship was further extended to a multi-substrate Monod equation. Except for this multi-substrate Monod equation, the logistic equation was used for the growth rate determination in order to achieve a satisfying match of experimental and computed model data [39].

The application of novel precursors, organic acids (for copolymer production), or industrial waste materials (to reduce the production costs) often negatively influences growth kinetics because of the toxic properties of such compounds. That was the reason for introducing substrate inhibition terms in some model rate equations. As examples, the sigmoidal term from enzyme kinetics reported by Hill [40], the term with negative exponential behavior reported by Aiba et al. [41], and the terms expressed by Webb, Andrews, and Noack, as well as by Yanno et al. (all presented in reference [18]) can be highlighted here. Analyzing the implementation of substrate inhibition terms in the literature, two cases of successful application can be found:

1) When primary or secondary C-sources (organic molecules) have a negative impact on the growth rate (e.g. the solvent effect of methanol or glycerol, the pH-value influence of organic acids; the osmotic pressure impact in the case of sugars and salts, or multi-substrate broth composition expressing a catabolic repression effect).

2) Nutritionally balanced concentration of N- and/or P- sources, needed in the exponential growth phase for biomass growth, was modeled as an inhibiting factor of the specific PHA production rate. When depleted, the critically low concentrations of these compounds were used in models as the "triggering variable in charge", which starts (or enhances) PHA synthesis. This is a typical example of formal-kinetic approaches. Formal mathematical models use the mentioned essential nutrients concentration as the only decisive factor triggering PHA biosynthesis, although that is not always entirely true, and often not in accordance with real metabolic mechanisms, which are much more complex.

Increasing pressure for PHA price reduction has intensified the search for, and investigation of novel wild-type microbial production species, and genetic manipulation of established and novel wild-type strains. This resulted in a broader spectrum of convertible substrates and better kinetic performances in PHA biosynthesis. The impact on mathematical modeling was unavoidable: more sophisticated approaches were requested, for example an introduction of maintenance energy terms in model equations, consideration the steric hindrance of PHA synthesis when a certain mass fraction biomass is achieved (the mobility of cytoplasm is hindered by high quantity of inclusion bodies), dissolved oxygen influence on growth and PHA synthesis, as well as the development of low structured models based on simplified metabolic pathways and on enzyme- or formal kinetic principles. Taking this into account, the Guthke equation version [18, 19] of the Monod type, which describes the dependence of the maintenance energy rate on C-source concentration, the dissolved oxygen concentration as a factor influencing specific growth and specific production rates (Tohiyama et al. [42]), the product inhibition terms established by Aiba et al. (1968) [41] as well as those proposed by Luong ([43, 44]) for steric hindrance were reported as a useful tool. Further, as an example of the necessity of establishing low structured mathematical models, the work published by Koller et al. can be specified [45]. These authors published a pure formal kinetic type of model for the haloarchaeon *Haloferax mediterranei* cultivated in batch and fed-batch mode; the terpolyester poly(3-hydroxybutyrate-*co*-3-hydroxyvalerate-*co*-4-hydroxybutyrate) was produced using a multi-substrate medium containing glucose, galactose, γ-butyrolactone. Here, it should be emphasized that in this case, biomass and the terpolyester were structured according to the two-compartments presumptions by Sonnleitner et al. [32] and Heinzle and Lafferty [33], any structuring of metabolic pathways was absent. For this process, the simple formal kinetic model type was sufficient for modeling the cultivation. On the contrary, in the case of homopolymer poly(3-hydroxybutyrate) (PHB) biosynthesis by *Pseudomonas hydrogenovora* (today: *Burkholderia fungorum*), the pure formal kinetic modeling approach was unsatisfying, and a certain structuring of metabolic space by including feedback control mechanisms was unavoidable, hence, a low-structured model was applied. The different modeling approaches just described were grounded on the fact that the archaeal galactose catabolic pathway(s) for *Hfx. mediteranei* were completely unknown. In general, for this haloarchaeon, all metabolic pathways and basic regulatory mechanisms were not described in detail in those days, so the lack of metabolic information was the reason for choosing the formal kinetic approach. In the second case (i.e. *P. hydrogenovora*), either a set of multiple subsequent metabolic chain reactions of special metabolic pathways was merged in one kinetically defined (representative) reaction, or the slowest reaction in a chain (or the most sensitive on the concentration of regulating factors/signaling molecules) was acknowledged as a "representative". This way, the complex metabolic network was simplified as much as possible, and, additionally, the intracellular pools of key metabolites were integrated into the model. That means that excepting previously described biomass structuring, an extremely simplified structure of the metabolic network is present in models of the low-structured formal kinetic type. This way, the model of PHA biosynthesis by *P. hydrogenovora* has been defined by the presence of the native biomass space (residual biomass Xr + PHA), the metabolic space (intracellular pools of metabolites and biocatalysts), and the broth space (substrates and excreted metabolites). In Table 2.5, the modeling presumptions for both cultivations (*H. mediteranei* and *P. hydrogenovora*) are presented.

Formal kinetic, unstructured, and structured mathematical models were successfully applied for investigation of substrate feeding strategies, production of heteropolymers, and process parameter optimizations, and for studying one-, two-, and multistage continuous cultivations. An example of such application of mathematical models is the modeling of a five-step continuous cultivation of *C. necator* DSM 545 performed by Atlić et al. [46]; this process was modeled by Horvat et al. [47]. The main idea of developing a five-stage cultivation process was to establish the different nutritional and process conditions in each stage of the five-stage cascade, having different substrates, concentrations, temperatures, pH-values, and dissolved oxygen [DO] levels in each stage; these factors should be decisive for "formation" of new types of biopolyesters (blocky structured chains, alternating polymer segments, desired molecular mass distribution). The first stage of the five-stage cascade was foreseen to ensure the nutritionally balanced biomass growth at high cell density, the second stage was planned to be the nitrogen-source depletion step (hence, the "PHB synthesis triggering step"), while the last three subsequent reactors acted as the PHA producers and "adjusters" of polymer composition. The modeling principles used in this work were:

- Partially growth-associated PHB synthesis in N-limited conditions (Luedeking-Piret's relation).

TABLE 2.5

Comparison of Modeling Presumptions for *H. mediteranei* and *P. hydrogenovora* Cultivations Performed for Production of Poly(3-hydroxybutyrate-*co*-3-hydroxyvalerate-*co*-4-hydroxybutyrate) Copolymer and Poly(3-hydroxybutyrate) Homopolyester, Respectively (Koller et al. [45])

	H. mediteranei	*P. hydrogenovora*
C-sources	glucose, galactose, γ-butyrolactone (GBL)	glucose, galactose
N-sources	yeast extract	NH₄⁺, complex-N (casein aminoacids)
Products (P)	Heteropolymer, Poly(3HB-*co*-3HV-*co*-4HB)	Monopolymer, poly-(3-HB); α-ketoglutarate; unknown extracellular compound (P22)
Intracellular pools	none	PHB, Acetyl-CoA, PHB-polymerase
Modeling assumptions	Independent synthesis of residual biomass (Xr, catalytically active part) from each C-source (glucose, galactose, γ-butyrolactone) and from yeast extract as additional complex N-source. The kinetic term: Monod relation. No kinetic influence of one C source on the conversion rate of the others.	Ac-CoA, stemming from the key metabolic pathways (glucose and galactose metabolism) were the omnipresent precursor of PHB compound. The kinetic behavior: Monod relation for Ac-CoA generation from both sugars.
Modeling assumptions	4-hydroxybutyrate (4HB) is generated only from γ-butyrolactone); Both other PHA-building blocks (3-hydroxybutyrate (3HB) and 3-hydroxyvalerate (3HV) are supposed to stem from both glucose and galactose.	Ac-CoA is the substrate for two competitive reactions: PHB synthesis (the thiolase reaction) and generation of precursors for biomass biosynthesis (citrate synthase reaction). Casamino acids were accepted to lead to Ac-CoA (deamination and degradation)
Modeling assumptions	PHA synthesis is inhibited by increasing PHA contents in cells, ("steric disturbing effect"). The kinetic term: Luong relation.	Negative feedback control by Ac-CoA in its synthesis was applied, (Luong inhibition type). Ac-CoA is metabolized for: • formation of biomass, • energy supply (inclusive maintenance energy), • NADPH generation (occurs in TCA cycle), • PHB production (3HB as β-thiolase substrate), • α-ketoglutarate (excreted), • unknown compound (excreted), • synthesis of the PHB-polymerase enzymatic system.

(Continued)

TABLE 2.5 (CONTINUED)

Comparison of Modeling Presumptions for *H. mediteranei* and *P. hydrogenovora* Cultivations Performed for Production of Poly(3-hydroxybutyrate-*co*-3-hydroxyvalerate-*co*-4-hydroxybutyrate) Copolymer and Poly(3-hydroxybutyrate) Homopolyester, Respectively (Koller et al. [45])

	H. mediteranei	*P. hydrogenovora*
Modeling assumptions		Ac-CoA production from glucose and galactose is inhibited (but not completely terminated) by casamino acids. Kinetic term used: Jerusalimsky type of inhibition. Casamino acids are separate C- and N-source for synthesis of proteins (a certain part of non-PHA biomass is directly synthesized from this source).
Modeling assumptions		Synthesis rate of PHB polymerase (generated from Ac-CoA and N-sources) is defined by multiple Michaelis-Menten kinetics. Its particular, degradation rate (protein turnover) is assumed. The PHB synthesis rate is proportional to the concentration of intracellular polymerase. High concentration of the complex N-source is set as the direct inhibitor of PHB polymerase. Kinetic term used: Jerusalimsky type (Mechanistically, there is no direct inhibition, the lack of knowledge about triggering regulatory mechanism for PHB synt. forced to this approach)
Modeling assumptions		No consumption of inorganic nitrogen source (NH_4^+) is allowed if casamino acids are present. α-ketoglutarate and P22 were set as a negative feedback controller of its own synthesis. NH_4^+ and PHB polymerase were assumed as inhibiting for its biosynthesis (Jerusalimsky type equation).
Model type	Formal kinetic	Formal kinetic, metabolically low structured

- Specific growth rate in accordance with Megee et al. [48] and Mankad-Bungay relations [49] (foreseen for a double substrate /C-and N-source/ limited growth).
- Bioreactor R1 of the cascade modeled as a continuous flow-in flow-out system for nutritionally balanced biomass formation.
- Bioreactor R2 considered as obeying to double substrates (C- and N-source) dependent growth kinetics.
- Reactors R3–R5 loaded with the C-source, strictly N-source deficient with expressed maximal possible accumulation of PHB.

Computed *in silico* simulations (after optimization) have resulted in the strong indication that PHB productivity can be significantly higher if certain dilution rates (D) and fine-tuned concentration of C- and N-source in the feed streams are provided. An alternative feeding strategy for "smooth" switching from batch to continuous mode during the start of the process was also suggested in order to prevent substrate inhibition.

The plurality of metabolic behaviors of microbial species, combined with available cultivation techniques (even among the strains belonging to the same species) was the main reason why diverse modeling strategies were performed. Below in Table 2.6, an overview of PHA producers, for which unstructured and structured mathematical models were applied, is given. Principles of formal-kinetic and low-structured modeling are presented in Figures 2.1 and 2.2.

More detailed comments and descriptions of the majority of the above-cited models are available in reviews published by Patnaik [28] and Horvat et al. [22].

2.3.2 Monoseptic and Mixed Culture Dynamic Models in PHA Biosynthesis

Detailed investigations of microorganisms and cultivation techniques carried out in the past have pointed out that environmental circumstances (e.g., substrate concentrations, temperature, pH-value, etc.) are not the only factors that determine growth kinetics. In long term industrial cultivations (performed starting from pure laboratory cultures through several upscaling steps to final industrial scale), biological and genetic factors have been expressed as the "generators" of kinetic changes. In extensively growing microbial populations, the number of cell divisions is incredibly high, so the possibility for "popping out" of some cells (in the sense of occurrence of natural mutations, losing of biological "power", senescence and degeneration) is increased. That leads to the plurality of attributes of individual cells, and these cell individualisms (among the whole population) are factors that additionally influence the characteristics of microbial processes. Such "individuality" factors cannot be expressed by unstructured formal kinetic models, which are only able to handle the steady-state type of microbial kinetic (e.g., stable exponential growth, nutrient balanced conditions, nutrients supplying in continuous cultivations, stationary phase of growth). Unfortunately, such models are very limited or unable to predict the dynamic behavior of the microbial culture under real cultivation conditions. This insufficiency is caused by missing consideration

TABLE 2.6
PHA-Producing Microorganisms and Applied Unstructured and Low-Structured Mathematical Models

Microorganism, cultivation type and definition of specific growth rate (μ)	Biomass growth rate eq. $\dfrac{dX}{dt}$	C-source consumption rate eq. $\dfrac{dS}{dt}$	N-source consumption rate eq. $\dfrac{dN}{dt}$	Product synthesis rate eq. $\dfrac{dP}{dt}$	Ref.
Azohydromonas sp.:					
Azohydromonas australica DSM 1124 $\left[\mu_m * \dfrac{S_1}{S_1 + K_{S1}} * \dfrac{(N)^{n1}}{N^{n1} + K_N^{n1}}\right]$ $* \left\{1 - \left[\dfrac{(N)^a}{(N_{max})^a}\right]\right\} * \left[\dfrac{K_I}{K_I + S_1}\right]$	$(\mu - D) * X$	$-\left(\mu / Y_{X/S} + m_S\right) * X + D * (S_0 - S_1)$	$K_1 * \mu * X - D * P$		[50]
Azohydromonas lata DSM 1123 $\left\{\mu_{m,s} * \dfrac{(N/S)*R_{O2,X}}{(N/S)*R_{O2,X} + k_m}\right\}$ $* \left\{1 + \dfrac{(P/(P+X))^{\alpha1}}{\left(\left(P_{max}/(P_{max}+X_{max})\right)\right)^{\alpha1}}\right\}^{-1}$	$\mu * X - \delta * (t - t_1) * \dfrac{F_{out}}{V} * X$	$-(R_{xx} * \mu * X + R_{C,SX} * X)$ $-C_{SP} * (k_1 * \mu * X + k_2 * X)$ $+\dfrac{F_{in}}{V} * S - \delta * (t - t_1) * * \dfrac{F_{out}}{V} * S$	$-(R_{NX} * \mu * X + R_{C,NX} * X)$ $+\dfrac{F_{in}}{V} * N - \delta * (t - t_1) * \dfrac{F_{out}}{V} * N$	$(k_1 * \mu * X + k_2 * X)$ $+\delta * (t - t_1) * \dfrac{F_{out}}{V} * P$	[51]

(Continued)

TABLE 2.6 (CONTINUED)
PHA-Producing Microorganisms and Applied Unstructured and Low-Structured Mathematical Models

Microorganism, cultivation type and definition of specific growth rate (μ)	Biomass growth rate eq. $\dfrac{dX}{dt}$	C-source consumption rate eq. $\dfrac{dS}{dt}$	N-source consumption rate eq. $\dfrac{dN}{dt}$	Product synthesis rate eq. $\dfrac{dP}{dt}$	Ref.
Azohydromonas australica DSM 1124 (upper equations: batch; lower equations: fed-batch) $$\left\{\mu_{max}\left[S_1(S_1+K_{S1})\right]\left[\frac{(S_2)^{a1}}{(S_2)^{a1}+(K_{S2})^{a1}}\right]\left[\frac{K_I}{K_I+S_1}\right]\left[1-\left(\frac{S_2}{S_m}\right)^a\right]\right\}X$$ $$\left\{\mu_{max}\left[S_1(S_1+K_{S1})\right]\left[\frac{(S_2)^{a1}}{(S_2)^{a1}+(K_{S2})^{a1}}\right]\left[\frac{K_I}{K_I+S_1}\right]\left[1-\left(\frac{S_2}{S_m}\right)^a\right]\right\}X$$	$$\left\{\mu_{max}\left[S_1(S_1+K_{S1})\right]\left[\frac{(S_2)^{a1}}{(S_2)^{a1}+(K_{S2})^{a1}}\right]\left[\frac{K_I}{K_I+S_1}\right]\left[1-\left(\frac{S_2}{S_m}\right)^a\right]\right\}X$$ $$\left\{\mu_{max}\left[S_1(S_1+K_{S1})\right]\left[\frac{(S_2)^{a1}}{(S_2)^{a1}+(K_{S2})^{a1}}\right]\left[\frac{K_I}{K_I+S_1}\right]\left[1-\left(\frac{S_2}{S_m}\right)^a\right]\right\}X$$	$$-\left[\frac{1}{Y_{X/P}^{Si}}\mu+m_{Si}\right]X$$ $$-\left[\frac{1}{Y_{X/P}^{Si}}\mu+m_{Si}\right]X+D\left(\frac{S_{0i}}{S_i}\right)$$	—	$$[K_1\mu]X$$ $$[K_1\mu]X-DP$$	[52]
Azotobacter sp.					
Azotobacter vinelandii UWD $$\mu*X*\left(1-\frac{X}{X_m}\right)$$ $$\left\{\mu_m*\frac{S}{S+K_S}\right\}$$		$$-\frac{1}{Y_{X/S}}*\frac{dX}{dt}-\frac{1}{Y_{P/S}}*\frac{dP}{dt}-k_e*X$$	—	$$\alpha*\frac{dX}{dt}+\beta*X$$	[53]

(Continued)

TABLE 2.6 (CONTINUED)
PHA-Producing Microorganisms and Applied Unstructured and Low-Structured Mathematical Models

Microorganism, cultivation type and definition of specific growth rate (μ)	Biomass growth rate eq. $\dfrac{dX}{dt}$	C-source consumption rate eq. $\dfrac{dS}{dt}$	N-source consumption rate eq. $\dfrac{dN}{dt}$	Product synthesis rate eq. $\dfrac{dP}{dt}$	Ref.
Bacillus sp.					
Bacillus megaterium	$(\mu_1 - k_4) * X$	$-(\alpha * \mu_1 + \gamma) * X$	$-\dfrac{\mu_1 * X}{Y_{x/N}}$	$k1 * \mu_1 * \Phi + k2 * X$	[54]
$\mu_1 = \mu_{\max} * \left\{ \dfrac{S}{S + K_{SS}} \right\} * \left\{ \dfrac{N}{K_{NS} + N} \right\}$					
$\mu_2 = \mu_m * \left\{ \dfrac{N}{K_{NS} + N} \right\}$					
$\mu_3 = \mu_m * \left\{ \dfrac{N}{K_{NS} + N} \right\} * \left\{ 1 - \dfrac{H}{H_m} \right\}$					
$\mu_4 = \mu_m * \left\{ \dfrac{N}{K_{NS} + N} \right\} * \left\{ 1 + \dfrac{k_{3H}}{H} + k_{2H} * H \right\}^{-1}$					
Bacillus megaterium BBST4	—	$\dfrac{dS_1}{dt} = -\left(k_3 \dfrac{dR}{dt} + k_4 \dfrac{dP}{dt} + k_5 R \right)$	—	$(\mu k_1 + k_2) R$	[55]
$\mu_{\max} \left(\dfrac{S_2}{K_{S2} + S_2} \right)$		$\dfrac{dS_2}{dt} = -\left[\dfrac{\mu}{Y_{\frac{R}{S2}}} + m_{S2} \right] R$			

(Continued)

TABLE 2.6 (CONTINUED)
PHA-Producing Microorganisms and Applied Unstructured and Low-Structured Mathematical Models

Microorganism, cultivation type and definition of specific growth rate (μ)	Biomass growth rate eq. $\dfrac{dX}{dt}$	C-source consumption rate eq. $\dfrac{dS}{dt}$	N-source consumption rate eq. $\dfrac{dN}{dt}$	Product synthesis rate eq. $\dfrac{dP}{dt}$	Ref.
Bacillus subtilis ssp. *subtilis* ATCC 6051 $\mu = \left(\dfrac{1}{X}\right)\dfrac{dX}{dt} = \mu_{max}\left(\dfrac{S}{S+K_S}\right)\left(\dfrac{K_I}{K_I+S}\right)$		$q_s = \left(\dfrac{1}{X}\right)\dfrac{dS}{dt} = -\left(\dfrac{1}{Y}\mu + m\right)$	—	$q_P = \left(\dfrac{1}{X}\right)\dfrac{dP}{dt} = k_1\mu + k_2$	[56]
Burkholderia sp.					
Pseudomonas hydrogenovora DSM 1749 (today: *Burkholderia fungorum*) $\mu_{HM} = \left[\left[\mu_{max1}\left(\dfrac{S_1}{S_1+K_{S1}}\right) +\mu_{max2}\left(\dfrac{S_2}{S_2+K_{S2}}\right)\right]\left(\dfrac{H}{H+K_H}\right)\right]$	$\dfrac{dXr}{dt} = \mu_{HM} \times M(Xr)$	$\dfrac{dS1}{dt} = -q_{pr1m}*\dfrac{S_1}{S_1+K_{s1}}\dfrac{K_{iNK}}{K_{iNK}+N}$ $\left(1-\dfrac{IPR}{IPR_m}\right)*X_1+\Sigma\left(\dfrac{F_{1S1,0}}{V}\right)$ $\dfrac{dS2}{dt} = -q_{pr2m}*\dfrac{S_2}{S_2+K_{s2}}\dfrac{K_{iNK}}{K_{iNK}+N}$ $\left(1-\dfrac{IPR}{IPR_m}\right)*X_1+\Sigma\left(\dfrac{F_{1S2,0}}{V}\right)$	$-\mu_2*\dfrac{N}{N+K_N}\dfrac{X1}{Y_{ymk}} -$ $-q_{prNm}*\dfrac{N}{N+K_N}\left(1-\dfrac{IPR}{IPR_m}\right)*X_1$	$\dfrac{dPR}{dt} = q_{pr1m}*\dfrac{S_1}{S_1+K_{s1}}\dfrac{K_{iNK}}{K_{iNK}+N}\left(1-\dfrac{IPR}{IPR_m}\right)*X_1$ $+q_{pr2m}*\dfrac{S_2}{S_2+K_{s2}}\dfrac{K_{iNK}}{K_{iNK}+N}\left(1-\dfrac{IPR}{IPR_m}\right)*X_1$ $-\mu_{m1m}\dfrac{IPR}{IPR+K_{ppr}}X_1-\dfrac{\mu_{m1Km}}{Y_{x1pr}}\dfrac{N}{N+K_N}\dfrac{IPR}{IPR+K_{qpr}}X_1$ $-q_m\dfrac{IPR}{IPR+K_{qpr}}\dfrac{(IEK)}{Y_{ppr}}X_1$ $-q_{21m}\dfrac{IPR}{IPR+K_{qpr}}\dfrac{K_{zN}}{K_{qN}+N}\dfrac{K_{ipol}}{K_{ipol}+IEK}\left(1-\dfrac{P_{21}}{P_{21m}}\right)\dfrac{X1}{Y_{p21yr}}$ $-q_{22m}\dfrac{IPR}{IPR+K_{qpr}}\dfrac{K_{zN}}{K_{qN}+N}\dfrac{K_{ipol}}{K_{ipol}+IEK}\left(1-\dfrac{P_{21}}{P_{21m}}\right)\dfrac{X1}{Y_{p22yr}}$	

(Continued)

TABLE 2.6 (CONTINUED)
PHA-Producing Microorganisms and Applied Unstructured and Low-Structured Mathematical Models

Microorganism, cultivation type and definition of specific growth rate (μ)	Biomass growth rate eq. $\dfrac{dX}{dt}$	C-source consumption rate eq. $\dfrac{dS}{dt}$	N-source consumption rate eq. $\dfrac{dN}{dt}$	Product synthesis rate eq. $\dfrac{dP}{dt}$	Ref.
				$\dfrac{dPOL}{dt} = a\dfrac{IPR}{IPR+K_{spr}}\dfrac{K_{iNK}}{K_{iNK}+NK}\dfrac{N}{N+N_K}$ $-b\dfrac{IEK}{IEK+K_{POL}}$ $\dfrac{dP}{dt} = q_{im}\dfrac{IPR}{IPR+K_{ppr}}IEK*X_1$ $\dfrac{dP_{21}}{dt} = q_{21m}\dfrac{IPR}{IPR+K_{spr}}\dfrac{K_{iN}}{K_{iN}+N}$ $\dfrac{K_{iPOL}}{K_{iPOL}+IEK}(1-\dfrac{P_{21}}{P_{21m}})*X_1$ $\dfrac{dP_{22}}{dt} = q_{22m}\dfrac{IPR}{IPR+K_{spr}}\dfrac{K_{iN}}{K_{iN}+N}$ $\dfrac{K_{iPOL}}{K_{iPOL}+IEK}(1-\dfrac{P_{21}}{P_{21m}})*X_1$	[45]
Burkholderia cepacia B27 $\mu = \mu_{max}\left(1-\left(\dfrac{S}{S_i}\right)^{a1}\right)\left(1-\left(\dfrac{N}{Ni}\right)^{a2}\right)$	$\dfrac{dX}{dt} = \left[\mu_{max}\left(\dfrac{S}{K_S+S}\right)\left(\dfrac{N}{K_N+N}\right)\left(1-\left(\dfrac{S}{S_i}\right)^{a1}\right)\left(1-\left(\dfrac{N}{Ni}\right)^{a2}\right)\right]X$	$\dfrac{dS}{dt} = -\left[\left(\dfrac{1}{Y_{\frac{X}{S}}}*\mu+m_S\right)\right]X$	$\dfrac{dN}{dt} = -\left[\left(\dfrac{1}{Y_{\frac{X}{N}}}*\mu+m_N\right)\right]X$	$\dfrac{dP}{dt} = [\alpha*\mu]X$	[57]

(Continued)

TABLE 2.6 (CONTINUED)
PHA-Producing Microorganisms and Applied Unstructured and Low-Structured Mathematical Models

Microorganism, cultivation type and definition of specific growth rate (μ)	Biomass growth rate eq. $\dfrac{dX}{dt}$	C-source consumption rate eq. $\dfrac{dS}{dt}$	N-source consumption rate eq. $\dfrac{dN}{dt}$	Product synthesis rate eq. $\dfrac{dP}{dt}$	Ref.
Cupriavidus necator					
Alcaligenes eutrophus ATCC 17697 (today *Cupriavidus necator*) $\left\{\mu_m * \dfrac{S}{S+K_S}\right\} * \left[1 - \left(\dfrac{S}{Sm}\right)^a\right]^n$	$\mu * X$	$-k4 * \dfrac{dX}{dt} - k5 * X$	$-k3 * \dfrac{dX}{dt}$	$k1 * \dfrac{dX}{dt} + k2 * X$	[36]
Alcaligenes eutrophus NCIMB 11 599 (today *Cupriavidus necator*) $\mu_m * \dfrac{S}{S+K_S+\dfrac{S^2}{K_{S,I}}} * \dfrac{N}{N+K_N+\dfrac{N^2}{K_{N,I}}}$	$\mu * X$	$F*S - \dfrac{1}{Y_{X/S}} * \dfrac{dX}{dt} - \dfrac{1}{Y_{P/S}} * \dfrac{dP}{dt} - m_e * X$	$F*N - Y_{N/X} * \mu * X$	$\pi_m * \left(1 - \dfrac{P/X}{(P/X)_m}\right) * \dfrac{S}{S+K_{PS}+\dfrac{S^2}{K_{PSI}}} * \dfrac{N+v_P}{N+K_{PN}+\dfrac{N^2}{K_{N,I}}}$	[58]
Ralstonia eutropha ATCC 17697 (today *Cupriavidus necator*) $\mu_{ml} * \dfrac{N/S}{N/S+K_{N/S}} * \left[1 - \dfrac{(N/S)^m}{(N_m/S_m)^m}\right]$	$\mu * X * \dfrac{F}{V}$	$-k4 * \dfrac{dX}{dt} - k5*X + \dfrac{F_F}{V} - \dfrac{dV}{V*dt}*S$	$-k3 * \dfrac{dX}{dt} + \dfrac{F_n}{V} - \dfrac{dV}{V*dt}*N$	$k1 * \dfrac{dX}{dt} + k2 * X - \dfrac{F}{V}*P$	[59]

(Continued)

TABLE 2.6 (CONTINUED)
PHA-Producing Microorganisms and Applied Unstructured and Low-Structured Mathematical Models

Microorganism, cultivation type and definition of specific growth rate (μ)	Biomass growth rate eq. $\dfrac{dX}{dt}$	C-source consumption rate eq. $\dfrac{dS}{dt}$	N-source consumption rate eq. $\dfrac{dN}{dt}$	Product synthesis rate eq. $\dfrac{dP}{dt}$	Ref.
Lactobacillus delbrueckii IAM1928 $\mu_1 = \left\{ \dfrac{\mu_{m1}(DO)*S_1}{S_1+K_{S1}} * \left(1 - \dfrac{S_2}{S_{2MAX}}\right) \right\}$	$\mu_1 * X_1 - \dfrac{F}{V} * X_1$ $\mu_2 * X_2 - \dfrac{F}{V} * X_2$	$\dfrac{dS_1}{dt} = \dfrac{-1}{Y_{X1/S_1}} * \mu_1 * X_1$ $+ \dfrac{\alpha\mu_1 + \beta}{Y_{Lac/S1}} * X_1 + \dfrac{F}{V}*(S_F - S_1)$ $\dfrac{dS_2}{dt} = \dfrac{\alpha\mu_1 + \beta}{Y_{Lac/S1}} * X_1$ $+ \dfrac{-1}{Y_{X_2/S2}} * \mu_2 * X_2 - \dfrac{F}{V} * S_2$	$-Y_{N/X2} * \mu_2(DO)* X_2 - \dfrac{F}{V} * N$	$q_m\left(\dfrac{k_n}{k_n + N}\right) * X_2 - \dfrac{F}{V} * N$	[42]
Ralstonia eutropha H16 (ATCC17699) (today Cupriavidus necator) $\mu_2 = \left\{ \dfrac{\mu_{m2}(DO)*S_2}{S_2 + \dfrac{S_2^2}{K_{S2}} + K_N} * \dfrac{N}{K_N + N} \cdot \dfrac{1}{K_I} \right\}$					
Ralstonia eutropha ATCC 17697 (today Cupriavidus necator) $\left\{ \mu_{m1} * \dfrac{N/S}{N/S + K_{N/S}} * \left[1 - \dfrac{(N/S)^m}{(N_m/S_m)^m}\right] \right\}$	$\mu * X - \dfrac{F}{V} * X$	$-k4 * \dfrac{dX}{dt} - k5*X + \dfrac{F_s}{V} - \dfrac{dV}{V*dt} * S$	$-k3 * \dfrac{dX}{dt} + \dfrac{Fn}{V} - \dfrac{dV}{V*dt} * N$	$k1 * \dfrac{dX}{dt} + k2*X - \dfrac{F}{V} * P$	[60]
Ralstonia eutropha ACM1296 (today Cupriavidus necator) $\left\{ \mu_{m1} * \dfrac{N/S}{N/S + K_{N/S}} * \left[1 - \dfrac{(N/S)^m}{(N_m/S_m)^m}\right] \right\}$	$\tau\dfrac{\partial X}{\partial t} = \dfrac{1}{Pe}\dfrac{\partial^2 X}{\partial\omega^2} - \dfrac{\partial X}{\partial\omega} - \tau * r_X$	$\tau\dfrac{\partial S}{\partial t} = \dfrac{1}{Pe}\dfrac{\partial^2 S}{\partial\omega^2} - \dfrac{\partial S}{\partial\omega} - \tau * r_S$	$\tau\dfrac{\partial N}{\partial t} = \dfrac{1}{Pe}\dfrac{\partial^2 N}{\partial\omega^2} - \dfrac{\partial N}{\partial\omega} - \tau * r_N$	$\tau\dfrac{\partial P}{\partial t} = \dfrac{1}{Pe}\dfrac{\partial^2 P}{\partial\omega^2} - \dfrac{\partial P}{\partial\omega} - \tau * r_P$	[61]
Ralstonia eutropha NCIMB11599 (today Cupriavidus necator) $\mu_m = \mu_{m1} * \dfrac{S1}{S1 + K_{S1}} * \dfrac{K1*S2}{S2 + K_{S2}}$	$\mu * X$	$F*S_1 - \dfrac{1}{Y_{X/S}} * \dfrac{dX}{dt} - \dfrac{1}{Y_{P/S}} * \dfrac{dP}{dt} - Ym * X$ $k_{Phosphate/X} * \mu * X$	—	$\beta_{m1} * \dfrac{S_1}{S_1 + k_S + \dfrac{S_2^2}{K_{S,I}}} * \dfrac{k_{p2}}{k_{p3} + S_2} * \left(1 - \dfrac{P}{P_{max}}\right) * X$	[62]

(Continued)

TABLE 2.6 (CONTINUED)
PHA-Producing Microorganisms and Applied Unstructured and Low-Structured Mathematical Models

Microorganism, cultivation type and definition of specific growth rate (μ)	Biomass growth rate eq. $\dfrac{dX}{dt}$	C-source consumption rate eq. $\dfrac{dS}{dt}$	N-source consumption rate eq. $\dfrac{dN}{dt}$	Product synthesis rate eq. $\dfrac{dP}{dt}$	Ref.
Cupriavidus necator DSM 545 $$\mu_1 = \left\{\mu_{max}\left(\frac{S_n}{S_n+K_S}\right)\left(\frac{N_n}{N_n+K_N}\right)\right\}$$ $$\mu_n = \left\{\mu_{max}\left(\frac{S_n}{S_n+K_S}\right)\left(\frac{N_n}{N_n+K_N}\right)\left(\frac{2K_S K_N}{K_S N_n + K_N S_n}+1\right)\right\}$$	$$\frac{dX_{r1}}{dt}=\left(\frac{F_{Lin}}{V_1}\right)X_{r0}$$ $$-\left(\frac{F_{Lout}}{V_1}\right)X_{r1}+\mu_1 X_{r1}$$ $$\frac{dX_{rn}}{dt}=\left(\frac{F_{n-Lin}}{V_n}\right)X_{rn-1}$$ $$-\left(\frac{F_{n,out}}{V_n}\right)X_{rn}+\mu_n X_{rn}$$	$$\frac{dS_1}{dt}=\left(\frac{F_{Lin}}{V_1}\right)S_0-\left(\frac{F_{Lout}}{V_1}\right)S_1$$ $$-\mu_1\left(\frac{X_{r1}}{m_{X,S}}\right)-(\mu_1 m_{P,X}+a_1)\left(\frac{X_{r1}}{m_{P,S}}\right)$$ $$-b_1 X_{r1}$$ $$\frac{dS_n}{dt}=\left(\frac{F_{n-Lin}}{V_n}\right)S_{n-1}-\left(\frac{F_{n,out}}{V_n}\right)S_n$$ $$+\left(\frac{F_{2n-1}}{V_n}\right)S_f-\mu_n\left(\frac{X_{rn}}{m_{X,S}}\right)$$ $$-(\mu_n m_{P,X}+a_n)\left(\frac{X_{rn}}{m_{P,S}}\right)-b_n X_{rn}$$	$$\frac{dN_1}{dt}=\left(\frac{F_{Lin}}{V_1}\right)N_0$$ $$-\left(\frac{F_{Lout}}{V_1}\right)N_1$$ $$-\mu_1\left(\frac{X_{r1}}{m_{X,N}}\right)$$ $$\frac{dN_n}{dt}=\left(\frac{F_{n-Lin}}{V_n}\right)N_{n-1}$$ $$-\left(\frac{F_{n,out}}{V_n}\right)N_n$$ $$-\mu_n\left(\frac{X_{rn}}{m_{X,N}}\right)$$	$$\frac{dP_1}{dt}=\left(\frac{F_{Lin}}{V_1}\right)P_0-\left(\frac{F_{Lout}}{V_1}\right)P_1$$ $$+(\mu_1 m_{P,X}+a_1)X_{r1}$$ $$\frac{dP_n}{dt}=\left(\frac{F_{n-Lin}}{V_n}\right)P_{n-1}-\left(\frac{F_{n,out}}{V_n}\right)P_n$$ $$+(\mu_n m_{P,X}+a_n)X_{rn}$$	[47]
Ralstonia eutropha DSM 545 (today *Cupriavidus necator*) $$\mu_{XS}=\left\{\frac{\mu_m S}{S+K_S+S^2/K_{IS}}*\frac{N}{K_N+N+N^2/K_{IN}}\right.$$ $$\left.*\left[1-\left(\frac{X}{Xm}\right)^\alpha\right]\right\}$$	$$(\mu_{XS}+\mu_{XP})$$ $$-D(t)*X$$	$$\frac{F(t)}{V(t)}*S_F-D(t)*S-$$ $$\left(\frac{\mu_{XS}}{Y_{XS}}+\frac{\mu_{XP}}{Y_{PS}}+m_s\right)*X$$	$$\frac{F_s(t)}{V(t)}*N_F-D(t)*N$$ $$-\left(\frac{\mu_{XS}+\mu_{XP}}{Y_{XN}}\right)*X$$	$$-D(t)*P+\left(\mu_{PS}-\frac{\mu_{XP}}{Y_{XP}}\right)*X$$	[63]

(Continued)

TABLE 2.6 (CONTINUED)
PHA-Producing Microorganisms and Applied Unstructured and Low-Structured Mathematical Models

Microorganism, cultivation type and definition of specific growth rate (μ)	Biomass growth rate eq. $\dfrac{dX}{dt}$	C-source consumption rate eq. $\dfrac{dS}{dt}$	N-source consumption rate eq. $\dfrac{dN}{dt}$	Product synthesis rate eq. $\dfrac{dP}{dt}$	Ref.
$\mu_{XP} = \left\{ \dfrac{\mu_{mP} * f_{PHB}}{f_{PHB} + K_{PHB}} * \dfrac{N}{K_N + N + N^2 / K_{IN}} \right\}$ $* \left\{ 1 - \left(\dfrac{X}{Xm} \right)^\alpha \right\}$ $\mu_{PS} = \left\{ \dfrac{\mu_m S}{S + K_S + S^2 / K_{IS}} \right\}$ $* \left\{ 1 - \left(\dfrac{f_{PHB}}{f_{PHB\,max}} \right)^\beta \right\} * \dfrac{K_{PIN}}{K_{PIN} + N}$					
Alcaligenes eutrophus H 16 (today *Cupriavidus necator*) $\left\{ \mu_{m1} \dfrac{S}{S + K_{S1}} + \mu_{m2} * \dfrac{(S / K_{S2})^a}{1 + (S / K_{S2})^a} \right\}$	$\mu * X$	$\dfrac{-1}{Y_{X/S}} * \mu * X$	—	$Y_{P/X} * \mu * X$ $+ \dfrac{K_I}{K_I + S} * (-k_1 * P + k_2 * X)$	[34]
Alcaligenes eutrophus B4383 (today *Cupriavidus necator*) $\left\{ \mu_{m1} * \dfrac{S}{S + K_{S1}} + \mu_{m2} * \dfrac{(S / K_{S2})^a}{1 + (S / K_{S2})^a} \right\}$ $* \left\{ 1 - \dfrac{(S_N / S_F)^m}{(Sm)^m} \right\}$	$\mu * X - \dfrac{F}{V} * X$	$-\left(a * \dfrac{dX}{dt} + b * \dfrac{dP}{dt} + c * X \right)$ $+ \dfrac{F_2}{V} * S_{F0} - \dfrac{F}{V} S$	$-d * \dfrac{dX}{dt} + \dfrac{F_1}{V} * N_{F0} - \dfrac{F}{V} N$	$\{k_1 * (X - X_{mo}) - k2 * P\} * \dfrac{K_S}{K_S + N}$ $+ e * \dfrac{dX}{dt} - \dfrac{F}{V} * P$	[64]

(Continued)

TABLE 2.6 (CONTINUED)
PHA-Producing Microorganisms and Applied Unstructured and Low-Structured Mathematical Models

Microorganism, cultivation type and definition of specific growth rate (μ)	Biomass growth rate eq. $\dfrac{dX}{dt}$	C-source consumption rate eq. $\dfrac{dS}{dt}$	N-source consumption rate eq. $\dfrac{dN}{dt}$	Product synthesis rate eq. $\dfrac{dP}{dt}$	Ref.
Cupriavidus necator ATCC 17699 $\left[\mu_m * \dfrac{S}{S+K_S}\right]$	$\mu * X$	$a*\mu*X*A + b*X*A$ $\left[\dfrac{dA}{dt} = K_A*S*(X*A)\right.$ $\left.\dfrac{dA}{dS} = -\dfrac{K_A S}{(\alpha*\mu*S/(S+K_S))+b}\right]$	—	$a*\dfrac{dX}{dt} + \beta*X$	[65]
Ralstonia eutropha NRRL 14690 (today *Cupriavidus necator*) $\left\{\mu_{m1} * \dfrac{S}{S+K_{S1}} + \mu_{m2}*\dfrac{(S/K_{S2})^n}{1+(S/K_{S2})^n}\right\}$ $*\left\{1-\dfrac{(N/S)^m}{(Sm)^m}\right\}$	$\mu * X$	$-\left(a*\dfrac{dX}{dt}+b*\dfrac{dP}{dt}+c*X\right)+\dfrac{F_2}{V}*S_{F0}-\dfrac{F}{V}S$	$-f*\dfrac{dX}{dt}+\dfrac{F_i}{V}$ $*N_{F0}-\dfrac{F}{V}N$	$\{k_1*(X-X_{min})-k2*P\}$ $*\dfrac{K_S}{K_S+N}+e*\dfrac{dX}{dt}-\dfrac{F}{V}*P$	[67]
Ralstonia eutropha NRRL B14690 (today *Cupriavidus necator*) $\mu_m*\dfrac{S_1^{a1}}{S_1^{a1}+K_{S1}^{a1}}*\dfrac{(S_2)^{v2}}{S_2^{n2}+K_{S2}^{n2}}$ $*\left\{1-\left(\dfrac{(S1)^{a1}}{(S_{m1})^{a1}}\right)^{v2}\right\}\left\{1-\left(\dfrac{(S2)^{v2}}{(S_{m2})^{v2}}\right)^{v2}\right\}$	$\mu*X-\dfrac{F}{V}*X$	$-(a*\mu+\gamma)*X+\dfrac{F_i}{V}*S_1-\dfrac{F}{V}S_1$ $-(\mu/Y_{X/S2}+m_S)*X+\dfrac{F_2}{V}*S_2-\dfrac{F}{V}S_2$	$\dfrac{\mu*X}{Y_{x/N}}-m_S*X$ $+\dfrac{F_2}{V}*N-\dfrac{F}{V}N$	$k1*\dfrac{dX}{dt}-k2*X-\dfrac{F}{V}*P$	[68–71]

(Continued)

TABLE 2.6 (CONTINUED)
PHA-Producing Microorganisms and Applied Unstructured and Low-Structured Mathematical Models

Microorganism, cultivation type and definition of specific growth rate (μ)	Biomass growth rate eq. $\dfrac{dX}{dt}$	C-source consumption rate eq. $\dfrac{dS}{dt}$	N-source consumption rate eq. $\dfrac{dN}{dt}$	Product synthesis rate eq. $\dfrac{dP}{dt}$	Ref.
Cupriavidus necator DSM 545 $$\mu_{S1} = \mu_{S1,max}\,\frac{S1}{S1+k_{S1}}\,\frac{N1}{N1+k_{N,G}}$$ $$\mu_{S2} = \mu_{S2,max}\,\frac{S2}{S2+k_{S2}}\,\frac{N1}{N1+k_{N,G}}\,\frac{k_i}{k_i+G}$$ $$q_{S1} = q_{S1,max}\,\frac{S1}{S1+k_{S1}}\,\frac{k_{qP,S1}}{k_{qP,S1}+N1}$$ $$q_{S2} = q_{S2,max}\,\frac{S2}{S2+k_{GLY}}\left(\frac{k_{s,S1}}{S1+k_{s,S1}}\right)$$ $$b_1 = b_{1,max}\,\frac{S1}{S1+k_{S1}}$$	$\mu_{S1}\times m(X_{r1})$ $+\,\mu_{S2}\times m(X_{r1})$	$\dfrac{dm(S1)}{dt} = -\mu_{S1}\,\dfrac{m(X_{r1})}{Y_2} - \mu_{S1}\,\dfrac{m(X_{r1})\times Y_1}{Y_3}$ $-\,q_{S1}\,\dfrac{m(X_{r1})}{Y_3} - b_1\times m(X_{r1})$ $-\,\sum\text{sampling} + \sum\text{feed}$ $\dfrac{dm(S2)}{dt} = -q_{S2}\,\dfrac{m(X_{r1})}{Y_4} - \mu_{S2}\,\dfrac{m(X_{r1})}{Y_5}$ $-\,\sum\text{sampling} + \sum\text{feed}$	$-\mu_{S1}\,\dfrac{m(X_{r1})}{Y_6} - \mu_{S2}\,\dfrac{m(X_{r1})}{Y_7}$ $+\,\mu_{S1}\times m(X_{r1})\times C$ $-\,\sum\text{sampling} + \sum\text{feed}$	$\mu_{S1}\times m(X_{r1})\times Y_1 + q_{S1}\times m(X_{r1})$ $+\,q_{S2}\times m(X_{r1}) - \sum\text{sampling}$	[72]
Cupriavidus necator DSM 545 $$\mu_{FAME} = \mu_{FAME,max}\,\frac{FAME}{FAME+k_{FAME}}\,\frac{N_2}{N_2+k_{N,2}}$$ $$q_{FAME} = q_{FAME,max}\,\frac{FAME}{FAME+k_{FAME}}\,\frac{k_{qN,FAME}}{k_{qN,FAME}+N_{FAME}}$$ $$q_{VA} = q_{V2,max}\,\frac{VA}{VA+k_{VA}}$$ $$b_2 = b_{2,max}\,\frac{VA}{VA+k_{VA}}$$ $$z_{VA} = z_{VA,max}\,\frac{VA}{VA+k_{VA}}$$	$\mu_{FAME}\times m(X_{r2})$ $-\,k_d\times m(X_{r2})$ $-\,\sum\text{sampling}$	$\dfrac{dm(FAME)}{dt} = -\mu_{FAME}\,\dfrac{m(X_{r2})}{Y_9} - q_{FAME}\,\dfrac{m(X_{r2})}{Y_{10}}$ $-\,b_2\times m(X_{r2}) - \mu * m(X_{r2})\times\dfrac{Y_8}{Y_{10}}$ $-\,\sum\text{sampling} + \sum\text{feed}$ $\dfrac{dm(VA)}{dt} = -q_{VA}\,\dfrac{m(X_{r2})}{Y_{11}} - z_{VA}\times m(X_{r2})$ $-\,\sum\text{sampling} + \sum\text{feed}$	$\dfrac{dm(N_2)}{dt} = -\mu_{VA}\,\dfrac{m(X_{r2})}{Y_{12}}$ $-\,\sum\text{sampling}$ $+\,\sum\text{feed}$	$\dfrac{dm(HB)}{dt} = \mu_{FAME}\times m(X_{r2})\times Y_8$ $+\,q_{FAME}\times m(X_{r2})\times \sum\text{sampling}$ $\dfrac{dm(3HV)}{dt} = q_{VA}\times m(X_{r2})$ $-\,\sum\text{sampling}$	[73]

(Continued)

TABLE 2.6 (CONTINUED)
PHA-Producing Microorganisms and Applied Unstructured and Low-Structured Mathematical Models

Microorganism, cultivation type and definition of specific growth rate (μ)	Biomass growth rate eq. $\dfrac{dX}{dt}$	C-source consumption rate eq. $\dfrac{dS}{dt}$	N-source consumption rate eq. $\dfrac{dN}{dt}$	Product synthesis rate eq. $\dfrac{dP}{dt}$	Ref.
Cupriavidus necator DSM 545 $\mu_1 = \mu_{max} \cdot \dfrac{S}{S+k_S} \cdot \dfrac{N}{N+k_N}$ $\mu_2 = \mu_{max} \cdot \dfrac{S}{S\left(1+(S/k_{i,GLY})^2\right)+k_S} \cdot \dfrac{N}{N+k_N}$ $\mu_3 = \mu_{max} \cdot \dfrac{S+S^2/k_{iS}}{S+k_S+S^2/k_{GLY}} \cdot \dfrac{N}{N+k_N}$ $\mu_4 = \mu_{max} \cdot \dfrac{S}{(S+k_S)(1+S/k_{i,GLY})} \cdot \dfrac{N}{N+k_N}$ $\mu_5 = \mu_{max} \cdot \dfrac{S}{S+k_S} \cdot e^{\left(S/k_{i,GLY}\right)} \cdot \dfrac{N}{N+k_N}$ $q = q_{max} \cdot \dfrac{S}{S+k_S} \cdot \dfrac{k_N}{k_N+N}$ $b = b_{max} \cdot \dfrac{S}{S+k_S}$	$\mu \cdot m(Xr)$ $-k_d \cdot m(Xr) \cdot \dfrac{K_{iN}}{K_{iN}+N}$ $-\sum sampling$	$-\mu_i \cdot \dfrac{m(Xr)}{f_{Xi/S}} - q_i \cdot \dfrac{m(Xr)}{f_{Pi/S}}$ $-b \cdot m(Xr) - \mu_i \cdot m(Xr) \cdot \dfrac{f_{Pi/Xr}}{f_{Pi/S}}$ $-\sum sampling + \sum feed$	$-\mu_i \cdot \dfrac{m(Xr)}{f_{Xr/N}}$ $-\sum sampling + \sum feed$	$\mu_i \cdot m(Xr) \cdot f_{Pi/Xr} + q_i \cdot m(Xr)$ $-\sum sampling$	[73]
Cupriavidus necator $\mu_X = \mu_{XS} + \mu_{Xp}$ $\mu_{XS} = \mu_{XS}^{max.} \cdot \left(\dfrac{S}{K_S+S+\dfrac{S^2}{K_{iS}}}\right)\left(\dfrac{N}{K_N+N+\dfrac{N^2}{K_{iN}}}\right)$ $\left(\dfrac{O_2}{K_{XO2}+O_2}\right)\left[1-\dfrac{X}{X_m}\right]^{\alpha}$	$\mu_X = \mu_{XS}+\mu_{Xp}$	$q_S = \dfrac{\mu_{XS}}{Y_{XS}} + \dfrac{q_{Pher.}}{Y_{PS}}$	$\dfrac{dN(t)}{dt} = \dfrac{F_N(t)N_F}{V(t)}$ $-D(t)N - \mu_N X(t) = 0$ $q_N = \dfrac{\mu_{XS}+\mu_{Xp}}{Y_{XN}}$	$q_{PS_{bat.}} = q_{PS}^{max.} \cdot \left(\dfrac{S}{K_{PS}+S+\dfrac{S^2}{K_{iPS}}}\right)^{\beta}$ $\left[1-\left(\dfrac{f_{PHB}}{f_{PHB(max.)}}\right)\right]\left[\dfrac{K_{PiN}}{N+K_{PiN}}\right]$ $\left(\dfrac{O_2}{K_{PO_2}+O_2+\dfrac{O_2^2}{K_{PiO_2}}}\right)$	

(Continued)

TABLE 2.6 (CONTINUED)
PHA-Producing Microorganisms and Applied Unstructured and Low-Structured Mathematical Models

Microorganism, cultivation type and definition of specific growth rate (μ)	Biomass growth rate eq. $\frac{dX}{dt}$	C-source consumption rate eq. $\frac{dS}{dt}$	N-source consumption rate eq. $\frac{dN}{dt}$	Product synthesis rate eq. $\frac{dP}{dt}$	Ref.
$\mu_{XP} = \mu_{XP}^{max.} \left(\dfrac{f_{PHB}}{K_{PHB} + f_{PHB}} \right) \dfrac{N}{K_N + N + \dfrac{N^2}{K_{iN}}}$ $\left[\dfrac{O_2}{K_{XO2} + O_2} \right] \left[1 - \dfrac{X}{X_m} \right]^\alpha$				$q_{PSaut.} = q_{PS}^{max.}{}_{aut.} \left(\dfrac{H_2}{K_{PH2} + H_2} \right) \left[\dfrac{O_2}{K_{pO2} + O_2 + \dfrac{O_2^2}{K_{PiO2}}} \right]$ $\left(\dfrac{CO_2}{K_{PCO2} + CO_2} \right) \left[1 - \dfrac{f_{PHB}}{f_{PHB(max.)}} \right]^\beta \dfrac{K_{PiN}}{N + K_{PiN}}$ $\dfrac{dP(t)}{dt} = \dfrac{dP_{het.}(t)}{dt} + \dfrac{dP_{aut.}(t)}{dt}$ $= (q_{PSaut} + q_{PShet.}) X - D(t) P_{het.}(t)$	[74]
Cupriavidus necator DSM 545 $\mu = \dfrac{\mu_{max.} S}{S + k_S}$	$\mu_{Xr} X_r$	$-\dfrac{dS}{dt} = \left(\dfrac{1}{Y_{\frac{Xr}{S}}} \mu_{Xr} + \dfrac{1}{Y_{\frac{P}{S}}} q_P \right) X_r$	$-\dfrac{dN}{dt} = \dfrac{1}{Y_{\frac{Xr}{N}}} \mu X_r$	$\mu_r X_r$	[75]
Cupriavidus necator DSM 545 $\mu = \dfrac{\mu_{max.} S}{S + k_S}$	$\dfrac{dX_R}{dt} = (\mu_{XR} - k_d) X_R$	$\dfrac{dS}{dt} = -\dfrac{1}{Y_{\frac{Xr}{S}}} \dfrac{dX_r}{dt} - \dfrac{1}{Y_{\frac{P}{S}}} \dfrac{dP}{dt}$	$\dfrac{dN}{dt} = -\dfrac{1}{Y_{\frac{Xr}{N}}} \dfrac{dX_r}{dt}$	$\dfrac{dP}{dt} = -Y_{\frac{P}{S}} \dfrac{dS}{dt} = q_P X_r$	[76]
Cupriavidus necator DSM 545 $\mu = \dfrac{\mu_{max.} S}{S + k_S}$	$\mu_{max} \left\{ \dfrac{S_1}{S_1 + KS_1} \right\} \left\{ \dfrac{(S_2)^{n1}}{(S_2)^{n1} + (KS_2)^{n1}} \right\}$ $\left[1 - \left(\dfrac{S_1}{S_{m1}} \right)^a \right] \left[1 - \left(\dfrac{S_2}{S_{m2}} \right)^b \right] X - DX$	$\dfrac{dS_i}{dt} = -\left\{ \dfrac{1}{Y_{\left(\frac{X+P}{Si}\right)}} + m_{Si} \right\} X + D(S_{0i} - Si)$	—	$\dfrac{dP}{dt} = [K_1 \mu] X - DP$	[77]

(Continued)

TABLE 2.6 (CONTINUED)

PHA-Producing Microorganisms and Applied Unstructured and Low-Structured Mathematical Models

Microorganism, cultivation type and definition of specific growth rate (μ)	Biomass growth rate eq. $\dfrac{dX}{dt}$	C-source consumption rate eq. $\dfrac{dS}{dt}$	N-source consumption rate eq. $\dfrac{dN}{dt}$	Product synthesis rate eq. $\dfrac{dP}{dt}$	Ref.
Cupriavidus necator DSM 545 $\mu = \dfrac{\mu_{max} \cdot S}{S + k_S}$	$\dfrac{dX}{dt} = \mu_{max}\, X\left(1 - \dfrac{X}{X_{max}}\right)$	$\dfrac{dS}{dt} = -\left(\dfrac{1}{Y_{X_S}}\right)\left(\dfrac{dX}{dt}\right) + m_S X$	—	$\dfrac{dP}{dt} = (\alpha\mu + \beta)X$	[78]
Cupriavidus necator DSM 545 $\mu = \dfrac{Y^{Clim.}_{X/S_1}}{Y^{Clim.}_{X/S_1} - \dfrac{Y^{Clim.}_{P}}{P} \, / \, e^{\mu_{mR} \cdot t}} \cdot \dfrac{X}{S_1} \cdot \mu_R$	$\mu_R = \mu_{mR} * \dfrac{S1}{K_{S1} + S1 + \dfrac{S1^2}{K_{i2}}}$ $* \dfrac{S2}{K_{S2} + S2 + \dfrac{S2^2}{K_{i2}}}$	$q^{Clim.}_{S,R} = \dfrac{1}{Y^0_{R/S_1}}\,\mu_R + m$ $q^{Nlim.}_{S,R} = \dfrac{1}{\rho} * Y^0_{R/S_1}\,\mu_R + \dfrac{m}{\rho}$	$q_{S,R} = \dfrac{\mu_R}{Y_{R/S_2}}$ $(S_2 = N)$	$q^{Clim.}_{P} = \dfrac{Y^{Clim.}_{P/S_1}}{Y^0_{R/S_1}}\,\mu_R + m * Y^{Clim.}_{P/S_1}$ $q^{Clim.}_{PR} = \alpha^{Clim.} * \mu_R + \beta^{Clim.}$ $q^{Nlim.}_{P} = \dfrac{Y^{Nlim.}_{P/S_1}}{\rho * Y^0_{R/S_1}}\,\mu_R + \dfrac{m * Y^{Nlim.}_{P/S_1}}{\rho}$ $q^{Nlim.}_{PR} = \alpha^{Nlim.} * \mu_R + \beta^{Nlim.}$	[79]
Haloferax mediterranei *Haloferax mediterranei* DSM 1411 $\mu_{HM} = \mu_{max1}\left(\dfrac{S_1}{S_1 + K_{S1}}\right)*\left(\dfrac{H}{H + K_H}\right)$ $+ \mu_{max2}\left(\dfrac{S_2}{S_2 + K_{S2}}\right)*\left(\dfrac{H}{H + K_H}\right)$ $+ \mu_{max3}\left(\dfrac{S_3}{S_3 + K_{S3}}\right)*\left(\dfrac{H}{H + K_H}\right)$	$\dfrac{dXr}{dt} = \mu_{HM} \times M(Xr)$	$\dfrac{dMS_1}{dt} = -\left\{\dfrac{\mu_{max}}{Y_{X/S1}} * \dfrac{S_1}{S_1 + K_{S1}}\,\dfrac{H}{H + K_H}\right\} * MX_r$ $+ \left\{\dfrac{q_{P1S1max}}{Y_{P1S1}} * \dfrac{S_1}{S_1 + K_{P1S1}} + \dfrac{q_{P2S2max}}{Y_{P2S1}} * \dfrac{S_2}{S_2 + K_{P1S2}}\right\}$ $\left\{1 - \dfrac{(P_f)}{(P_{f\,max})}\right\} * MX_r + \Sigma(F_i S_{0,i})$	$-\dfrac{\mu_{HM}}{Y_{X/ComplexN}} \times M(Xr)$	$\dfrac{dMP_1}{dt} = \left\{q_{P1S1max} * \dfrac{S_1}{S_1 + K_{P1S1}} + q_{P1S2max} * \dfrac{S_2}{S_2 + K_{P1S2}}\right\}$ $*\left\{1 - \dfrac{(P_f)}{(P_{f\,max})}\right\} * MX_r$ $\dfrac{dMP_2}{dt} = \left\{q_{P2S1max} * \dfrac{S_1}{S_1 + K_{P2S1}} + q_{P2S2max} * \dfrac{S_2}{S_2 + K_{P1S2}}\right\}$ $*\left\{1 - \dfrac{(P_f)}{(P_{f\,max})}\right\} * MX_r$	

(Continued)

TABLE 2.6 (CONTINUED)
PHA-Producing Microorganisms and Applied Unstructured and Low-Structured Mathematical Models

Microorganism, cultivation type and definition of specific growth rate (μ)	Biomass growth rate eq. $\frac{dX}{dt}$	C-source consumption rate eq. $\frac{dS}{dt}$	N-source consumption rate eq. $\frac{dN}{dt}$	Product synthesis rate eq. $\frac{dP}{dt}$	Ref.
		$\dfrac{dMS_2}{dt} = -\left\{\dfrac{\mu_{max}}{Y_{X/S2}} * \dfrac{S_2}{S_2+K_{S2}}\dfrac{H}{H+K_H}\right\} * MX_r$ $+\left\{\dfrac{q_{P2S2max}}{Y_{P2S1}} * \dfrac{S_2}{S_2+K_{P1S1}}\right.$ $+\left.\dfrac{q_{P1S2max}}{Y_{P1S1}} * \dfrac{S_2}{S_2+K_{P1S2}}\right\}$ $\left\{1-\dfrac{(P_1)}{(P_{max})}\right\} * MX_r + \Sigma(F_i S_{0,2})$ $\dfrac{dMS_3}{dt} = -\left\{\dfrac{\mu_{max}}{Y_{X/S3}} * \dfrac{S_3}{S_3+K_{S3}}\dfrac{H}{H+K_H}\right\} * MX_r$ $+\left\{\dfrac{q_{P3S3max}}{Y_{P3S3}} * \dfrac{S_3}{S_3+K_{P3S3}}\left(1-\dfrac{(P_1)}{(P_{max})}\right)\right\}$ $* MX_r + \Sigma(F_i S_{0,2})$		$\dfrac{dMP_3}{dt} = \left\{q_{P2S1max} * \dfrac{S_3}{S_3+K_{P3S1}}\right.$ $\left.*\left[1-\dfrac{(P_1)}{(P_{max})}\right]\right\} * MX_r$	[45]
Haloferax mediterranei DSM 1411 $\mu = \mu_{max.}\left(\dfrac{S}{K_S+S}\right)\left(\dfrac{N}{K_N+N}\right)X$	$\dfrac{dX}{dt} = \mu_{max.}\left(\dfrac{S}{K_S+S}\right)\left(\dfrac{N}{K_N+N}\right)X$	$\dfrac{dS}{dt} = q_{max.}\left(\dfrac{S}{K_S+S}\right)X$	—	$\dfrac{dP}{dt} = \dfrac{dS}{dt} - \dfrac{dX}{dt} - k_d X$	[80]
Haloferax mediterranei DSM 1411 $\mu = \mu_{max.}\left(\dfrac{S}{K_S+S}\right)X - k_x X$	$\dfrac{dX}{dt} = \mu_{max.}\left(\dfrac{S}{K_S+S}\right)X - k_x X$	$\dfrac{dS}{dt} = q_{max.}\left(\dfrac{S}{K_S+S}\right)X$	—	$\dfrac{dP}{dt} = q_P\left(\dfrac{S}{K_S+S}\right)X - k_P X$	[81]
Halomonas sp.:					

(Continued)

TABLE 2.6 (CONTINUED)
PHA-Producing Microorganisms and Applied Unstructured and Low-Structured Mathematical Models

Microorganism, cultivation type and definition of specific growth rate (μ)	Biomass growth rate eq. $\dfrac{dX}{dt}$	C-source consumption rate eq. $\dfrac{dS}{dt}$	N-source consumption rate eq. $\dfrac{dN}{dt}$	Product synthesis rate eq. $\dfrac{dP}{dt}$	Ref.
Halomonas bluephagenesis TD01 (CGMCC 4353)	$N_{cell} = [2.32 + 550 / (0.9987 + 0.3738 e^{-0.2885 x})^{36.32}] \times 10^8$				[82]
Pseudomonas sp.					
Pseudomonas putida KT2440 $\mu = \mu_{max} \left(\dfrac{S}{S + K_{S1}} \right) \left(\dfrac{N}{N + K_{N1}} \right) X$	$\mu_S = \dfrac{X}{X + K_S} S$ $\mu_N = \dfrac{X}{X + K_N} N$	$\dfrac{dS}{dt} = -k_4 q_s$	$\dfrac{dN}{dt} = -k_4 q_N$	$\dfrac{dP}{dt} = k_3 \mu + k_4 q_P$	[83]
Pseudomonas putida GPo1 (ATCC 29347) $\mu = \dfrac{Y_{X/S_1}^{Clim.}}{Y_{X/S_1}^{Clim.} - Y_P^{Clim.}/S_1} / e^{m_R R t} * \mu_R$	$\mu_R = \mu_{mR} * \dfrac{S_1}{K_{S1} + S_1 + \dfrac{S_1^2}{K_{I2}}}$ $* \dfrac{S_2}{K_{S2} + S_2 + \dfrac{S_2^2}{K_{I2}}}$	$q_{S,R}^{Clim.} = \dfrac{1}{Y_{X/S_1}^{Clim.}} \mu_R + m$ $q_{S,R}^{Nlim.} = \dfrac{1}{\rho * Y_{R/S_2}^0} \mu_R + \dfrac{m}{\rho}$	$q_{S,R} = \dfrac{\mu_R}{Y_{R/S_2}}$ $(S_2 = N)$	$q_P^{Clim.} = \dfrac{Y_{P/S_1}^{Clim.}}{Y_{R/S_1}^0} \mu_R + m * Y_{P/S_1}^{Clim.}$ $q_{PR}^{Clim.} = \alpha^{Clim.} * \mu_R + \beta^{Clim.}$ $q_P^{Nlim.} = \dfrac{Y_{P/S_1}^{Nlim.}}{\rho * Y_{R/S_1}^0} \mu_R + \dfrac{m * Y_{P/S_1}^{Nlim.}}{\rho}$ $q_{PR}^{Nlim.} = \alpha^{Nlim.} * \mu_R + \beta^{Nlim.}$	[84]
Mixed cultures: $\mu_S = Y_{X/S} \left\{ q_{S,max} * \dfrac{S}{S + K_S} * \left[1 - \left(\dfrac{P}{P_{max}} \right)^{\alpha} \right] * \dfrac{P}{P + K_{P,S}} \right\} * \dfrac{N}{K_{N,PHB} + N}$ $\mu_{PHB} = \left\{ \mu_{PHB,max} * \dfrac{P}{P + K_{P,S}} \right\} * \dfrac{N}{K_{N,PHB} + N}$	$(\mu_s + \mu_{PHB}) * X$	$-1 * \left\{ q_{S,max,PHB} * \dfrac{S}{S + K_S} * \left[1 - \left(\dfrac{P}{P_{max}} \right)^{\alpha} \right] \right\}$ $+ \left\{ q_{S,max} X * \dfrac{S}{S + K_S} \left[1 - \left(\dfrac{P}{P_{max}} \right)^{\alpha} \right] \left(\dfrac{N}{K_{N,PHB} + N} \right) + m_s \right\} * X$	$-Y_{N/X} * (\mu_s) * X$ $-Y_{N/X} * (\mu_{PHB}) * X$	$(Y_{P/X} * \mu_S - Y_{P/X} * \mu_{PHB} - m_{PHB})$ $* X - (\mu_S + (\mu_{PHB})) * P$	[85]

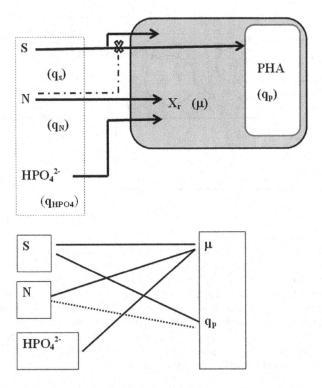

FIGURE 2.1 *Upper part:* Mass flux diagram for the formal-kinetic modeling of PHA biosynthesis. *Legend:* S, C-source; N, nitrogen source; HPO_4^{2-} phosphorus source; X_r, residual biomass; PHA, biopolyester; μ, specific growth rate; q_s, q_N, q_{HPO4}, specific consumption rate for C-, N-, and phosphorus source, respectively; q_p specific rate of product (PHA) formation; cross-stitch (X), inhibition of PHA synthesis at high nitrogen concentration. *Lower part:* Diagram of kinetic influences. Full lines, positive influence on the reaction rate; dotted line, inhibition of reaction rate.

of cell segregation, inhibitory effect of intracellular PHA (steric hindrances), metabolic regulatory mechanisms, and the dynamic responses to environmental conditions. To overcome these shortcomings, dynamic models were developed. They are "constructed" to be able to handle different transient states present in microbial cultivations (for example the transition from lag phase to exponential growth phase, or from exponential to stationary growth phase). In other words, dynamic models are appropriate for describing changes in the system's state during the entire cultivation. They are typically represented by a system of interconnected differential equations, which address the structuring of biomass (producers/non-producers, young/old cells, plasmids harboring/non-harboring, PHA sterically hindering/not hindering, gene expressed/non-expressed, biologically active/nonactive), bioreactor compartments (well mixed/plug flow, well aerated/non-aerated), metabolites (intracellular/extracellular), and products (homopolymer/heteropolymer, intracellular/extracellular). Here, it is necessary to state that a certain degree of simplification (if compared with the real state) is always present; however, these

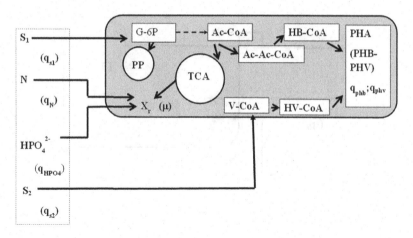

FIGURE 2.2 Mass flux diagram for the low-structured modeling of PHA (PHB-PHV) copolymer biosynthesis. *Legend*: S_1: C- and energy source (glucose), S_2: valeric acid (PHV precursor), N: nitrogen source, HPO_4^{2-}: phosphorus source, X_r: residual biomass, PHA: biopolyester, G-6P: glucose-6-phosphate, Ac-CoA, acetyl-coenzyme A; Ac-Ac-CoA, acetoacetyl-coenzyme A, HB-CoA: hydroxy-butyryl coenzymeA, V-CoA: valeryl-coenzyme A, HV-CoA, 3-hydroxyvaleryl-coenzyme A, TCA: (Krebs cycle), PP: pentose phosphate pathway, μ: specific growth rate, q_s: q_N, q_{HPO4}, specific consumption rate for C-, N-, and phosphorus source, respectively, q_{phb}: specific rate of product (PHA-PHB part) formation, q_{phv}: specific rate of product (PHA-PHV part) formation.

dynamic models are much more complicated than "black box" models, but they are far away from being models of the "white box" type. To build up a dynamic model, the kinetic properties of basic enzymatic reactions, stoichiometry of metabolic pathways as well as other cellular processes (i.e., protein-protein, DNA-protein, PHA-protein interactions) should be known. The dynamic responses of genetically structured models and the metabolic network on changes of extracellular substrates were published earlier ([86–89]). These works and other publications similar in nature published later on were the fundaments for the development of dynamic models dealing with PHA biosynthesis. Taking into consideration the dynamic models adjusted for describing PHA biosynthesis, the basic prime example is probably the one published by [90], developed to model the behavior of wastewater-activated sludge microorganisms. A pure culture of *Paracoccus pantotrophus* LMD 94.21 was used as the model organism for studying the response of wastewater treatment plant microorganisms subjected to a feast/famine regime in light of biosynthesis of storage polymers (glycogen and PHA). For that purpose, the mentioned organism was first cultivated in a steady-state C-limited chemostat process; later, the cultivation switched to batch mode by pulse-addition of acetate. The presence of this external substrate was the basis for microbial growth and PHB accumulation. After acetate depletion, intracellular PHB degradation took place to further enable biomass growth. The mass-balanced PHB content was correlated to the difference in metabolic rates of total acetate uptake and that required for biomass growth. A metabolically structured model that encompasses the observed kinetics of acetate consumption and PHB formation/consumption has been developed and successfully tested.

A highly intriguing cultivation strategy of co-culture of *Lactobacillus delbrueckii* and *A. eutrophus* (today: *C. necator*) was investigated by Katoh and coworkers [58]. Here, sugars were metabolized to lactate by *L. delbrueckii*, and lactate was utilized as a C-source by *A. eutrophus* for growth and PHB synthesis. In this paper, the dynamic modeling of a co-culture in one single fermenter ("one-pot-reaction") was carried out. Two cultivation phases (biomass growth and PHB production) were distinguished, and metabolic flux distributions were computed. When NH_4^+ was in surplus, the isocitrate dehydrogenase reaction in the tricarboxylic acid (TCA) cycle generated NADPH, which was predominantly the reactant for the α-ketoglutarate-to-glutamate reaction. If the NH_4^+ concentration was reduced to a certain critical level (or completely exhausted), NADPH was available for the acetoacetyl-CoA reductase reaction in the PHB production pathway. The authors reported that the model offered possibilities for the optimization of process control strategy, C-, N-sources feeding optimization, and investigation of process dynamics.

Mantzaris et al. [91] developed a population balance model (PBM) for productivity control. Later, Roussos and Kiparissides published a bivariate population balance model [92], which intended to cover the dynamic behavior of a PHB-producing microbial batch culture. In this work, three different numerical methods (continuous finite elements, discrete-continuous finite elements, and discretized) for the solution of the model's equation system, were tested in order to find an accurate one. It was proven that, when the accumulated PHB inhibits the biomass growth, the bivariate population balance modeling approach is necessary to be in accordance with dynamic behaviors of the microbial culture. The advantages of PBMs are that they consider that the biophase consists of a number of individual cells, different to some extent. This allows a high degree of detailing, so the PBMs are successful and accurate tools that comprehensively reflect cell growth, nutrient uptake, and product formation phenomena. For example, in this type of model it is possible to solve the mother-cell division process considering the cell space partition (with intracellular PHB redistribution into new daughter cells). In short, instead of the average cell properties used in formal kinetic models, the PBM type allows the differentiation and distribution of an interesting property over the total cell population.

In wastewater treatment technology, the settling of activated sludge is an important step, necessary to achieve a specific degree of purification. The presence and the content of "storage" materials (i.e., glycogen and PHA) are decisive factors that influence the settling velocity and the sludge volume index. The possibility of PHA isolation from the activated sludge and the settling attributes of sludge were the main reasons for the investigation and modeling in this field. Insel et al. [93] studied the simultaneous growth of cells and PHA storage kinetics in aerobic heterotrophic biomass cultivated under unsteady feast conditions. The authors have investigated the disturbed feeding conditions (short-term variations) as well as its consequences on growth and PHA storage kinetics of activated sludge. In this study, the multi-component biodegradation model was applied, inclusive of a multi-response procedure and identifiability analysis. If a lush quantity of external substrate was present, the heterotrophic biomass responded by an increase of growth activity and by simultaneous reduction of substrate storage capability. Furthermore, the reduction of sludge age (SRT) from 10 to 2 days consequently led to the increase of the specific growth rate from 3.9 to 7.0 [d^{-1}], but did not affect the maximum of storage rate. The direct kinetic consequence was the elevation of value of the half-saturation constant for growth from 5 to 25 mg L^{-1} chemical oxygen demand (COD). This was a strong indication of an altered species distribution among the activated sludge microbial consortium. An increase of primary growth metabolism followed by the reduction of the PHA storage rate was a tool for the validity test of the applied model: the system's response was tested under two different sludge ages combined with the availability of external substrate. Altered feeding conditions were found to have different storage and growth kinetics. Modifications in the model structure were highlighted as a benefit in the reduction of the laborious tasks dealing with frequent recalibration efforts (because of various process conditions).

The PHA biosynthesis from chemically defined substrates was also the object of dynamic modeling. Wu et al. [94] have worked on adaptive control mechanisms applicable to fed-batch cultivation of *R. eutropha* (today: *C. necator*). Attention was dedicated to the detection of "uncertainties of very sensitive kinetic parameters". The specific optimal adaptive control strategy was established with the help of the constrained discrete-time optimization algorithm and the genetic algorithm solver originated from the "Matlab" software package. Applying simplified kinetics, the feed flow control methods were implemented using stepwise changes of variables. The stability analysis proposed by Lyapunov for parametric uncertainties and state estimation errors was performed. Using simulation procedures, an output tracking with good correlations to the experiment was achieved applying the "two-input control" configuration.

Chen et al. [95] presented a mathematical model adapted for an aerobic dynamic discharge (ADD) process of PHA biosynthesis by mixed microbial cultures (MMCs). This process was intended to improve the selection of PHA-accumulating mixed microbial cultures of high efficiency. The model included the substrate uptake, PHA degradation, synthesis, and inhibition. Satisfactory agreement between experimental and simulated data has been obtained, so the model was able to represent the dynamic processes (mixed culture selection phase and transition to the pseudo steady state PHA production). Two variables applied to quantitatively express the "physical selective pressure" on the PHA-accumulating characteristic of MMCs were useful for the process control. The model assumptions were: in the feast phase, acetate is absorbed and activated by the expense of adenosine triphosphate (ATP) (1:1 stoichiometric requirement), the carbon source is converted to acetyl-CoA, CO_2 and NADH, acetyl-CoA and NADH are the reactants for PHA synthesis. During the famine phase, PHA is degraded to acetyl-CoA and NADH. Intracellularly, the substrate is decomposed into acetyl-CoA; one fraction of acetyl-CoA is stored in PHA, while the rest is directly metabolized into NADH and CO_2. Moreover, oxidative phosphorylation occurs in the famine phase; here, intracellular NADH is oxidized by oxygen and ATP is generated, which allows biomass growth. In addition, the model takes cell reproduction into account; in the absence of substrate, the microorganisms are in a maintenance state with gradual consumption of ATP.

A tabular review of dynamic oriented models of PHA production is given in Table 2.7.

TABLE 2.7
Properties of Dynamic Models Dealing with PHA Biosynthesis

Microorganism	Number of:			Substrates and products		Ref.
	Mass balance equations	Intracellular reactions	Intracellular metabolites	Mass-balanced extracellular substrates	Mass-balanced products	
Paracoccus pantotrophus	7	7	6	Acetate, NH_4^+	CO_2, X, PHA	[90]
Lactobacillus delbrueckii, Alcaligenes eutrophus (today *C. necator*)	30	23	25	Lactate, NH_4^+	CO_2, X, PHA	[58]
MMC	Bivariate population balance model			S	PHA, X	[92]
MMC	According to ASM1 and ASM3 models	According to ASM1 and ASM3 models	According to ASM1 and ASM3 models	Acetate, DO	X, PHA	[93]
Ralstonia eutropha (today *C. necator*)	5	–	–	Acetate, NH_4^+	X, PHA	[94]
MMC	5	7	5	Acetate, NH_4^+, HPO_4^{2-}	X, PHA	[95]

[1] at 1 bar, [2] at 15°C, [3] at 200 bar; DO: dissolved oxygen; MMC: mixed microbial culture

2.3.3 Metabolic Models

Metabolic models are the expression of scientific aspirations to have a mathematical tool able to describe real biological systems, metabolic networks, its responses to changes in the environment (transient states), and the biological adaptability. Such metabolic models have been developed as the consequence of the limited ability of formal kinetic and low structured models to mathematically encompass the broad spectrum of activities in biological systems. For the basic principles and methods of metabolic modeling, the reader can consult a comprehensive article published by Giersch [96].

Concerning the model complexity, the main general differences between different model types should be explained here. There is no exact distinction between the individual types of mathematical models used in literature. However, some general agreements are accepted by the scientific community active in "biological modeling", although these agreements should be considered "unwritten rules". In principle, three categories respecting their characteristics can be differentiated:

1. If metabolic reactions and intracellular kinetics are not included, and the extracellular concentration of substrates as well as products, and the biomass concentration itself are the key variables, we talk about *formal kinetic models.*
2. When the extracellular and some intracellular molecules are considered, a restricted number of intracellular reactions is more or less merged in "representative" reactions, and the reaction space is divided into a few compartments, but the biomass is seen as being "without the presence of genetic regulations", the *low-structured model* type can be proclaimed.
3. If interrelated, highly networked metabolic pathways exist in the model, but the intracellular space is divided into compartments in accordance with the cell topology, and/or the genetic pathway regulation is present, we can talk about *high-structured metabolic models.*

The metabolic models are very divergent in their purpose; in simple cases, the metabolic pathway with two or a few enzymatic reactions is investigated. Nevertheless, they may also encompass the complete catabolic and anabolic routes (and respective genetic regulations) of the cellular metabolism. To represent the highly organized metabolism of living cells, the metabolic control and regulatory mechanisms must be incorporated into the model. Here, a typical problem often occurs, regulatory mechanisms and complete metabolisms are usually not completely elucidated and known (except for some "model organisms"). Therefore, at the present level of scientific knowledge, it is not possible to form *"fully mechanistic models"* able to simulate the whole cellular metabolism and/or the biological cell lifecycle. For that reason, metabolic models must contain restrictions and simplifications. Some of them are reduced enzyme kinetics, lumped metabolic pathway reactions, simplified regulating patterns, or simplified transcriptions. Therefore, instead of "white box" models, "grey box" models are mostly presented in the literature. To be more precise, one should emphasize that there are huge scientific efforts ongoing to upgrade such "grey box" models to the complexity of "white box" models. These efforts are strongly dependent on the state of the art of advanced analytical techniques (e.g. determination of regulating factors, signal molecules, measurement of intracellular metabolite concentrations, and intracellular/membrane-bound enzyme activities), and on the improvement of speed and processing power of computers. Here, one should mention the development of "metabolic oriented" software tools enabling facilitated solving of very complex non-linear systems of differential equations related to complicated metabolic networks. Among the high number of metabolic models, kinetic, stoichiometric, and cybernetic models can be distinguished. Moreover, in the scientific and professional literature, hybrid types of metabolic models, which combine at least two of the aforementioned types, are published.

Kinetic metabolic models are based on extracellular and intracellular enzyme kinetics (and/or microbial growth kinetics) respecting the stoichiometry of metabolic reactions. Theoretically, provided that the whole metabolic network with the information about all enzymatic reactions for an organism is known, it is no problem to establish a detailed metabolic model, which agrees with experimental results regarding the simulation and prediction of dynamic changes in the cells and in the cultivation broth. Unfortunately, the usually available in vitro kinetic data, which are often not the same as those in vivo, and the scarce knowledge about general regulatory mechanisms are the bottlenecks of such models. Some authors therefore organize the metabolic models in accordance with certain biological situations, i.e. resting cell metabolism, metabolic steady-state fluxes for exponential phase, or secondary product formation in the stationary phase.

Stoichiometric models are based on the time invariant characteristics of metabolic networks, and an application of strict stoichiometric relationships considering extracellular substrates, products, and intracellular metabolites is unavoidable for these models. The structure of a model understands that each macroscopic or molecular species, both intracellular and extracellular, undergoes mass balancing, a tool for the metabolic flux analysis (MFA) necessary to quantify central metabolic intracellular fluxes. With the help of MFA, the fluxes of different branches in the metabolic network are determined (either measured or calculated). After that, the stoichiometric matrix has to be formatted. In the next step, the stoichiometric matrix and vectors of reaction rates are multiplied to define the mass balance equations. This is followed by the algebraic manipulation of the stoichiometric matrix. In this context, five different procedures were published by different authors (depending on the objective of the analysis):

- Imposing of constraints (e.g. measured flux data) and then solving the system using linear algebra (flux balancing).
- The combination of flux balancing with labeled metabolites balancing and application of numerical methods.
- If the system cannot be constrained in any way, the linear optimization is targeted to the determination of the maximum or minimum of a chosen objective function (OF).
- Convex analysis implementation for the elementary flux modes (EFMs) determination.
- Computed biologically important vectors for the null space of the stoichiometric matrix.

The above-mentioned procedures, if applied, can yield a picture of the metabolic situation. To some degree, limited predictive power, caused by the lack of regulatory and genetic information, is the main shortcoming of stoichiometric models.

Cybernetic models, as a third main group among metabolic models, are based on the optimal nature of microbial processes and on the metabolic regulation that is solved with the help of a cybernetic framework, which has to be discriminated from the pure kinetic procedures. As mentioned before, it is unlikely to have all the data necessary to organize a regulatorily exact mechanistic model with the help of kinetic equations. A modus operandi to create the regulatory mechanisms is perhaps cybernetic modeling (Dhurjati et al. [97]; Varner and Ramkrishna [98–101]). Using this method, the created cybernetic variables replace the unknown regulatory and reaction mechanisms of the cell, e.g. activation, inhibition, induction, and repression. This model type needs objective functions (OF) because of the assumption that the metabolic system operates in an optimal way with a physiologically specific goal, e.g. maximal growth rate or maximal product yield. These goal-functions are the subject of optimization procedures. A comprehensive review of the above methods is given by Patnaik [102].

Metabolic mathematical models applied to the cellular metabolism of PHA producers have been of great importance in the investigation of metabolism, transmembrane substrate transporting, and in the improvement and development of new strains.

2.3.3.1 Metabolic Models Adjusted to Activated Sludge Mixed Microbial Cultures

Activated sludge MMCs are an interesting source of PHA. Despite some constraints, e.g. unfavorable molecular mass distribution, they are the subject of intensive investigation. This was especially intensified after the realization that molecular mass distribution of PHA can be influenced by the regulation of feeding in feast and famine phases of running wastewater treatment plants (WWTPs). In order to overcome the difficulties of interactions of biomass and complex substrates from wastewater, these systems were usually studied in laboratories with the help of artificial media containing acetate, propionate, succinate, or glucose as the C-source, both by MMCs and pure cultures (Table 2.8). Some early works from this field were summarized in a publication by Gernaey et al. [103]. Yu and Wang [104] have studied a detoxification mechanism related to acetic acid for *R. eutropha* (today *C. necator*) using a metabolic cell model that encompassed five fluxes (for mass transfer and acetyl-CoA synthesis) and the formation of three products (active biomass, PHB, and CO_2). Diverse conditions were imposed on the system for measuring metabolic fluxes of interest. These experiments and model simulations led to the conclusion that at high biomass concentration, increased extracellular acetate concentrations are the result of acetate detoxification. The magnitude of detoxification fluxes correlated with the intracellular acetate pool.

Biological phosphorus and nitrogen removal and its relation to PHA synthesis/degradation were modeled for an activated sludge full-scale plant by Veldhuizen et al. [105].

TABLE 2.8

Characteristic Properties of Metabolic Models Dealing With PHA Production by Mixed Microbial (MMC) and by Pure Cultures

Micro-organisms	Substrates/Products		Number of:			Ref.
	Mass-balanced extracellular substrates	Mass-balanced products	Mass balance equations	Intracellular reactions	Intracellular metabolites	
MMC						
MCC	$COD, NH_4^+, NO_3^-, PO_4^{3-}$	Xr, PHB, Glycogen	Based on ASM1 and ASM2, 22 stoichiometric and 42 kinetic parameters model			[105]
MCC	Acetate	PHB, PHBV	13 (GAO) 17 (PAO)	6 (GAO) 7 (PAO)	12 (GAO) 15 (PAO)	[107, 108]
MCC	Acetate, NH_4^+	CO_2, Xr, PHB	13	7	9	[85]
MCC	Propionate, NH_4^+	CO_2, Xr, PHB, PHBV	15	8	9	[110]
MCC	Acetic acid, Propionate, NH_4^+	CO_2, PHB, PHBV, PH2Mv	14	8	10	[111]
MCC	Acetate, NH_4^+	CO_2, Xr, PHB	12	6	7	[112]
MCC	Acetic acid, Propionate, NH_4^+	CO_2, Xr, PHB, PHBV	9	9	9	[113]
R. jostii RHA1 (iMT1174) and E. coli K-12 (i AF1260)	Glucose, Acetate, Succinate	PHB, Glycogen, Triacyl-glycerols	Genome scale metabolic models for Rhodococcus jostii RHA1 (iMT1174) and Escherichia coli K-12 (iAF1260)			[114]
MMC	Volatile fatty acids	CO_2, Xr, PHA	13	7	10	[115]
MMC	Acetate, Methanol, DO	PHA, Xr	Based on [112, 115, 117]			[116]

(Continued)

TABLE 2.8 (CONTINUED)

Characteristic Properties of Metabolic Models Dealing With PHA Production by Mixed Microbial (MMC) and by Pure Cultures

	Substrates/Products		Number of:			
Micro-organisms	Mass-balanced extracellular substrates	Mass-balanced products	Mass balance equations	Intracellular reactions	Intracellular metabolites	Ref.
Plasticicumulans acidivorans + PHB nonstoring heterotroph	Acetate, NH_4^+	PHA, Xr, CO_2	5	6	7	[117–120]
MMC	Acetate, Propionate, Butyrate, Valerate	PHB/PHBV, Ac-CoA	7	11	5	[121]
MMC	Molasses, Volatile fatty acids (Acetate, Butyrate, Propionate, Valerate, Even and odd numbered fatty acids), NH_3	Xr, PHA	17 (depends on substrate number)	10 (plus 9 transport reaction)	7	[148]
Plants; Enzyme kinetic, MCA, SCAMP/ Sauro [123]	-	PHBV	9	6	8	[122]
Pseudomonas sp.						
Pseudomonas stutzeri 1317	Acetate, Butyrate, Hexanoate, Octanoate, Decanoate, Dodecanoate, Tetradecanoate	*mcl*-PHA	6–10		5–8	[126]

(Continued)

TABLE 2.8 (CONTINUED)

Characteristic Properties of Metabolic Models Dealing With PHA Production by Mixed Microbial (MMC) and by Pure Cultures

Micro-organisms	Substrates/Products			Number of:			Ref.
	Mass-balanced extracellular substrates	Mass-balanced products	Mass balance equations	Intracellular reactions	Intracellular metabolites		
Pseudomonas putida KT 2440	Glucose	Xr, PHB	889	877	886		[129]
Genome-scale network reconstruction, FBA, FVA, C13-IFD, gene knock-out, MILP							
Pseudomonas putida KT 2440	Glucose, Glycerol, Pyruvate	Xr, PHA	1046	1071 total; 958 metabolic	1044		[130]
Genome-scale network, *In silico* FA, 13C-IFD							
Escherichia coli							
E. coli	Acetate, Propionate	PHBV	11 (for monomer formation only)	See Ref. [97]	See Ref. [97]		[127]
Model modules:							
• Enzyme kinetic (monomer formation),							
• Artificial genetic network (regulation),							
• Polymerization (PHBV formation)							

(Continued)

TABLE 2.8 (CONTINUED)

Characteristic Properties of Metabolic Models Dealing With PHA Production by Mixed Microbial (MMC) and by Pure Cultures

	Substrates/Products		Number of:			
Micro-organisms	Mass-balanced extracellular substrates	Mass-balanced products	Mass balance equations	Intracellular reactions	Intracellular metabolites	Ref.
Escherichia coli DH5α	Glucose, Acetate	PHB, Formate	13	14 (3 transport reaction included)	12	[137]
Methylobacterium sp.						
Methylobacterium extorquens AM1 Kinetic model, generic enzymatic rate equations	Methanol	Xr, PHB	80	80	80 (total); 56 (related to central C-flux only) 20 related to biomass precoursors)	[128]
Azohydromonas sp.						
Azohydromonas lata DSM 1123 (metabolic and polymerisation model) *C. necator*	Sucrose	Xr, PHB, CO₂	See Ref. [102–106]	See Ref. [102–106]	See Ref. [102–106]	[132–136]
Ralstonia eutropha (today *C. necator*)	Acetate	CO₂, Xr, H⁺	10	5	5	[104]

(Continued)

TABLE 2.8 (CONTINUED)

Characteristic Properties of Metabolic Models Dealing With PHA Production by Mixed Microbial (MMC) and by Pure Cultures

	Substrates/Products		Number of:			
Micro-organisms	Mass-balanced extracellular substrates	Mass-balanced products	Mass balance equations	Intracellular reactions	Intracellular metabolites	Ref.
Ralstonia eutropha H16 (today *C. necator*)	Fructose	Xr, PHB, CO_2	1121	1391 total; 229 transport reaction included	1117	[142]
Cupriavidus necator DSM 545 Kinetic model, EFM, YSA, QP	Glucose, NH_3	Xr, PHB	48	45 (inclusive 2 transport reaction)	44	[143]
Cupriavidus necator DSM 545 Kinetic model, EFM, YSA, QP	Glycerol, NH_3	Xr, PHB	48	48 (inclusive 3 transp. reactions)	44	[144]
Cupriavidus necator DSM 545 MFA, CIM, NLWL	Butyrate	Xr, PHB	77	53 + 45 (anabolic + katabolic)	74	[150]

(Continued)

TABLE 2.8 (CONTINUED)

Characteristic Properties of Metabolic Models Dealing With PHA Production by Mixed Microbial (MMC) and by Pure Cultures

| | Substrates/Products | | | Number of: | | |
Micro-organisms	Mass-balanced extracellular substrates	Mass-balanced products	Mass balance equations	Intracellular reactions	Intracellular metabolites	Ref.
Ralstonia eutropha (today *C. necator*) CFD + metabolic enzyme kinetic model; GAMBIT / FLUENT, SIMPLE method of pressure-velocity coupling	Glucose	PHB	Accord. Tetra-Hybrid/ Hex Core Mesh space	5	13	[151]

GAO: Glycogen-accumulating microorganisms, PAO: Phosphate-accumulating microorganisms, PH2Mv: poly(2-methyl-3 hydroxyvalerate), Xr: residual biomass, DO: dissolved oxygen, PHBV: copolymer poly(3-hydroxybutyrate-*co*-3-hydroxyvalerate), MMC: mixed microbial culture, MCA: metabolic control analysis, FBA: flux balance analysis, FVA: flux variability analysis, IFD: internal flux distribution, MILP: mixed integer linear programming, FA: flux analysis, EFM: elementary flux modes, YSA: yield space analysis, QP: quadratic programming, MMC: mixed microbial culture, CIM: consistency index method, NLWL: non-linear weighted least squares, CFD: computational fluid dynamics

This work was based on a previously published "activated sludge model" (ASM2d) developed for the removal of COD, N, and P (Kuba et al. [106]). The model developed contained 22 stoichiometric and 42 kinetic parameters, has been handled by the SIMBA 3.0 software package, and tested (validated) using experimental data from aerobic and anoxic conditions. Yagci and associates [107, 108] have formed a metabolic model focusing on phosphate- and glycogen-accumulating organisms (PAOs and GAOs, respectively) from a microbial sludge sample that synthetizes/degrades PHA. The acetate uptake was investigated under aerobic and anaerobic conditions. Similar to the above, this model also originated from the ASM2d model (published by Henze et al. 1999 [109]). In this work, the glyoxylate pathway was assumed as the reaction producing reductive agents that in PAOs participate in PHA synthesis. Experimental values and model predictions were in high agreement for glycogen uptake, P release, PHA production, and PHA composition.

In three subsequent papers, Dias and coworkers have presented the application of elementary flux modes (EFMs), metabolic flux analysis (MFA), and flux balance analysis (FBA) for modeling of feeding conditions, culture enrichment, and quality of PHA [85, 110, 111]. First, the authors developed a two-compartmental metabolic model for PHA; thereafter, they decomposed a whole metabolic network into elementary flux modes and establish the dynamic (hybrid/semi-parametric) model and used propionic acid as a C-source in experimental setups. After that, the copolymer production from mixtures of acetic and propionic acid was modeled in a metabolic model. Material and energy balances were established on the basis of previously elucidated metabolic pathways. Two types of feast-and-famine culture-enrichment strategies were tested by using either acetate or propionate. The equations were formed with the help of theoretical yields of biomass and PHA production (related to the oxidative phosphorylation efficiency, i.e. P/O ratio).

PHB-producing MMCs cultivated under aerobic conditions in sequencing batch and fed-batch reactors were investigated by Johnson and coworkers [112]. The whole process was assumed to be of a two-step type: (I) biomass propagation, and (II) PHB accumulation. For model development, these authors used an advanced procedure consisting of a few steps:

(i) Measuring of experimental data for processes in bioreactors (including exhausting-gas concentrations).
(ii) Incorporation of data corrections for the influence of pH control, substrate concentration control (adding of solutions causing volume change), and sampling.
(iii) Determination of oxygen uptake and CO_2 evolution rates.
(iv) Mass balancing.
(v) Metabolic model structuring and evaluation of measurements.

The feast phase was assumed to encompass PHB-affected acetate uptake, biomass growth, maintenance energy, PHB production (affected by feedback control), CO_2 generation, oxygen, and NH_4^+ uptakes. Biomass growth, PHB degradation, CO_2 evolution, maintenance, oxygen, and NH_4^+ uptakes were accounted to the famine phase.

Apart from PHA homopolymers, the biosynthesis of poly(3-hydroxybutyrate-*co*-3-hydroxyvalerate) (PHBV) copolymers was also the subject of mathematical modeling. Using the above-citied work of Dias and coworkers, Jiang and associated scientists have developed a metabolic mathematical model for the production of PHBV copolyester. The C-sources were acetate and propionate metabolized by MMCs with the dominant fraction being *P. acidivorans* [113]. Some improvements were incorporated into the model:

1. Famine phase-specific metabolic active reactions were accounted for.
2. Acetyl-CoA and propionyl-CoA were the sources for the biomass synthesis and were separately specified.
3. ATP requirements for the maintenance energy were accounted for (estimated parameter under a fixed P/O ratio).
4. When propionate was the sole C-source, the TCA cycle was accounted as inactive.
5. Acetate and propionate uptake were regulated simultaneously dependent on the acetate-propionate ratio.

Tajparast and Frigon [114] studied the "production of storage compounds during feast-famine cycles of activated sludge" including foam formation, bulking, and process optimization by two genome-scale metabolic models for *Rhodococcus jostii* RHA1 (iMT1174) and for *Escherichia coli* K-12 (iAF1260). The production potential of value-added byproducts (e.g., PHA) was the target. The models were able to simulate growth on glucose, acetate, and succinate with PHB, glycogen, and triacylglycerols (TAGs) as products. In this work, special attention was dedicated to the development of proper OFs suitable for the application in the model to predict the production of storage compounds in both feast/famine cycle conditions. With the help of flux balance analysis (FBA), the main OF was established to maximize the growth rate, but it was insufficient for the time interval between feast and famine phases. Two additional sub-OFs were necessary, one related to minimization of biochemical fluxes, and the second one to minimize the metabolic adjustments (MoMA). When simulating glucose and acetate as the sole carbon sources in the process, all (sub-)OFs pointed out the identical substrate-storage "combinations" (i.e. acetate-PHB; glucose-glycogen). Interestingly, if succinate was used as substrate, the predictions were dependent on the metabolic network structure. While the metabolic model adjusted for *E. coli* has predicted glycogen accumulation, the model for *R. jostii* targeted on PHB accumulation. Therefore, the authors have concluded that new modeling strategies addressing the population ecology must be developed to properly predict the metabolism.

Model upgrades were introduced for the case of mixed substrate uptake related to the microbial growth, to the switching between feast and famine phases, and to PHA degradation and accumulation (Tamis et al. [115]). These authors have described the concepts for a generalized, "more predictive", model, focusing on the switching of feast/famine phases.

Korkakki et al. [116] reported that a distinction exists between substrates "that select for PHA-producers and for biomass growth". In that view, the volatile fatty

acids were underlined as stimulating PHA-producers, while methanol affected biomass growth. Mixed substrate was applied, and a sedimentation step was introduced
after acetate exhaustion; the supernatant with unconsumed methanol was discharged.
This action resulted in an increased PHA mass fraction from 48 to 70 wt.-% of biomass. Applied process configuration was simulated by modeling, resulting in the
realization that the "length of the pre-settling period and the discharged supernatant
quantity are determining for the elimination of the side population". The conversions
of acetate and methanol were assumed to be the carbon and energy source for two
independent microbial groups: PHA-producers, i.e. *P. acidivorans* and methanol-
growers. Acetate was assumed to be initially converted to PHB and then to biomass,
while methanol was directly converted to biomass according to Monod kinetics. In
addition, two different populations (distinguished by kinetic properties) were underlined as decisive for the success of the described strategy.

Marang et al. [117–120] studied the conditions that favor the enrichment of mixed
cultures with PHA producers. These authors have adopted and upgraded the model
proposed by Johnson et al. [112] and applied it for modeling of processes performed
in SBR and CSTR reactors. The aim was to elucidate the influence of the reactor type
and diverse feeding conditions (i.e. feast-famine ratio, substrate uptake rate) on the
enrichment of PHA-producing bacteria. This mathematical model for the growth of
P. acidivorans (model organism for PHA producers) as a competitor to a PHA non-
storing heterotroph species has been able to consider macroscopic behaviors, lumped
biomass, individual cells, the effect of residence time distribution, and the PHA content distribution among the microbial community. The model used encompasses the
typical intracellular reactions of given stoichiometry:

(i) Acetate uptake, activation: $HAc + ATP \rightarrow Ac\text{-}CoA$

(ii) Catabolism: $Ac\text{-}CoA \rightarrow CO_2 + 2\,NADH$

(iii) PHB production: $Ac\text{-}CoA + 0.25\,NADH \rightarrow PHB$

(iv) Anabolism: $1.267\,Ac\text{-}CoA + 0.2\,NH_3 + 2.16\,ATP \rightarrow$
$CH_{1.8}O_{0.5}N_{0.2} + 0.267\,CO_2 + 0.434\,NADH$

(v) PHB consumption: $PHB + 0.25\,ATP \rightarrow Ac\text{-}CoA + 0.25\,NADH$

(vi) Oxidative phosphorylation: $NADH + 0.5\,O_2 \rightarrow \delta ATP$

and six process parameters, six stoichiometric yields, eight kinetic parameters for
P. acidivorans as well as four kinetic parameters for the non-PHB-storing heterotroph species. The δ-factor represented the mean of the oxidative phosphorylation
efficiency (assumed: 2.0 mol ATP/mol NADH for all species). Furthermore, the
kinetic equations for *P. acidivorans* related to acetate uptake, growth on acetate,
growth on PHB, maintenance on acetate, maintenance on PHB, PHB production,
PHB degradation, and ammonium uptake were established by the use of Monod
terms and inhibition terms. Similarly, for the non-PHB-storing heterotroph, the
kinetic terms for acetate uptake, growth on acetate, maintenance on acetate, biomass
decay, and ammonium uptake were incorporated.

Wang et al. [121] have been engaged in the investigation of PHBV copolyester
production from a mixture of volatile fatty acids by MMC. A metabolic model was
developed to describe the existent substrate competition among acetate, propionate,

butyrate, and valerate present in the broth. Interestingly, the authors have introduced the inhibition parameter that regulates the inhibition of acetate and propionate uptake by butyrate and valerate. This model offered the understanding of multiple substrate uptakes for PHA production, prediction of copolyster composition, and optimization of the process efficiency. The metabolic model constructed in this work was focused on the uptake of VFAs via two pathways respecting the even and odd number of C-atoms, either converted to acetyl-CoA or to acetyl-CoA + propionyl-CoA. In this metabolic model, the uptake of acetate, propionate, butyrate, and valerate was assumed to end as corresponding acyl-CoAs. After the active transport uptake, the assumption that butyrate and valerate are directly converted to 3HB and 3-hydroxyvalerate (3HV) monomers was considered. In addition, acetate and propionate were assumed to be converted to acetyl-CoA and propionyl-CoA, respectively. Furthermore, the propionyl-CoA fraction was considered to be decarboxylated to acetyl-CoA and CO_2. Acetyl-CoA that was not consumed via the TCA cycle but was assumed to be the precursor for PHA synthesis. Furthermore, the PHA composition was the result of propionyl-CoA being completely shifted towards 3HV synthesis, and acetyl-CoA was partially converted to 3HV (after coupling with propionyl-CoA) and partially to 3HB. ATP was assumed to be formed by oxidative phosphorylation with contribution to maintenance energy.

The model encompasses a total of eleven reactions, five intracellular intermediates, five basic substrates (acetate, propionate, butyrate, valerate, O_2), and three products.

In general, mathematical modeling of PHA production by microbial cultures from activated sludge is tightly connected to the modeling of the entire wastewater treatment processes. Metabolic models related to PHA biosynthesis in activated sludge cultures and their characteristic properties are summarized in Table 2.8.

2.3.3.2   Metabolic Models Targeted to Industrial Scale Modeling and to Development of High Productive Strains

One of the most ground-breaking ideas related to improvements in PHA production was the transfer of microbial PHA synthesis genes into plants. The intention was to direct carbon captured by photosynthesis to the synthesis of PHA molecules instead of to the usual products, e.g., starch, inulin, etc. Daae and coworkers [122] have structured a metabolic model for PHA biopolyester biosynthesis in the plastids of genetically modified plants. The model established was based on the kinetic properties of the main enzymes responsible for the PHA anabolism: 3-ketothiolase, acetoacetyl-CoA reductase, and PHB synthase. For that purpose, the authors resorted to the Michaelis-Menten equation for non-reversible systems, corrected it by a competitive inhibition term, and to the reversible sequential ordered mechanism of the "BiBi" type and "ping-pong BiBi" reversible mechanism type. The metabolic pathway simulation was performed by a simulation software package published by Sauro [123] (SCAMP – system for mathematical simulation). To calculate the flux and the concentration control coefficients for assumed steady-state conditions, metabolic control analysis (MCA) was incorporated [124, 125]. This model was able to simulate dynamics in production rates and steady-state concentrations of key substances. The illuminated and dark phases were an important factor that influenced the 3HB/3HV ratio as well as the rate of PHBV copolyester biosynthesis. Most importantly, the natural differences in the concentrations of substrates are decisive for the production rate as well as for the PHA composition.

Xu and coworkers [126] have applied a metabolic mathematical model for the "tuning" of monomer composition in medium-chain-length copolymers (*mcl*-PHA), a group of heteropolyesters synthesized by some species of the pseudomonades. A simplified metabolic network of irreversible and reversible reactions that operate under pseudo-steady-state conditions was established. For example, PHA synthases from Class II encoded on the *PhaC1* and *PhaC2* operons preferring 3-hydroxyoc-tanoyl-CoA and 3-hydroxydecanoyl-CoA as substrates were, because of similar sub-strate specificity, treated as a "one synthase kinetic" in the model. The model was used to simulate the "outputs" caused by different "inputs": the influence of substrate chain length, usage of a sole and mixed C-source, the effect of enzyme concentration on copolyester composition, the selectivity of the enzymatic systems, and metabolic pathways constructed by genetic engineering. Additionally, the model was used to elucidate the biosynthesis of short-chain-length PHA (*scl*-PHA) copolyesters and the formation of "mixed" *scl-co-mcl*-PHA. As an interesting result it was under-lined that an engineering of two simultaneously operated, independent, metabolic pathways for precursor synthesis is an opportunity to regulate monomers for PHA copolymer synthesis.

Biosynthesis of the PHBV copolyester was the object of the work by Iadevaia and Mantzaris [127]. These authors have established a mathematical model based on three sub-steps:

- The control of enzyme expression levels by artificial genetic network (aim: synthesis of monomers).
- Kinetic properties of monomer synthesis pathways.
- Polymerization dynamics of PHA synthesis.

For that purpose, the genetic switch module (activated when a certain expression level exists) was interconnected with the driver module of monomer synthesis. The biosynthesis of monomers was set to be leaning on monomer polymerization with the rates of initiation, elongation, and transition being taken into account. Additionally, under the simulated steady state, both the 3HB mass fraction as well as the polydis-persity ($Đ$) of PHBV were strongly influenced by levels and ratios of chain elonga-tion rates. The authors stated that the model is a powerful instrument for bioprocess design dedicated to the production of PHBV copolyesters of the desired structure.

Methylobacterium extorquens AM1 was the object of a large-scale kinetic meta-bolic model developed by Ao et al. [128]. These investigators have used a set of steady-state fluxes, combined with metabolite levels, with the intention to simplify the parametric range of biocatalytic reactions. This way, the number of parameters related to biocatalytic rate equations was reduced, but still represented the key char-acteristics. Furthermore, if enzyme equilibrium- and saturation-constants were not determined through experiments, the values of physiological concentrations were chosen instead. The metabolic network has reflected sugar metabolism (gluconeo-genesis, pentose phosphate pathway), acetyl-CoA (synthesis and consumption),

Krebs (TCC) and serine cycles, formaldehyde incorporation, and PHB synthesis and respiration (energy generation). The related kinetic model has encompassed 80 computational (metabolic) species. A very stable model that follows the C-source fluctuations was the result. However, although the PHA modeling was not the primary goal of this study, this approach could be a successful way to develop metabolic models to study PHA biosynthesis.

Puchałka and coworkers [129] performed the genomic reestablishment of metabolic network for *P. putida* KT2440. Metabolic network formation and flux balance analysis (FBA) have enabled establishing important metabolic functions, discovering signaling gaps, and gene structure improvements. This comprehensive work has encompassed flux variability analysis (FVA), intracellular flux distributions (C^{13}-analysis), FBA, continuous cultivation technique, the knockout of key genes, and the mathematical model validation.

The resulting finalized model network consisted of 62 extracellular compounds, 877 individual biochemical reactions, and 824 intracellular substances. Altogether, 6% of the reactions were not gen-associated. The authors have reported that the metabolic network structure was one of the most decisive factors for model accuracy. As Ac-CoA was set to be the main PHA precursor, two methods were modeled to in silico enhance its intracellular pool concentration: the action of pyruvate dehydrogenase (PDH) was improved, and an additional reaction was introduced for the simultaneous production of CoA.

The above-mentioned organism *P. putida* KT2440 was the object of research performed by Sohn et al. [130]. The metabolic model of genome scale type (termed as PpuMBEL1071) was applied to investigate the degradation of compounds containing aromatic ring, as well as the PHA accumulation capacity. This model was reliant on 900 genes (16.6% of the whole genome), and contained 113 transport reactions; in total, the metabolic network encompassed 958 reactions and 1044 metabolic species. Depending on the metabolic function, biochemical reactions were organized into 11 classes, the largest were related to amino acid metabolism (ca. 25% of reactions), to sugar metabolism (16%), and approximately 10% were tightly bounded to about 300 protein transporters. Metabolic fluxes through the network were determined with the help of biochemical thermodynamics, mass balancing, and linear programming. The irreversible reaction fluxes were identified and labeled as positive; in contrast, a negative sign was given to fluxes of reversible reactions for the opposite direction. More detailed consideration of the genome-scale metabolic modeling principles as well as in silico calculated metabolic flux analysis, the reader can find, for example, in the work by Thiele and Palsson [131].

In literature, a series of articles, all devoted to PHA production and related to the metabolism of *Alcaligenes latus* (today: *Azohydromonas lata*), can be found [132–136]. Established mathematical models comprise dynamics of metabolic regulation, metabolic kinetics, and PHA biosynthesis. They encompass two sub-model units, one responsible for the simulation of polymerization kinetics (chain length and time scale), and a second one that reconsiders the metabolic situation. A metabolic connection between two sub-models was set: 3-hydroxybutyryl-CoA (the substrate of PHA synthase) concentration. Furthermore, the biomass was divided into three compartments: the non-PHA biomass (residual biomass, Xr), the PHB precursor

3HB, and the product (P), i.e. PHB. Except for C- and N-sources, all substrates were chosen to be in surplus, but the heterogeneity of the microbial population and mass-transfer phenomena were not considered.

This model was successful in the prediction of C-source consumption rate dynamics, molar mass, and concentration of PHA, PHA distribution respecting cultivation time, and the simultaneous biomass and biopolyester formation. Furthermore, thanks to separate compartments related to bioreactor space, the model was used to study the influence of cultivation procedures (batch and fed-batch type) on the synthesized PHA and biomass.

In scientific investigation of improvement of PHA production, the genes from natural PHA producers were transferred to non-producers in a manner to use benefits of the new host strain, e.g. higher growth rate. By gene manipulations and transfer of plasmids from R. eutropha, the recombinant E. coli DH5 strain was constructed by Carlson et al. [137]. To explain some unusual capabilities of this construct, a metabolic network model was set up for axenic conditions. Experimental situations (regarding PHA synthesis and metabolite excretion) were analyzed using elementary flux modes (EM) [138, 139]. In making this procedure, 202 modes related to anaerobic PHB-formation were identified, 98 modes of them were characterized by sole PHB accumulation (i.e. without growth of non-PHA biomass). The anaerobic biopolyester synthesis was found to be possible during active growth as well as under non-growth conditions.

Interestingly, one of the first examined PHA producers, R. eutropha (today C. necator), in particular its wild-type strain H16, was not able to utilize glucose, but it was capable of metabolizing fructose and N-acetyl-glucosamine (NAG) [140]. Performed genetic manipulations have yielded mutants (i.e. G^{+1} and H_1G^{+3}) able to utilize glucose [141]. From the genetic point of view, the mentioned R. eutropha H16 genome consists of two individual chromosomes and one separate "megaplasmid pHG1". A sequence analysis deciphered chromosome 1 (4,052,032 bp), chromosome 2 (2,912,490 bp), and the "megaplasmid" (452,156 bp). These data were the basis for the development of the RehMBEL1391 genomic metabolic model (Park and colleagues [142]) that encompasses 1171 metabolites, 229 transport reactions, and 1391 reactions in total. As reported by the authors, "the metabolic flux distribution and the changes of metabolic fluxes under several perturbed conditions were examined by flux variability". For that purpose, the metabolic fluxes were varied computationally and the minimal and maximal flux values for each reaction were applied to build an objective function (i.e. maximum cell growth rate, or maximal production rate, for example). Using this procedure (including "metabolic adjustment method" /MOMA/ and quadratic programming /QP/), the authors have discovered the necessary gene knockouts to increase the synthesis of 2-methylcitric acid as well as the strategies leading to enhanced PHB production (optimal pH-value and C/N-source ratio).

Among others, Lopar et al. investigated in silico the strain C. necator DSM 545. A five-stage continuous process of PHB biosynthesis on glucose in a bioreactor cascade was analyzed [143]. The same method, based on kinetic, structured, metabolic model, elementary flux modes (EM), and two-dimensional yield space, was applied to investigate the glycerol metabolism for this microorganism [144]. For that purpose the established model contained 43 mass balance equations related to external substrates and intracellular metabolite pools [143]. EMs as well as the stoichiometric matrix were processed using the Metatool 5.1 program (Pfeiffer and coworkers [145]) upgraded by von Kamp and Schuster [146], and the metabolic yield analysis (including weighting factors) was handled by the Matlab program ("fmincon" function applied). Metabolic fluxes were computed from the experimentally obtained yield data (for residual biomass and PHB) and for two different feeding strategies for C-source supply with the help of the quadratic programming approach [147] and the minimized sum of squared weighting factors. The target of this work was to ensure a more efficient approach for optimizing biopolyester and non-PHA biomass yields. An excellent agreement of experimental data and in silico results was reported, including the growth-associated PHA synthesis phase. Experimental yields for Xr and PHA were compared with the in silico data for all five reactors in the cascade. To analyze the metabolic pathways of *C. necator* DSM 545 for the cultivation of glycerol, a byproduct from biodiesel production, as C-source, the model was slightly improved. A metabolic network of 48 biochemical reactions was created for that purpose [144]. Performing this work, the authors wanted to answer a key question: is the flux throughout metabolic network itself the limiting factor for biological growth and PHB synthesis on glycerol, or it is the glycerol transport system? High attention was given to the intersections of metabolic pathways active in the carbohydrate metabolism (the direction of reversible enzymatic reactions), to metabolic flux in gluconeogenesis, and to pools of NAD^+ or FAD^+, coenzymes possibly engaged in the glycerol-3-P-dehydrogenase (GLY-3-P DH) reaction. Four EM sets were the outcome of this work, considering the pairs NAD/NADH or FAD/FADH$_2$ that can contribute to the GLY-3-P DH reaction. In addition, the possible presence or absence of 6-phosphogluconate dehydrogenase (6-PG DH) was tested. It was interesting to see that the simplified metabolic network applied in this work has resulted in multiple solutions for the parallel synthesis of PHB and biomass. Additionally, the number of possible solutions went up if either the pairs NAD/NADH or FAD/FADH$_2$ were accounted to be the reactants in the GLY-3-P DH reaction. Furthermore, it was shown in silico that the experimental yields of biomass and PHB on glycerol can be achieved in two very different metabolic situations: either if the Entner-Doudoroff pathway (ED) is predominant in comparison to glycolysis (Embden-Meyerhof-Parnas pathway, EMP) or and vice versa. The findings indicated that the organism can be functional in multiple ways when utilizing glycerol for PHB biosynthesis.

Pardelha and associated authors have used MMCs to produce PHA on cane molasses and on VFAs [148]. In this work, the authors have applied the metabolic modeling systems described for MMCs from activated sludge, establishing the dynamic metabolic model able to consider the consumption of mixtures of VFAs. Flux balance analysis was part of the procedure used. This way, the scientists have reported that the minimal TCA cycle activity is the most effective metabolic condition for PHA accumulation [149], no matter if considering feast or famine regimes.

C. necator DSM 545 (H$_1$ G^{+3} mutant) was the object of investigations performed by Grousseau et al. [150]. The objective of this work was to elucidate the role of NADPH in PHA producers. Using knowledge about metabolic pathways from databases and bibliographic studies about enzyme kinetics and the stoichiometry of biochemical reactions, metabolic models were prepared. The two equation systems were as follows:

1. Related to an anabolic network of 53 reactions, intended to represent the synthesis of main cellular components. In addition, 28 intracellular molecular species were applied to calculate the overall stoichiometry.
2. Consisting of 45 reactions (46 metabolic intermediates) considering the glyoxylate shunt, the TCA cycle, gluconeogenesis, and PHB biosynthesis, was established to represent the catabolism.

The work was concentrated on growth control and NADPH generation as well as on the resulting impact on PHA synthesis in the presence of butyric acid as the C-source.

It was assumed that butyric acid enters the cells as a non-ionized molecule (protonized form) that is thereafter activated by butyrate kinase and phosphate butyryltransferase. Butyryl-CoA formed this way is the substrate for the oxidative conversion to acetoacetyl-CoA that is subsequently transformed into acetyl-CoA (3-ketothiolase-catalyzed), which is the precursor for biomass, energy, and PHB. To simplify the metabolic (and mathematic) situation, these authors have also performed the mass balancing of pools (respecting the pseudo steady state) to solve the system of linear equations. Results reported in this work warn that the anabolic requirements allow the NADPH generation via the ED pathway (theoretical C yield for PHB of 0.89 mol/mol). The metabolic situation without residual biomass formation requires the only possible NADPH regeneration in the TCA (the action of isocitrate dehydrogenase with theoretical carbon yield 0.67 mol/mol). Furthermore, it was pointed out that "the highest specific NADPH production rate was maximal at maximal growth rate" in all conditions. It turned out that "sustaining a controlled residual growth improves the PHB specific production rate without altering production yield".

Except for complicated metabolic networks, the reactor space was the object of mathematical modeling. Computational fluid dynamics and highly advanced software packages (e.g., FLUENT 6.3.26) were necessary to develop a 3D simulation. Afterwards, this simulation was connected to appropriately describe PHB synthesis by a metabolic model (Mavaddat and coworkers [151]). The hydrodynamics and PHA biosynthesis in an airlift-bioreactor system was investigated. Gas-phase to liquid-phase interactions were modeled by a Eulerian approach where this system, having the effects of coalescence and bubble ruptures, was mathematically presented using a "population" balance model. The biosynthesis of PHB itself was limited by the maximal rates (flux values) of 3-ketothiolase, reductase, and synthase reaction. From the macroscopic point of view, the liquid velocity vectors, the gas hold-up, the shear forces stress, and the volumetric oxygen transfer coefficient were investigated to achieve the PHA and substrate concentration profiles.

The basic properties of metabolic models devoted to development of PHA production are summarized in Table 2.8, while the structure of metabolic network model is presented in Figure 2.3.

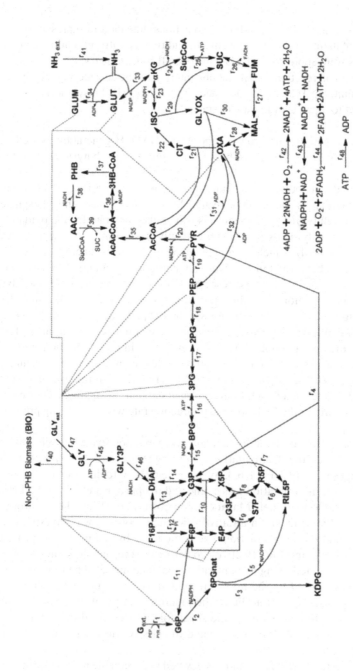

FIGURE 2.3 Metabolic network of with specified reaction rates (r) as the basis for the building up of high structured metabolic model for glucose and glycerol metabolism of PHA-producing *C. necator* DSM 545. Reaction rates definitions (r) are of the enzyme kinetic nature (adapted from reference [144]).

2.3.4   CYBERNETIC MODELS IN PHA BIOSYNTHESIS

Despite the great advance in the development of genome-scale metabolic models which include more than thousand metabolic reactions (encompassing more than 95% of mass related metabolic fluxes), at today's level of scientific knowledge, it is still not realistic to expect that we can get an exact picture of cell metabolism by metabolic mechanistic models. The main reasons for that are deficient kinetic data (usually in vitro estimated kinetic constants), and the insufficient knowledge of regulatory mechanisms of the cell cycle. One mode among others to "build in" the regulatory mechanisms in the metabolic model is to introduce the cybernetic way of thinking described earlier [152–156]. That means that the appropriately constructed cybernetic variables inserted into the mathematical model functionally replace the role of regulatory and mechanistic factors of the investigated cell system. For that purpose, a suitable OF must be created, usually tightly bounded to crucial cellular processes, e.g. gene activation, enzyme induction, metabolic repression, feedback inhibition. A physiologically specific target must be reflected in an OF, assuming that the biological cell metabolism is constructed in a manner to ensure the optimal final result, e.g. maximal growth or product yield. This strategy is known in the modeling literature as "the optimality hypothesis" that assumes that the synthesis is regulated in the cells in such a way that a nutritional target (the goal) is optimally fulfilled. In one model, several OFs can exist, often with an established "metabolic hierarchy". These goal-functions should be optimized by mathematical optimization procedures [157]. The above briefly described strategy was not the only one applied in cybernetic models. Hatzimanikatis and coworkers used mixed-integer linear optimization and discrete variables for modeling the activity/inactivity (presence/absence) of regulatory loops [158]. These authors declared the model to be a "mixed integer linear programming optimization problem (MILP)". The verification of a qualitative model/experiment was done by (log) linear approximation for a structurally "non-linear model". If experimental data were affordable in high quantity and the mechanistic details are insufficient, the fuzzy logic-based models [159, 160] or neural networks [161] can be organized to simulate the metabolic system. Among others, one of the first cybernetic models dedicated to PHA biosynthesis was the one published by Yoo and Kim [162], who built up their model on two basic presumptions:

(i) Two compartment principle, i.e. the residual cell mass (Xr, non-PHB part) and PHB (storage material);

(ii) C-source is allocated to enzyme synthesis, assuming a certain flexibility under N-limitation, and the pathway of PHB synthesis is under transcriptional control induced by environmental stress (i.e. nitrogen limitation). The degree of metabolic flexibility in catabolic processes was achieved by choosing the optimization of acetyl-CoA utilization as the objective for this purpose.

The key protein synthesis state equation was considered as dependent on the nonlinear control variable. Yoo and Kim did not use the standard, conventional cybernetic approach in this work where "if many resources are to be allocated to many activities (not necessarily an equal number), the fractional allocations are proportional to the fractional returns". The authors reported that the applied singular optimality criterion has made "the model too stiff to integrate numerically". Therefore, these authors introduced a nonsingular strategy and applied the macroeconomic Herrnstein's law [163] (rarely used in the biotechnology field). The resulting model predicted well the "mixed-growth-associated synthesis of PHB" for the overall fermentation range.

A very instructive paper about the model's utility was published by Varner and Ramkrishna [164]. They have tested by simulation different metabolic situations (blocking of two unproductive pathways and an over-expressing a key enzyme that catalyzes PHB synthesis) to improve productivity. This model was a generic type devoted to storage pathways investigation. It comprised of pathways related to two substrates and seven enzymes. Accomplished model simulations suggested that the nutritional circumstances are the important factors that determine the metabolic pathway (network) performances. The investigated genetic alterations and the cybernetic criteria (for over-expressing and blocking two unproductive pathways) resulted in a PHB content of 90% in total biomass, which was much higher than the experimentally (fed-batch and continuous process) achieved 65–75%. What remained unclear was if this difference was the problem of the model's shortcomings because of a "steric inhibition factor" (not incorporated in the model!).

A more complicated cybernetic model (PHA synthesis by *A. eutrophus*; today *C. necator* DSM 545) with a metabolic network consisting of 25 equations, including 53 parameters in total, among them 14 fixed and 39 adjustable, was structured by Ferraz et al. [165]. According to the authors, the preliminary adjustable parameter determination was performed by application of linearization and simplification/comparison of experimental results. After that, the advanced model parameters estimation using flexible polyhedron search (a flexible geometric simplex method, Nelder and Mead [166]) was introduced in the procedure. To have an effective application of this technique, the preliminarily estimated values of parameters were used as the "initial guess" for the searching of optimal values. In this model, the simulation of PHBV production and residual biomass growth encompass the substrate feedings, the sampling of medium for analysis, and the evaporation of fluids. As the authors stated, the two-phase cultivation process was correctly represented by the model; however, in the tested range, the model was less efficient for transient states.

Gadkar and colleagues [167] have also investigated the applicability of a "four units" cybernetic model for the cultivation and PHA synthesis of *R. eutropha* (today: *C. necator*). This cybernetic metabolic model consisted of four metabolic parts (subnetworks): sugar degradation (pentose phosphate and glycolysis pathways), biomass synthesis (i.e. production biomass precursors by the TCA cycle), PHB synthesis, and PHB degradation (depolymerization). The authors have established the substrate (glucose) concentration as a triggering factor for switching from PHB synthesis to PHB degradation. This way, the PHB degradation pathway (i.e. depolymerization) was set to be activated if the glucose concentration drops and becomes the growth rate limiting step. In the above cited paper, Gadkar et al. have compared the simulation results achieved by their model with the model formerly developed by Yoo and Kim [162]. These authors have applied a multi-rate predictive control algorithm for the simulation of a continuous flow bioreactor with a cell recycling system based on a microfiltration unit for cell separation; this algorithm addressed the PHB productivity and the permeate flux through the microfilter.

Another cybernetic model was originally developed for optimizing penicillin production and was later reorganized and adapted for PHB and PHBV production by *A. eutrophus* (today: *C. necator*) [168]. The authors, Riascos and Pinto, applied the orthogonal collocation method for a simultaneous optimization procedure. In this approach, the optimal control problem (OCP) was set to be the focus of optimization. The authors practiced the orthogonal collocation to discretize the optimization problem on finite elements (to solve the singularity problem). According to the authors, "the discretization of differential-algebraic equation systems (DAE) by orthogonal collocation in finite elements efficiently transforms dynamic optimization problems into nonlinear programming (NLP) problems". Their model considered nitrogen source, glucose, fructose, and propionic acid as C-sources, 2–4 control profiles, and 11 states, (all feed rates were chosen to be the control variables). This model of continuous PHB production by *A. eutrophus* was further upgraded and aligned with an industrial two-stage continuous process [169]. Three metabolic sub-networks were taken into consideration: the glycolysis and the pentose phosphate pathway for sugar metabolism, as well as the Krebs cycle (for the synthesis of biomass precursors). Additionally, formation and depolymerization of PHB were also included in the metabolic network. The cybernetic variables were formed on two levels: the "local" and the "global" level. As stated by the authors, the local cybernetic variables were related to the "individual strands of the metabolic pathways (production of growth precursors, production of PHB, etc.)". At the global level, the cybernetic variables were formed to present the "competition between possible metabolic pathways for synthesis of certain metabolite", such as between biomass and PHB precursors, respectively. This solution enabled a higher quantity of precursors to be directed towards PHB under the excess of glucose in the broth medium. Alternatively, during C-limiting growth, the competition between the usage of metabolites originating either from PHB degradation or from glucose catabolism was incorporated. The cultivation procedure in the experimental part of this study was performed in a series of two bioreactors with a separation device attached. For the first reactor, the growth on glucose and NH_4^+ under non-limiting conditions was foreseen for enhanced biomass growth. The outlet stream from this reactor was the inflow stream for the second reactor, which was additionally fed by glucose only. Any NH_4^+ present in the second reactor originated only from the part not used in the first reactor and was totally consumed in the second step. Hence, in the second stage, bacteria were exposed to excess glucose that was dominantly transformed into PHB. In addition, the output from the second reactor was subjected to a separation procedure (PHB extraction from biomass). Interestingly, the remaining residual biomass was degraded by different methods into carbohydrates that were added as a feed stream to both reactor steps. The dynamics of the process in bioreactors were analyzed with the help of a cybernetic model and by bifurcation analysis, which resulted in better insight into the steady states and limit cycles, thus helping to better design and control such cultivation systems.

Elementary flux modes interfering with cybernetic modeling has resulted in the development of a hybrid model published by Kim et al. [170]; this hybrid model was applied for the metabolism of *E. coli*, and was characterized by its high number of cybernetic solutions. Later on, PHA synthesis on carbohydrates by *R. eutropha* (today: *C. necator*) was investigated by Franz et al. [171] in a similar way. Here, these authors used elementary mode analysis for the "systematic derivation of the model equations". They have accounted that PHB is an intracellular metabolite with "slow dynamics" and that all other intracellular metabolites were assumed to be in quasi-steady state. The vector related to fluxes through EFMs was set as a "controlling vector of cybernetic variables vector". Furthermore, the enzymes included in the metabolic network were treated as inducible or constitutive, respectively. Separate cybernetic variables were formed to be the control factors of the vectors of inducible enzymes. In addition, the established two-compartmental principle (PHB and "residual non-PHB biomass") was again applied with the maximum carbon source uptake as an OF. Metabolic yield analysis was performed to reduce the number of EFMs. Following the stoichiometry of the network, it was determined that the concentration of the non-PHB part of cells (Xr) is linearly correlated with the quantity of used nitrogen source. The consequence was that the yield space can be treated as a two-dimensional one. Using the described approach for non-quasi-stationary metabolites (i.e. PHB), the validation of the model was performed by separate experiments. For that purpose, five metabolically different model configurations were developed to investigate the ratio of metabolic fluxes between biomass and PHB synthesis. Furthermore, an identification of nonlinear phenomena (oscillations and multiple steady states) was done by nonlinear analysis of continuous cultivation [172, 173]. Multiple steady states were discovered for the continuous bioreactor operation under a given range of substrate loading, depending on a given dilution rate. According to the authors, the steady states multiplicity region was of a narrow range, and the multiple steady states are realistically not very likely to occur in the real system.

Concluding this section, it is important to say that cybernetic models are formed respecting the concepts of natural evolution. If compared with mechanistic models, they are physiologically more exact, and also more feasible. Still, it should not be forgotten that cybernetic models display some disadvantages. One of the most significant ones is the complex structure of cybernetic models as a consequence of the high complexity of the represented intracellular regulatory or active transport systems. The second problem refers to the multiple cybernetic goals and the difficulties in recognizing which of them is the most relevant. Especially in MMCs, a conflict of modeling aims leads to a conflict of results.

2.3.5 MODELING OF PHA BIOSYNTHESIS USING NEURAL NETWORKS AND HYBRID MODELS

Non-biological neural networks (NNs), better to say artificial neural networks (ANNs), are systems which mimic biological neural networks present in living systems, especially in brains. It is important to state that such systems consist of a frame of many different learning algorithms. So, it is wrong to call NNs "algorithms". ANNs are created to be able to "learn" the specific tasks comparing the submitted examples and real data. Interestingly, they do not need to be programmed in accordance with the desired tasks. In mathematical (or kinetical) function recognition, for instance, they learn the graphical or equational form of functions using added examples that are marked as a certain "function-type" or, if different, as "no-type". Thereafter, the identification of such functions can be performed in analyzed sources, because the knowledge about functions is generated from the learning material.

The complexity of ANNs is expressed by the number of "inputs", "nodes", and "outputs", respectively, and additionally, by the number of interconnections between nodes. Another name for nodes in the literature is "artificial neuron". In biological systems, a signal is received by neurons and transmitted through synapses to other members of the network. In ANNs, the function of synapses is assigned to connections between nodes. These connections transmit a signal from one node to (an)other node(s). After the signal is received, it is processed in the node and split up to other artificial neurons. In ANNs, the "signal" is a real number, but the outputs of nodes are computed by a non-linear function, which depends on the sum of all its inputs. Usually a "weighting" factor is integrated to adjust a learning process by increasing or decreasing the signal strength. Sometimes, "limiting factors" are incorporated to establish the threshold under/above which the signal should/should not be sent. Interestingly, in neural network models, the mathematical formulation of biological or technological occurrences is not obligate. ANNs handle existing problems originating from the structure of cybernetic and mechanistic models. It can happen that more than one neural network gives an optimal result for the investigated situation. An assembling of different types of hybrid models can help to spot the best one. For that purpose, mechanistic and/or cybernetic mathematical models, either partially or fully developed, are hybridized with the neural networks.

In Table 2.9, an overview of literature dealing with PHA oriented neural networks and hybrid modeling is given. Figure 2.4 represents the structure of an artificial neural network.

2.4 METABOLIC PATHWAY AND FLUX ANALYSIS METHODS IN MODELING OF PHA BIOSYNTHESIS

The investigations of cellular metabolisms using both qualitative and quantitative analysis of metabolic pathway fluxes, genes activity, signaling molecules synthesis, and regulating mechanisms, are indispensable tools for the progress in biotechnological productivity; PHA synthesis is no exception in this context. The knowledge about metabolic networks dedicated to PHA synthesis is unavoidable during the optimization of microbial industrial production processes in this field. Experimental results supported by mathematical simulation of metabolic networks can improve the productivity and yields of a given production strain, providing deeper insights into its physiology, molecular mechanisms, and cell cycle. This knowledge is applicable for genetic improvements targeted on enzyme properties, metabolic flux bottlenecks, gene expression/knockouts, hosting of foreign genes, or plasmids as well as the change of plasmid copy numbers. Metabolic network reconstruction is based on the known genome structure (gene sequence) and on developed bioinformatic tools [183, 184]. For a certain organism, this reconstruction includes a cumbersome analysis and collecting of metabolic data registered in related databases (e.g., BioCyc,

TABLE 2.9

PHA-Producing Microorganisms and Applied Neural Networks and Hybrid Models

Microorganism	Cultivation type	Methods	Model type	Ref.
MMC		EFM	hybrid	[111]
MMC	SBR	bootstrap aggregated neural networks	ANNs	[174]
MMC	SBR	EM (Expectation-Maximization) algorithm	hybrid: "first principles models" + modular ANNs	[175]
Ralstonia eutropha (C. necator)	Fed-batch	back-propagation, generalized regression type, and recurrent Elman and Hopfield type algorithm; Lyapunov exponents, noise filters: auto-associative neural, cumulative sum control chart-(CUSUM), extended Kalman, Butterworth	feed-forward ANNs, hybrid	[176–179]
Ralstonia eutropha (C. necator)	Batch	EFM, Metatool; CellNettAnalyser	Hybrid (ANNs+cybernetic)	[171]
Azohydromonas lata MTCC 2311	SF STR	Genetic algorithm, Response surface method, Central composite design	ANNs hybrid	[180–182]

ANNs: artificial neural networks, EFM: elementary flux modes, MMC: mixed microbial cultures, SBR: sequencing batch reactor, SF: shaking flasks, STR: stirred tank reactor

BiGG, BRENDA, BioSilico, EcoCyc, ExPASy, GeneDB, IMG, KEGG, MetaCyc, metaTIGER, UMBBD, Transport DB), as well as searching published results in the scientific literature.

When reconstructed, such complicated genome-scale metabolic models resulted in the development of complex mathematical equation systems that are "mathematical pictures" (models) of a biological system. This system of differential equations is solvable by the application of specialized computing software using numerical integration techniques and appropriate simplifications. Metabolic network analysis by bioinformatic methods originates in the early works dealing with chemical open reaction systems [185], steady states and dynamics of open reaction systems [186], and convex stoichiometric analysis of reaction networks [187, 188]. After that, more specialized algorithms were developed for biological reaction networks (cf. ref. [189, 190]). The great advance in this field was achieved with the publishing of methods appropriate for quantitative analysis of complex metabolic networks, in short:

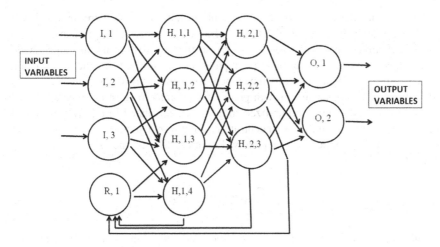

FIGURE 2.4 Artificial neural network structure. *Legend:* I, input layer (nodes); H, hidden layer (nodes); O, output layer (nodes), R, recurrent node; arrow, direction of information flow. The recurrent neurons (if many) are present in certain types of networks (i.e. Hopfield, Elman types). Networks differ in the neuron processing functions, the information flow directions, and the neuron numbers and positions.

- Extreme pathways approach [191–193],
- Minimal metabolic behaviors [194],
- Elementary (flux) mode (EFM) analysis [195, 196],
- Flux balance methods [197–199] based on linear programming as flux coupling [200], flux balance [197, 201], and flux variability analysis, respectively [202–204],
- Dynamic simulation and parameter estimation methods [205],
- Synthetic accessibility [206].

All the listed approaches targeted to the metabolic network analysis have a foundation in the mass conservation law as the basic principle of chemistry. The biological cell is assumed to be a "biochemical microreactor" with a complicated reaction network having a high number of intracellular metabolites (transformed through reversible and irreversible intracellular, enzymatic, chain reactions). Additionally, the system is considered open, which means that an exchange of mass and energy with the external space, hence, the transport of products and substrates, is considered. An intracellular metabolite can be mass-balanced using an equation of a general form (Equation 2.1):

$$\frac{dC_i}{dt} = ST_M * R_V - \left(\mu * C_i\right) \tag{2.1}$$

where C_i is the internal metabolite concentration vector, R_v is the reaction rate (flux) vector related to biochemical reactions (conversion of internal metabolites), ST_M is the stoichiometry coefficients matrix, and μ is the specific growth rate. A working metabolic network allows cell growth, changing the cell mass and the cell volume, describing the "dilution" of internal metabolites caused by the increase of cell volume. Fortunately, in a microbial cell, this "dilution" by growth is significantly slower than biochemical reactions (μ x C_i << R_v), so the negative term can be neglected. This allows some simplifications in computational handling of the above equation. Under some circumstances (fully developed exponential growth, biochemical transformations in the stationary growth phase, conversions in substrate limited growth or by resting cells, cells immobilized in gel or encapsulated in a polymer matrix), the steady state assumption (concentrations of internal metabolite are practically constant) can be applied. Consequently, Equation 2.1 becomes simpler and linear (Equation 2.2).

$$\frac{dC_i}{dt} = ST_M * R_V = 0 \tag{2.2}$$

If fully developed exponential growth is considered, some kinetic consequences become significant: the concentrations of essential substrates (carbon-, nitrogen-, phosphate sources, etc.) are significantly higher than the corresponding saturation constants (Ks). If not, exponential growth will not occur. Additionally, they must be lower than the respective inhibitory levels, so, no growth limitation for any reason is possible. For other purposes, the main substrates/co-substrates/precursor concentrations should be at a level that ensures saturation of the respective biocatalysts, i.e. enzymes should work at the maximum reaction rate. Furthermore, some reactions in metabolic networks are practically irreversible from the thermodynamic point of view, so the related rate in vector $\underline{R}v$ is of positive value (Equation 2.3):

$$R_i > 0 \tag{2.3}$$

Considered constraints (3), the equation system (2) becomes underdetermined (characterized by a multiplicity of reactions differing to the number of metabolites). An experimental work is of help in such cases: if some of the metabolic fluxes can be determined (measured, e.g., by C^{13} isotope labeling), the reaction flux vector $\underline{R}v$ is to be divided into sub-vectors, where R_e contains measured, and R_c unmeasured metabolic fluxes. R_c can be calculated, provided that a minimal fraction of the whole set of reaction fluxes is experimentally determined. Fortunately, combining the experimental and the computing procedures, metabolic flux analysis (MFA), flux balance analysis (FBA), and metabolic pathway analysis (MPA) can handle the system consisting of the matrix ST_M and the experimentally measured vector R_e. For further details, the reader can consult the review [30] and the original literature presented therein.

2.4.1 EFM in the Microbial PHA Biopolyester Synthesis

Apart from other purposes, elementary flux modes (EFMs) were also applied for the analysis of metabolic networks related to PHA synthesis. Peculiarly, PHB homopolyester was investigated using this method. Two research areas should be underlined: PHB's role and synthesis in activated sludge from wastewater treatment processes, and PHB formation from renewable raw materials. Among others, Carlson et al. [207] have published an article that was focused on rather exotic PHB producer, namely genetically engineered *Saccharomyces cerevisiae*. These authors have investigated PHB carbon yields from different substrates by application of elementary modes analysis (EMA). In this work, the two-compartments strategy (PHA and non-PHA part of biomass [Xr]) was combined with a network of biochemical reactions encompassing their reversibility or irreversibility, respectively. Nine molecular species (glucose, acetate, glycerol, ethanol, PHB, CO_2, succinate, and ATP) were considered as substrates and products, respectively. First, the wild-type *S. cerevisiae* system was analyzed by EMA, resulting in 241 unique reaction combinations. After that, the authors changed the metabolic network by the introduction of the recombinant PHB synthesis/accumulation pathway. This action resulted in 20 unique EMs; among them, seven were producing PHB. Comparing the calculated theoretical PHB carbon yield, 0.67 g/g PHB per g carbon source was achieved for an optimal EM. Furthermore, the modifications of the metabolic network were tested using EMA to check eventual positive effects. For example, incorporation of the ATP citrate-lyase reaction, which is absent in the *S. cerevisiae* wild-type strain, in the metabolic network of the PHB producing recombinant strain has resulted in an enormous increase of the unique modes from 20 to 496. Simultaneously, the number of PHB-generating EMs reached 314. Applied modification consequently influenced the theoretical PHB carbon yield, which increased from 0.67 to 0.83 g/g. Incorporation of a transhydrogenase reaction in the model network resulted in a theoretical PHB carbon yield of 0.71 g/g. Introducing both the transhydrogenase and the ATP citrate-lyase reaction, a yield of 0.84 g/g was achieved. Anaerobic PHB synthesis in an *S. cerevisiae* system was also investigated by this work; this resulted in two EMs capable of producing PHB and ethanol. Similar investigations of anoxic PHB synthesis by the same group of authors was performed for recombinant *Escherichia coli* [137–139]. It was reported that this genetically modified bacteria have the potential to accumulate PHB mass fractions exceeding 0.5 g per g cell dry mass. Such a result was possible both for the cultivation in a nutritionally balanced medium (growth-associated synthesis) as well as for an unbalanced medium. The achieved maximum of specific PHB production rate ($q_{p\,max}$) for the growth-associated part of the process amounted to 2.3 ± 0.2 mmol PHB per g of residual dry biomass per hour. Interestingly, the excreted byproducts were different comparing the PHB-synthesizing recombinant construct and the control strain. Under balanced and unbalanced cultivation conditions, PHB production occurred in parallel with acetate excretion. The PHB-producing recombinant culture excreted less acetate (33%) than a control culture under unbalanced cultivation conditions. The authors used the above results for the development of a theoretical, biochemical, metabolic network model to study the capability of this organism in anoxic cultivation procedures. A maximum theoretical PHB carbon yield for anoxic PHB synthesis of 0.8 g/g was reported. In addition, Wlaschin and co-authors [208] have introduced the weighted sum of all EMs to present the cell's metabolic system. They have underlined that the high variety of EMs caused difficulties in the identification of the individual weights. Two deletions of genes were introduced to determine the weighting factors. Combined with growth conditions, this action reduced the original number of EMs from 4374 to 40 for a PHB synthesizing strain, and to 24 for a non-PHB synthesizing control organism. Further, the modes were divided into five groups, depending on the overall stoichiometry. As the authors stated, this resulted in a reduction of complexity, so the weighting factors of each EMs group were estimated from measured metabolite production rates. Moreover, the inverse correlation of estimated individual weights and the related entropy (generated through the function of the pathways defined in EMs) was underlined. This is perhaps Mother Nature's way of making available those regulatory patterns that favor the pathways which generate less entropy. The excellent correlation between experimentally determined metabolic fluxes and in silico results based on reaction entropies of EMs was demonstrated. According to the authors, "interpretation of the obtained set of EMs in analogy to a metabolic ensemble of quantum states of a macroscopic system" is a useful tool.

The naturally PHA-producing bacterial strain R. eutropha (today C. necator) was investigated with the help of EMs and yield space analysis (YSA) by Franz and colleagues [171]. This model was discussed above in the part of this chapter related to the hybrid cybernetic models (vide supra, Section 3.4), so for more details the reader should refer to the mentioned section.

The five-step continuous process of PHB synthesis by C. necator DSM 545 on glucose reported by Atlić et al. [46], where a bioreactor cascade was considered as the process-engineering substitute for a tubular plug flow reactor, was an object of mathematical modeling performed with the help of EFMs and YSA [143]. Two different metabolic situations occurring alongside the cascade and the transient state between them were covered by this modeling approach, namely nutritionally balanced growth and nitrogen limited PHB synthesis. Lopar and associated colleagues have used two-dimensional YSA to improve the feeding strategies for the five-step continuous chemostat. Using both the mentioned methods (EFMs and YSA), the same group of authors also investigated the possibilities of improvement of C. necator metabolic network for PHB biosynthesis on glycerol as a carbon source [144]. For a more detailed description of both investigations, the reader should consult the "metabolic models" section, especially the subsection "metabolic models targeted to industrial-scale modeling and to development of high productive strains" in this chapter.

EFM analysis was also applied for the research of PHA biosynthesis by Pseudomonas putida on glycerol (Poblete-Castro and colleagues [209]). Focusing on possible "hot points" of genetic engineering, i.e., amplification, attenuation, and deletion, in silico experiments were performed to achieve a deeper insight into metabolic networks. For that purpose, the P. putida KT2440 genome-scale network, containing the anabolic routes to biomass, PHA, and byproducts, the central catabolism, and energy generating reactions, was used for the EFM analysis. In this work, the structuring of cell space was assumed; it was foreseen that byproducts, i.e. gluconate and α-ketogluconate, are synthesized in the periplasm, while other metabolic reactions take place in the cytoplasm. The modes obtained were evaluated using yields, maximal theoretical yields, and relative metabolic fluxes throughout pathways. Concerning the results of metabolic flux correlation coefficients, strain improvement by metabolic engineering was suggested based on analyzing the ratio of metabolic fluxes towards PHA and other fluxes. For example, the gcd-encoded glucose dehydrogenase was underlined as the major target for gene deletion. The resulting mutant (*P. putida Δgcd*) revealed the two-fold PHA accumulation capacity compared to its parental strain, without undesirable side effects caused by modification. On the contrary, the 6-phosphogluconolactonase-lacking mutant (*P. putida Δpgl*) was inferior in PHA synthesis, and both situations were exactly predicted in silico. The potential of *P. putida Δgcd* mutant for PHA production was tested under controlled batch cultivation, where the PHA yield was increased by 80%. Further, the double PHA concentration was achieved, with a 50% increase of PHA mass fraction in cell dry matter without reduction of the growth rate.

Besides the work of Poblete-Castro et al., an article published by Borrero de Acuna and colleagues [210] also attracted the attention of peers. Here, the overproduction of enzymes to overcome "bottlenecks" in *mcl*-PHA precursor formation was investigated. The researchers have established a genome-based in silico model with the intention to improve *mcl*-PHA production on glucose. The model prediction was successful; both the wild type (*P. putida* KT2440) and the mutant (*P. putida Δgcd*) showed enhanced PHA production if the pyruvate dehydrogenase subunit was overexpressed. PHA production was increased by 33% (wild-type strain), and by 121% (*P. putida Δgcd*), respectively. Interestingly, an overexpression of 6-phospho-gluconolactonase encoding gene (*pgl*) did not cause better PHA biosynthesis. In addition, it turned out that NADPH has to be quantitatively consumed to achieve optimum PHA synthesis. An overview of software tools applied in metabolic pathway analysis can be seen in Table 2.10.

Other software tools that contain computing routines for elementary modes are EMPATH, JARNAC, In-Silico-DiscoveryTM, ScrumPy, PySCeS, and, for on-line computation, pHpMetatool.

2.5   CONCLUSIONS AND OUTLOOK

Mathematical modeling of PHA biosynthesis turned out to be a powerful tool for the prediction of different biological, biochemical, genetic, metabolic, or biotechnological situations. This approach is more successful if the predicted variables lie in the range of data that were determined when the model was structured (interpolation) than in the case of extrapolation. Today's scientific state of knowledge does not allow to clearly state which model type is the definitely best option for handling diverse problems related to PHA biosynthesis, such as mass transfer phenomena, bioreactor performance, metabolic, and genetic engineering, influence of cultivation conditions, and biochemical kinetics. This impossibility emerges from the plurality and complexity of biological (and physical) systems. Indeed, all models presented in this chapter have a certain range of validity, and, depending on applied adjustable parameters, they possess some range of adjustability, i.e., the "interpolation/extrapolation power of model". Of course, there is no doubt that a certain model type is more appropriate for the simulation/description of some biological systems than another one. That is because the models are meant and structured to be the direct mathematical pictures of the natural systems. This is especially valid for genome-scale and for metabolic kinetic models. Exceptions of that statement are neural network models, which are not constructed to be a "direct picture" of biological systems. It is

TABLE 2.10

The Overview of Software Tools Applied in Metabolic Pathway Analysis (adapted from [30])

Software	Application field	Basic language and Software	Operating software	Main refs
GEPASI	dynamics, steady states and control of biochemical systems	C	Linux	[211]http://www.gepasi .org/
COPASI	metabolic networks, cell-signaling pathways, regulatory networks	C++	Linux, Mac OS Microsoft Windows	[212]http://www.copasi .org/
METATOOL	nullspace matrix, elementary modes	C GNU Octave Matlab	Linux, Windows, Intel Mac	[145]http://pinguin.biolo gie.unijena.de/bioinformá tik/networks/metatool/ metatool5.1/metatool5.1 .html
FluxAnalyzer	structure, pathways, flux distributions, elementary modes	Matlab	Windows	[213]http://www.mpi -magdeburg.mpg.de/ projects/fluxanalyzer
CellNetAnalyser	structural and functional analysis of biochemical networks	Matlab LP (M)ILP	Linux, Windows	[214]http://www.mpi -magdeburg.mpg.de/ projects/cna/manual/toc _frame.htm
YANA	elementary modes flux distributions		Linux, Windows, Mac OS	[215] http://yana.bioapps. biozentrum.uni-wuerz burg.de
YANA square	elementary modes flux distributions	Java	Windows, Linux, General Unix, Mac OS	[216, 217] http://yana.bio apps.biozentrum .uni-wuerzburg.de
ExPa	extreme pathways	C++	Linux, Mac OS	[218]; http:// systemsbiology.ucsd.edu/ Downloads/ ExtremePathwayAnalysis
SNA		Mathematica Matlab	Linux	[219, 220] http://bioinfor matics.org/project/?group _id=546

worth saying that they are intended to achieve results "as exact and compliant as possible" compared with experiments. Mathematical models presented in this chapter are applied for very different tasks: mixing and homogenization of bioreactor space, metabolic pathways, and networks, cell cycles, genetic, and metabolic engineering. Respectively to that, they have highly diverse structures. The main question here is: how to know when and for which explicit purpose to apply a specific type of model? More than one answer applies to this question! Some authors are convinced that "standard microbial laboratory cultivations and everyday industrial praxis" can be simulated/predicted by simple formal kinetic models and by "low-structured" models. Despite their simplicity, these models are accurate enough for that purpose, and they have a low computational demand, hence, they can be performed by commercial software on PCs. Advanced scientific purposes and industrial equipment development, i.e. real natural or technical systems, require highly structured, multilevel-organized hybrid models. For instance, computational fluid dynamics (CFD) is an excellent tool for the modeling (computational characterization) of concentration/temperature/momentum fields in bioreactors and in up- and down-stream equipment. Further, genome-scale models, encompassing enzyme kinetic, are more or less able to reflect the real cell metabolism. It seems to be that the combination of latter types of models (hybrid-type in its nature) can be the solution for the coupling of biological and physical properties in one mathematical system. In addition, the cybernetic or/and neural network models, if hybridized with kinetic/genome-scale metabolic systems or with CFD models, can be the solution for computational simulation of complex biotechnological systems, encompassing the cellular physiology, broth rheology, and related mass transfer effects. Here, it is necessary to state that cellular metabolic systems are still a demanding task for mathematical modeling.

Approximately two decades ago, elementary modes and elementary flux modes (EMs /EFMs) calculation procedures were established as a powerful tool for the analysis of metabolic networks. This was great support for the analysis of fully developed, large metabolic networks. Unfortunately, these methods frequently result in an overproduction of solutions. To overcome this problem, additional constraints are integrated, e.g., YSA, null space application, etc., intending to reduce the number of possible solutions to a convenient level for handling. A few years ago, powerful computational algorithms, i.e. conversion flux cone encapsulated in the parental flux cone, network decomposition, and parallel computing, were developed for the handling of full genome scale metabolic networks. As many other methods, as well as EMs/EFMs are of limited use. They are not applicable for many transient states happening in the cell's lifecycle. Metabolic, kinetic and genetic properties of cells limit the number of states suitable for such work on steady state fluxes through networks of reactions, enzymes knock out/build in investigations, exponential growth, steady state continuous cultivations, transformations by resting cells, and on the synthesis of extracellular or intracellular metabolites under restricted or terminated growth. Fortunately, other similar methods can be combined with EMs/ EFMs, such as flux balance analysis, flux variability analysis, or extreme pathways, helping to achieve the desired results. In addition, the development of computing algorithms (e.g., simple rank/nullity test for the distinguishing of extreme pathways from elementary modes using the invariant stoichiometric matrix) greatly facilitates the solution of complete genome-scale metabolic networks. In general, the constraint is now on the biological side. In a nutshell, despite fully explained genomes for a certain number of microorganisms, there is still missing knowledge about interactions on the genomic/proteomic/signaling level. Once the biological facts are known in total, mathematical modeling of biotechnological processes will be of indispensable help for the development of processes and production strains. In the meantime, mathematical modeling and biological investigations (genomic, proteomic, kinetic, signaling, etc.) are convicted to support each other in parallel to proceed and enhance "step by step" by the exchange of information and cross-linked testing of results and hypotheses.

REFERENCES

1. Oeding V, Schlegel HG. Beta-ketothiolase from *Hydrogenomonas eutropha* H16 and its significance in the regulation of poly-beta-hydroxybutyrate metabolism. Biochem J 1973; 134(1): 239–248.
2. Haywood GW, Anderson AJ, Chu L, Dawes EA. Characterization of two 3-ketothiolases possessing differing substrate specificities in the polyhydroxyalkanoate synthesizing organism *Alcaligenes eutrophus*. FEMS Microbiol Lett 1988; 52: 91–96.
3. Haywood GW, Anderson AJ, Chu L, Dawes EA. The role of NADH- and NADPH-linked acetoacetyl-CoA reductases in the poly-3-hydroxybutyrate synthesizing organism *Alcaligenes eutrophus*. FEMS Microbiol Lett 1988; 52: 259–264.
4. Belova LL, Sokolov AP, Sidorov IA, Trotsenko YA. Purification and charactereization of NADPH-dependent acetoacetyl-CoA reductase from *Methylobacterium extorquens*. FEMS Microbiol Lett 1997; 156: 275–279.
5. Rehm BHA. Polyester synthases: natural catalysts for plastics. Biochem J 2003; 376: 15–33.
6. Jendrossek D, Pfeiffer D. New insights in the formation of polyhydroxyalkanoate granules (carbonosomes) and novel functions of poly(3-hydroxybutyrate). Environ Microbiol 2014; 16: 2357–2373.
7. Mezzolla V, D'Urso OF, Poltronieri P. Role of PhaC Type I and Type II Enzymes during PHA Biosynthesis. Polymers 2018; 10(8): 910–922.
8. Anonymus 1. Available from: https://www.brenda-enzymes.org/enzyme.php?ecno=3.1.1.75 (Accessed 21. 01. 2019).
9. Abe T, Kobayashi T, Saito T. Properties of a novel intracellular poly(3-hydroxybutyrate) depolymerase with high specific activity (PhaZd) in *Wautersia eutropha* H16. J Bacteriol 2005; 187: 6982–6990.
10. Gebauer B, Jendrossek D. Assay of poly(3-hydroxybutyrate) depolymerase activity and product determination. Appl Environ Microbiol 2006; 72: 6094–6100.
11. Jendrossek D, Handrick R. Microbial degradation of polyhydroxyalkanoates. Annu Rev Microbiol 2002; 56: 403–432.
12. Uchino K, Saito T, Gebauer B, Jendrossek D. Isolated poly(3-hydroxybutyrate) (PHB) granules are complex bacterial organelles catalyzing formation of PHB from acetyl Coenzyme A (CoA) and degradation of PHB to acetyl-CoA. J Bacteriol 2007; 189(22): 8250–8256.
13. Jendrossek D. Extracellular polyhydroxyalkanoate depolymerases: the key enzymes of PHA degradation. In: Y. Doi and A. Steinbüchel (ed.), Biopolymers, vol. 3b. Polyesters II. Wiley-VCH, Weinheim, Germany, 2002, p. 41–83.
14. Doi Y, Segawa A, Kawaguchi Y, Kunioka M. Cyclic nature of poly(3-hydroxyalkanoate) metabolism in *Alcaligenes eutrophus*. FEMS Microbiol Lett 1990; 55: 65–169.

15. Taidi B, Mansfield D, Anderson AJ. Turnover of poly(3-hydroxybutyrate) (PHB) and its influence on the molecular mass of the polymer accumulated by *Alcaligenes eutrophus* during batch culture. FEMS Microbiol Lett 1995; 129: 201–206.

16. Reusch RN. Physiological importance of poly-(*R*)-3-hydroxybutyrates. Chem Biodivers 2012; 9: 2343–2366.

17. Reusch RN. The role of short-chain conjugated poly-(*R*)-3-hydroxybutyrate (cPHB) in protein folding. Int J Mol Sci 2013; 14: 10727–10748.

18. Moser A. Kinetics of batch fermentations. In: H.J. Rehm, G. Reed (eds.), Biotechnology, vol. 2, Verlagsgesellschaft GmbH, Weinheim, 1985, pp. 243–283.

19. Moser A. Bioprocess Technology Kinetics and Reactors. Springer-Verlag, New York, 1988, pp. 197–306.

20. Gaden EL. Fermentation process kinetics. J Biochem Microb Technol Eng 1959; 11(4): 413–429; cited from: Gaden EJ Jr. Fermentation process kinetics. Biotechnol Bioeng 2000; 67: 629–635.

21. Braunegg G, Genser K, Bona R, Haage G, Schellauf F, Winkler E. Production of PHAs from agricultural waste material. Macromol Symp 1999; 144: 375–383.

22. Horvat P, Koller M, Braunegg G. Kinetic aspects and mathematical modelling of PHA biosynthesis. In: M. Koller (ed.), Recent Advances in Biotechnology, Volume 2, Microbial Polyester Production Performance and Processing, Bioengineering, Characterization, and Sustainability. Bentham Science Publishers, 2016, pp. 111–214.

23. Koller M, Maršálek L, Miranda de Sousa Dias M, Braunegg G. Producing microbial polyhydroxyalkanoate (PHA) biopolyesters in a sustainable manner. New Biotechnol 2017; 37: 24–38.

24. Braunegg G, Lefebvre G, Renner G, Zeiser A, Haage G, Loidl-Lanthaler K. Kinetics as a tool for polyhydroxyalkanoate production optimization. Can J Microbiol 1995; 41: 239–248.

25. Hartmann R, Hany R, Pletscher E, Ritter A, Witholt B, Zinn M. Tailor-made olefinic medium-chain-length poly[(R)-3-hydroxyalkanoates] by *Pseudomonas putida* GPo1: batch versus chemostat production. Biotechnol Bioeng 2001; 93: 737–746.

26. Hartmann R, Hany R, Geiger T, Egli T, Witholt B, Zinn M. Tailored Biosynthesis of olefinic medium-chain-length poly[(R)-3-hydroxyalkanoates] in *Pseudomonas putida* GPo1 with improved thermal properties. Macromolecules 2004; 37: 6780–6785.

27. Koller M. A review on established and emerging fermentation schemes for microbial production of polyhydroxyalkanoate (PHA) biopolyesters. Fermentation 2018; 4: 1–30; (2018).

28. Patnaik PR. Perspectives in the modeling and optimization of PHB production by pure and mixed cultures. Crit Rev Biotechnol 2005; 25: 153–171.

29. Novak M, Koller M, Braunegg G, Horvat P. Mathematical modelling as a tool for optimized PHA production. Chem Biochem Eng Q 2015; 29(2): 183–220.

30. Horvat P, Koller M, Braunegg G. Recent advances in elementary flux modes and yield space analysis as useful tools in metabolic network studies. World J Microbiol Biotechnol 2015; 31(9): 1315–1328.

31. Kossen NWF, Oosterhuis NMG. Modeling and scaling-up of bioreactors. In: H. J. Rehm, G. Reed (eds.), Biotechnology (2nd ed.) Vol. 2., VCH Verlag, Weinheim, Germany, 1985.

32. Sonnleitner B, Heinzle E, Braunegg G, Lafferty RM. Formal kinetics of poly-β-hydroxybutyric acid (PHB) production in *Alcaligenes eutrophus* H 16 and *Mycoplana rubra* R 14 with respect to the dissolved oxygen tension in ammonium-limited batch cultures. Eur J Appl Microbiol Biotechnol 1979; 7(1): 1–10.

33. Morinaga Y, Yamanaka S, Ishizaki A, Hirose Y. Growth characteristics and cell composition of *Alcaligenes eutrophus* in chemostat culture. Agric Biol Chem 1978; 42(2): 439–444.

34. Heinzle E, Lafferty RM. A kinetic model for growth and synthesis of poly-β-hydroxybutyric acid (PHB) in *Alcaligenes eutrophus* H 16. Eur J Appl Microbiol Biotechnol 1980; 11(1): 8–16.

35. Asenjo JA, Suk JS. Kinetics and models for the bioconversion of methane into an intracellular polymer, poly-β-hydroxybutyrate (PHB). Biotechnol Bioeng Symp 1986; 15: 225–234.

36. Mulchandani A, Luong JH, Groom C. Substrate inhibition kinetics for microbial growth and synthesis of poly-β-hydroxybutyric acid by *Alcaligenes eutrophus* ATCC 17697. Appl Microbiol Biotechnol 1989; 30(1): 11–17.

37. Luedeking R, Piret EL. A kinetic study of the lactic acid fermentation. Batch process at controlled pH. J Biochem Microbiol Technol Eng 1959; 1(4): 393–412.

38. Megee RD III, Drake JF, Fredrickson AG, Tsuchiya HM. Studies in intermicrobial symbiosis *Saccharomyces cerevisiae* and *Lactobacillus casei*. Can J Microbiol 1972; 18(11): 1733–1742.

39. Dhanasekar R, Viruthagiri T, Sabarathinam PL. Poly(3-hydroxybutyrate) synthesis from a mutant strain *Azotobacter vinelandii* utilizing glucose in a batch reactor. Biochem Eng J 2003; 16(1): 1–8.

40. Hill AV. The possible effects of the aggregation of the molecules of haemoglobin on its dissociation curves. J Physiol 1910; 40: 4–7.

41. Aiba S, Shoda M, Nagatani M. Kinetics of product inhibition in alcohol fermentation. Biotechnol Bioeng 1968; 10(6): 845–864.

42. Tohyama M, Patarinska T, Qiang Z, Shimizu K. Modeling of the mixed culture and periodic control for PHB production. Biochem Eng J 2002; 10(3): 157–173.

43. Luong JH. Kinetics of ethanol inhibition in alcohol fermentation. Biotechnol Bioeng 1985; 27(3): 280–285.

44. Luong JH. Generalization of monod kinetics for analysis of growth data with substrate inhibition. Biotechnol Bioeng 1987; 29(2): 242–248.

45. Koller M, Horvat P, Hesse P, Bona R, Kutschera C, Atlić A, Braunegg G. Assessment of formal and low structured kinetic modeling of polyhydroxyalkanoate synthesis from complex substrates. Bioprocess Biosyst Eng 2006; 29(5–6): 367–377.

46. Atlić A, Koller M, Scherzer D, Kutschera C, Grillo-Fernandes E, Horvat P, Chiellini E, Braunegg G. Continuous production of poly([R]-3-hydroxybutyrate) by *Cupriavidus necator* in a multistage bioreactor cascade. Appl Microbiol Biotechnol 2001; 91(2): 295–304.

47. Horvat P, Vrana Špoljarić I, Lopar M, Atlić A, Koller M, Braunegg G. Mathematical modelling and process optimization of a continuous 5-stage bioreactor cascade for production of poly[–(R)-3-hydroxybutyrate] by *Cupriavidus necator*. Bioprocess Biosyst Eng 2013; 36: 1235–1250.

48. Megee RD III, Drake JF, Fredrickson AG, Tsuchiya HM. Studies in intermicrobial symbiosis *Saccharomyces cerevisiae* and *Lactobacillus casei*. Can J Microbiol 1972; 18(11): 1733–1742.

49. Mankad T, Bungay HR. Model for microbial growth with more than one limiting nutrient. J Biotechnol 1988; 7(2): 161–166.

50. Gahlawat G, Srivastava AK. Development of a mathematical model for the growth associated Polyhydroxybutyrate fermentation by *Azohydromonas australica* and its use for the design of fedbatch cultivation strategies. Bioresour Technol 2013; 137: 98–105.

51. Chatzidoukas C, Penloglou G, Kiparissides C. Development of a structured dynamic model for the production of polyhydroxybutyrate (PHB) in *Azohydromonas lata* cultures. Biochem Eng J 2013; 71:72–80.

52. Gahlawat G, Srivastava AK. Model-Based nutrient feeding strategies for the increased production of polyhydroxybutyrate (PHB) by *Alcaligenes latus*. Appl Biochem Biotechnol 2017; 183: 530–542.

53. Dhanasekar R, Viruthagiri T, Sabarathinam PL. Poly(3-hydroxybutyrate) synthesis from a mutant strain *Azotobacter vinelandii* utilizing glucose in a batch reactor. Biochem Eng J 2003; 16(1): 1–8.
54. Faccin DJ, Correa MP, Rech R, et al. Modeling P(3HB) production by *Bacillus megaterium*. J Chem Technol Biotechnol 2012; 87: 325–333.
55. Porras MA, Ramos FD, Diaz MS, Cubitto MA, Villar MA. Modeling the bioconversion of starch to P(HBco-HV) optimized by experimental design using *Bacillus megaterium* BBST4 strain. Environ Technol 2018; 40: 1185–1202.
56. Panda I, Balabantaray S, Sahoo SK, Patra N. Mathematical model of growth and polyhydroxybutyrate production using microbial fermentation of *Bacillus subtilis*. Chem Eng Commun 2018; 205:2, 249–256.
57. Mendez DA, Cabezab, IO, Moreno, NC, Riascos CAM. Mathematical modelling and scale-up of batch fermentation with *Burkholderia cepacia* B27 using vegetal oil as carbon source to produce polyhydroxyalkanoates. Chem Eng Trans 2016; 49: 277–282.
58. Katoh T, Yuguchi D, Yoshii H, Shi H, Shimizu K. Dynamics and modeling on fermentative production of poly (β-hydroxybutyric acid) from sugars via lactate by a mixed culture of *Lactobacillus delbrueckii* and *Alcaligenes eutrophus*. J Biotechnol 1999; 67(2–3): 113–134.
59. Shahhosseini S. Simulation and optimisation of PHB production in fed-batch culture of *Ralstonia eutropha*. Process Biochem 2004; 39: 963–969.
60. Patnaik PR. Neural network designs for poly-β-hydroxybutyrate production optimization under simulated industrial conditions. Biotechnol Lett 2005; 27(6): 409–415.
61. Patnaik PR. Dispersion optimization to enhance PHB production in fed-batch cultures of *Ralstonia eutropha*. Bioresour Technol 2006; 97(16): 1994–2001.
62. Shang L, Fan DD, Kim M, et al. Modeling of poly(3-hydroxybutyrate) production by high cell density fed-batch culture of *Ralstonia eutropha*. Biotechnol Bioproc Eng 2007; 12: 417–423.
63. Mozumder MS, Goormachtigh L, Garcia-Gonzalez L, De Wever H, Volcke EI. Modeling pure culture heterotrophic production of polyhydroxybutyrate (PHB). Bioresour Technol 2014; 155: 272–280.
64. Raje P, Srivastava AK. Updated mathematical model and fed-batch strategies for poly-β-hydroxybutyrate (PHB) production by *Alcaligenes eutrophus*. Bioresour Technol 1998; 64: 185–192.
65. Yu J, Si Y, Wong WK. Kinetics modeling of inhibition and utilization of mixed volatile fatty acids in the formation of polyhydroxyalkanoates by *Ralstonia eutropha*. Process Biochem 2002; 37: 731–738.
66. Wang J, Yue ZB, Sheng GP, Yu HQ. Simulation and optimisation of PHB production in fed-batch culture of *Ralstonia eutropha*. Appl Microbiol Biotechnol 2007; 75(4): 871–878.
67. Patwardhan PR, Srivastava AK. Model-based fed-batch cultivation of *R. eutropha* for enhanced biopolymer production. Biochem Eng J 2004; 20: 21–28.
68. Khanna S, Srivastava AK. A simple structured mathematical model for biopolymer (PHB) production. Biotechnol Prog 2005; 21(3): 830–838.
69. Khanna S, Srivastava AK. Computer simulated fed-batch cultivation for over production of PHB: a comparison of simultaneous and alternate feeding of carbon and nitrogen. Biochem Eng J 2006; 27: 197–203.
70. Khanna S, Srivastava AK. Optimization of nutrient feed concentration and addition time for production of poly(β-hydroxybutyrate). Enzyme Microb Technol 2006; 39: 1145–1151.
71. Khanna S, Srivastava AK. Continuous production of poly-β-hydroxybutyrate by high-cell-density cultivation of *Wautersia eutropha*. J Chem Technol Biotechnol 2008; 83: 799–805.

72. Vrana Špoljarić I, Lopar M, Koller M, et al. Mathematical modeling of poly[(R)-3-hydroxyalkanoate] synthesis by *Cupriavidus necator* DSM 545 on substrates stemming from biodiesel production. Bioresour Technol 2013; 133: 482–494.

73. Vrana Špoljarić I, Lopar M, Koller M, et al. *In silico* optimization and low structured kinetic model of poly[(R)-3-hydroxybutyrate] synthesis by *Cupriavidus necator* DSM 545 by fed-batch cultivation on glycerol. J Biotechnol 2013; 168(4): 625–635.

74. Mozumder MSI, Garcia-Gonzalez L, De Wever H, Volcke EIP. Model-based process analysis of heterotrophic-autotrophic poly(3-hydroxybutyrate) (PHB) production. Biochem Eng J 2016; 114: 202–208.

75. Pérez Rivero C, Sun C, Theodoropoulos C, Webb C. Building a predictive model for PHB production from glycerol. Biochem Eng J 2016; 116: 113–121.

76. Triguerosa DEG, Hinterholz CL, Fioresea ML, et al. Statistical evaluation and discrimination of competing kinetic models and hypothesis for the mathematical description of poly-3(hydroxybutyrate) synthesis by *Cupriavidus necator* DSM 545. Chem Eng Sci 2017; 160: 20–33.

77. Das M, Grover A. Fermentation optimization and mathematical modeling of glycerol-based microbial poly(3-hydroxybutyrate) production. Proc Biochem 2018; 71: 1–11.

78. Marudkla J, Lee WC, Wannawilai S, et al. Model of acetic acid-affected growth and poly(3-hydroxybutyrate) production by *Cupriavidus necator* DSM 545. J Biotechnol 2018; 268: 12–20.

79. Vega R, Castillo A. A simplemathematical model capable of describing the microbial production of poly(hydroxyalkanoates) under carbon- and nitrogen-limiting growth conditions. J Chem Technol Biotechnol 2018; 93: 2564–2575.

80. Cui YW, Zhang HY, Ji SY, Wang ZW. Kinetic analysis of the temperature effect on polyhydroxyalkanoate production by *Haloferax mediterranei* in synthetic molasses wastewater. J Polym Environ 2017; 25:277–285.

81. Cui YW, Shi YP, Gong XY. Effects of C/N in the substrate on the simultaneous. production of polyhydroxyalkanoates and extracellular polymeric substances by *Haloferax mediterranei* via kinetic model analysis. RSC Adv 2017; 7: 18953–18961.

82. Ye J, Huang W, Wang D, et al. Pilot scale-up of poly(3-hydroxybutyrate-co-4-hydroxy butyrate) production by *Halomonas bluephagenesis* via cell growth adapted optimization process. Biotechnol J 2018; 13: 1800074 (1–10).

83. Torres-Cerna CE, Alanis AY, Poblete-Castro I, Hernandez-Vargasc EA. Batch cultivation model for biopolymer production. Chem Biochem Eng Q 2017; 31(1): 89–99.

84. Vega R, Castillo A. A simple mathematical model capable of describing the microbial production of poly(hydroxyalkanoates) under carbon- and nitrogen-limiting growth conditions. J Chem Technol Biotechnol 2018; 93: 2564–2575.

85. Dias JM, Serafim LS, Lemos PC, et al. Mathematical modelling of a mixed culture cultivation process for the production of polyhydroxybutyrate. Biotechnol Bioeng 2005; 92(2): 209–222.

86. Lee SB, Seressiotis A, Bailey JE. A kinetic model for product formation in unstable recombinant populations. Biotechnol Bioeng 1985; 27(12): 1699–1709.

87. Nielsen J, Emborg C, Halberg K, Villadsen J. Compartment model concept used in the design of fermentation with recombinant microorganisms. Biotechnol Bioeng 1989; 34(4): 478–486.

88. Grosz R, Stephanopoulos G. Physiological, biochemical, and mathematical studies of micro-aerobic continuous ethanol fermentation by *Saccharomyces cerevisiae*. I: hysteresis, oscillations, and maximum specific ethanol productivities in chemostat culture. Biotechnol Bioeng 1990; 36(10): 1006–1019.

89. Rizzi M, Baltes M, Theobald U, Reuss M. *In vivo* analysis of metabolic dynamics in *Saccharomyces cerevisiae*: II. Mathematical model. Biotechnol Bioeng 1997; 55: 592–608.

90. Van Aalst-van Leeuwen MA, Pot MA, van Loosdrecht MC, Heijnen JJ. Kinetic modeling of poly(β-hydroxybutyrate) production and consumption by *Paracoccus pantotrophus* under dynamic substrate supply. Biotechnol Bioeng 1997; 55(5): 773–782.
91. Mantzaris NV, Srienc F, Daoutidis P. Nonlinear productivity control using a multi-staged cell population balance model. Chem Eng Sci 2002; 57: 1–14.
92. Roussos AI, Kiparissides C. A bivariate population balance model for the microbial production of poly(3-hydroxybutyrate). Chem Eng Sci 2012; 70: 45–53.
93. Insel G, Yavaşbay A, Ozcan O, Cokgor EU. Modeling of simultaneous growth and storage kinetics variation under unsteady feast conditions for aerobic heterotrophic biomass. Bioprocess Biosyst Eng 2012; 35(8): 1445–1454.
94. Wu W, Lai SY, Jang MF, Chou YS. Optimal adaptive control schemes for PHB production in fedbatch fermentation of *Ralstonia eutropha*. J Process Contr 2013; 23: 1159–1168.
95. Chen Z, Guo Z, Wena Q, Huang L, Bakke R, Du M. Modeling polyhydroxyalkanoate (PHA) production in a newly developed aerobic dynamic discharge (ADD) culture enrichment process. Chem Eng J 2016; 298: 36–43.
96. Giersch C. Mathematical modelling of metabolism. Curr Opin Plant Biol 2000; 3(3): 249–253.
97. Dhurjati P, Ramkrishna D, Flickinger MC, Tsao GT. A cybernetic view of microbial growth: modelling of cells as optimal strategists. Biotechnol Bioeng 1985; 27(1): 1–9.
98. Varner J, Ramkrishna D. Application of cybernetic models to metabolic engineering: investigation of storage pathways. Biotechnol Bioeng 1989; 58(2–3): 282–291.
99. Varner J, Ramkrishna D. Metabolic engineering from a cybernetic perspective. 1. Theoretical preliminaries. Biotechnol Prog 1999; 15(3): 407–425.
100. Varner J, Ramkrishna D. Metabolic engineering from a cybernetic perspective. 2. Qualitative investigation of nodal architechtures and their response to genetic perturbation. Biotechnol Prog 1999; 15(3): 426–438.
101. Varner J, Ramkrishna D. Metabolic engineering from a cybernetic perspective: aspartate family of amino acids. Metab Eng 1999; 1(1): 88–116.
102. Patnaik PR. Microbial metabolism as an evolutionary response: the cybernetic approach to modeling. Crit Rev Biotechnol 2001; 21: 155–175.
103. Gernaey KV, van Loosdrecht MC, Henze M, Lind M, Jørgensen SB. Activated sludge wastewater treatment plant modelling and simulation: state of the art. Environ Model Softw 2004; 19: 763–783.
104. Yu J, Wang J. Metabolic flux modeling of detoxification of acetic acid by *Ralstonia eutropha* at slightly alkaline pH levels. Biotechnol Bioeng 2001; 73(6): 458–464.
105. Van Veldhuizen HM, Van Loosdrecht MC, Heijnen JJ. Modelling biological phosphorus and nitrogen removal in a full scale activated sludge process. Water Res 1999; 33: 3459–3468.
106. Kuba T, Murnleitner E, van Loosdrecht MC, Heijnen JJ. A metabolic model for biological phosphorus removal by denitrifying organisms. Biotechnol Bioeng 1996; 52(6): 685–695.
107. Yagci N, Artan N, Cokgör EU, et al. Metabolic model for acetate uptake by a mixed culture of phosphate- and glycogen-accumulating organisms under anaerobic conditions. Biotechnol Bioeng 2003; 84(3): 359–373.
108. Yagci N, Insel G, Tasli R, et al. A new interpretation of ASM2d for modeling of SBR performance for enhanced biological phosphorus removal under different P/HAc ratios. Biotechnol Bioeng 2006; 93(2): 258–270.
109. Henze M, Gujer W, Mino T, et al. Activated sludge model No.2D, ASM2D. Water Sci Technol 1999; 39: 165–182.
110. Dias JM, Oehmen A, Serafim LS, et al. Metabolic modelling of polyhydroxyalkanoate copolymers production by mixed microbial cultures. BMC Syst Biol 2008; 2: 59.

111. Dias JM, Lemos P, Serafim L, et al. Development and implementation of a nonparametric/metabolic model in the process optimisation of PHA production by mixed microbial cultures. In: European Symposium on Computer Aided Process Engineering, ESCAPE17. Elsevier, Amsterdam; conference held in Bucharest, 2017, pp. 995–1000.
112. Johnson K, Kleerebezem R, van Loosdrecht MC. Model-based data evaluation of polyhydroxybutyrate producing mixed microbial cultures in aerobic sequencing batch and fed-batch reactors. Biotechnol Bioeng 2009; 104(1): 50–67.
113. Jiang Y, Hebly M, Kleerebezem R, et al. Metabolic modeling of mixed substrate uptake for polyhydroxyalkanoate (PHA) production. Water Res 2011; 45(3): 1309–1321.
114. Tajparast M, Frigon D. Genome-scale metabolic modeling to provide insight into the production of storage compounds during feast-famine cycles of activated sludge. Water Sci Technol 2013; 67(3): 469–476.
115. Tamis J, Marang L, Jiang Y, et al. Modeling PHA-producing microbial enrichment cultures-towards a generalized model with predictive power. N Biotechnol 2014; 31(4): 324–334.
116. Korkakaki E, van Loosdrecht MCM, Kleerebezem R. Survival of the fastest: selective removal of the side population for enhanced PHA production in a mixed substrate enrichment. Bioresource Technol 2016; 216: 1022–1029.
117. Marang L, Jiang Y, van Loosdrecht MCM, Kleerebezem R. Impact of non-storing biomass on PHA production: an enrichment culture on acetate and methanol. Int J Biol Macromol 2014; 71: 74–80.
118. Marang L, van Loosdrecht MCM, Kleerebezem R. Modeling the competition between PHA-producing and non-PHA-producing bacteria in feast-famine SBR and staged CSTR systems. Biotechnol Bioeng 2015; 112:2475–2484.
119. Marang L, van Loosdrecht MCM, Kleerebezem R. Combining the enrichment and accumulation step in non-axenic PHA production: Cultivation of *Plasticicumulans acidivorans* at high volume exchange ratios. J Biotechnol 2016; 231:260–267.
120. Marang L, van Loosdrecht MCM, Kleerebezem R. Enrichment of PHA-producing bacteria under continuous substrate supply. New Biotechnol 2018; 41: 55–61.
121. Wang X, Carvalhoa G, Reis, MAM. Oehmena, A. Metabolic modeling of the substrate competition among multiple VFAs for PHA production by mixed microbial cultures. J Biotechnol 2018; 280: 62–69.
122. Daae EB, Dunnill P, Mitsky TA, et al. Metabolic modeling as a tool for evaluating polyhydroxyalkanoate copolymer production in plants. Metab Eng 1999; 1(3): 243–254.
123. Sauro HM. SCAMP: a general-purpose simulator and metabolic control analysis program. Comput Appl Biosci 1993; 9(4): 441–450.
124. Fell DA. Understanding the Control of Metabolism. Portland Press, London, 1997.
125. Fell DA. Metabolic control analysis: a survey of its theoretical and experimental development. Biochem J 1992; 286(Pt 2): 313–330.
126. Xu J, Guo B, Zhang Z, et al. A mathematical model for regulating monomer composition of the microbially synthesized polyhydroxyalkanoate copolymers. Biotechnol Bioeng 2005; 90(7): 821–829.
127. Iadevaia S, Mantzaris NV. Genetic network driven control of PHBV copolymer composition. J Biotechnol 2006; 122(1): 99–121.
128. Ao P, Lee LW, Lidstrom ME, Yin L, Zhu X. Towards kinetic modeling of global metabolic networks: *Methylobacterium extorquens* AM1 growth as validation. Sheng Wu Gong Cheng Xue Bao 2008; 24(6): 980–994.
129. Puchałka J, Oberhardt MA, Godinho M, et al. Genome-scale reconstruction and analysis of the *Pseudomonas putida* KT2440 metabolic network facilitates applications in biotechnology. PLOS Comput Biol 2008; 4(10): e1000210.
130. Sohn SB, Kim TY, Park JM, Lee SY. *In silico* genome-scale metabolic analysis of *Pseudomonas putida* KT2440 for polyhydroxyalkanoate synthesis, degradation of aromatics and anaerobic survival. Biotechnol J 2010; 5(7): 739–750.

131. Thiele I, Palsson BO. A protocol for generating a high-quality genome-scale metabolic reconstruction. Nat Protoc 2010; 5(1): 93–121.
132. Penloglou G, Roussos AI, Chatzidoukas C, Kiparissides C. Thessaloniki, Preprints, Bioproduction 2nd Annual Meeting, –"Digital BioProduction", 229–235, Greece.
133. Penloglou G, Roussos A, Chatzidoukas C, Kiparissides C. A combined metabolic/polymerization kinetic model on the microbial production of poly(3-hydroxybutyrate). N Biotechnol 2010; 27(4): 358–367.
134. Penloglou G, Chatzidoukas C, Roussos A, Kiparissides C. 21st European Symposium on Computer Aided Process Engineering – ESCAPE 21, Elsevier B.V, Amsterdam; conference held in Chalkidiki, Greece, 2011.
135. Penloglou G, Chatzidoukas C, Kiparissides C. Microbial production of polyhydroxybutyrate with tailor-made properties: an integrated modelling approach and experimental validation. Biotechnol Adv 2012; 30(1): 329–337.
136. Chatzidoukas C, Penloglou G, Kiparissides C. Development of a structured dynamic model for the production of polyhydroxybutyrate (PHB) in *Azohydromonas lata* cultures. Biochem Eng J 2013; 71: 72–80.
137. Carlson R, Wlaschin A, Srienc F. Kinetic studies and biochemical pathway analysis of anaerobic poly-(R)-3-hydroxybutyric acid synthesis in *Escherichia coli*. Appl Environ Microbiol 2005; 71(2): 713–720.
138. Carlson R, Srienc F. Fundamental *Escherichia coli* biochemical pathways for biomass and energy production: identification of reactions. Biotechnol Bioeng 2004; 85(1): 1–19.
139. Carlson R, Srienc F. Fundamental *Escherichia coli* biochemical pathways for biomass and energy production: creation of overall flux states. Biotechnol Bioeng 2004; 86(2): 149–162.
140. König C, Sammler I, Wilde E, Schlegel HG. Konstitutive Glucose-6-phosphat-Dehydrogenase bei Glucose verwertenden Mutanten von einem kryptischen Wildstamm. Arch Mikrobiol 1969; 67(1): 51–57.
141. Schlegel HG, Gottschalk G. Utilization of glucose by a mutant of *Hydrogenomonas* H16. Biochem Z 1965; 341: 249–259.
142. Park JM, Kim TY, Lee SY. Genome-scale reconstruction and in silico analysis of the *Ralstonia eutropha* H16 for polyhydroxyalkanoate synthesis, lithoautotrophic growth, and 2-methyl citric acid production. BMC Syst Biol 2011; 5: 101.
143. Lopar M, Vrana Špoljarić I, Atlić A, et al. Five-step continuous production of PHB analyzed by elementary flux, modes, yield space analysis and high structured metabolic model. Biochem Eng J 2013; 79: 57–70.
144. Lopar M, Špoljarić IV, Cepanec N, Koller M, Braunegg G, Horvat P. Study of metabolic network of *Cupriavidus necator* DSM 545 growing on glycerol by applying elementary flux modes and yield space analysis. J Ind Microbiol Biotechnol 2014; 41(6): 913–930.
145. Pfeiffer T, Sánchez-Valdenebro I, Nuño JC, Montero F, Schuster S. METATOOL: for studying metabolic networks. Bioinformatics 1999; 15(3): 251–257.
146. von Kamp A, Schuster S. Metatool 5.0: fast and flexible elementary modes analysis. Bioinformatics 2006; 22(15): 1930–1931.
147. Schwartz JM, Kanehisa M. A quadratic programming approach for decomposing steady-state metabolic flux distributions onto elementary modes. Bioinformatics 2005; 21 (Suppl. 2): 204–205.
148. Pardelha F, Albuquerque MG, Reis MA, Oliveira R, Dias JM. Dynamic metabolic modelling of volatile fatty acids conversion to polyhydroxyalkanoates by a mixed microbial culture. N Biotechnol 2014; 31(4): 335–344.
149. Pardelha F, Albuquerque MG, Reis MA, Dias JM, Oliveira R. Flux balance analysis of mixed microbial cultures: application to the production of polyhydroxyalkanoates from complex mixtures of volatile fatty acids. J Biotechnol 2012; 162(2–3): 336–435.

150. Grousseau E, Blanchet E, Déléris S, Albuquerque MG, Paul E, Uribelarrea JL. Impact of sustaining a controlled residual growth on polyhydroxybutyrate yield and production kinetics in *Cupriavidus necator.* Bioresour Technol 2013; 148: 30–38.

151. Mavaddat P, Mousavi SM, Amini E, et al. Modeling and CFD-PBE simulation of an airlift bioreactor for PHB production. Asia-Pac J Chem Eng 2014; 9: 562–573.

152. Varner J, Ramkrishna D. Metabolic engineering from a cybernetic perspective. 1. Theoretical preliminaries. Biotechnol Prog 1999; 15(3): 407–425.

153. Varner J, Ramkrishna D. Metabolic engineering from a cybernetic perspective. 2. Qualitative investigation of nodal architectures and their response to genetic perturbation. Biotechnol Prog 1999; 15(3): 426–438.

154. Varner J, Ramkrishna D. Metabolic engineering from a cybernetic perspective: aspartate family of amino acids. Metab Eng 1999; 1(1): 88–116.

155. Dhurjati P, Ramkrishna D, Flickinger MC, Tsao GT. A cybernetic view of microbial growth: modelling of cells as optimal strategists. Biotechnol Bioeng 1985; 27(1): 1–9.

156. Varner J, Ramkrishna D. Application of cybernetic models to metabolic engineering: investigation of storage pathways. Biotechnol Bioeng 1998; 58(2–3): 282–291.

157. Patnaik PR. Microbial metabolism as an evolutionary response: the cybernetic approach to modeling. Crit Rev Biotechnol 2001; 21(3):155–175.

158. Hatzimanikatis V, Floudas CA, Bailey JE. Analysis and design of metabolic reaction networks via mixed-integer linear optimization. AIChE J 1996; 42: 1277–1292.

159. Lee B, Yen J, Yang L, Liao JC. Incorporating qualitative knowledge in enzyme kinetic models using fuzzy logic. Biotechnol Bioeng 1999; 62(6): 722–729.

160. Babuška R, Verbruggen HB, van Can HJ. Fuzzy modeling of enzymatic penicillin-G conversion. Eng Appl Artif Intell 1999; 12: 79–92.

161. van Can HJ, Te Braake HA, Hellinga C, Luyben KC. An efficient model development strategy for bioprocesses based on neural networks in macroscopic balances. Biotechnol Bioeng 1997; 54(6): 549–566.

162. Yoo S, Kim WS. Cybernetic model for synthesis of poly-β-hydroxybutyric acid in *Alcaligenes eutrophus.* Biotechnol Bioeng 1994; 43(11): 1043–1051.

163. Herrnstein RJ. On the law of effect. J Exp Anal Behav 1970; 13(2): 243–266.

164. Varner J, Ramkrishna D. Application of cybernetic models to metabolic engineering: investigation of storage pathways. Biotechnol Bioeng 1998; 58(2–3): 282–291.

165. Ferraz L, Bonomi A, Piccoli RA, et al. Cybernetic structured modeling of the production of polyhydroxyalkanoates by *Alcaligenes eutrophus.* Braz J Chem Eng 1999; 16: 201–212.

166. Himmelblau DM. Applied Nonlinear Programming. McGraw-Hill, New York, 1972.

167. Gadkar KG, Doyle FJ III, Crowley TJ, Varner JD. Cybernetic model predictive control of a continuous bioreactor with cell recycle. Biotechnol Prog 2003; 19(5): 1487–1497.

168. Riascos CA, Pinto JM. Optimal control of bioreactors: a simultaneous approach for complex systems. Chem Eng J 2004; 99: 23–34.

169. Pinto MA, Immanuel CD. AIChE 2005 Annual meeting, food, pharmaceutical and bioengineering division, (479c). Available from: http://www3.aiche.org/Proceedings /Abstract.aspx?ConfID=Annual 2005&GroupID=1007&SessionID=1214&PaperID =25880 (Accessed 01. 03. 2019).

170. Kim JI, Varner JD, Ramkrishna D. A hybrid model of anaerobic *E. coli* GJT001: combination of elementary flux modes and cybernetic variables. Biotechnol Prog 2008; 24(5): 993–1006.

171. Franz A, Song HS, Ramkrishna D, Kienle A. Experimental and theoretical analysis of poly(β-hydroxybutyrate) formation and consumption in *Ralstonia eutropha.* Biochem Eng J 2011; 55: 49–58.

172. Weber J, Kayser A, Rinas U. Metabolic flux analysis of *Escherichia coli* in glucose-limited continuous culture. II. Dynamic response to famine and feast, activation of the methylglyoxal pathway and oscillatory behaviour. Microbiology 2005; 151(3): 707–716.

173. Disli I, Kremling A, Kienle A. Proceedings MATHMOD 09 Vienna – full papers CD volume, 2009.
174. Ganjian A, Zhang J, Dias JM, Oliveira R. Modelling of a sequencing batch reactor for producing polyhydroxybutyrate with mixed microbial culture cultivation process using neural networks and operation regime classification. Chem Eng Trans 2013; 32: 1261–1266.
175. Peres J, Oliveira R, Serafim LS, et al. Hybrid modelling of a PHA production process using modular neural networks. In: A. Barbosa-Póvoa, H. Matos (eds.), European Symposium on Computer-Aided Process Engineering-14, Elsevier, Amsterdam; conference held in Lisbon, 2004, pp. 733–738.
176. Patnaik PR. Neural network designs for poly-β-hydroxybutyrate production optimization under simulated industrial conditions. Biotechnol Lett 2005; 27(6): 409–415.
177. Patnaik PR. Enhancement of PHB biosynthesis by *Ralstonia eutropha* in fed-batch cultures by neural filtering and control 1. Food Bioprod Process 2006; 84: 150–156.
178. Patnaik PR. Application of the lyapunov exponent to evaluate noise filtering methods for a fed-batch bioreactor for PHB production. Bioautomation 2008; 9: 1–14.
179. Patnaik PR. Design considerations in hybrid neural optimization of fed-batch fermentation for PHB production by *Ralstonia eutropha*. Food Bioprocess Technol 2010; 3: 213–225.
180. Zafar M, Kumar S, Kumar S, Dhiman AK. Optimization of polyhydroxybutyrate (PHB) production by *Azohydromonas lata* MTCC 2311 by using genetic algorithm based on artificial neural network and response surface methodology. Biocatal Agric Biotechnol 2012; 1: 70–79.
181. Zafar M, Kumar S, Kumar S, Dhiman AK. Modeling and optimization of poly(3hydr oxybutyrate-co-3hydroxyvalerate) production from cane molasses by *Azohydromonas lata* MTCC 2311 in a stirred tank reactor: effect of agitation and aeration regimes. J Ind Microbiol Biotechnol 2012; 39(7): 987–1001.
182. Zafar M, Kumar S, Kumar S, Dhiman AK. Artificial intelligence based modeling and optimization of poly(3-hydroxybutyrate-co-3-hydroxyvalerate) production process by using *Azohydromonas lata* MTCC 2311 from cane molasses supplemented with volatile fatty acids: a genetic algorithm paradigm. Bioresour Technol 2012; 104: 631–641.
183. Caspi R, Foerster H, Fulcher CA, Hopkinson R, Ingraham J, Kaipa P, Krummenacker M, Paley S, Pick J, Rhee SY, Tissier C, Zhang P, Karp PD MetaCyc: a multiorganism database of metabolic pathways and enzymes. Nucl Acids Res 2006; 34: D511–D516.
184. Kanehisa M, Araki M, Goto S, Hattori M, Hirakawa M, Itoh M, Katayama T, Kawashima S, Okuda S, Tokimatsu T, Yamanishi Y. KEGG for linking genomes to life and the environment. Nucleic Acids Res 2008; 36: D480–D484.
185. Milner PC. The possible mechanisms of complex reactions involving consecutive steps. J Electrochem Soc 1964; 111: 228–232.
186. Feinberg M, Horn FJM. Dynamics of open chemical systems and the algebraic structure of the underlying reaction network. Chem Eng Sci 1974; 29: 775–787.
187. Clarke BL. Complete set of steady states for the general stoichiometric dynamical system. J Chem Phys 1981; 75: 4970–4979.
188. Clarke BL. Stoichiometric network analysis. Cell Biophys 1988; 12:237–253.
189. Stephanopoulos GN, Aristidou AA. Nielsen J Metabolic Engineering. Principles and Methodologies. Academic Press, San Diego, 1998, pp. 464–580.
190. Gombert AK. Nielsen J. Mathematical modelling of metabolism. Curr Opin Biotechnol 2000; 11: 180–186.
191. Papin JA, Price ND, Palsson BØ. Extreme pathway lengths and reaction participation in genome-scale metabolic networks. Genome Res 2002; 12: 1889–1900.
192. Papin JA, Stelling J, Price ND, Klamt S, Schuster S, Palsson BØ. Comparison of network-based pathway analysis methods. Trends Biotechnol 2004; 22: 400–405.

193. Price ND, Reed JL, Papin JA, Wiback SJ, Palsson BØ. Network-based analysis of metabolic regulation in the human red blood cell. J Theor Biol 2003; 225: 185–194.
194. Larhlimi A, Bockmayr A. A new constraint-based description of the steady-state flux cone of metabolic networks. Discrete Appl Math 2009; 157: 2257–2266.
195. Schuster S, Fell DA, Dandekar T. A general definition of metabolic pathways useful for systematic organization and analysis of complex metabolic networks. Nat Biotechnol 2000; 18: 326–332.
196. Schuster S, Hilgetag C. On elementary flux modes in biochemical reaction systems at steady state. J Biol Syst 1994; 2: 165–182.
197. Edwards JS, Ramakrishna R, Schilling CH, Palsson BØ. Metabolic flux balance analysis. In: S.Y. Lee, E.T. Papoutsakis (eds.), Metabolic Engineering. Marcel Deker, New York, 1999, pp. 13–57.
198. Kauffman KJ, Prakash P, Edwards JS. Advances in flux balance analysis. Curr Opin Biotechnol 2003; 14: 491–496.
199. Price ND, Reed JL, Palsson BØ. Genome-scale models of microbial cells: evaluating the consequences of constraints. Nat Rev Microbiol 2004; 2: 886–897.
200. Burgard AP, Nikolaev EV, Schilling CH, Maranas CD. Flux coupling analysis of genome-scale metabolic network reconstructions. Genome Res 2004; 14: 301–312.
201. Orth JD, Thiele I, Palsson BØ. What is flux balance analysis? Nat Biotechnol 2010; 28: 245–248.
202. Bushell ME, Sequeira SIP, Khannapho C, Zhao H, Chater KF, Butler MJ, Kierzek AM, Avignone-Rossa CA. The use of genome scale metabolic flux variability analysis for process feed formulation based on an investigation of the effects of the zwf mutation on antibiotic production in *Streptomyces coelicolor*. Enzyme Microb Technol 2006; 39: 1347–1353.
203. Gudmundsson S, Thiele I. Computationally efficient flux variability analysis. BMC Bioinf 2010; 11: 489–591.
204. Mahadevan R, Schilling CH. The effects of alternate optimal solutions in constraint-based genome-scale metabolic models. Metab Eng 2003; 5: 264–276.
205. Dräger A, Kronfeld M, Ziller MJ, Supper J, Planatscher H, Magnus JB, Oldiges M, Kohlbacher O, Zell A. Modeling metabolic networks in *C. glutamicum*: a comparison of rate laws in combination with various parameter optimization strategies. BMC Syst Biol 2009; 3: 5–28.
206. Wunderlich Z, Mirny L. Using the topology of metabolic networks to predict viability of mutant strains. Biophys J 2006; 91: 2304–2311.
207. Carlson R, Fell D, Srienc F. Metabolic pathway analysis of a recombinant yeast for rational strain development. Biotechnol Bioeng 2002; 79: 121–134.
208. Wlaschin AP, Trinh CT, Carlson R, Srienc F. The fractional contributions of elementary modes to the metabolism of *Escherichia coli* and their estimation from reaction entropies. Metab Eng 2006; 8: 338–352.
209. Poblete-Castro I, Binger D, Oehlert R, Rohde M. Comparison of mcl-Poly (3-hydroxyalkanoates) synthesis by different *Pseudomonas putida* strains from crude glycerol: citrate accumulates at high titer under PHA-producing conditions. BMC Biotechnol 2014; 14: 962–972.
210. Borrero-de Acuña JM, Bielecka A, Häussler S, Schobert M, Jahn M, Wittmann C, Poblete-Castro I. Production of medium chain length polyhydroxyalkanoate in metabolic flux optimized *Pseudomonas putida*. Microb Cell Fact 2014; 13: 88–102.
211. Mendes P. GEPASI: a software package for modelling the dynamics, steady states and control of biochemical and other systems. Comput Appl Biosci CABIOS 1993; 9: 563–571.

212. Hoops S, Sahle S, Gauges R, Lee C, Pahle J, Simus N, Singhal M, Xu L, Mendes P, Kummer U. COPASI-a complex Pathway simulator. Bioinformatics (2006) 22: 3067–3074.
213. Klamt S, Stelling J, Ginkel M, Gilles ED. FluxAnalyzer: Exploring structure, pathways, and flux distributions in metabolic networks on interactive flux maps. Bioinformatics 2003; 19: 261–269.
214. Klamt S, Saez-Rodriguez J, Gilles ED. Structural and functional analysis of cellular networks with Cell NetAnalyzer. BMC Syst Biol 2007; 1: 2.
215. Schwarz R, Musch P, von Kamp A, Engels B, Schirmer H, Schuster S, Dandekar T YANA-a software tool for analyzing flux modes, gene-expression and enzyme activities. BMC Bioinf 2005; 6: 135.
216. Schwartz J-M, Gaugain C, Nacher JC, de Daruvar A, Kanehisa M. Observing metabolic functions at the genome scale. Genome Biol 2007; 8: R123–R139.
217. Schwartz R, Liang C, Kaleta C, Kuhnel M, Hoffmann E, Kuznetsov S, Hecker M, Griffiths G, Schuster S, Dandekar T. Integrated network reconstruction, visualization and analysis using YANAsquare. BMC Bioinf 2007; 8: 313–322.
218. Bell SL, Palsson BØ. Expa: a program for calculating extreme pathways in biochemical reaction networks. Bioinformatics 2005; 21: 1739–1740.
219. Urbanczik R. SNA-a toolbox for the stoichiometric analysis of metabolic networks. BMC Bioinf 2006; 7: 129.
220. Urbanczik R, Wagner C. Functional stoichiometric analysis of metabolic networks. Bioinformatics 2005; 21: 4176–4180.

Part II

Environmental and Stress Factors

3 Interconnection between PHA and Stress Robustness of Bacteria

Stanislav Obruca, Petr Sedlacek, Iva Pernicova,
Adriana Kovalcik, Ivana Novackova,
Eva Slaninova and Ivana Marova

CONTENTS

3.1 IMPORTANCE OF STRESS ROBUSTNESS FOR BACTERIA

In their natural habitats, bacteria are constantly challenged by the environment since they are exposed to rapidly changing conditions such as fluctuations in temperature, pH or salinity and availability of nutrients or radiation. Therefore, to cope with such challenging conditions, bacteria developed numerous strategies which help them to face various stressors. These scenarios differ in their complexity and include relatively simple strategies such as the escape from local environments exhibiting stress nature to less harmful locations, but also more sophisticated actions such as the rearrangement of transcriptome as a response to stress conditions [1]. Actually, the effectiveness of a stress response strategy is directly associated with the efficiency of adaptation to changes in environmental conditions and therefore, the stress response can be considered as one of the key features determining the success or failure in the process of natural selection in highly competitive environments.

In general, microorganisms produce various proteins, biopolymers and other metabolites to enhance their robustness and resistance against various stress factors. Polyhydroxyalkanoates (PHA) are microbial polyesters which are accumulated in the form of intracellular granules by various prokaryotic microorganisms when carbon substrate is present in excess. The primary biological function of these materials is the storage of carbon, energy and reducing power which is utilized when the external carbon source is exhausted [2]. As storage materials they have for decades been considered being predominantly associated with survival in starvation periods, and their protective roles against other stress factors were overlooked. Nevertheless, according to numerous recent studies, the biological role of PHA is surprisingly much more complex, since it was observed that the presence of PHA granules in bacterial cells enhances their resistance against numerous stress factors such as high or low temperature [3, 4], repeated freezing and thawing [5], osmotic [6, 7] and oxidative pressure [8, 9] or UV-irradiations [10]. It was also observed that genes encoding for enzymes involved in the PHA biosynthesis are localized within a genomic island suggesting that the horizontal transfer of PHA-related genes might be the mechanism of adaptation to stress conditions [11]. Obviously, apart from their primary storage function, PHA provide bacteria with additional benefits as non-specific protectants against numerous stressors and, herein, they enhance the stress robustness of bacteria. Therefore, this chapter summarizes current understanding on the protective function of PHA against various stress factors.

3.2   PHA AND STRESS INDUCED BY HIGH TEMPERATURE

Temperature is one of the most important environmental factors for life in microorganisms. All metabolic actions are very sensitive to temperature and its changes [12]. The life of microorganisms is related to a wide range of temperature limits. Some species of bacteria and archaea live and thrive at temperatures above 45°C, some of them are capable of surviving the temperatures exceeding values in excess of 110°C. Higher temperature environments include hot springs, hydrothermal vents in lakes or oceans, volcanic and geothermal deposits or solfataric fields. In addition to natural resources, they also include artificial biotopes such as activated sludge, compost or smoldering piles of coal waste and hot effluents from geothermal plants [13, 14].

Temperature stress is a common factor in natural conditions. The exposure to higher temperatures presents a risk primarily for three types of key biomolecules, namely nucleic acids, proteins and membrane lipids. High temperatures increase the fluidity of cell membranes which can affect the cell integrity. Simultaneously, high temperature disrupts supramolecular structures of nucleic acids and proteins which are crucial for their proper biological functions [15].

Nevertheless, microorganisms are able to partially adapt to the increase in temperature. The fluidity of membranes is ensured by adjusting the amount and type of lipid – saturated versus unsaturated. [16, 17]. Moreover, even in microorganisms adapted to higher temperatures, there may be a sudden heat shock. A sudden temperature shock leads to protein denaturation which results in the formation of protein aggregates and the disruption of cell metabolism. Therefore, at a sudden rise in temperature, microorganisms induce a group of genes called heat shock genes which encode for so-called heat shock proteins (HSP). HSP are a convenient strategy to protect against temperature stress, leading to the removal of denaturation of the protein and to the maintenance of homeostasis [18]. HSP include chaperones and prostheses necessary to overcome the effect of temperature shock for other cellular proteins. They are responsible for the correct folding of proteins which have lost their correct higher structures as a consequence of their exposure to higher temperatures. The regulation of HSP expression is based on alternative sigma factors. In *Escherichia coli*, for example, the high temperature change response is controlled by a specific sigma factor 32, this factor is coded in the gene *rpoH*. The factor binds to promotors located in front of the heat shock proteins and enhances the level of their expression [19].

It should be noted that the chaperoning activity, apart from HSP, was also observed in PHA granule associated proteins called phasins (PhaP). Due to their amphiphilic nature, the primary function of these proteins is to serve as an interphase between the hydrophobic core of PHA granules and the hydrophilic cytoplasm environment. Nevertheless, their indisputable chaperoning effect was observed in many studies. It was observed that the poly(3-hydroxybutyrate) (PHB) production causes the stress in recombinant *E. coli* which was evident from enhanced expression of chaperones, sigma factors and other stress-related genes [20, 21] but in the presence of PhaP proteins, a dramatic reduction in the expression of stress-related genes was determined. Unexpectedly, PhaP also protected *E. coli* strains incapable of PHA biosynthesis under both normal and stress conditions. The expression of PhaP resulted in the reduction in HSP levels, and increased the growth and the resistance to heat shock [22]. It was observed that PhaP can protect the enzyme from spontaneous thermal denaturation and enable its refolding after chemical denaturation [23], which implies that PhaP might play an important role as chaperones and also support HSP during the exposure to temperature shock in naturally PHA-producing bacteria.

In addition to HSP production, a complementary microbial strategy of how to protect against stress induced by high temperatures is the production of small molecules called osmolytes, which can serve as chemical chaperones. These molecules help increase the stability of native proteins and support the refolding of polypeptides which were misfolded as a consequence of exposure of microbial cells to thermal shock. It should be noted that the monomer of PHB, 3-hydroxybutyrate (3HB), can also function as an osmolyte or chemical chaperone. The accumulation of 3HB monomer is associated with the prevention of protein aggregation under thermal stress [24]. Although the PHB compound is used predominantly as the source of carbon and energy by bacteria, its monomer, 3HB, can serve as a chemical chaperone that can protect enzymes from denaturation. Actually, biosynthesis and degradation occur in bacterial cells simultaneously, therefore PHA-accumulating cells possess a high intracellular pool of 3HB. The denaturation of lipase and lysozyme by the increase in temperature in the presence or absence of 3HB was monitored by dynamic light scattering (DLS) and differential scanning calorimetry (DSC). The experiment revealed the significant protective effect of 3HB, moreover, the protective effect of 3HB was comparable to other well-known chemical chaperones such as trehalose and hydroxyectoine [25].

In addition to PHB, thermal stress protection may also be provided by a copolymer of poly(hydroxybutyrate-*co*-hydroxyhexanoate) (PHBHx). *Aeromonas hydrophila* 4AK4 produces a PHA copolymer composed precisely of 3-hydroxybutyrate and

3-hydroxyhexanoate (P(3HB-*co*-3HHx). It is the accumulation of PHA that leads to the ability to survive adverse conditions. In addition to the strain *Aeromonas hydrophila*, a mutant unable to produce PHA was also used to detect the protective effect of PHA. The series of experiments revealed that the wild-type strain was more resistant to heat shock, compared to the mutant strain unable to produce PHA. The stress-related sigma factor level was also lower in the mutant CQ4 strain than in the 4AK4 strain. P(3HB-*co*-HHx) is also likely to have a protective effect through increased expression of sigma factors that activate other gene products related to a protective mechanism against stress [26]. It was observed that the degradation of PHA storage is associated with an increased concentration of guanosine tetraphosphate (ppGpp) in *Pseudomonas oleovorans* [27]. ppGpp serves as an intracellular alarmone and induces the expression of various genes involved in the stress response, therefore it is likely that a lowered level of sigma stress factor in PHA deficient cells of *A. hydrophila* was the consequence of the decrease of ppGpp levels under particular conditions.

Thermophiles are a group of heat-loving microorganisms. They live and thrive at temperatures higher than 45°C [28]. However, their optimum temperature may also be higher. Thermophiles can be divided into thermophiles with optimum growth temperature above 45°C, extreme thermophiles with optimum growth temperature above 65°C, and hyperthermophiles with optimum growth temperature above 80°C [29]. Similar to the mesophilic microorganisms, thermophiles can produce PHA as well. Their use in biotechnology provides several advantages, for example higher resistance to contamination due to high cultivation temperature and allowing the process to be conducted in semi-sterile or even non-sterile conditions [30].

Chelatococcus daeguensis TAD1, isolated from the biofilm of a biofilter used to remove NOx from a coal-fired power plant, is one of the thermophilic bacteria producing PHA. This thermophilic bacterium is capable of producing PHB up to 84% of cell dry mass at 45°C when glucose is used as the carbon source. It can also produce high PHB titers at higher temperatures, above 50°C. In addition to glucose, cheaper substrates, such as starch or glycerol, can also be used to produce PHB, reaching 80% of dry weight of cells on glycerol [31].

Another interesting thermophilic producer of PHA is thermophilic bacteria identified as *Aneurinibacillus* sp. XH2, which has been isolated from the oil field. The optimum growth temperature is about 55°C. Using glucose, peptone and yeast extract, PHA production has been detected using Nile red and a transmission electron microscope. Using gas chromatography with mass spectrometry (GS-MS), PHA was identified as a copolymer of 3-hydroxybutyrate with 3-hydroxyvalerate [32].

Extreme thermophiles producing PHA include bacteria of the genus *Thermus thermophilus*. The cultivation temperature is 75°C and when sodium gluconate (1.5% w/v) or sodium octanoate (10 mM) are used as solo carbon sources, *Thermus thermophilus* can produce PHA at approximately 35 to 40% dry weight of biomass. By gas chromatography analysis, the production on gluconate was found to give a polyester composed mainly of 3-hydroxydecanoate (3HD) with the molar fraction of up to 64 mol-%. Other components in polyester, in addition to 3HD, were 3-hydroxyoctanoate (3HO), 3-hydroxyvalerate (3HV) and 3-hydroxybutyrate (3HB). However, the polyester produced when grown on octanoate as the only carbon source was composed of more than one monomer unit, namely 35.4 mol-% 3-hydroxyundecanoate (3HUD), 24.5 mol-% 3HB, 14.6 mol-% 3HD, 12.3 mol-% 3-hydroxynonanoate (3HN) and 7.8 mol-% 3-hydroxydodecanoate [33].

In addition to thermophilic heterotrophic bacteria, there are also thermophilic cyanobacteria, which are also able to produce PHA. One was isolated from volcanic rock and identified as *Synechococcus* sp. MA19. It is a thermophilic cyanobacterium that is capable of autotrophic accumulation of PHB. This strain is able to accumulate more than 20% of PHB in dry mass of cells in the presence of nitrogen atmosphere with 2% CO_2. Its optimal cultivation temperature is about 50°C. The production of PHB in dark cultivations with nitrogen deficiency led to the increase in PHB to 27% by dry weight of cells. However, the change in cultivation conditions led to glycogen degradation, the second storage product accumulated by cyanobacteria, resulting in a decrease in cell weight [34].

The production of PHA is also described for other thermophilic bacterial species such as *Caldimonas* [35, 36], for example *Caldimonas taiwanensis* [37], which can produce P(3HB-*co*-3HV) copolymer or members of genus *Geobacillus* [38] bacteria capable of producing PHB at 60°C [39]. However, although the use of thermophiles in biotechnology offers many benefits, the thermophilic producers are currently underexplored compared to the PHA mesophilic producers. [40].

3.3 PROTECTIVE FUNCTIONS OF PHA AGAINST LOW TEMPERATURE AND FREEZING

Approximately 80% of our planet's biosphere is permanently cold with average temperatures below 5°C and even in the remaining regions the temperature fluctuates wildly, occasionally decreasing close to, or even below 0°C [41]. Therefore, many bacteria developed sophisticated strategies to endure low temperatures. These include the ability to produce and accumulate cryoprotectants, the substances which are able to protect cells from the adverse effects of freezing and low temperature. There are numerous low molecular weight solutes, e.g. amino acids and their derivatives, sugars, ectoines and their derivatives, etc., but also high molecular weight substances such as proteins and polysaccharides, which are produced by bacteria and exhibit cryoprotective activity [42].

The effect of low temperatures on microbial cell depends upon the fact of whether or not the temperature decreases below 0°C. Above this breakpoint the bacterial cells are usually capable of active defense against the stress conditions which typically include the formation of cold shock proteins and other specific metabolites which enable them to cope with low temperatures. Nevertheless, when the temperature drops to values where water starts freezing, the response of most prokaryotes is passive, frequently leading slowly to the death of cells [43]. The growth of extracellular ice crystals increases the osmotic pressure in the medium since the excluded solutes are concentrated in a decreasing volume of water. This effect leads to so-called "freeze dehydration", and has harmful consequences for challenged cells. Further, the cells are subsequently damaged by the formation of intracellular ice crystals which might damage membranes and organelles, and cause the formation of gas bubbles. Beside freeze dehydration and intracellular ice, cells can also be damaged by reactive oxygen species (ROS) formed in cells during freezing [44] and the decreasing volume of the bacteria-inhabited channels of unfrozen liquid surrounded by growing ice crystals can lead to mechanical injury of the cells [45]. Mesophiles, unlike psychrophiles, do not tolerate cold conditions well. The exposure of mesophiles to near-freezing temperatures usually results in reduced enzyme activity, decreased membrane fluidity, altered transport of nutrients and waste products, decreased rates of transcription, translation and cell division, protein cold-denaturation, inappropriate protein folding and intracellular ice formation [41]. On the other hand, the microorganisms, which can grow and maintain their vital metabolic functions in cold environments ranging from −20°C to +10°C, are known as psychrophiles. Different strategies could contribute to the accommodation of microorganisms to a cold environment, such as environmental selection, and modifications at the molecular or physiological level [46].

Lower sensitivity of several PHA-positive strains against cold environments has recently been heavily reported. Recent study of the behavior of bacteria producing PHA and their PHA deficient mutants showed higher capability of bacteria with PHA inclusions to endure unfavorable low temperatures or even freezing when compared to PHA – deficient mutant strain [26]. The explanation can be the relationship between the PHA metabolism and ppGpp briefly described in the part dedicated to heat shock. It was found that the presence of PHA in bacteria may be related to the synthesis of nucleotides, including ppGpp and alarmone which activates the expression of the *rpoS* gene encoding for alternative sigma factor of RNA polymerase. *RpoS* mediates the formation of protective proteins, and thus increases the resistance of bacteria against various stresses including low temperatures [27, 47]. It was observed that PHA as carbon storage materials alleviates the oxidative stress induced in bacteria by cold environments. Principally, the PHA metabolism is a dynamic process which combines simultaneous biosynthesis and degradation. Hence, it was experimentally confirmed that after cold shock PHA is degraded and the NADH/NAD+ ratio is maintained to keep the redox state in balance and to adapt bacteria to cold environments [3]. The investigation of some Antarctic strains (e.g. *Pseudomonas* sp. 14-3 also classified as *Pseudomonas extremaustralis* DSM 17835T) showed that the PHA synthesis and degradation significantly contribute to the increased motility and survival of these bacteria under icy conditions and high pressure [3, 48]. Also, in another psychrophilic bacterium *Sphingopyxis chilensis* PHA plays a crucial role when cells are exposed to freezing [49]. Nevertheless, cold protective efficiency of PHA against low temperatures has also been detected in the case of mesophiles such as model strain of PHA metabolism – *Cupriavidus necator* (formerly *Ralstonia eutropha*, *Wautersia eutropha* and *Alcaligenes eutrophus*). Nowroth et al. reported that PHA reservoirs in *C. necator* wild-type strain protected cells from entering the viable but not cultivable physiological state when exposed to 5°C since the PHA-deficient strain was much more sensitive to the exposition to low temperature [4]. Furthermore, the survival of *C. necator* was also studied when exposed to sub-zero temperatures. Again, it was observed that wild-type strains of *C. necator* revealed higher tolerance to repeated freezing-thawing cycles than the PHA-deficient mutant. This might be a consequence of higher intracellular concentration of PHA monomers in wild-type strains since 3HB demonstrated strong cryoprotective properties.

In addition, it was demonstrated that the PHA granules maintain high flexibility even under extremely low temperatures, which suggests that the PHA granules might protect bacterial cells against injury from extracellular ice. Further, the presence of PHA granules modifies the adhesive forces between water and cellular components and thus protects bacterial cells at low temperatures and inhibits the formation of cytoplasmatic ice [5].

The importance of PHA for bacteria inhabitating cold environments was also confirmed by numerous studies dealing with the composition of microflora occurring in niches constantly exposed to low and even sub-zero temperatures. Numerous PHA-accumulating bacteria have been identified in Antarctic soil [50], subarctic sea ice in Greenland [51], Antarctic freshwater [52], the Baltic Sea [53] and the Pangi-Chamba trans-Himalayan region [54] suggesting that the PHA biosynthesis is an efficient strategy to cope with the challenges of cold environments.

3.4 OSMOPROTECTIVE FUNCTION OF PHA GRANULES

The changes in external osmolarity are very common challenges experienced by prokaryotes in their natural habitats. Soil bacteria, for instance, are frequently exposed to quick fluctuations in external salinity depending upon the weather [55]. When bacterial cells are exposed to hypertonic conditions caused by high extracellular concentration of salts (or other soluble substances), water goes out of the cells which results in quick dehydration of the cytoplasm. Moreover, since the volume of cytoplasm is reduced as a consequence of water efflux, the cytoplasm membrane shrinks to cover the lower volume of the cytoplasm and separates from outer layers of the cell envelope (especially in Gram-negative bacteria). This effect is known as plasmolysis. Massive plasmolysis might represent a serious problem for challenged cells since the cytoplasmic membrane is not capable of shrinking more than 2–5%. Therefore, serious plasmolysis usually results in damage and even collapse of the cytoplasmic membrane [56]. Conversely, when bacterial cells are exposed to hypotonic conditions, the influx of water "blows" the cell up which might disrupt the integrity of the cytoplasm membrane and, therefore, also the entire cell. The disruption of cells due to the application of osmotic down-shock is known as a hypotonic lysis [57].

To defend against the harmful effect of osmotic fluctuations, bacterial cells, regardless of their salinity preferences, developed sophisticated protective mechanisms. Bacterial cells are able to detect the variances in external osmotic pressure by the action of various membrane-associated mechanosensitive channels and osmotic transporters. The osmosensing capability enables cells to react to fluctuations in osmolarity. The exposure of the cells to osmotic up-shock induces the synthesis of osmolytes (also called compatible solutes). These small organic molecules such as trehalose, ectoines, glutamate, glycine betaine, etc. compensate the extracellular osmotic pressure and therefore protect bacterial cells from the harmful effect of osmotic up-shock. Apart from their balancing function, these small molecules also act as chemical chaperones and therefore protect various sensitive biomolecules from denaturation and loss of biological activity [58]. When the opposite situation occurs and cells are introduced into a hypotonic environment, water as well as osmolytes are pumped out of the cells by so-called "mechanosensitive channels" to protect the cell from hypotonic lysis [59, 60].

Nonetheless, the presence of PHA also protects cells against severe consequences of osmotic fluctuations as was observed, for instance, in the experiments performed by Zhao et al. who compared the survival rate of *Aeromonas hydrophila* and its PHA synthase knock-out mutant. When exposed to various stress factors including osmotic challenge, the wild-type strain capable of PHA synthesis demonstrated higher resistance to osmotic pressure [26]. Comparable results were obtained by Kadouri et al. who compared the wild-type strain of *Azospirillum brasilense* and its PHA synthase deletion mutants [61] and intracellular PHA depolymerase deletion mutants [62]. Similarly, transgenic *Escherichia coli* capable of PHA synthesis was more robust against osmotic challenge than the wild-type strain [63].

The reports mentioned above confirmed that the osmoprotective function of PHA granules is a general feature that can be identified in numerous microorganisms. The protective mechanism of PHA granules is most likely complex. For instance, *Rhizobium leguminosarum* TA-1 and *Rhizobium meliloti* SU-47 responded to osmotic up-shock by the mobilization of intracellular PHA granules to cover the energetic and carbon demands of osmolytes biosynthesis [64]. Again, it should be pointed out that the PHA monomers such as 3HB are considered potent compatible solutes [24, 25]. The protective function of PHA granules against osmotic up-shock was also recently investigated in *Cupriavidus necator* and it was observed that the PHA-accumulating wild-type strain survived osmotic up-shock much better than the PHA non-accumulating mutant. Nevertheless, the osmotic up-shock did not induce any intracellular PHA degradation. The PHA granules demonstrated a scaffold-like effect and prevented the PHA accumulating cells against massive plasmolysis which, oppositely, occurred in the up-shocked cells of PHA non-accumulating mutant as was observed by transmission electron microscopy and by thermal analysis. Moreover, the PHA granules revealed unique and unexpected liquid-like behavior. When the plasmolysis occurred in the very close vicinity of PHA granules, it was possible to observe that PHA partially stabilized membranes by plugging small gaps. Therefore, it is likely that the osmoprotective effect of PHA granules is at least partially enabled by unique biophysical properties of intracellular amorphous PHA granules. As a consequence, the level of dehydration and the fluctuations in intracellular pH value was also substantially lower in PHA-containing cultures than in their PHA negative counterparts [6]. In the follow-up study, the same team explored the fate of the bacterial cells when exposed to subsequent osmotic down-shock. Again, the PHA-accumulating wild type of *Cupriavidus necator* performed much better than the PHA negative mutant. Actually, the PHA non-accumulating mutant underwent massive hypertonic lysis, but the PHA-accumulating wild type was capable of maintaining the cell integrity when suddenly transferred from 200 g/L of NaCl to distilled water. The microphotographs obtained by electron microscopy are provided in Figure 3.1. It is very likely that this effect is a result of a different level of membrane damage during the osmotic up-shock. However, the same response was also observed in PHA accumulating halophilic bacterium *Halomonas halophila* which is adapted to high salinity and, therefore, did not experience any osmotic up-shock during the experiment (transfer from 66 g/L to distilled water). Also, in this microorganism, sudden

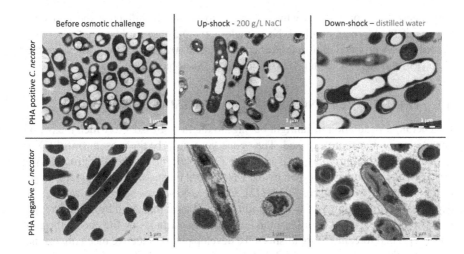

FIGURE 3.1 Effect of osmotic challenge consisting of osmotic up- and subsequent down-shock on a wild type of *C. necator* and its PHA-negative mutant.

exposure of the cells to hypotonic conditions resulted in a massive lysis of PHA-poor cells and considerable capability of keeping the cell integrity and viability in PHA-rich cells. The osmotic down-shock protective mechanism of PHA for the halophilic strain, which is prone to hypotonic lysis due to high intracellular concentration of osmolytes, deserves further investigation. It is likely that, since hydrophobic PHA granules represent a substantial portion of the intracellular volume, the total amount of osmolytes per cell will be considerably lower in PHA-rich cells. Furthermore, it is possible that the presence of PHA granules, due to their capability to protect bacterial cells from massive plasmolysis, reduces the intracellular concentration of osmolytes; nevertheless, these explanations need further experimental approval [7].

The study described above with halophilic PHA producer emphasizes the importance of PHA for prokaryotes adapted to high osmolarity. Actually, the fact, that the PHA biosynthetic capability is very common among halophiles including extremely halophilic *Arachae* such as *Haloferax mediterranei* [65], *Halogeometricum borinquense* [66] or *Natrinema ajinwuensis* [67], indicates that the PHA accumulation might be, due to osmoprotective consequences, the additional adaptation strategy to high salinity environments exhibited by many prokaryotes.

The fact that many halophiles accumulate PHA has also a biotechnological importance since halophiles are considered as auspicious candidates for industrial production of PHA as high salt concentration reduces the risk of undesired microbial contamination and, therefore, the PHA production can be performed under semi-sterile or non-sterile conditions even in a continuous mode of operation; the recycling of medium salts further increases the economic aspects of the process [68]. Furthermore, cheap seawater can be directly used for the preparation of cultivation media [69], and PHA can be efficiently recovered from halophiles by simple hypotonic lysis of bacterial cells, which substantially reduces the cost of the isolation process [70]. In addition, compatible solutes such as ectoine or hydroxyectoine represent very interesting co-products which might also positively influence the economic aspects of PHA production [71]. Nevertheless, the purpose of this chapter is not to describe PHA production by halophilic bacteria and *Archaea* in detail since another chapter of this book (Chapter 4) is dedicated to this topic.

3.5 PROTECTIVE FUNCTION OF PHA AGAINST RADIATION

In general, UV radiation in sunlight belongs to the first stressors to which the organisms living on Earth are exposed. The ozone layer absorbs the most harmful UV-C radiation (100–295 nm) but UV-B radiation (295–320 nm) reaches the ground. And this UV-B irradiation is absorbed by the DNA of organisms and causes fatal changes in the molecular structure of DNA [72]. Due to a lower ozone concentration, it is expected that the intensity of UV-B irradiation will have increased by around 5–15% in the mid-latitudes and by more than 20% at polar latitudes by the end of the century [73]. Therefore, microorganisms must develop photoprotective mechanisms that minimize the damage caused by UV-B radiation especially from the viewpoint of natural selection.

Among common lesions induced by UV-B radiation are pyrimidine dimers such as pyrimidone and cyclobutene pyrimidine dimers, which are covalently bonded from two adjacent pyrimidines. However, the fatal damage corresponding with UV radiation is not related only to DNA but also to RNA, proteins and enzymes and other biomolecules due to oxidative consequences [72, 74]. As was already mentioned several times, for living organisms it is essential for life to adapt and make progress in protective mechanisms. Therefore, living organisms developed various mechanisms to decrease or even prevent fatal changes in DNA.

Generally, microorganisms surviving in extreme conditions are called extremophiles. In the literature, the term for ultraviolet radiation-resistant (UVR) microorganisms which are known for the production of various metabolites such as pigments, mycosporine-like amino acids, melanin, pannarin and sphaerophorin, can be found [75, 76]. Another group of UVR microorganisms possesses the capability to repair already-changed DNA due to their specific enzymes, e.g. photolyases, which directly break covalent bonding between pyrimidine dimers to form original pyrimidine monomers. This process is known as photoreactivation and it requires light in the near U/blue light (300–500 nm) region as an energy source [72, 77]. Another enzymatic DNA repair pathway includes the DNA glycosylase base excision repair (BER) where the glycosyl bond between damaged base and deoxyribose is hydrolyzed. Next, DNA repair mode is based on the repair enzyme called UV-damage endonuclease (UVDE) which can recognize the photoproducts and cut them out immediately. The last of the most frequently used DNA repair pathways which was identified is the nucleotide excision repair (NER) which belongs to the main defensive strategies against UV radiation. This strategy consists of the removal of a damaged oligonucleotide [72, 77, 78].

Nevertheless, living microorganisms developed alternative protective mechanisms against ultraviolet radiation applying the PHA granules also as a protectant against UV-B radiation. There are numerous reports on the relationship between the protective effect of PHA granules in microbial cells and UV-B protection. For instance, a positive effect of PHA granules was observed in a wild type of rhizobac-terium *Azospirillum brasilense* producing high levels of PHB in comparison with its PHA synthase deletion mutant. The exposure of both strains to UV irradiation showed that wild-type cells exhibited higher resistance to UV than *A. brasilense phbC* null mutant. Even in this experiment, where short wave UV (254 nm) was used, 20% of wild-type cells remained alive in contrast to 2% of viable cells of *phbC* mutant cells after 60 s of exposure [61].

The protective effect of PHA granules was also investigated on the PHA depoly-merase deletion PhaZ mutant of *Azospirillium brasilense* that lost the ability to hydrolyze PHA. While wild-type cells depolymerase PHB into D-3-hydroxybu-tyrate oligomers by PhaZ depolymerase, the D-ß-hydroxybutyrate is oxidized by NAD dependent D-ß-hydroxybutyrate dehydrogenase to acetoacetate. Acetoacetate can be indirectly transferred into acetoacetyl-CoA *via* acetoacetyl-CoA synthetase or directly using succinyl-CoA as CoA donor via the activity of CoA transferase. Afterwards, activated acetoacetyl-CoA is hydrolyzed into two acetyl-CoA which can be integrated into the Krebs cycle. The results of the viability of wild type and *phaZ* mutant are also significantly different, as in previous cases with *phbC* mutant cells. The greatest difference between bacterial strains after 40 s of UV exposure (254 nm) was that 47.8% of wild type cells remained viable compared to 11% of *phaZ* deletion mutant cells [62].

Similarly, another study observed greater stress resistance of PHA producing a wild type of *Aeromonas hydrophila* in comparison with its PHA synthase-negative mutant. However, the protective effect of PHA granules was not as high as previ-ous studies demonstrated. After the exposure to UV (254 nm) for 10 s, the viability decreased rapidly for both bacterial strains. The rest of the living wild-type strain was about 1.15% and for the mutant it was 0.17%. A further decrease until 120 s was slow. At the end of measurement (120 s of UV exposure), the survival ratio of the wild type was two times higher than in the mutant strain [26].

Furthermore, another study used genetically modified *Escherichia coli* to imitate the properties of natural PHA producer. Wang et al. used the genetically modified *E. coli* strain DH5α (pSCP-CAB) that contains genes able to biosynthe-size PHA and *E. coli* DHα5 (pQWQ2/pSCP-CAB) that contains genes capable of both PHA biosynthesis and hydrolysis. The tolerance of UV irradiation (254 nm) increased with the introduction of the only gene encoding for PHA synthase as well as with the introduction of genes encoding both for PHA synthase and PHA depolymerase [63].

The mechanism of a UV protective effect of microbial PHA granules was inves-tigated in a study which employed *Cupriavidus necator* H16 as a model PHA accu-mulating bacterial strain (74% of cell dry mass) and its mutant strain PHB-4 without the ability to produce PHA. The first experimental part confirmed the UV protective properties by decreased viability of bacterial strain exposed to harmful UV irradia-tion (400–320 nm), also introduced in previous studies. To reveal the mechanism of the UV protective action of PHA granules, several spectroscopic techniques were used such as UV-Vis spectroscopy with regular transmission mode and spatially integrating mode and nephelometry. Based on the results from UV-Vis spectroscopy, it could be assumed that the cells of PHA accumulating *C. necator* H16 are capable of efficient scattering of UV radiation. This assumption was also supported by neph-elometry measurement. Moreover, it was suggested that the shielding effects of PHA granules could be enhanced by specific binding of PHA granules to DNA through simultaneous attachment of protein PhaM to PHA synthase and to DNA. Moreover, the presence of PHA granules protected the cells from UV radiation inducing the formation of oxygen radicals. Therefore, the PHA granules serve as great UV shield-ing with high scattering efficiency in the wavelengths close to DNA absorption max-ima compared to PHA-poor bacterial strain [10].

3.6   OXIDATIVE STRESS AND PHA

Oxidative stress is also a very common bacterial stressor in numerous environments and metabolic situations. Molecular oxygen (O_2) is the most abundant oxidative agent on Earth. This property is utilized in energetically effective aerobic metabo-lism in which O_2 acts as a final electron acceptor. Nevertheless, aerobic oxidative metabolism also generates substantial amounts of reactive oxygen species (ROS) such as superoxide (O_2-), hydroxyl radical ($\cdot$OH) or hydrogen peroxide (H_2O_2), which are very reactive and may, therefore, harm numerous cellular metabolites including but not limited to proteins and nucleic acids which cause their irreversible and dev-astating damage leading potentially to cellular death [79]. Apart from endogenous ROS generated by their own respiratory metabolism, microbes can be also exposed to oxidative pressure during chemical oxidation of reduced species at oxic-anoxic interfaces, as a consequence of H_2O_2 production by lactic acid bacteria. During the oxidative burst of phagocytes, pathogens are exposed to ROS during the inflamma-tion, and also some antimicrobial substances such as antibiotics induce the oxidative pressure [80]. Therefore, all cells including prokaryotes produce enzyme machin-ery composed of catalases, peroxiredoxins and superoxide dismutases, superoxide reductase or hydrogen peroxide reductase which are capable of catalyzing the reac-tions with harmful oxidants and to detoxify them before they can cause damage to crucial cellular components [81]. Moreover, to improve the efficiency of ROS detoxi-fying apparatus, cells experiencing oxidative pressure tend to increase the intracellu-lar level of reduced coenzymes, in particular NADPH, which serves as the reduction power utilized by ROS detoxifying enzymes to eliminate ROS [82].

There are numerous reports indicating that the presence of PHA granules enhances the stress robustness of bacteria against oxidative pressure. For instance, Kadouri et al. observed that the wild-type strain of *Azospirillum brasilense* reveals substantially improved survival rates when exposed to various stressors including external H_2O_2 compared to knock-out mutants incapable of PHA biosynthesis [61] or intracellular PHA degradation [62]. Similarly, it was observed that the presence of PHA granules protects *Delftia acidovorans* against oxidative stress induced by photo-activated titanium dioxide (TiO_2) [83]. Zhao et al. observed that a wild type *Aeromonas hydrophila* 4AK4 is capable of surviving the H_2O_2 challenge substan-tially better than the PHA synthase deletion mutant unable to biosynthesize PHA. The authors attributed the protective function of PHA to the relationship between the PHA metabolism and the expression of stress-related alternative sigma factor *rpoS* since the mutant strain revealed downregulated expression of the *rpoS* gene compared to wild-type strains [26]. Affirmative results were also revealed by Ayub et al., who observed that the PHA metabolism is crucial for psychrophilic *Pseudomonas* sp. 14-3 when exposed to low temperatures. Under cold conditions, aerobic metabolism generates substantially higher amounts of ROS as a consequence of enhanced O_2 solubility and damage of integrity of membrane in which the respiratory chain is located. Nevertheless, the PHA metabolism helped the investigated strain maintain the redox potential and eliminate generated ROS considerably better than in the PHA synthase deletion mutant which was, unlike the wild-type strain, completely unable to grow at 10°C [3]. Further, Batista et al. investigated the biological function of PHA in diazotrophic bacterium, *Herbaspirillum seropedicae*. To achieve the goal, the authors constructed a PHA-negative mutant by the deletion of PHA synthase. The absence of PHA in the mutant strain not only perturbs the redox balance and increases the oxidative stress, but also influences the activity of the redox-sensing Fnr transcription regulators. The transcriptional profiling of the mutant reveals that the loss of PHB synthesis affects the expression of many genes, including approximately 30% of the Fnr regulon. It can be stated that hampering the PHA biosynthesis has a negative impact on cell robustness and fitness [9]. However, despite numerous evidence that PHA helps bacteria to cope with oxidative stress, molecular mechanisms supporting a positive association between PHA metabolism and oxidative stress tolerance are not yet fully understood.

One of the possible mechanisms of how the PHA granules and PHA related metabolism protect against oxidative pressure is that the presence of PHA granules simply protects cells from the ROS formation. As was described above, it was observed that the PHA granules efficiently scatter UV-irradiation which in turn protects bacterial cells against ROS generated as a consequence of undesirable photochemical reactions [10]. Further, PHA monomers such 3HB as potent chemical chaperones protect sensitive biomolecules not only from the harmful effect of high temperature but also from the oxidation by ROS. It was observed that 3HB protected lipase as a model enzyme against oxidative damage by Cu^{2+} and H_2O_2 [25]. A very interesting contribution to the potential ROS protective mechanism of PHA was recently provided by Koskimäki et al. The authors looked into the PHA metabolism of pine endophyte *Methylobacterium extorquens* and observed that the PHA biosynthesizing strain produces methyl-esterified dimers and trimers of 3-hydroxybutyrate. These have three times greater hydroxyl radical-scavenging activity than glutathione and 11 times greater activity than vitamin C or the monomer 3-hydroxybutyric acid. The authors suggested that the PHA reserves are mobilized for the synthesis of methyl-esterified 3HB oligomers in bacteria cells when exposed to hydroxyl radical stress induced by infected plants [8].

It should be also pointed out that controlled application of oxidative pressure seems to be a promising strategy to enhance the PHA yields during the biotechnological production of PHA. For instance, it was observed by Obruca et al. that the application of an appropriate amount of H_2O_2 raises the PHA yields in *C. necator* H16 by about 30% [84]. The reasons causing the enhancement of PHA accumulation by H_2O_2 were further investigated in the follow-up study. The introduction of H_2O_2 induced an oxidative stress response which is, besides other cellular processes, associated with raised activity of NAPDH-generating enzymes, such as the pentose phosphate cycle;

consequently, the intracellular ratio of NADPH/NADP$^+$ is increased. This metabolic effect further stimulated the flow of acetyl-CoA towards the PHA biosynthesis rather than to the TCA cycle. Therefore, the application of hydrogen peroxide can be considered as an interesting strategy to improve PHA productivity by a well dose-tuned application of an inexpensive substance [85]. The fact that the oxidative stress supports PHA accumulation in bacteria was also observed during the metabolic analysis of PHA over-producing a mutant of *C. necator* which was obtained by random chemical mutagenesis. The PHA over-producing mutant demonstrated substantially higher activities of NADPH-producing enzymes involved in the oxidative stress response, such as glutamate dehydrogenase, NADP-dependent isocitrate dehydrogenase, glucose-6-phosphate dehydrogenase or malic enzymes. Furthermore, the mutant strain not only reached higher yields of PHA, it was also capable of more efficient incorporation of 3-hydroxyvalerate (3HV) precursors into the P(3HB-*co*-3HV) copolymer structure than observed for the wild-type strain. The incorporation of 3HV into the copolymer structure substantially lowers the crystallinity and melting temperature of the resulting material and, therefore, improves the mechanical and technological properties of the copolymer [86]. The application of increased oxidative pressure to boost the PHA production was also studied by Follonier et al., who enhanced the oxygen transfer rate in a bioreactor by the application of elevated atmospheric pressure in the bioreactor. Such a strategy tripled the PHA yields by *Pseudomonas putida* [87]. A similar positive effect of application of high atmospheric pressure during the cultivation on PHA biosynthesis was also observed in *Paracoccus denitrificans* when glycerol was used as a carbon source. The application of very high pressures (10–50 MPa) inhibited the growth of bacteria and, therefore also lowered the PHA titers, but enhanced the PHA content in bacterial biomass and also affected the monomer composition of produced PHA [88].

3.7 STRESS INDUCED BY HEAVY METALS AND OTHER XENOBIOTICS AND PHA METABOLISM

The pollution of air, water and soil is directly related to the constantly increasing population, leading to urbanization and industrialization. This brings heavy metal and other xenobiotics as anthropogenic pollutants. The problem lies not only in acute pollution of the environment, e.g. in cases of accidents, for the future the main problem lies in the bioaccumulation of heavy metals (due to their small size) and other xenobiotics (dyes, solvents and other organic and inorganic compounds), which become integrated with food chains. That makes this stress a global problem for all living systems including plants, animals, aquatic ecosystems and the human population. The decrease in the diversity of microorganisms in areas characterized by the presence of heavy metals and other xenobiotics has already been observed. As already mentioned, environments polluted by heavy metals and xenobiotics are located near industrial plants and factories associated with tannery, electronics and metal processing industries, mining of ores, distilleries, agricultural environments, etc. [89, 90].

Generally, heavy metals are classified as metals with a density higher than 5 g/cm^3. Some of them, (e.g., nickel, iron, copper, zinc and manganese) are necessary for the reactions involved in metabolic pathways in the form of cations in low amounts because of the ability to form complex compounds; these metals belong to the essential trace elements. A group of toxic heavy metals, (e.g. mercury, chromium, lead, silver, arsenic and cadmium) do not have any biological role in almost any metabolisms, however they are well known due to the toxicity and harmful effects connected with mutagenic and carcinogenic impact on living ecosystems. Nevertheless, most heavy metals, regardless of the biological functions, are toxic for the microorganisms at higher concentrations [91–95].

Microorganisms showing higher capacity of survival in the presence of heavy metals have been isolated from different niches. Considering the ability to survive in the presence of heavy metals there are two options, namely resistance and tolerance. The term resistance can be used when a specific mechanism of detoxification can be described and it has been described for many bacteria similar to antibiotic-resistant strains, whereas tolerance is a more general term when the mechanism of survival of cells has not been known. There are also other factors that influence the toxic effect of heavy metal, e.g. the presence and concentration of substances forming chelates, competitive interactions, pH value, etc. The toxic effect of higher amounts of heavy metals lies in the displacement of essential nutritional minerals by the formation of unspecific complexes in living systems leading to the disruption of their vital function. That is the reason why the homeostasis of heavy metals in cells should be maintained [89, 96–98].

The flux of toxic ions into a cell is often mediated by transporting systems for ions essential for the growth of cells of very similar structure due to insufficient impermeability. When a high concentration of heavy metal is present in cells due to unspecific transport systems, despite higher intracellular amounts other ions are transported into the cytoplasm. It is caused by constitutively expressed unspecific transporters which ensure the open gate for the transport of them which explains one of the causes of the toxic effect of heavy metals. Within the cell, different ions bind to various functional groups of proteins, nucleic acids etc., leading to the disruption of their biological function. Because of the incapability of degradation of heavy metal ions there are three possible mechanisms of how to transform them into a harmless form. The first one lies in active efflux from the cell, the second mechanism is based on the formation of complex compounds from molecules containing thiol groups and the third one involves the reduction of ions into a less toxic oxidation form. The metabolism of heavy metals in cells is therefore based on the transport predominantly. For many metals, the combinations of these mechanisms have been described [89, 99–101].

Xenobiotics represent a group of compounds of foreign origin and moreover, they often have harmful effects on cells. Generally, almost any substance apart from heavy metals, with an at least partial inhibition effect could be considered as xenobiotics. The most common organic pollutants originating from the pharmaceutical, chemical and petroleum industries and also from agricultural substances and coal refining include aromatic compounds represented by benzene and its derivatives, phenol, phenolic compounds and man-made compounds such as chlorobiphenyls. Xenobiotics are present at least in water and soil and some of them, as well as heavy metals, could be accumulated in the environment. In contrast to heavy metals some xenobiotic compounds could serve as substrates for growth of microorganisms *via* specialized enzyme systems found in adapted strains [102–106].

The toxicity of aromatic compounds correlates with their hydrophobicity, as they preferably dissolve in biological membranes of cells. This causes the increase in membrane fluidity even at low concentrations of xenobiotics leading to the loss of important intracellular metabolites such as ions, ATP and to the disruption of motion of protons to keep proton motive force and respiratory systems going. The general stress response deals with mainly effective efflux pump to get rid of compounds, with the production of heat-shock proteins and also with the modification of biologic membrane to decrease its hydrophobicity together with the enhancement of rigidity. Other mechanisms include increased production of membrane-repair enzymes and enzymes able to inactivate solvents. The oxygen-limiting or anoxic conditions are typical for environments polluted by aromatic compounds which lead to the requirement for their anoxic degradation by microorganisms. The microbial strains able to grow in the presence of xenobiotics causing environmental pollution are considered to be proper candidates for decontamination within bioremediation processes [102–106].

Despite many reports describing the isolation of PHA-producing microorganisms from contaminated environments, only a few studies deal with PHA production as a response to stress caused by heavy metals [107].

The study by Kamnev et al. deals with the comparison of endophytic and non-endophytic strains of *Azospirillum brasilense* considering the influence of heavy metals (Co^{2+}, Cu^{2+} and Zn^{2+}) on the PHA metabolism. For the experiments, the concentrations up to 0.2 mM were tested. It was observed that a non-endophytic, *A. brasilense* Sp7, showed the induction of PHB production unlike endophytic strain *A. brasilense* Sp245. The dissimilarities between these strains could be related to their different adaptation abilities considering their biological role in the ecosystem [108].

In another study by Chien et al. the mechanism of resistance of *Cupriavidus taiwanensis* against cadmium was described. This strain is known for the production of a significant amount of PHA, the same as the phylogenetically related strain *Cupriavidus necator*. A cadmium-resistant strain *C. taiwanensis* EJ02 was isolated from the sediments taken from the Er-Jen River in southern Taiwan which is well known for contamination from local industrial activities such as electroplating factories and smelters. The laboratory strains of *C. taiwanensis* from the collections of microorganisms do not exhibit resistance to heavy metals. The aim of the work was to study the PHA metabolism under stress conditions caused by high concentrations of heavy metals represented by cadmium. Except for the determination of PHB content in cells cultivated in the presence of different amounts of cadmium, the expression of genes involved in the PHA metabolism (namely *phaA*, *phaB* and *phaC*) was also observed by real-time polymerase chain reaction (PCR). The PHA content was increased when the excess of carbon source was added into the media. *C. taiwanensis* EJ02 was able to grow in the medium with 5 mM cadmium chloride without adding the excess of carbon whereas the non-resistant strains grew only up to a concentration of 1.5 mM. Considering the expression of genes involved in the PHA metabolism, there was no overexpression of genes *phaA*, *phaB* and *phaC* in the presence of cadmium in growth media despite an increased amount of PHA in

C. taiwanensis EJ02 cells when exposed to cadmium stress. Other genes participating in the PHA metabolism of *C. necator* and *C. taiwanensis* were identified, e.g. the second gene for PHA synthase (PhaC2) or isologs of enzymes PhaA and PhaB. The activities of these genes, as well as the activity of PHB depolymerase coded by *phaZ*, could also influence the PHA metabolism of *C. taiwanensis* under the stress caused by heavy metals [109].

As mentioned above there is a substantial difference between the metabolization of heavy metals and potential xenobiotics. Whereas heavy metals are usually excluded from microbial cells or incorporated into chelates, xenobiotics could be metabolized when serving as carbon and energy sources.

The capability of phenol degradation was described for the strain *Alcaligenes* sp. d_2 isolated from soil contaminated by detergent. Besides strain *Alcaligenes* sp. d_2 this ability was described also for other *Alcaligenes* strains. The metabolization of phenol and other phenolic compounds led to acetyl-CoA which is transformed via a well-known metabolic PHA biosynthetic pathway into PHA. It was observed that the stress caused by phenols serving as a sole carbon source in a growth medium manifests in an early production of PHA when after eight hours of cultivation 60% of phenol was degraded. After 24 hours of cultivation all phenol was degraded and the amount of intracellular PHA was the highest. Subsequently, PHA was decomposed when it served as an energy source. The optima for biodegradation of phenol and more significant PHA production consist of approximately neutral pH-values, the concentration of phenol of about 15 mg per 100 mL and the presence of a nitrogen source [110].

Besides water soluble xenobiotics such as phenol, there are also water insoluble hydrophobic compounds such as benzene, toluene, ethylbenzene, xylenes and others. Their toxic effect is directly related to their hydrophobicity. Almost all these substances occur naturally in crude oil, and also serve as starting compounds for chemical synthesis, while accidents during the processing, storage or transport could lead to their release into the environment. The study by Trautwein et al. deals with bacterial strain *Aromatoleum aromaticum* EbN1 and its ability to utilize aromatic compounds under nitrate-reducing conditions in the sense of stress response against selected aromatic compounds in semi-inhibitory concentration. An oxygen limitation or totally anoxic environment is often related to the pollution caused by hydrophobic compounds, therefore for the purpose of biodegradation it is necessary to use the microorganisms able to grow and metabolize during oxygen limitation. Individual solvents and their mixtures were tested, and the additive effect was described. The most toxic effect, which totally inhibited the growth of cells, was observed when ethylbenzene was added into the media (0.6 mM), the inhibiting concentration of the least toxic phenol was 8 mM, and the concentration of mixture of solvents was reduced to 3 mM. These concentrations are high enough to use the strain in the process of decontamination of ground water from aromatic compounds. The biosynthesis of PHA is connected with the production of phasins (PhaP), the proteins that are associated with PHA granules. There cannot be free phasins in cells, therefore a higher expression of them could correlate with higher PHA production. The analysis of proteome of *A. aromaticum* EbN1 showed that in the presence of toluene enzymes PhbB, PhaC and phasin-likes proteins (EbA1323, EbA6852 and

EbA5033) were over-expressed as well as in the presence of ethylbenzene, but not in cases when the compounds were not catabolized. The similarities between these phasins and phasins *Cupriavidus necator* PhaP1 and PhaP4 were observed. The up-regulation of phasins determined through 2D electrophoresis corresponded with the amount of PHA in cells, when only the homopolymer PHB was observed. With regards to PHA production a reductive force is necessary. That is a possible explanation why only in the presence of ethylbenzene PHA was detected when it served as a reduction equivalent for the degradation of substrate under nitrate-reducing conditions. Moreover, the hydrophobic granules of PHA could serve as a trap for aromatic compounds as well as for hydrophobic fluorescence probes such as Nile red. The aromatic compounds also had an influence on the expression of other proteins associated with oxidative stress and with general stress response [102].

Mezzina et al. studied the involvement of PhaP protein isolated from soil bacterium *Azotobacter* sp. FA8 into the stress response of recombinant *Escherichia coli* during the exposure to organic solvents. It was observed that the over-expression of PhaP led to the enhanced growth and biosynthesis of PHB together with the reduction of forming of inclusions during the PhaP overproduction. The solvents were represented by biotechnology-relevant chemicals presented in biofuels such as ethanol, butanol and 1,3-propanediol. The PhaP phasin was characterized as a chemical chaperone based on the results of experiments comparing the growth of recombinant a PhaP overproducing strain with a wild type resulting in higher resistance against selected solvents. The protective effect was also described during the production of ethanol or 1,3-propanediol when the recombinant strain was able to produce a higher amount of these which significantly improved the production characteristics. Besides the stress caused by the presence of solvents, the influence of PhaP was also observed during general stress response. The reduction of expression of heat shock related genes, lower level of sigma factor RpoH (main regulator under exposure to high temperature), increased growth of cells and generally higher resistance against heat shock and oxidative stress were described in connection with phasin PhaP overproduction [111].

The presence of xenobiotics inhibiting cell growth in substrates and originating from renewable resources is quite frequent. One of these could be levulinic acid which naturally occurs in hydrolysates of lignocellulosic biomass. Based on the results of a study dealing with the adaptation of *Cupriavidus necator* H16 to levulinic acid it was described that the adaptation led to the changes in the metabolism of cells. Generally, adapted strains showed better growth in the presence of levulinic acid than the wild-type strain. Some of them also showed more effective utilization of levulinic acid leading to the production of poly(3-hydroxybutyrate-*co*-3-hydroxyvalerate) associated with enhanced propionyl-CoA metabolism. More effective utilization of levulinic acid was associated with higher respiration activity and increased NADPH production considering the possibility of oxidative stress mitigation due to more intensive respiration and consequent enhanced PHA biosynthetic pathway [112].

The microorganisms which are able to grow with a high concentration of heavy metals are known as metallotolerants. There are no microbial strains that require the presence of heavy metals for their growth but some of their production metabolic pathways could be directly caused or at least enhanced by the presence of heavy metal ions in the solutions where cells occur. A typical example could be seen for the PHA model microorganism *Cupriavidus necator* when, moreover, nitrogen limitation together with an excess of a carbon source led to a higher PHB amount in cells. Also, the presence of heavy metal ions such as copper, iron or nickel led to enhanced production of PHA. The resistance against high concentrations of different heavy metals including copper, nickel, zinc, cadmium, lead, arsenic, mercury and others was described for the microbial strain *Cupriavidus metallidurans* CH34 isolated from the metal processing factory. All defense mechanisms were described for this strain. The resistance against heavy metals could be associated with genetic determinants occurring within plasmids, transposons or chromosomal DNA when the transfer of genes causing the resistance is used by many microorganisms as a strategy of how to survive this stress. Enzymes of the PHA biosynthetic pathway were characterized and it was observed that more copies of individual genes are located in different places in the genome in contrast with the genome of *C. necator*. Moreover, some of them showed dissimilarities considering the substrate specificity which could lead to the ability of utilization of different carbon sources for PHA production as well as an ability to incorporate a wider range of monomers [113–115].

The toxitolerants represent a group of microorganisms able to survive in the presence of some xenobiotics when some of them (e.g. aromatic compounds) could advantageously serve as carbon and energy sources. An example of these could be the bacterial strain *Pseudomonas putida* which, apart from many other interesting products, is able to produce *mcl*-PHA in quite significant amounts [116].

3.8 CONCLUSIONS AND OUTLOOK

As described above, the protective function of PHA against a wide range of stress factors was described and proved for various PHA-producing microorganisms including both natural PHA producers as well as recombinant strains equipped with PHA biosynthetic capability. The protective mechanisms of PHA granules are very complex. They involve many modes of action from passive biophysical consequences of the presence of highly amorphous and flexible materials representing substantial volumetric portion of bacterial cells, through the chaperoning potential of PHA granules associated proteins and PHA monomers to the impact of PHA on the re-transformation of the gene expression profile under stress conditions. Of course, PHA cannot compete in the efficiency of the shielding effect with specific stress response metabolites; nevertheless, they can provide additional sheltering and increase the robustness and survival rate of prokaryotes suddenly exposed to various stress factors. Therefore, PHA-accumulating capability is widely distributed among microbes inhabiting extreme conditions, in particular PHA biosynthesis is common, mainly among the psychrophiles and halophiles, which underlines its evolutionary importance with respect to extreme conditions. The fact that PHA substantially enhances the stress robustness of bacteria could be, of course, considered in various biotechnological processes employing bacterial cells such as the *in situ* bioremediation and removal of anthropogenic pollutants, agricultural inoculant preparation, wastewater treatment or the process of PHA production.

ACKNOWLEDGMENT

This work was supported by project GA19-20697S of the Czech Science Foundation (GACR). Further, Ivana Novackova is a Brno Ph.D. Talent Scholarship Holder – Funded by the Brno City Municipality. This work was also funded through the project SoMoPro (project No. 6SA18032). This project has received funding from the European Union's Horizon 2020 research and innovation program under the Marie Skłodowska-Curie action, and it is co-financed by the South Moravian Region under grant agreement No. 665860. The authors confirm that the content of this work reflects only the authors' views and that the EU is not responsible for any use that may be made of the information it contains.

REFERENCES

1. Marles-Wright J, Lewis RJ. Stress responses of bacteria. *Curr Opin Struc Biol* 2007; 17(6): 755–760.
2. Pötter M, Steinbüchel A. Poly(3-hydroxybutyrate) granule-associated proteins: Impacts on poly(3-hydroxybutyrate) synthesis and degradation. *Biomacromolecules* 2005; 6(2): 552–560.
3. Ayub ND, Tribelli PM, López NI. Polyhydroxyalkanoates are essential for maintenance of redox state in the Antarctic bacterium *Pseudomonas* sp. 14-3 during low temperature adaptation. *Extremophiles* 2009; 13(1): 59–66.
4. Nowroth V, Marquart L, Jendrossek D. Low temperature-induced viable but not culturable state of *Ralstonia eutropha* and its relationship to accumulated polyhydroxybutyrate. *FEMS Microbiol Lett* 2016; 363(23): fnw249.
5. Obruca S, Sedlacek P, Krzyzanek V, *et al.* Accumulation of poly(3-hydroxybutyrate) helps bacterial cells to survive freezing. *Plos One* 2016; 11(6): e0157778.
6. Obruca S, Sedlacek P, Mravec F, *et al.* The presence of PHB granules in cytoplasm protects non-halophilic bacterial cells against the harmful impact of hypertonic environments. *New Biotechnol* 2017; 39: 68–80.
7. Sedlacek P, Slaninova E, Koller M, *et al.* PHA granules help bacterial cells to preserve cell integrity when exposed to sudden osmotic imbalances. *New Biotechnol* 2019; 49: 129–136.
8. Koskimäki JJ, Kajula M, Hokkanen J, *et al.* Methyl-esterified 3-hydroxybutyrate oligomers protect bacteria from hydroxyl radicals. *Nat Chem Biol* 2016; 12(5): 332–338.
9. Batista MB, Teixeira CS, Sfeir MZT, *et al.* PHB biosynthesis counteracts redox stress in *Herbaspirillum seropedicae*. *Front Microbiol* 2018; 9: 472.
10. Slaninova E, Sedlacek P, Mravec F, *et al.* Light scattering on PHA granules protects bacterial cells against the harmful effects of UV radiation. *Appl Microbiol Biot* 2018; 102(4): 1923–1931.
11. Ayub ND, Pettinari MJ, Méndez BS, *et al.* The polyhydroxyalkanoate genes of a stress resistant Antarctic *Pseudomonas* are situated within a genomic island. *Plasmid* 2007; 58: 240–248.
12. Aragno M. Responses of microorganisms to temperature. *Physiol Plant Ecol* 1981; 1081: 339–369.
13. Thummes K, Schäfer J, Kämpfer P, *et al.* Thermophilic methanogenic Archaea in compost material: Occurrence, persistence and possible mechanisms for their distribution to other environments. *Syst Appl Microbiol* 2007; 30(8): 634–643.
14. Stetter KO. Hyperthermophiles in the history of life. *Philos T R Soc B* 2006; 361(1474): 1837–1843.

15. Jaenicke R. Stability and folding of ultrastable proteins: Eye lens crystallins and enzymes from thermophiles. *FASEB J* 1996; 10(1): 84–92.
16. Rothschild LJ, Mancinelli RL. Life in extreme environments. *Nature* 2001; 409(6823): 1092–1101.
17. Marguet E, Forterre P. Protection of DNA by salts against thermodegradation at temperatures typical for hyperthermophiles. *Extremophiles* 1998; 2(2): 115–122.
18. Schumann W. Regulation of bacterial heat shock stimulons. *Cell Stress Chaperon* 2016; 21(6): 959–968.
19. Maleki F. Bacterial heat shock protein activity. *J Clin Diagn* 2016; 10(3): BE01–BE03.
20. Han M, Yoon SS, Lee SY. Proteome analysis of metabolically engineered *Escherichia coli* producing poly(3-hydroxybutyrate). *J Bacteriol* 2001; 183(1): 301–308.
21. Han M, Park SJ, Lee JW, *et al*. Analysis of poly(3-hydroxybutyrate) granule-associated proteome in recombinant *Escherichia coli*. *J Microbiol Biotechnol* 2006; 16(6): 901–910.
22. de Almeida A, Catone MV, Rhodius VA, *et al*. Unexpected stress-reducing effect of PhaP, a poly(3-hydroxybutyrate) granule-associated protein, in *Escherichia coli*. *Appl Environ Microbiol* 2011; 77(18): 6622–6629.
23. Mezzina MP, Wetzler DE, de Almeida A, *et al*. A phasin with extra talents: A polyhydroxyalkanoate granule-associated protein has chaperone activity. *Environ Microbiol* 2015; 17(5): 1765–1776.
24. Soto G, Setten L, Lisi C, *et al*. Hydroxybutyrate prevents protein aggregation in the halotolerant bacterium *Pseudomonas* sp. CT13 under abiotic stress. *Extremophiles* 2012; 16(3): 455–462.
25. Obruca S, Sedlacek P, Mravec F, *et al*. Evaluation of 3-hydroxybutyrate as an enzyme-protective agent against heating and oxidative damage and its potential role in stress response of poly(3-hydroxybutyrate) accumulating cells. *Appl Microbiol Biot* 2016; 100(3): 1365–1376.
26. Zhao YH, Li HM, Qin LF, *et al*. Disruption of the polyhydroxyalkanoate synthase gene in *Aeromonas hydrophila* reduces its survival ability under stress conditions. *Fems Microbiol Lett* 2007; 276(1): 34–41.
27. Ruiz JA, López NI, Fernández RO, *et al*. Polyhydroxyalkanoate degradation is associated with nucleotide accumulation and enhances stress resistance and survival of *Pseudomonas oleovorans* in natural water microcosms. *Appl Environ Microbiol* 2001; 67(1): 225–230.
28. Panda AK, Bisht SS, De Mandal S, *et al*. Microbial diversity of thermophiles through the lens of next generation sequencing. Microbial diversity in the genomic era 2019; 217–226.
29. Noll KM. Thermophilic bacteria. *Encyclopedia of Genetics* 2001; 1961–1963.
30. Ibrahim MHA, Steinbuchel A. High-cell-density cyclic fed-batch fermentation of a poly(3-Hydroxybutyrate)-accumulating thermophile, *Chelatococcus* sp. strain MW10. *Appl Environ Microb* 2010; 76(23): 7890–7895.
31. Xu F, Huang S, Liu Y, *et al*. Comparative study on the production of poly(3-hydroxybutyrate) by thermophilic *Chelatococcus daeguensis* TAD1: A good candidate for large-scale production. *Appl Microbiol Biot* 2014; 98(9): 3965–3974.
32. Xiao Z, Zhang Y, Xi L, *et al*. Thermophilic production of polyhydroxyalkanoates by a novel *Aneurinibacillus* strain isolated from Gudao oilfield, China. *J Basic Microb* 2015; 55(9): 1125–1133.
33. Pantazaki AA, Tambaka MG, Langlois V, *et al*. Polyhydroxyalkanoate (PHA) biosynthesis in *Thermus thermophilus*: Purification and biochemical properties of PHA synthase. *Mol Cell Biochem* 2013; 254(1/2): 173–183.

34. Miyake M, Erata M, Asada Y. A thermophilic cyanobacterium, *Synechococcus* sp. MA19, capable of accumulating poly-β-hydroxybutyrate. *J Ferment Bioeng* 1996; 82(5): 512–514.
35. Chen W-M, Chang J-S, Chiu C-H, *et al. Caldimonas taiwanensis* sp. nov., a amylase producing bacterium isolated from a hot spring. *Syst Appl Microbiol* 2005; 28(5): 415–420.
36. Hsiao L-J, Lee M-C, Chuang P-J, *et al.* The production of poly(3-hydroxybutyrate) by thermophilic *Caldimonas manganoxidans* from glycerol. *J Polym Res* 2018; 25(4).
37. Sheu D-S, Chen WM, Yang JY, *et al.* (2009). Thermophilic bacterium *Caldimonas taiwanensis* produces poly(3-hydroxybutyrate-*co*-3-hydroxyvalerate) from starch and valerate as carbon sources. *Enzyme Microb Tech* 2009; 44(5): 289–294.
38. Arena A, Gugliandolo C, Stassi G, *et al.* An exopolysaccharide produced by *Geobacillus thermodenitrificans* strain B3-72: Antiviral activity on immunocompetent cells. *Immunol Lett* 2009; 123(2): 132–137.
39. Giedraitytė G, Kalėdienė L. Purification and characterization of polyhydroxybutyrate produced from thermophilic *Geobacillus* sp. AY 946034 strain. *Chemija* 2015; 26(1): 38–45.
40. Chen GQ, Jiang XR. Next generation industrial biotechnology based on extremophilic bacteria. *Curr Opin Biotech* 2018; 50: 94–100.
41. De Maayer P, Anderson D, Cary C, *et al.* Some like it cold: Understanding the survival strategies of psychrophiles. *EMBO Rep* 2014; 15(5): 508–517.
42. Hubalek Z. Protectants used in the cryopreservation of microorganisms. *Cryobiology* 2003; 46(3): 205–229.
43. Panoff JM, Thammavongs B, Gueguen M, *et al.* Cold stress responses in mesophilic bacteria. *Cryobiology* 1998; 36(2): 75–83.
44. Baek KH, Skinner DZ. Production of reactive oxygen species by freezing stress and the protective roles of antioxidant enzymes in plants. *J Agric Chem Environ* 2012; 1(1): 34–40.
45. Mazur P. Cryobiology: The freezing of biological systems. *Science* 1970; 168(3934): 939–949.
46. Mocali S, Chiellini C, Fabiani A, *et al.* Ecology of cold environments: New insights of bacterial metabolic adaptation through an integrated genomic-phenomic approach. *Scientific Rep* 2017; 7: 839.
47. Kadouri D, Burdman S, Jurkevitch E, *et al.* Identification and isolation of genes involved in poly (β-hydroxybutyrate) biosynthesis in *Azospirillum brasilense* and characterization of a phbC mutant. *Appl Environ Microbiol* 2002; 68(6): 2943–2949.
48. Tribelli PM, Lopez NI. Poly(3-hydroxybutyrate) influences biofilm formation and motility in the novel Antarctic species *Pseudomonas extremaustralis* under cold conditions. *Extremophiles* 2011; 15(5): 541–547.
49. Pavez P, Castillo JL, González C, *et al.* Poly-β-hydroxyalkanoate exert a protective effect against carbon starvation and frozen conditions in *Sphingopyxis chilensis*. *Curr Microbiol* 2009; 59(6): 636–640.
50. Goh YS, Tan IKP. Polyhydroxyalkanoate production by antarctic soil bacteria isolated from Casey Station and Signy Island. *Microbiol Res* 2012; 167(4): 211–219.
51. Kaartokallio H, Søgaard DH, Norman L, *et al.* Short-term variability in bacterial abundance, cell properties, and incorporation of leucine and thymidine in subarctic sea ice. *Aquat Microb Ecol* 2013; 71: 57–73.
52. Ciesielski S, Górniak D, Możejko J, *et al.* The diversity of bacteria isolated from Antarctic freshwater reservoirs possessing the ability to produce Polyhydroxyalkanoates. *Curr Microbiol* 2014; 69(5): 594–603.
53. Pärnänen K, Karkman A, Virta M, *et al.* Discovery of bacterial polyhydroxyalkanoate synthase (PhaC)-encoding genes from seasonal Baltic Sea ice and cold estuarine waters. *Extremophiles* 2015; 19(1), 197–206.

54. Kumar V, Thakur V., Ambika, *et al*. Bioplastic reservoir of diverse bacterial communities revealed along altitude gradient of Pangi-Chamba trans-Himalayan region. *FEMS Microbiol Lett* 2018; 365(14).
55. Morbach S, Kramer R. Body shaping under water stress: Osmosensing and osmoregulation of solute transport in bacteria. *ChemBioChem* 2002; 3(5): 384–397.
56. Schwarz H, Koch AL. Phase and electron microscopic observation of osmotically induced wrinkling and the role of endocytotic vesicles in the plasmolysis of the Gram-negative bacteria. *Microbiol* 1995; 141(12): 3161–3170.
57. Brown RB, Audet J. Current techniques for single-cell lysis. *J R Soc Interface* 2008; 5(2): 131–138.
58. Roberts MF. Organic compatible solutes of halotolerant and halophilic microorganisms. *Saline Systs* 2005; 1(1): 1–5.
59. Kouwen TRHM, Antelmann H, van der Ploeg R., *et al*. MscL of Bacillus subtilis prevents selective release of cytoplasmic proteins in a hypotonic environment. *Proteomics* 2009; 9(4): 1033–1043.
60. Bialecka-Fornal M, Lee HJ, Phillips R. The rate of osmotic downshock determines the survival probability of bacteria mechanosensitive channel mutants. *J Bacteriol* 2015; 197(1): 231–237.
61. Kadouri D, Jurkevitch E., Okon Y. Involvement of the reserve material Poly- beta-hydroxybutyrate in *Azospirillum brasilense* stress endurance and root colonization. *Appl Environ Microbiol* 2003a; 69(6): 3244–3250.
62. Kadouri D, Jurkevitch E, Okon Y. Poly beta-hydroxybutyrate depolymerase (PhaZ) in *Azospirillum brasilense* and characterization of a phaZ mutant. *Arch Microbiol* 2003b; 180(5): 309–318.
63. Wang Q, Yu H, Xia Y, *et al*. Complete PHB mobilization in *Escherichia coli* enhances the stress tolerance: A potential biotechnological application. *Microb Cell Fact* 2009; 8(1): 47.
64. Breedveld MW, Dijkema C, Zevenhuizen LPTM, *et al*. Response of intracellular carbohydrates to a NaCl shock in *Rhizobium leguminosarum* biovar trifolii TA-1 and *Rhizobium meliloti* SU-47. *J Gen Microbiol* 1993; 139(12): 3157–3163.
65. Mahler N, Tschirren S, Pflügl S, *et al*. Optimized bioreactor setup for scale-up studies of extreme halophilic cultures. *Biochem Eng J* 2018; 130: 39–46.
66. Salgaonkar BB, Bragança JM. Utilization of sugarcane bagasse by *Halogeometricum borinquense* strain E3 for biosynthesis of poly (3-hydroxybutyrate-*co*-3-hydroxyvalerate). *Bioeng* 2017; 4(2): 50.
67. Mahansaria R, Dhara A, Saha A, *et al*. Production enhancement and characterization of the polyhydroxyalkanoate produced by *Natrinema ajinwuensis* (as synonym) ≡ *Natrinema altunense* strain RM-G10. *Int J Biol Macromol* 2018; 107: 1480–1490.
68. Koller M. Recycling of waste streams of the biotechnological poly(hydroxyalkanoate) production by *Haloferax mediterranei* on whey. *Int J Polym Sci* 2015; 1–8.
69. Takahashi RYU, Castilho NAS, Silva MACD, *et al*. Prospecting for marine bacteria for polyhydroxyalkanoate production on low-cost substrates. *Bioeng* 2017; 4(3): 60.
70. Koller M, Chiellini E, Braunegg G. Study on the production and re-use of poly(3-hydroxybutyrate-*co*-3-hydroxyvalerate) and extracellular polysaccharide by the archaeon *Haloferax mediterranei* strain DSM 1411. *Chem Biochem Eng Q* 2015; 29(2): 87–98.
71. Van-Thuoc D, Guzmán H, Quillaguamán, *et al*. High productivity of ectoines by *Halomonas boliviensis* using a combined two-step fed-batch culture and milking process. *J Biotechnol* 2010; 147(1): 46–51.
72. Goosen N, Moolenaar, GF. Repair of UV damage in bacteria. *DNA Repair* 2008; 7(3): 353–379.

73. Zepp RG, Erickson DJ 3rd., Paul ND, *et al.* Effects of solar UV radiation and climate change on biogeochemical cycling: Interactions and feedbacks. *Photochem Photobiol Sci* 2011; 10(2): 261–279.
74. Mouret S, Philippe C, Gracia-Chantegrel J, *et al.* UVA-induced cyclobutane pyrimidine dimers in DNA: A direct photochemical mechanism? *Org Biomol Chem* 2010; 8(7): 1706–1711.
75. Koller M, Muhr A, Braunegg G. Microalgae as versatile cellular factories for valued products. *Algal Res* 2014; 6: 52–63.
76. Gabani P, Singh OV. Radiation-resistant extremophiles and their potential in biotechnology and therapeutics. *Appl Microbiol Biotechnol* 2013; 97(3): 993–1004.
77. Takahashi A, Ohnishi T. Molecular mechanisms involved in adaptive responses to radiation, UV light, and heat. *J Radiat Res* 2009; 50(5): 385–393.
78. Singh OV, Gabani P. Extremophiles: Radiation resistance microbial reserves and therapeutic implications. *J Appl Microbiol* 2011; 110(4): 851–861.
79. Ranawat P, Rawat S. Stress response physiology of thermophiles. *Arch Microbiol* 2017; 199(3): 391–414.
80. Imlay JA. Where in the world do bacteria experience oxidative stress? *Environ Microbiol* 2019; 21(2): 521–530.
81. Ezraty B, Gennaris A, Barras F, *et al.* Oxidative stress, protein damage and repair in bacteria. *Nat Rev Microbiol* 2017; 15(7): 385–396.
82. Sigler K, Chaloupka J, Brozmanova J, *et al.* Oxidative stress in microorganisms – I. Microbial vs. higher cells – damage and defenses in relation to cell aging and death. *Folia Microbiol* 1994; 44(6): 587–624.
83. Goh LK, Purama RK, Sudesh K. Enhancement of stress tolerance in the Polyhydroxyalkanoate producers without mobilization of the accumulated granules. *Appl Biochem Biotechnol* 2014; 172(3): 1585–1598.
84. Obruca S, Marova I, Svoboda Z, *et al.* Use of controlled exogenous stress for improvement of poly(3-hydroxybutyrate) production in *Cupriavidus necator*. *Folia Microbiol* 2010; 55(1): 17–22.
85. Obruca S, Marova I, Stankova M, *et al.* Effect of ethanol and hydrogen peroxide on poly(3-hydroxybutyrate) biosynthetic pathway in *Cupriavidus necator* H16. *World J Microbiol Biotechnol* 2010; 26(7): 1261–1267.
86. Obruca S, Snajdar O, Svoboda Z, *et al.* 2013. Application of random mutagenesis to enhance the production of polyhydroxyalkanoates by *Cupriavidus necator* H16 on waste frying oil. *World J Microbiol Biotechnol* 2013; 29(12): 2417–2428.
87. Follonier S, Henes B, Panke S, *et al.* Putting cells under pressure: A simple and efficient way to enhance the productivity of medium-chain-length polyhydroxyalkanoate in processes with *Pseudomonas putida* KT2440. *Biotechnol Bioeng* 2012; 109(2): 451–461.
88. Mota MJ, Lopes RP, Mário RQ, *et al.* Effect of high pressure on paracoccus denitrificans growth and polyhydroxyalkanoates production from glycerol. *Appl Biochem Biotechnol* 2019; 188(3): 810–823.
89. Marzan LW, Hossain M, Mina SA, *et al.* Isolation and biochemical characterization of heavy-metal resistant bacteria from tannery effluent in Chittagong city, Bangladesh: Bioremediation viewpoint. *Egypt J Aquat Res* 2017; 43(1): 65–74.
90. Robin RS, Muduli PR, Vardhan KV, *et al.* Heavy metal contamination and risk assessment in the marine environment of Arabian Sea, along the Southwest Coast of India. *Am J Chem* 2012; 2(4): 191–208.
91. Trevors JT, Oddie KM, Belliveau BH. Metal resistance in bacteria. *FEMS Microbiol Rev* 1985; 32(1): 39–54.
92. Deborah, S, Raj JS. Bioremediation of heavy metals from distilleries effluent using microbes. *J Appl Adv Res* 2016; 1(2): 23–28.

93. Mustapha, MU, Halimoon N. Screening and isolation of heavy metal tolerant bacteria in industrial effluent. *Procedia Environ Sci* 2015; 30: 33–37.
94. Nies, DH. Microbial heavy-metal resistance. *Appl Microbiol Biotechnol* 1999; (51): 730–750.
95. Adriano, DC. *Trace Elements in Terrestrial Environments: Biogeochemistry, Bioavailability, and Risks of Metals* (2001); Springer.
96. de Lima SAA, de Carvalho MA, de Souza SA, *et al.* Heavy metal tolerance (Cr, Ag and Hg) in bacteria isolated from sewage. *Braz J Microbiol* 2012; 43(4): 1620–1631.
97. Duxbury T. Microbes and heavy metals: An ecological overview. *Microbiol Sci* 1986; 3(11): 330–333.
98. Sterritt RM, Lester JN. Interactions of heavy metals with bacteria. *Scie Total Environ* 1980; 14(1): 5–17.
99. Nies DH, Silver S. Ion efflux systems involved in bacterial metal resistances. *J Ind Microbiol* 1995; 14(2): 186–199.
100. François F, Lombard C, Guigner J-M, *et al.* PignolIsolation and characterization of environmental bacteria capable of extracellular biosorption of mercury. *Appl Environ Microbiol* 2012; 78(4): 1097–1106.
101. Hrynkiewicz K, Baum C. Application of microorganisms in bioremediation of environment from heavy metals. *Environ Deterioration Human Health* 2014; 215–227.
102. Trautwein, K, Kuhner S, Wohlbrand L, *et al.* Solvent stress response of the denitrifying bacterium "*Aromatoleum aromaticum*" strain EbN1. *Appl Environ Microbiol* 2008; 74(8): 2267–2274.
103. van der Meer JR, de Vos WM, Harayama S, *et al.* Molecular mechanisms of genetic adaptation to xenobiotic compounds. *Microbiol Rev* 1992; 56(4): 677–694.
104. Keweloh, H, Weyrauch G, Rehm HJ. Phenol induced membrane changes in free and immobilized Escherichia coli. *Appl Microbiol Biotechnol* 1990; 33: 66–71.
105. Sikkema J, Bont JAM, Poolman B. Mechanisms of membrane toxicity of hydrocarbons. *Microbiol Rev* 1995; 59: 201–202.
106. de Lima SAA, Pereira MP, Silva FRG, *et al.* Utilization of phenol in the presence of heavy metals by metal-tolerant nonfermentative gram-negative bacteria isolated from wastewater. *Rev Latinoam Microbiol* 2007; 49(3–4): 68–73.
107. Pal A, Paul AK. Accumulation of polyhydroxyalkanoates by rhizobacteria underneath nickel-hyperaccumulators from serpentine ecosystem. *J Polymer Environ* 2012; 20(1): 10–16.
108. Kamnev AA, Tugarova AV, Antonyuk LP, *et al.* Effects of heavy metals on plant-associated rhizobacteria: Comparison of endophytic and non-endophytic strains of *Azospirillum brasilense*. *J Trace Elem Med Biol* 2005; 19(1): 91–95.
109. Chien Ch-Ch, Wang L-J, Lin W-R. Polyhydroxybutyrate accumulation by a cadmium-resistant strain of *Cupriavidus taiwanensis*. *J Taiwan Inst Chem Eng* 2014; 45(4): 1164–1169.
110. Nair IC, Pradeep S, Ajayan MS, *et al.* Accumulation of intracellular polyhydroxybutyrate in *Alcaligenes* sp. d2 under phenol stress. *Appl Biochem Biotechnol* 2009; 159(2): 545–552.
111. Mezzina MP, Álvarez DS, Egoburo DE, *et al.* A New player in the biorefineries field: Phasin PhaP enhances tolerance to solvents and boosts ethanol and 1,3-propanediol synthesis in escherichia coli. *Appl Environ Microbiol* 2017; 83(14).
112. Novackova I, Kucera D, Porizka J, *et al.* Adaptation of *Cupriavidus necator* to levulinic acid for enhanced production of P(3HB-*co*-3HV) copolyesters. *Biochem Eng J* 2019; 151.
113. Manasi, Rajesh N, Rajesh V. Evaluation of the genetic basis of heavy metal resistance in an isolate from electronic industry effluent. *J Genet Eng Biotechnol* 2016; 14(1): 177–180.

114. Passantha P, Esteves SR, Kedia G, *et al*. Increasing polyhydroxyalkanoate (PHA) yields from *Cupriavidus necator* by using filtered digestate liquors. *Bioresour Technol* 2013; 147: 345–352.
115. Janssen PJ, van Houdt R, Moors H, *et al*. The complete genome sequence of *Cupriavidus metallidurans* strain CH34, a master survivalist in harsh and anthropogenic environments. *PLoS One* 2010; 5(5): e10433.
116. Loeschcke A, Thies S. *Pseudomonas putida* – a versatile host for the production of natural products. *Appl Microbiol Biotechnol* 2015; 99(15): 6197–6214.

4 Linking Salinity to Microbial Biopolyesters Biosynthesis

Polyhydroxyalkanoate Production by Haloarchaea and Halophilic Eubacteria

*Martin Koller, Stanislav Obruca,
and Gerhart Braunegg*

CONTENTS

4.1   INTRODUCTION

Regarding the selection of powerful microbial polyhydroxyalkanoate (PHA) production strains, robust species, which are resistant to microbial contamination and which possess a broad substrate spectrum and a well-studied genome, proteome and metabolome are currently investigated by the scientific community all over the world. In general, extremophilic microbes inhabit challenging environments, and are adapted to high concentrations of heavy metals like chromium, cobalt, copper, nickel, cadmium, or lead ("metalophiles"), extreme pH-value conditions ("acidophiles" and "alkaliphiles"), extreme temperature ("thermophiles" on the one, and "kryophiles" or "psychrophiles" on the other side), or osmotic pressure ("osmophiles" including salt-requiring "halophiles" and "saccharophiles" thriving well under elevated sugar concentrations) (reviewed by [1, 2]. While using thermophilic PHA production strains, which prefer a cultivation temperature of 50°C and higher, only now starts to outgrow its infancy [3], particularly the application of halophiles, typically isolated from saline brines and salterns [4], salt lakes [5], or deep-sea marine environments [6, 7], is currently a strongly emerging research route. As a major advantage, the elevated or often even extreme concentration of sodium chloride (NaCl) needed by these strains for optimum growth and PHA biosynthesis, inhibits or completely prevents the growth of competing microbial species, hence, halophilic strains can be cultivated in simple, flexible devices, sometimes even under completely unsterile conditions in open systems, as it is known from algal cultivation in open racing ponds [8]. It is self-evident that this approach saves energy normally needed to create and maintain strictly sterile and monoseptic conditions [1–3, 9, 10]. In addition, salinity is a crucial parameter when operating wastewater treatment plants, where anaerobic digestion of organic waste by mixed microbial cultures is coupled with PHA accumulation, as was demonstrated by Wen and colleagues, who reported an optimum NaCl concentration of 5 g/L and low organic load for the conversion of food waste fermentation leachate into microbial biomass and PHA. Phylogenetic analysis of the microbial consortium revealed that, under the given process conditions, the stability of the system was caused by the enrichment of halophilic strains belonging to the genera *Paracoccus* and *Thauera* [11].

The use of such halophilic organisms for biotechnological purposes opens up several options: first, the high salt concentration in saline culture media reduces the risk of microbial contamination by unwanted foreign germs, which in turn results in a dramatic increase in the stability of the production batch [10]. From a mechanistic point of view, the cultivation of halophilic organisms in saline media leads to the intracellular accumulation of salt and to the partial removal of the salt load from the medium [9]. This is a decisive advantage when using salt-rich waste streams as a culture medium, which otherwise would have to be disposed of in a cost-intensive manner. Furthermore, various high-carbon waste streams are converted by hydrochloric acid (HCl)-catalyzed hydrolysis into biotechnologically accessible substrates. After hydrolysis, neutralization by caustic soda (NaOH) is necessary, which in turn produces salt. Such processes have recently been demonstrated by the example of using acid-catalyzed whey permeate, straw, bagasse, sawdust, or coffee stock to extract the microbially accessible raw material for PHA biosynthesis. Halophilic production strains used therein require the salt formed in the subsequent neutralization step as an essential media component and are therefore the ideal candidates for such bioprocesses [12–14].

4.2 HALOPHILIC MICROBES PRODUCING PHA

Table 4.1 provides a compilation of the halophilic PHA production strains covered in this chapter.

TABLE 4.1

Overview of Selected PHA Production Processes by Halophiles

Microbial species (Type of microbe)	Origin of the microorganism	Salinity and applied carbon source(s)	Monomeric composition of produced PHA	Production scale, feeding regime and output (product concentration and volumetric productivity)	Ref.
Haloarchaea					
Haloferax mediterranei DSM 1411	Salt pond located at the Costa Blanca near Alicante, Spain	Extreme halophile: 15% NaCl; Glucose plus yeast extract	Copolyester: PHBHV	10 L bioreactor; fed-batch feeding 13 g/L, 0.21 g/(L·h)	[103]
"	"	Extreme halophile: 15% NaCl; CGP; CGP plus GBL	Copolyesters: PHBHV P(3HB-*co*-3HV-*co*-4HB)	42 L / 10 L bioreactor fed batch process; 16.2 g/L, 0.12 g/(L·h) 11.1 g/L, 0.10 g/(L·h),	[91]
"	"	Extreme halophile: 20% NaCl; Hydrolyzed whey permeate Hydrolyzed whey permeate plus GBL	Copolyesters: PHBHV P(3HB-*co*-3HV-*co*-4HB)	42 L bioreactor fed batch process; 0.09 g/(L·h), 12.2 g/L PHBHV) 0.14 g/(L·h), 14.7 g/L (3HB-*co*-3HV-*co*-4HB)	[112]
"	"	Extreme halophile: 15.6% NaCl; Hydrolyzed whey permeate, elevated trace element concentration	Copolyester: PHBHV (below 2% 3HV in PHBHV)	2 L bioreactor; batch feeding 8 g/L, 0.17 g/(L·h)	[113]
"	"	Extreme halophile: 20% NaCl; Hydrolyzed whey permeate, spent fermentation broth of previous whey-based processes	Copolyester: PHBHV	10 L bioreactor batch process 0.04 g/(L·h), 2.28 g/L PHA	[123]

(Continued)

TABLE 4.1 (CONTINUED)

Overview of Selected PHA Production Processes by Halophiles

Microbial species (Type of microbe)	Origin of the microorganism	Salinity and applied carbon source(s)	Monomeric composition of produced PHA	Production scale, feeding regime and output (product concentration and volumetric productivity)	Ref.
?	?	Extreme halophile: 20% NaCl; Waste stillage from rice-based ethanol production; parallel recovery and re-use of medium salts	Copolyester: PHBHV	Shaking flask setups; batch feeding 16.4 g/L PHA, 0.7 g/g PHA in CDM, 0.17 g/(L·h)	[117]
?	?	Extreme halophile: 23.4% NaCl; Native cornstarch treated via enzymatic reactive extrusion plus yeast extract	Copolyester: PHBHV	6 L bioreactor pH-stat fed-batch process 0.28 g/(L·h), 0.508 g/g PHA in CDM; 20 g/L PHA	[115]
?	?	Extreme halophile: 15.6% NaCl; Glucose plus yeast extract	Copolyester: PHBHV	Shaking flask cultivations; batch feeding NH_4^+ as nitrogen source: 10.7 g/L CDM, 0.046 g/g PHA in CDM; NO_3^- as nitrogen source: 5.6 g/L CDM, 0.093 g/g PHA in CDM	[108]
?	?	Extreme halophile: 156 g/L NaCl; Glucose; varying phosphate concentrations	Copolyester: PHBHV (22.36 wt.-% 3HV)	500 mL shaking flasks, batch 0.95 g/L PHA, 0.156 g/g PHA in CDM; 0.007 g/(L·h) with optimum phosphate concentration 0.5 g/L KH_2PO_4	[109]

(Continued)

TABLE 4.1 (CONTINUED)
Overview of Selected PHA Production Processes by Halophiles

Microbial species (Type of microbe)	Origin of the microorganism	Salinity and applied carbon source(s)	Monomeric composition of produced PHA	Production scale, feeding regime and output (product concentration and volumetric productivity)	Ref.
"	"	Extreme halophile: 156 g/L NaCl; Mixes of butyric and valeric acid; addition of emulsifier Tween80	Copolyester: PHBHV (43 mol-% 3HV at feed mix butyric/valeric acid of 56/44)	Fed-batch bioreactor cultivation 4.01 g/L PHA, 0.59 g/g PHA in CDM; 0.01 g/(L·h) (37°C)	[129]
"	"	Extreme halophile: 156 g/L NaCl; Odd- and even numbered fatty acids	PHBHV Copolyester: Acetic acid: 10 mol-% 3HV Valeric acid: ~90 mol-% 3HV	Shaking flask; parallel or sequential feeding; max. 1.5 g/L PHA; max. 0.25 g/g PHA in CDM	[130]
"	"	Extreme halophile: 23.4% NaCl; Extruded rice bran plus extruded cornstarch	Copolyesters: PHBHV	5 L bioreactor; pH-stat feeding strategy; 24.2 g/L PHA	[115]
"	"	Extreme halophile: 25% marine salts Starch (20 g/L) Glucose (10 g/L)	Copolyesters: PHBHV (in article: "PHB homopolyester")	Stable, monoseptic continuous cultivation over 3 months in 1.5 L bioreactor; T = 38°C 6.5 g/L PHA on starch 3.5 g/L on glucose	[95]
"	"	Extreme halophile: 14.4% NaCl; Hydrolyzed *Ulva* sp. (macroalgae) as carbon source	Copolyester: PHBHV (8 mol-% 3HV in PHBHV)	Shaking flask scale, batch cultivation; 2.2 g/L, 0.58 g/g PHA in CDM, 0.035 g/(L·h)	[121]

(Continued)

TABLE 4.1 (CONTINUED)
Overview of Selected PHA Production Processes by Halophiles

Microbial species (Type of microbe)	Origin of the microorganism	Salinity and applied carbon source(s)	Monomeric composition of produced PHA	Production scale, feeding regime and output (product concentration and volumetric productivity)	Ref.
*	*	Extreme halophile: 220 g/L NaCl Native and dephenolized olive oil wastewater	Copolyester: PHBHV (6.5 mol-% 3HV in PHBHV)	Shaking flask scale, batch cultivation; 43 wt.% PHA in CDM (concentration and productivity data not consistent in original publication)	[119]
*	*	Extreme halophile: 200 g/L NaCl 25–50% (v/v) pre-treated vinasse from ethanol production	Copolyester: PHBHV (12.4–14.1 mol-% 3HV in PHBHV)	Shaking flask scale, batch cultivation; 19.7 g/L, 0.70 g/g PHA in CDM, 0.21 g/(L·h)	[120]
Natrinema (Nnm.) ajinwuensis (altunense) RM-G10	Indian salt production pans	Extremely halophile: 20% NaCl Glucose	Copolyesters: PHBHV (3HV: 14 mol-%)	Repeated batch cultivations in shaking flasks PHA content in CDM 0.61 g/g PHA in CDM; 0.21 g/(L·h) PHA	[124]
Halobiforma (Hbf.) haloterrestris	Hypersaline soil samples from Aswan, Egypt	Extremely halophile: 22% NaCl	Homopolyester PHB	Shaking flask scale; 40 wt.% PHB in CDM on butyric acid, 0.15 g/g PHB in CDM on complex medium	[31]

(Continued)

TABLE 4.1 (CONTINUED)
Overview of Selected PHA Production Processes by Halophiles

Microbial species (Type of microbe)	Origin of the microorganism	Salinity and applied carbon source(s)	Monomeric composition of produced PHA	Production scale, feeding regime and output (product concentration and volumetric productivity)	Ref.
Halogeometricum (Hgm.) borinquense TN9	Marakkanam solar salterns, India	Extreme halophile: 20% NaCl; Glucose	Homopolyester PHB	Shaking flask cultivations 0.14 g/g PHA in CDM; ca. 3 mg/(L·h) PHA	[138]
Halogeometricum (Hgm.) borinquense E3	"	Extreme halophile: 20% NaCl; Glucose	Copolyesters: PHBHV (3HV: 21.5 mol-%)	Shaking flask cultivations PHA content in CDM 0.74 g/g; 0.21 g/(L·h) PHA	[139]
"	"	Extreme halophile: 20% NaCl; 25 and 50% sugarcane bagasse hydrolysates	Copolyesters: PHBHV (13.3% 3HV)	Shaking flask cultivations; between 0.45 and 0.50 g/g PHA in CDM; 0.0113 g/(L·h) PHBHV on 25% sugar cane bagasse hydrolysate	[140]
"	"	Extreme halophile: 20% NaCl; Starch and carbon-rich fibrous waste (cassava bagasse)	Copolyesters: PHBHV (13.1% 3HV with starch, 19.7% 3HV with cassava waste)	Shaking flask cultivations in batch mode; Starch: 4.6 g/L PHA, 0.02 g/(L·h), 0.74 g/g PHA in CDM, Cassava bagasse: 1.52 g/L, 0.006 g/(L·h), 0.45 g/g PHA in CDM	[141]

(Continued)

TABLE 4.1 (CONTINUED)
Overview of Selected PHA Production Processes by Halophiles

Microbial species (Type of microbe)	Origin of the microorganism	Salinity and applied carbon source(s)	Monomeric composition of produced PHA	Production scale, feeding regime and output (product concentration and volumetric productivity)	Ref.
Halogranum (Hgr.) amylolyticum	Tainan marine solar salterns, PR China	Extreme halophile: 20% NaCl; Glucose	Copolyesters: PHBHV (20 mol.-% 3HV!)	7.5 L bioreactor; fed-batch feeding strategy; 0.074 vv, 14 g/L PHBHV	[143]
Natronobacterium (Nbt.) gregoryi NCMB 2189T	Soda salt lake liquors from the soda lake Magadi, Kenia	Extreme halophile: 20% NaCl; alkaliphile; Carbohydrate	Homopolyester PHB	Shaking flask scale; batch feeding; 0.4 wt.-% PHB	[144]
Haloarcula (Har.) sp. IRU1	Hypersaline Urmia salt lake, Iran	Extreme halophile: 250 g/L NaCl; Glucose (other carbon sources tested)	Homopolyester: PHB	Shaking flask cultivations; batch feeding; T = 42°C, pH = 7.0; 0.66 g/g PHB in CDM	[132]
"	"	Extreme halophile: 250 g/L NaCl; Glucose (other carbon sources tested)	Homopolyester: PHB	Shaking flask scale; 0.62 (glucose), 0.57 (starch), 0.56 (sucrose), 0.55 (fructose), 0.40 (acetate), 0.39 (palmitic acid) g/g PHB in CDM Max. PHA concentration and productivity: 0.98 g/L, 0.016 g/ (L·h) (glucose)	[133]

(Continued)

TABLE 4.1 (CONTINUED)
Overview of Selected PHA Production Processes by Halophiles

Microbial species (Type of microbe)	Origin of the microorganism	Salinity and applied carbon source(s)	Monomeric composition of produced PHA	Production scale, feeding regime and output (product concentration and volumetric productivity)	Ref.
"	"	Extreme halophile: 250 g/L NaCl 47°C (other T tested) Substrate: Petrochemical wastewater, tryptone	Homopolyester: PHB	Shaking flask scale, batch feeding; max. 0.47 g/g PHB in CDM (2% petrochemical wastewater, yeast extract, 47°C	[134]
"	"	Extreme halophile: 250 g/L NaCl 47°C (other T tested) Substrate: Crude oil, yeast extract (other N-sources tested)	Homopolyester: PHB	Shaking flask scale, batch feeding; max. 0.41 g/g PHB in CDM (2% crude oil, yeast extract, 47°C)	[135]
Haloarcula (Har.) marismortui	Dead Sea	Extreme halophile: 20% NaCl; Raw and charcoal-pretreated vinasse from bioethanol production	Homopolyester: PHB	Shaking flask cultivations; between 0.23 (10% non-detoxified vinasse) and 0.30 (100% charcoal-detoxified vinasse) g/g PHA content in CDM; 0.015 g/(L·h) (non-detoxified) and 0.02 (detoxified) g/(L·h) PHB (2.8 g/L and 4.5 g/L PHB, respectively)	[137]

(Continued)

TABLE 4.1 (CONTINUED)
Overview of Selected PHA Production Processes by Halophiles

Microbial species (Type of microbe)	Origin of the microorganism	Salinity and applied carbon source(s)	Monomeric composition of produced PHA	Production scale, feeding regime and output (product concentration and volumetric productivity)	Ref.
Natrinema pallidum JCM 8980 (= isolate 1KYS1)	Kayacik saltern, Turkey	Extreme halophile: 25% NaCl; Starch	Copolyesters: PHBHV (3HV: 25 mol-%)	Shaking flask cultivations; 0.53 g/g PHA in CDM; 0.3 mg/ (L·h) PHA	[142]
Halopiger (Hpg.) aswanensis 56	Samples collected from surface of hypersaline soil collected in Aswan, Egypt	Extreme halophile: 25% NaCl; Sodium acetate and butyric acid	Homopolyester: PHB	Corrosion-resistant 8 L bioreactor; batch feeding; 0.0045 g/(L·h), 0.53 g/g PHB in CDM, 4.6 g/L PHB, 0.018 g/ (L·h)	[55]
Halobiforma (Hbf.) haloterrestris	Samples collected from surface of hypersaline soil collected in Aswan, Egypt	Extremely halophile: 2.2 M NaCl (in addition: moderately thermophilic: optimum 42°C)	Homopolyester: PHB	Shaking flask scale 0.4 g/g PHB in CDM on butyric acid, 0.15 g/g PHB in CDM on peptone	[31]
Haloterrigena (Htg.) hispanica	Saltern crystallizer pond at Fuente de Piedra saline lake, province Malaga, southern Spain	Extreme halophile: 20% NaCl; Carrot waste; complex medium	Homopolyester: PHB	Bioreactor; batch feeding; dialysis device 0.0013 g/g PHB in CDM after five days of cultivation	[145]

(Continued)

TABLE 4.1 (CONTINUED)
Overview of Selected PHA Production Processes by Halophiles

Microbial species (Type of microbe)	Origin of the microorganism	Salinity and applied carbon source(s)	Monomeric composition of produced PHA	Production scale, feeding regime and output (product concentration and volumetric productivity)	Ref.
Eubacteria (Gram-negative)					
Halomonas boliviensis LC1	Soil samples from the Laguna Colorada lake, Bolivian Andeans	Moderate halophile: 4.5% NaCl (moreover: psychrophilic and alkaliphilic) Starch incompletely hydrolyzed	Homopolyester: PHB	2 L bioreactor 25 g/L PHB, 0.49 g/g PHB in CDM, 0.9 g/(L·h)	[158]
"	"	Moderate halophile: 4.5% NaCl Acetate and butyrate	Homopolyester: PHB	2 L bioreactor 1–2 g/L PHB, 0.88 g/g PHB in CDM	[37]
"	"	Moderate halophile: 4.5% NaCl Glucose as main carbon source plus sodium glutamate as growth supplement	Homopolyester: PHB	Fed-batch setups in 2 L bioreactor 20.7 g/L PHB, 0.90 g/g PHB in CDM, 1.15 g/(L·h) PHB (repeated feed of glutamate, phosphate, and NH$_4$Cl) 35.6 g/L PHB, 0.81 g/g PHB in CDM, 1.1 g/(L·h) PHB (supply of glutamate, phosphate, and NH$_4$Cl only at the beginning)	[159]

(Continued)

TABLE 4.1 (CONTINUED)
Overview of Selected PHA Production Processes by Halophiles

Microbial species (Type of microbe)	Origin of the microorganism	Salinity and applied carbon source(s)	Monomeric composition of produced PHA	Production scale, feeding regime and output (product concentration and volumetric productivity)	Ref.
"	"	Moderately halophile: 4.5% NaCl; Hydrolyzed wheat bran, acetate, butyrate	Homopolyester: PHB	Batch cultivation in 2 L bioreactor; 4 g/L PHB, 0.5 g/g PHB in CDM, 0.2 g/(L·h) PHB	[160]
Halomonas elongata 2FF	Hypersaline meromictic Fără fund lake, Romania	Moderately halophile: 10% NaCl; Glucose	Homopolyester: PHB	Shaking flask cultivation; 0.4 g PHB per g CDM; 0.95 g/L PHB; 0.02 g/(L·h)	[39]
Halomonas campaniensis LS21	Sludge and plant residues collected from a salt lake in PR China	Moderately halophile and alkaliphile: 27 g/L NaCl and pH-value 10; Alkaline seawater and artificial carbonaceous kitchen waste	Homopolyester: PHB	Open cultivation for 65 days 26% PHA in CDM (wild type strain), 70% PHA in CDM (engineered strain)	[169]
Yangia sp. ND199	Mangrove forest soil	Moderate halophile: 45 g/L NaCl; Fructose, glycerol, sugarcane molasses, and corn syrup)	Homopolyester PHB (glucose, fructose, glycerol, sugarcane molasses, and corn syrup; glutamate as N– source) PHBHV (glycerol plus yeast extract as N– source)	3 L bioreactor cultivations 53% PHA in CDM, 5.7 g/L PHA	[200]

(Continued)

TABLE 4.1 (CONTINUED)
Overview of Selected PHA Production Processes by Halophiles

Microbial species (Type of microbe)	Origin of the microorganism	Salinity and applied carbon source(s)	Monomeric composition of produced PHA	Production scale, feeding regime and output (product concentration and volumetric productivity)	Ref.
"	"	Moderate halophile: 30 g/L NaCl; Fructose		10 L Bioreactor cultivations; batch and fed-batch feeding Batch: 4 g/L PHA, 0.49 g/g PHA in CDM, 0.12 g/(L·h) Fed-batch: 20 g/L PHA, 0.68 g/g PHA in CDM, 1 g/(L·h)	[201]
Halomonas bluephagenesis TD01 (wild type)	Aydingkol salt lake, PR China	Halophile and alkaliphile: 60 g/L NaCl and pH-value 9; Glucose	Homopolyester: PHB	Open continuous fed-batch cultivation process for 56 h; 64 g/L PHA	[170]
Halomonas bluephagenesis (genetically engineered)	"	Halophile: 60 g/L NaCl; Glucose, propionic acid	Homopolyester PHB and copolyester PHBHV	Open continuous cultivation process for 56 h; 80 g/L PHB, 56 g/L PHBHV	[172]
Halomonas bluephagenesis (genetically engineered)	"	Halophile: 60 g/L NaCl; Glucose, GBL as 4HB precursor	Copolyester: Poly(3HB-*co*-4HB)	Open continuous cultivation process in 1 m³ vessel 51 g/L PHA, 0.61 g PHA per g CDM; 1.04 g/(L·h) PHA	[175]
Halomonas bluephagenesis (genetically engineered)	"	Halophile and alkaliphile: 60 g/L NaCl and pH-value 9; Waste gluconate and GBL as 4HB precursor	Copolyester: Poly(3HB-*co*-14%-4HB)	5 m³ bioreactor 100 g/L CDM, 0.6 g PHA per g CDM; 1.67 g/(L·h)	[176]

(Continued)

TABLE 4.1 (CONTINUED)
Overview of Selected PHA Production Processes by Halophiles

Microbial species (Type of microbe)	Origin of the microorganism	Salinity and applied carbon source(s)	Monomeric composition of produced PHA	Production scale, feeding regime and output (product concentration and volumetric productivity)	Ref.
Halomonas bluephagenesis (genetically engineered)	"	Halophile: 60 g/L NaCl; Glucose, GBL as 4HB precursor	Poly(3HB-*co*-25%-4HB)	Open continuous cultivation process in 1 m³ vessel 51 g/L PHA, 0.61 g PHA per g CDM; 1.04 g/(L·h) PHA	[178]
H. bluephagenesis TDHCD-R₃-8⁻₃ (genetically engineered)	"	Halophile and alkaliphile: 50 g/L NaCl and pH-value 8.5; Glucose	Homopolyester: PHB	7 L bioreactor cultivation 71 g/L PHA, 0.79 g/g PHA in CDM	[179]
Halomonas bluephagenesis (genetically engineered)	"	Halophile: 60 g/L NaCl; Glucose, acetate	Copolyesters: Poly(3HB-*co*-4HB) and PHBHV	7.5 L bioreactor; fed-batch feeding Up to 0.94 g/g PHA in CDM; up to 0.08 mol/mol 3HV in PHBHV, up to 0.12 mol/mol 4HB in Poly(3HB-*co*-4HB)	[180]
Halomonas halophila	Salt pond located at the Costa Blanca near Alicante, Spain	Halophile: 66 g/L NaCl; 20 g/L Glucose; in addition: saccharose, galactose, xylose and mannose converted. Different salinities tested	Homopolyester: PHB	Shaking flask cultivation; batch feeding; 0.61 g PHA per g CDM	[12]

(Continued)

TABLE 4.1 (CONTINUED)
Overview of Selected PHA Production Processes by Halophiles

Microbial species (Type of microbe)	Origin of the microorganism	Salinity and applied carbon source(s)	Monomeric composition of produced PHA	Production scale, feeding regime and output (product concentration and volumetric productivity)	Ref.
"	"	Halophile: 66 g/L NaCl; Hydrolyzed spent coffee grounds	Homopolyester: PHB	Shaking flask cultivation; batch feeding; 1 g/L PHB, 0.27 g PHA per g CDM, 14 mg/(L·h)	[187]
Halomonas hydrothermalis	West coast of India	Moderate halophile: 81 g/L Waste frying oil (WFO) WFO plus precursors	Homopolyester: PHB Copolyester: PHBHV (3HV up to 50 mol-%)	Shaking flask cultivation; batch feeding; 2.26 g/L PHB, 0.62 g/g PHB in CDM 1.61 g/L PHBHV, 0.69 g/g PHB in CDM	[192]
Halomonas marina HMA 103	Orissa solar saltern, India	Extreme halophile; 100 g/L NaCl Glucose Fatty acids (e.g., valerate)	Homopolyester: PHB Copolyester: PHBHV	0.59 g/g PHB in CDM 0.8 g/g PHA in CDM (two step cultivation)	[186]
Halomonas neptuinia	Deep-sea hydrothermal-vent environments	Moderate halophile: 81 g/L WFO	Homopolyester: PHB	Shaking flask cultivation; batch feeding; 0.67 g/L PHB, 0.23 g/g PHB in CDM	[192]
Halomonas venusta KT832796	Samples collected on the Indian coast	Moderate halophile: 15 g/L NaCl Glucose	Homopolyester: PHB	2 L bioreactor; fed-batch high concentration pulse feeding strategy (pH-stat or single pulse feeding, respectively) 33.4 g/L PHA, 0.88 g PHA per g CDM	[166]

(Continued)

TABLE 4.1 (CONTINUED)
Overview of Selected PHA Production Processes by Halophiles

Microbial species (Type of microbe)	Origin of the microorganism	Salinity and applied carbon source(s)	Monomeric composition of produced PHA	Production scale, feeding regime and output (product concentration and volumetric productivity)	Ref.
Halomonas profundus	Isolated from a shrimp collected next to a hydrothermal deep-sea vent field of the Mid Atlantic ridge	Moderate halophile: 27 g/L sea salts	Homopolyester PHB and Copolyester PHBHV when adding precursors	5 L bioreactor (batch feeding, two-stage process) Up to 0.27 g/L PHA and 0.8–0.9 g PHA per g CDM	[167]
Halomonas salina	Hypersaline microbial mat at Exportadora de Sal (ESSA) saltworks at Guerrero Negro, Baja California Sur, Mexico	Moderate halophile: 25 g/L NaCl Glucose	Homopolyester: PHB	Shaking flask cultivation: batch feeding 0.6 g/L PHB, 0.26 g/g PHB in CDM	[191]
Halomonas smyrnensis	Çamalti Saltern area (Aegean Region of Turkey)	Extreme halophile: 137.2 g/L sea salts Sucrose, glucose	Homopolyester PHB	Shaking flasks: 50 g/L sucrose: Up to 0.48 g/L PHB, 27% PHB in CDM, 0.003 g/(L·h) 20 g/L glucose: Up to 1.34 g/L PHB, 46% PHB in CDM, 0.01 g/(L·h)	[188]

(Continued)

TABLE 4.1 (CONTINUED)
Overview of Selected PHA Production Processes by Halophiles

Microbial species (Type of microbe)	Origin of the microorganism	Salinity and applied carbon source(s)	Monomeric composition of produced PHA	Production scale, feeding regime and output (product concentration and volumetric productivity)	Ref.
Rhodobacter sphaeroides U7 (mutant strain of strain *R. sphaeroides* ES16)	n.r.	Moderate halophile: 30 g/L NaCl	Homopolyester PHB and copolyester PHBHV when adding valerate	5 L bioreactor; aerobic-dark condition with different aeration rates, batch feeding Up to 2.5 g/L PHA and 0.65 g PHA per g CDM	[194]
Paracoccus sp. strain LL1	Soil and water samples collected from Lonar crater, Maharashtra, India	Slight halophile: 10 g/L NaCl Fungal hydrolyzed corn stover (glucose, xylose, cellobiose as main carbon substrates)	Homopolyester PHB	5 L bioreactor, batch cultivation; 10 g/L PHA and 0.72 g/g PHA in CDM, 0.14 g/(L·h)	[199]
"	"	Slight halophile: 10 g/L NaCl Glycerol	Copolyester PHBHV (up to 11 mol-% 3HV in PHBHV) (contrast to [197]!)	5 L bioreactor, batch cultivation: up to 3.77 g/L PHA, 0.04 g/(L·h), and 0.43 g/g PHA in CDM 5 L bioreactor with cell retention: up to 9.5 g/L PHA, 0.08 g/(L·h), and 0.39 g/g PHA in CDM	[200]
"	"	Slight halophile: 10 g/L NaCl Waste cooking oil, Tween 80 as surfactant	Copolyester PHBHV (3 mol-% 3HV in PHBHV)	2.5 L bioreactor, batch cultivation; 1 g/L PHA, 0.31 g/g PHA in CDM, 0.01 g/L·h)	[201]

(Continued)

TABLE 4.1 (CONTINUED)
Overview of Selected PHA Production Processes by Halophiles

Microbial species (Type of microbe)	Origin of the microorganism	Salinity and applied carbon source(s)	Monomeric composition of produced PHA	Production scale, feeding regime and output (product concentration and volumetric productivity)	Ref.
Vibrio proteolyticus	Seashore of the Korean peninsula	Halotolerant: 50 g/L NaCl Fructose	Homopolyester PHB and copolyester PHBHV when adding propionate	Shaking flask scale, batch cultivation. 2.7 g/L PHA, 54.7% PHA in CDM, 0.06 g/(L·h) Unsterile medium	[203]
Salinivibrio sp. M318	Fermenting shrimp paste in the Nam Dinh province, Vietnam	Moderate halophile: 30 g/L NaCl Carbon source: mixture of waste fish oil and glycerol; Nitrogen source: waste fish sauce	Homopolyester PHB with addition of precursors; production of PHBHV and PHB4HB copolyesters after precursor addition	Bioreactor, batch feeding: 10.7 g/L PHA, 0.57 g/g PHA in CDM, 0.22 g/(L·h) Bioreactor, fed-batch feeding: 35.6 g/L PHA, 0.515 g/g PHA in CDM, 0.46 g/(L·h)	[202]
Spirulina subsalsa	Gujarat coast samples, India	Moderate halophile: 25–50 g/L NaCl	Homopolyester: PHB	Shaking flask cultivations; batch feeding 25 g/L NaCl: 0.28 g/L PHA, 0.13 g/g PHA in CDM 50 g/L NaCl: 0.29 g/L PHA, 0.15 g/g PHA in CDM	[160]
Eubacteria (Gram-positive)					
Bacillus megaterium uyuni S29	Uyuni salt lake, Bolivia	Halotolerant: 45 g/L NaCl; Glucose	Homopolyester: PHB	Shaking flask cultivations; batch feeding 2.2 g/L PHB, 0.10 g/(L·h)	[215]

(Continued)

TABLE 4.1 (CONTINUED)
Overview of Selected PHA Production Processes by Halophiles

Microbial species (Type of microbe)	Origin of the microorganism	Salinity and applied carbon source(s)	Monomeric composition of produced PHA	Production scale, feeding regime and output (product concentration and volumetric productivity)	Ref.
"	"	Halotolerant: optimum 45 g/L NaCl; growth even at 100 g/L Glucose	Homopolyester: PHB	Shaking flask cultivations; batch feeding 5 g/L: max. ~1 g/L PHB, 0.33 g PHB per g CDM, 0.06 g/(L·h) 45 g/L NaCl: max.~2 g/L PHB, 0.41 g PHB per g CDM, 0.10 g/(L·h) 100 g/L: max. ~0.72 g/L PHB, 0.22 g PHB per g CDM, 0.03 g/(L·h)	[9]
"	"	Moderate halophile: 10 g/L NaCl; Desugarized sugar beet molasses, phosphate-limited	Homopolyester: PHB	Shaking flask cultivations; batch feeding 16.7 g/L PHB; 0.6 g/g PHB in CDM; 0.42 g/(L·h)	[218]
Bacillus megaterium H16	Solar salterns in Goa, India	Halotolerant: 50 g/L NaCl; Glucose	Homopolyester: PHB	Shaking flask cultivations; 0.39 g PHB per g CDM	[219]
Bacillus MG12	Soil samples from a mangrove area in Kannur district, North Kerala, India	Halotolerant: up to 90 g/L NaCl, cultivations carried out with 5 g/L Carbohydrates, glycerol, organic acids, hydrolyzed bagasse waste	PHBHV copolyesters via propionyl-CoA dependent and independent pathways; PHB homopolyester on acetate as sole substrate	Shaking flask scale, batch cultivation. 2.7 g/L PHA, 54.7% PHA in CDM, 0.06 g/(L·h)	[220]

CGP: Crude glycerol phase; GBL: γ-butyrolactone

4.2.1 DEFINITIONS TO CATEGORIZE HALOPHILES

The term "halophilicity" describes the adaptation of different life forms to habitats that are characterized by elevated salt concentrations. Halophile organisms are a phylogenetically diversified and highly heterogenous group of life forms, which require or prefer hypersaline media in which NaCl usually acts as the main salt constituent. The most common definition of "halophiles" by Oren defines such species to thrive optimally at NaCl concentrations of at least 50 g/L, and to tolerate at least 100 g/L [15]. More precisely, five different groups of halophilic microbes are distinguished in the literature based on the salt requirement and the salt level to which they are tolerant. The first group encompasses "halotolerant" microbes, which are insensitive towards up to 15% (2.5 M) NaCl but are not dependent on such high salinity (prototype organisms: halotolerant *Bacilli*; *cf.* Section 4.5 of this chapter). In contrast, "slightly halophile" and "moderately halophile" as the second and third group not only tolerate, but even require 1–3% (~0.3–1 M) or 3–15% (~1–2.5 M), respectively, NaCl to thrive well (prototype organisms: *Halomonas* sp.). The group of "borderline extreme halophiles" prospers best in media containing 1.5–4.0 M NaCl; prototype organisms of this group are members of the *Halobacteriaceae* and *Haloferacaceae* families. Finally, the fifth group, the so-called "extreme halophilic" microbes, require excessive salinity of 15–30% (2.5– 5 M NaCl) (reviewed by [9, 16, 17]). In contrast, well-established biotechnologically important microbes frequently used for PHA biosynthesis, such as the best-described PHA producer *C. necator*, were shown to be highly sensitive against elevated salt concentrations; as demonstrated by Mozumder and colleagues, biomass growth and PHA biosynthesis by *C. necator* were completely stopped at sodium ion concentrations in the medium in the range of 1 wt.-%. In fed-batch bioreactor cultivations, these authors revealed that prolonged biomass growth phases, characterized by the steady addition of NaOH as pH-corrective, results in intracellular sodium enrichment, and subsequently decreased product formation rates in the PHA accumulation phase [18]. However, reports on the impact of NaCl on PHA biosynthesis by osmo-mesophilic strains are somewhat contradictory. In contrast to the studies by Mozumder et al. [18], Obruca and colleagues have observed that mild salt stress even induces PHA biosynthesis in *C. necator* [19]; similar results were later obtained by Passanha and colleagues, who described the viability of NaCl addition as a tool to improve polyhydroxyalkanoate production by *C. necator* [20]. A first comprehensive review on PHA biosynthesis by halophiles was provided in 2010 by Quillaguamán and colleagues, who already differentiated the mechanisms involved in PHA production by haloarchaea on the one hand, and by halophilic eubacteria on the other [21]. Figure 4.1 illustrates the classification of microbes in terms of their halophilicity.

4.2.2 ADAPTATION STRATEGIES OF HALOPHILES TO COPE WITH HIGH SALINITY

4.2.2.1 Adaptation of Proteins and DNA to Salinity

Halophilicity occurs throughout the biosphere's entire lifecycle and is observed both among prokaryotes (*Archaea* and *Bacteria*) and *Eukarya*. Among halophiles, there are several strategies to compensate for osmotic pressure induced by high environmental

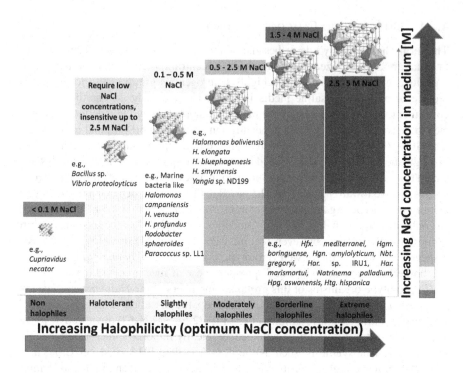

FIGURE 4.1 Schematic classification of PHA-accumulating microbes described in this chapter regarding their adaptation to salinity. NB: Transitions, especially between "slight halophiles" and "moderate halophiles" on the one hand, and "borderline halophiles" and "extreme halophiles", on the other hand, are often overlapping in literature.

salinity. Extreme halophiles accumulate a high concentration of KCl, therefore, the high intracellular salt concentration in extreme halophiles requires adaptation of universally conserved macromolecules like proteins and nucleic acids in order to enable functionality at excessive salinity. Hence, the genome of extreme halophiles is different from that of osmo-mesophiles (which do not accumulate KCl but small organic molecules called osmolytes to compensate osmotic pressure), and frequently needs particular editing, which is illustrated, e.g., by the high number of salt-resistant genes occurring in these microorganisms [22]. In addition, "halo-adaptation" of several functional proteins in haloarchaea has been reported, which makes them structurally compact and active exclusively at high salt concentrations. In this context, the genome of the haloarchaeon *Halobacterium* sp. NRC-1 (a.k.a. *Halobacterium salinarum* NRC-1) and its predicted proteome were analyzed by Kennedy and colleagues in 2001 by in silico techniques; these authors revealed adaptive features of the proteome, which is caused by the strain's life in extreme environments characterized by hypersalinity and excessive solar radiation. In particular, the proteome was highly acidic, mainly lacking basic proteins. This complies with a highly negative surface charge as the predominant adaptive mechanism of halo-proteins to make them work even in almost saturating salinity [23]. As shown by Bolhuis and colleagues, the genome of the haloarchaeon *Haloquadratum [Hqr.] walsbyi* displays a range of unique adaptive particularities that enable this organism to thrive in highly saline ecological niches [24]. In this context, the genome sequence of *Halobacterium* sp. NRC-1 also revealed several remarkable physical adaptations to highly saline environments, together with an acidic proteome, which is believed to be vital for circumventing salting-out of proteins in the hypersaline cytoplasm [23, 25]. Such a high surface negative charge of folded proteins as a tool to overcome the salting-out phenomenon in hypersaline cytoplasm was also described for the PHA-accumulating haloarchaeon *Haloarcula [Har.] marismortui* [26]. Mechanisms to accomplish this protein adaptation involve the formation of a surplus of acidic amino acids on the enzyme surface [27, 28]. As an example in this context, glucose dehydrogenase of the PHA-accumulating haloar- chaeon *Haloferax [Hfx.] mediterranei* has no flexible side chains on its surface, thus, it appears as a high-ordered multilayered solvation shell and is well adapted to resist attachment of ions [29, 30]. The genome of extreme halophiles is also different from that of osmo-mesophiles, and frequently needs particular editing, which is illustrated, e.g. by the high number of salt-resistant genes occurring in these microorganisms [22]. As shown by Bolhuis and colleagues, the genome of the haloarchaeon *Hqr. wals- byi* displays a range of unique adaptive particularities that enable this organism to thrive in highly saline ecological niches [24].

Hezayen and colleagues were the first researchers who reported on the particu- larities of haloarchaeal PHA synthase enzymes. Investigating "strain 56", an organ- ism isolated from Egyptian hypersaline soil samples near Aswan and cultivated with 250 g/L NaCl, these scientists discovered a covalently granule-bond PHA synthase enzyme. This new PHA synthase revealed particularities when compared to previously described PHA synthases of eubacteria; it showed thermostability up to a temperature of 60°C, and strongly increasing activity with increasing salinity, with special impact of Mg^{2+} ions. Moreover, this halophilic synthase showed a very narrow substrate spectrum, accepting only 3-hydroxybutyryl-CoA, but not 3-hydroxyvaleryl-CoA, 4-hydroxybutyryl-CoA, or 3-hydroxyhexanoyl-CoA. Most remarkably, no other PHB biosynthesis enzymes (3-ketothiolase, NADH/NADPH-dependent acetoacetyl–CoA reductase) as known from mesophilic PHA production strains were detected, which evidenced for the first time that haloarchaea might resort to a different metabolic route of PHA biosynthesis than known for eubacteria [31]. Years later, the organism, which forms motile, Gram-negative, pleomorphic pinkish rods, was categorized in depth and today is known as *Halopiger [Hpq.] aswanensis* DSM 13151. High salinity of 3.8–4.3 M NaCl, pH 7.5 (range: 6–9.2), and 40°C (maximum: 55°C) were reported as the optimum conditions for the cultivation of this extreme halophilic strain. Also, the rapid lysis of cells in distilled water was reported for this organism [32].

4.2.2.2 Pigment Formation in Halophilic PHA Producers

Such proteome adaptation and many other approaches developed by nature to help halophiles to get along with high salt loads impressively illustrate that this adapta- tion to high salinity is a widespread process among life forms, and displays a phe- nomenon well-optimized by nature (reviewed by [9, 16]). A well-known example is the adaptation of the pigment pattern of different microalgae strains as a reaction to fluctuating salinities [8]; analogously, in the case of the extreme halophilic PHA pro- ducer *Hfx. mediterranei*, overproduction of the pigment bacterioruberin, typically produced to cope with excessive UV-irradiation, was observed as a protective reaction to suboptimal salt concentrations, presumably for altering the flexibility of the cell membrane [33]. The feasibility of mass production of bacterioruberin, a C50 carotenoid *Hfx. mediterranei*, by triggering the salinity of the cultivation medium was later suggested by Chen et al., who obtained more than 0.5 g pigments per L of culture medium when cultivating the strain on inexpensive carbon sources extruded low-cost rice bran and corn starch at optimum salinity of 25 S/m (conductivity of the medium to quantify concentration of ions); these authors also confirmed the indirect relationship between salinity and pigment formation, and the direct relationship between salinity and PHA biosynthesis [34].

4.2.2.3   Solutes Formation by Halophilic PHA Producers

The most common approach for adaptation is based on the intracellular accumulation of dissolved compatible organic kosmotropic osmotic compounds (also called osmolytes) such as amino acids, betaine, ectoines, or trehalose; these solutes are accumulated to balance osmotic pressure. When using this strategy, it is not necessary for the microbes to adapt their enzymes specifically to the high salinity. The accumulation of these compounds occurs either intracellularly via biosynthesis, or they are taken up by the cells from the extracellular space. When the salinity/osmolarity of the environment drops, these compatible solutes are released from the cells' interior to balance inner- and outer-cellular osmotic pressure, which facilitates the isolation and purification of these products [35]. Because of their general protective role and activity as chemical chaperones, such compatible solutes find manifold uses in biotechnology, cosmetics (mainly skincare products), the food industry, or medicine [36].

In the context of halophilic PHA-synthesizing microbes, it is especially interesting to study how PHA production and formation of the described compatible solutes are interlinked by these organisms, and how these processes are associated with the organisms' strategies to cope with high external salinity. In 2006, Quillaguamán and colleagues noticed the sudden change of the morphology of PHA-accumulating *Halomonas boliviensis* LC1 cultures when challenged to sudden hypersalinity, and assumed a switch towards the formation of organic solutes to withstand the osmotic pressure prior to boosted PHA accumulation [37]. Later, Mothes et al. reported efficient parallel production of PHA and ectoines by the halobacterium *Halomonas elongata*. In the presence of 10% NaCl, the strain accumulated 0.5 g PHA per g biomass together with up to 14% of ectoines, which reinforced the validity of the concept of parallel PHA and ectoine formation. Similar findings were made in this study for parallel ectoine and poly(3-hydroxybutyrate) (PHB) accumulation for the strains *Halomonas halodenitrificans*, *Halomonas haloneurihalina*, and *Halomonas salina*. Mechanistically, it should be considered that the production of both products, ectoines and PHB, are favored by the same factors, namely nutrient limitation and osmotic stress, and that acetyl-CoA acts as common metabolite during biosynthesis of both products [38]. The closely related strain *Halomonas elongata* 2FF was tested by Cristea and colleagues in the presence of minimal media containing 0–25% NaCl and different carbon sources; the use of 10% NaCl and 10 g/L glucose turned out to result in the highest PHA productivity. After 48–72 h of cultivation, a

PHA biopolyester evidenced as PHB homopolyester via Fourier transform infrared spectroscopy (FT-IR), X-ray powder refraction analysis, Raman spectroscopy, proton nuclear magnetic resonance spectroscopy (^{1}H-NMR), and the spectroscopic crotonic acid assay according to Law and Slepecky, was obtained. Thermogravimetric analysis by differential scanning calorimetry (DSC) revealed a melting point T_m of this material of 162.5°C. The mass fraction of PHB in cell dry mass (CDM), PHB concentration, and volumetric productivity after 48 h amounted to 0.4 g/g, 0.95 g/L, and 0.02 g/(L·h), respectively [39]. Similar to *Halomonas elongata* ssp., Guzmán and colleagues investigated the parallel biosynthesis of PHA and ectoines in two fed-batch cultivation setups, using the bacterium *Halomonas boliviensis*. In the first cultivation, performed at 45 g/L of NaCl without nitrogen or phosphate limitation, the authors aimed at obtaining high biomass density, while in the second cultivation, the salinity of the medium was increased to 75 g/L NaCl in order to boost ectoine biosynthesis; here, nitrogen- and phosphate sources were supplied exclusively during the initial phase of the cultivation, and were later depleted in order to provoke conditions enhancing PHB formation. A PHB fraction in biomass of about 0.96 g/g, a volumetric productivity of about 1 g/(L·h), and an ectoine concentration and content of 4.3 g/L and 0.072 g/g, respectively, were achieved by this process [40]. Further, it was shown that the presence of PHA granules in cells exerted a protective function against hyperosmotic shock, as originally demonstrated by directly comparing the reaction of the halotolerant Gram-negative PHA-producer *Pseudomonas* sp. CT13 and PHA accumulating mutant strain to hypertonic up-shock. By this study carried out by Soto and colleagues, it was shown that the presence of PHA granules raises the intracellular level of PHB's monomer 3-hydroxybutyrate (3HB), which, in turn, inhibits protein agglomeration as a typical fatal consequence of exposure to high salt concentration and elevated temperature. As evidenced by boosted PHA productivity in parallel to increasing salinity, PHA and 3HB act as compatible solutes, which are essential for the bacteria to counterattack the hypertonic stress [41]. Obruca and coworkers recently confirmed and substantiated these findings by elucidating 3HB's outstanding chaperoning efficiency; it turned out that the protective effect of 3HB is competitive with that exerted by established compatible solutes like hydroxyectoine or trehalose. Mechanistically, PHA granules, water insoluble and highly polymerized in their nature and therefore not increasing the intracellular osmotic pressure, act as a pool of water-soluble compatible compounds, to which the cells can resort under hyperosmotic challenge and other environmental stress factors [42]. Later, it was shown by these authors that the presence of PHA granules in bacteria protects them from substantial plasmolysis, massive damage of the cytoplasmic membrane, and thus helps cells to prevent loss of cellular integrity under conditions of hypersalinity [43]. Therefore, it is very likely that a major reason for high amounts of PHA being so frequently found among halophilic organisms is that PHA biosynthesis constitutes a convenient strategy developed by them to manage osmotic up-shock [43].

4.2.2.4 Salt-In Strategy

The described accumulation of compatible solutes, a typical feature of halophilic eubacteria and *Eukarya*, is different from the alternative "salt-in" approach, which describes the intracellular accumulation of salts like KCl; this strategy is a typical feature of haloarchaea. In any case, this "salt-in" strategy requires the adaptation of enzymes by making them salt-resistant ("extremozymes", "halozymes") [27]. The changes in the intracellular concentrations of different ions during different phases of growth of a "*Halobacterium* sp." isolated from the Dead Sea, was described by Ginzburg and colleagues in 1970 [44]. According to later accomplished studies, this red-pigmented thermophilic strain (temperature optimum between 40 and 50°C), with forms of non-motile, disc-shaped cells, was a representative of the genus *Haloarcula [Har.]*, and classified as *Haloarcula marismortui* volcani [45]. Some haloarchaea, e.g. representatives of the genera *Natronococcus [Ncc.]*, *Natronobacterium [Nbt.]*, *Natrialba [Nab.]*, or *Natronomonas [Nmn.]*, are reported to replace part of KCl by 2-sulfotrehalose   (1-(2-O-sulfo-alpha-D-glucopyranosyl)-alpha-D-glycopyranose), when cultured under nutrient-limited conditions and high salinity [46].

4.2.2.5   Protective Role of PHA in Osmotic Up- and Down-Shock

The fact that PHA accumulation is a common feature among prokaryotes inhabiting high salinity environments, including extremely halotolerant *Archaea*, indicates that PHA biosynthesis could play a role in adaptation to osmotic pressure. As was mentioned above, it was proved that PHA protects non-halophilic bacteria against the harmful consequences of osmotic up-shock, since PHA-accumulating cells of *C. necator* were much more resistant to hypertonic challenge and demonstrated a substantially lower degree of plasmolysis when exposed to hypertonic environments than a PHA-negative mutant of *C. necator*. Massive plasmolysis results in damage to the cytoplasmic membrane and leakage of cytoplasmic content from the cells, therefore, PHA granules act as an internal scaffold and prevent bacterial cells from such harmful effects [43]. Apart from osmotic up-shock, PHA granules also provide a protective function against osmotic down-shock not only for the non-halophile *C. necator*, but also for the halophilic bacterium *Halomonas halophila*. The presence of PHA in bacterial cells substantially enhances their capability to retain cell integrity when suddenly exposed to osmotic down-shock. In the case of *C. necator*, the protective function of PHA granules was probably associated with a lower level of plasmolysis-induced cytoplasm membrane damage during osmotic up-shock, which enabled bacterial cells to survive following hypotonic shock. Unlike in PHA rich cells, a sudden induction of osmotic up- and subsequent down-shock resulted in massive hypotonic lysis of non-PHA-containing cells. In the case of the halophilic bacterium *Halomonas halophila*, challenged PHA-rich cells were capable of keeping cell integrity more effectively than their PHA-poor counterparts. *H. halophila* is a moderate halophile accumulating organic osmolytes to compensate for osmotic imbalances in extracellular and intracellular space. Under hyperosmotic conditions, intracellular concentration of these solutes can reach up to several moles per liter. Therefore, halophiles are much more sensitive to hypo-osmotic damage and hypotonic lysis than non-halophiles. Since hydrophobic PHA granules represent a substantial portion of the intracellular volume, the total amount of osmolytes per cell will be considerably lower in PHA-rich cells. Furthermore, it is very likely that the presence of PHA granules, due to their capability to protect bacterial cells from massive plasmolysis, reduces intracellular concentration of osmolytes. Hence, it seems that the presence of PHA granules provides numerous benefits to halophiles [47].

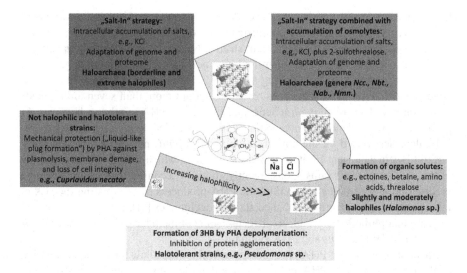

FIGURE 4.2 Adaptive strategies of halophilic PHA producers to cope with elevated salinity, and protective mechanisms provided to halotolerant strains based on intracellular presence of PHA.

Figure 4.2 illustrates adaptation strategies of halophilic PHA producers to cope with elevated salinity, and protective mechanisms provided to halotolerant strains based on the intracellular presence of PHA.

4.2.3 TECHNIQUES TO SPOT HALOPHILIC PHA PRODUCTION STRAINS

Various techniques were used in the past to detect PHA granules in new isolated microorganisms (reviewed by [48]). However, in the context of halophilic microbes, most established methods to identify PHA inclusions are not precise enough, and in many cases falsely show positive results. In this context, Nile red staining, the most frequently used staining technique to rapidly visualize PHA inclusions, also stains lipophilic inclusions other than PHA, e.g., wax esters or oil droplets; moreover, this method needs high intracellular PHA fractions to give positive results [49]. Only recently, a study by Fathima and colleagues demonstrated that a range of established tests is needed to unambiguously identify halophilic PHA producers by established techniques; these authors tested new isolates by Sudan black staining, the fluorescence dye Nile blue A, the spectroscopic determination after conversion to crotonic acid, and by FT-IR, before PHA producers were finally spotted and subjected to cultivation studies [50]. In 2015, Mahansaria and colleagues developed a new genetic method in order to better detect such strains, which was based on amplifying the *phaC* gene region that is highly conserved in halophilic PHA producing strains. This region is about 280–300 bp in size and encodes the Class-III PHA synthase present in many, albeit not all, halophiles [51]. The gene sequences *codehopCF* and *codehopCR*, originally developed in 2010 by Han and colleagues [52], acted as primers for the amplification process. This new method was tested using nine already known haloarchaea and halobacteria plus 28 new halophilic organisms isolated at the Indian coast. Twenty-eight of these 37 organisms showed *phaC*-positive, and eight strains *phaC*-negative response; for these eight strains, Nile red staining falsely gave positive results. 16S rRNA analysis identified nine new haloarchaea and nine new halobacteria as potential PHA-producing strains. Multiple sequence alignment of *phaC* gene-derived amino acid sequences revealed that not more than seven amino acids were conserved in all four *phaC*-encoded classes of PHA synthase; however, 61 amino acids were identical amongst the synthase encoded by *phaC* specific to studied halophiles. Based on the amplification test, all *phaC*-positive strains also produced PHA in a nutrient-limited cultivation medium, which was not the case for any of the *phaC*-negative strains. Consequently, this new technique can be considered a precise option for tracing new halophilic PHA producers by eliminating the risk of false positive results from established staining techniques [51]. However, it does not trace all halophilic PHA producers, because some of them have Class I (*Halomonas* sp.) or Class IV (*Bacillus* sp.) PHA synthase.

In addition to tracing PHA producers, new techniques allow for the determination of the concentration of PHA produced by halophiles during running bioprocesses [48]. In this context, a flow cytometry (FC) method was developed by García-Torreiro for the determination of the PHB content in *Halomonas boliviensis* LC1, isolated in the Bolivian Andes, in a fast and convenient fashion, which allows for a real time process control. This halophilic organism was stained with the fluorescence dye BODIPY 493/503 and analyzed by FC. In two-liter bioreactor cultivations with 45 g/L NaCl, PHB biosynthesis was induced by two different nutrient limitation regimes (nitrogen limitation and nitrogen limitation with parallel limited oxygen availability), and was monitored by established gas chromatography (GC) analysis and by FC. A high correlation between the data for PHB concentration obtained by the two methods was observed. Additionally, the FC technique turned out to be a viable tool to measure the heterogeneity of the microbial population regarding the distribution of the PHB content among individual cells; in this study, PHA contents in individual cells fluctuated between 0 and almost 0.9 g/g [53].

4.2.4 USING EXTREME HALOPHILES – CHALLENGES FOR THE BIOREACTOR EQUIPMENT

The high salinity of the cultivation media used for *Hfx. mediterranei* fermentation batches, which amounts to 20–25 g/L NaCl, considerably challenges the resistance of the bioreactor equipment and accompanying measuring electronics. Steel of the highest quality, borosilicate glass, ceramics, or high-performance polymers, e.g. poly(ethylene ether ketone) (PEEK), are possible materials to be used for designing bioreactors suitable for extremely halophile organisms to thrive [54]. Hezayen and colleagues constructed a corrosion-resistant bioreactor composed of PEEK, tech glass, and silicium nitrite ceramics for the cultivation of two new haloarchaeal isolates, aiming at production of PHB (using *Hpg. aswanensis* DSM 13151 "strain 56" at more than 200 g/L NaCl) and poly(γ-glutamic acid) (using *Natrialba* [*Nab.*] sp. at more than 200 g/L NaCl) as target products. PHB production by *Hpg. aswanensis* after 12 days batch cultivation on acetate and n-butyric acid as carbon sources amounted to 4.6 g/L and an intracellular PHB fraction of 0.53 g per g biomass [55]. Later, a hybrid bioreactor consisting of borosilicate glass and PEEK was designed by the company Labfors (CH) as a lab-scale bioreactor for the cultivation of extreme halophiles. Lorantfy and colleagues used this corrosion-resistant device to culture *Hfx. mediterranei* on diverse substrates (acetate, lactate, and glycerol); different cultivation regimes were used, together with a new technique for online biomass determination, which was based on the consumption rates for acid and hydroxide. In this study, it was shown that the carbon sources are consumed by the strain in a diauxic mode; moreover, kinetic process analysis revealed that growth of *Hfx. mediterranei* occurs in accordance with the Monod equations [54].

For pilot-scale cultivation of extreme halophilic PHA producers, a corrosion-resistant bubble column bioreactor was recently introduced by Mahler and associates. This bioreactor, 1.1 m in height and about 13 cm in diameter, was developed by an Austrian company and operated with a working volume of 15 L; the vessel was made of corrosion-resistant nickel-molybdenum-based alloy Hastelloy. Circulation of the fermentation broth was accomplished by aeration at a flow rate of 105 L/h. Moreover, a cell retention system based on a tangential flow microfiltration module with a poly(sulfone) membrane was implemented to obtain drastically higher volumetric productivities than normally achievable by the organism's metabolic potential. By direct comparison, this bubble column bioreactor design enables processing of higher energy efficiency than possible with typically applied stirred tank bioreactors, as evidenced by assessing physiological and hydro-dynamical parameters. In a cultivation medium containing 150 g/L NaCl, the maximum $k_L a$ value and oxygen transfer rate (OTR) in the bubble column at about 1 bar amounted to 84 L/h and 6 mmol O_2/(L·h), respectively. To reach the same OTR in a conventional stirred tank bioreactor studied in parallel, three times the energy input was needed. By this setup, a stable, continuous production process was operated with *Hfx. mediterranei* as the production strain. The process conditions were adjusted to mimic a process using brine with low glycerol concentration (0.27 g/L) operated at a high dilution rate (0.37 L/h), which surpasses $\mu_{max.}$ obtained in the stirred tank reactor by five times. Steady state conditions for biomass formation were maintained for 12 retention times (about 32 h). Unfortunately, no results were reported for PHA productivity. Most importantly, it was demonstrated that saline industrial waste streams can be used without pretreatment as a salt source in halophilic PHA production processes [56].

4.3 PHA PRODUCTION BY HALOPHILIC ARCHAEA ("HALOARCHAEA")

4.3.1 General Features of Haloarchaea

4.3.1.1 First Description of Haloarchaeal PHA Production

Halobacterium halobium, a strain isolated from the Dead Sea and first described in the 1930s, was the first haloarchaeon which revealed PHA accumulation when observed under the microscope. Similar findings were described in 1972 by Kirk and Ginzburg, who carried out morphological studies of another organism isolated from the Dead Sea by the freeze-fracture method, and noticed the presence of PHB inclusions, which were extracted by the authors and analyzed by X-ray diffractometry. Unfortunately, no quantitative data for PHB biosynthesis were reported. The authors suggested the classification of this organism as a species of *"Halobacterium marismortui"* or *"Halobacterium trapanicum"* according to the taxonomic nomenclature valid in those days [57]. Later publications classify this isolate as *Har. marismortui* [58].

4.3.1.2   Definition and Particular Products of "Haloarchaea"

Haloarchaea, also called "halophilic archaea" or, previously, "halophilic archaebacteria", represent a particular class within the phylum of the Euryarchaeota. Haloarchaea are typically found in aquatic habitats highly saturated with salt [59]. Importantly, they are classified as members of the *Archaea* domain, and should no longer be merged with the eubacterial group of "halobacteria", which was introduced into the scientific literature before the existence of the two different prokaryotic domains *Bacteria* and *Archaea* was established in science. Nowadays, microbiologists call halophilic archaea "haloarchaea" in order to unambiguously distinguish them from halophilic eubacteria. Metabolites produced by haloarchaeal species are typically stable in conditions of high salinity and temperature, which makes them useful for special industrial applications [60]. Haloarchaeal proteins and enzymes function at such salinity levels at which their eubacterial equivalents are no longer active, which makes haloarchaeal enzymes appropriate for salt-challenged processes and applications under dehydration conditions [61–63]. As an example, bacteriorhodopsin, a light-driven photon-pump found, e.g., in *Halobacterium halobium* [64, 65], or the purple membrane protein present in various haloarchaea [66, 67] are expected to become widely implemented in artificial retinas, photoelectric devices, or holograms. Moreover, thanks to their gas vesicles [68, 69] and S-layer glycoproteins [70], haloarchaea have exceptional potential to be utilized as drug delivery vehicles and as workhorses in the emerging field of nanobiotechnology [60]. Haloarchaea can also be implemented in bioremediation of hypersaline environments [71, 72] contaminated by, e.g. aromatic hydrocarbons [73], chlorate [74], or crude oil [75, 76]. PHA and extracellular polysaccharides (EPS) produced by such microbes are biodegradable, compostable, and biocompatible, and therefore have the potential to substitute established recalcitrant plastics and other polymers [60].

In any case, haloarchaea, as halophilic microorganisms, necessarily need elevated salt concentrations to thrive; the majority among them require more than 2.0 M NaCl for survival and cell multiplication. In most cases, haloarchaea live under aerobic conditions, with some exceptions, which resort to nitrate as terminal electron acceptors and are capable of denitrification (reviewed by [16]). Interestingly, haloarchaea are a special evolutionary subdivision among *Archaea*; they are characterized by the presence of lipids linked by ether groups and the lack of bacteria-specific murein layers in the cell wall. From the application-oriented point of view, haloarchaea produce a colorful range of marketable products as a reaction of their primary and secondary metabolism to their extreme cultivation conditions (reviewed by [16]). Besides PHA (reviewed by [77]), special polysaccharides [78, 79], halocins (bacteriocins synthesized by halophiles) [80, 81], or pigments [82, 83] are typical products of haloarchaea. Only recently, it was elucidated that haloarchaea also reveal particular resistance to heavy metal contaminated environments, as demonstrated for the high zinc resistance of members of the genera *Haloarcula [Har.]*, *Haloferax [Hfx.]*, *Halococcus [Hcc.]*, and *Halorubrum [Hrr.]* [84].

4.3.1.3 Genomic Studies Related to Haloarchaeal PHA Biosynthesis

In general, PHA synthase enzymes in haloarchaea belong to the Class III PHA synthases and are clustered separately [reviewed by 21]. To better understand and optimize PHA production by haloarchaea, genomic studies are needed. In this context, the haloarchaeal genes encoding for homologs of Class III PhaC synthase enzyme, which catalyze hydroxyacyl-CoA polymerization to PHA in bacteria like *Allochromatium vinosum*, were identified in the genomes of the haloarchaea *Har. marismortui* ATCC 43049 [27] and *Hqr. walsbyi* DSM 16790 [24]. In 2007, Han and associates investigated experimentally the expression profile of genes encoding haloarchaeal PHA synthesis enzymes for the first time [85]. *Har. marismortui*, a haloarchaeon isolated from the Dead Sea, was shown in this study to store PHB up to 0.21 g/g in biomass when cultivated in saline-minimal medium with high amounts of glucose as a sole carbon substrate. As a major outcome of this study, the genes $phaE_{Hm}$ and $phaC_{Hm}$ were identified by molecular characterization of the $phaEC_{Hm}$ operon; these two neighboring genes encode two Class III PHA synthase subunits and are directed by only one single promoter. It was shown by the authors that these genes are constitutively expressed, both under nutritionally balanced conditions in the microbial growth phase, and under nutrient limited conditions. Remarkably, in contrast to the non-granule associated $PhaE_{Hm}$ gene, $PhaC_{Hm}$ is strongly connected to the PHA granules. When $phaE_{Hm}$ or $phaC_{Hm}$ genes are transferred into the haloarchaeon *Haloarcula hispanica* (formerly known as "*Halobacterium hispanicum*"), which harbors highly homolog $phaEC_{Hh}$ genes, PHB synthesis considerably increased. Particularly co-expression of both genes results in best results for PHB production; the other way around, the knock-out of genes in *Har. hispanica* leads to complete termination of PHA synthesis. When $phaEC_{Hm}$ genes are transferred into such $phaEC_{Hh}$-knockout mutants, PHA synthase activity and PHA biosynthesis are fully restored. For the first time, it was demonstrated by these studies that *phaEC* genes are of high significance for PHA biosynthesis by haloarchaea [85]. In a similar way, Ding and coworkers reported on the sequencing of the *Har. hispanica* ssp. genome [86]. Surprisingly, substantial differences were noticed when comparing with the gene sequence of *Har. hispanica* ATCC 33960, the model strain used for above described molecular characterization studies by Han et al. [85]. Hence, the elucidation of the genomic background of haloarchaeal PHA synthesis still offers plenty of room for future studies.

4.3.2 *HALOFERAX MEDITERRANEI* – THE PROTOTYPE HALOARCHAEAL PHA PRODUCER

4.3.2.1 Isolation of *Hfx. mediterranei* and General Features

The first haloarchaeon, which was investigated in detail in order to elucidate its PHA-accumulation kinetics was *Hfx. mediterranei*, one out of 19 organisms originally enriched by Rodriguez-Valera and colleagues in a continuous cultivation process inoculated with samples collected on the Spanish coast from ponds of solar salterns [87]. As the most intriguing among these new isolates, the strain R-4 was labeled by the researchers as "*Halobacterium mediterranei*". The strain attracted attention due to its significantly different behavior to halobacteria known at that time; it turned out to use several defined substrates as sources of carbon and energy, display high hydrolytic activity to convert starch or lipids, have a thicker cell envelope than that described for halobacteria, and lack a peptioglucane layer (murein) in the cell wall. Already during these early studies, it was revealed that coloration of colonies of this organism decreases with increasing salinity [87]. In 1986, Fernandez-Castillo and colleagues identified PHA inclusions in this organism when investigating it together with other halophilic isolates, namely *Halobacterium volcanii*, *Halobacterium halobium*, *Halobacterium gibonsii*, and *Halobacterium hispanicum* in cultivations carried out in 10-liter aerated, magnetically stirred glass vessels [88]; these organisms are currently known as *Haloferax volcanii*, *Haloarcula marismortui*, *Haloferax gibbonsii*, and *Haloarcula hispanica* due to the new numerical taxonomy based on the composition of polar lipids of "halobacteria": in 1986, the old genus *Halobacterium* was split into genera *Halobacterium*, *Haloarcula*, and *Haloferax*, with the strain "*Halobacterium mediterranei*" (isolate R-4) being re-classified as "*Haloferax mediterranei*" [89]. Surprisingly, Fernandez-Castillo and colleagues reported the production of PHB homopolyester by all of these isolates, although ^{13}C-NMR measurements were carried out [88]. According to the present knowledge, at least *Hfx. mediterranei* should have had produced PHBHV copolyester under the applied cultivation conditions (excess availability of glucose or glycerol as carbon and energy sources). For *Hfx. mediterranei*, higher PHA fractions in biomass (0.17 g/g) were observed when using glucose than in the case of glycerol, citrate, or cellobiose. Only modest PHA accumulation was observed for the other isolates when cultivated in a medium containing 25% NaCl, 10 g/L glucose, and 1 g/L yeast extract; 0.07, 0.012, and 0.024 g PHA per g CDM were obtained for *Hfx. volcanii*, *Hfx. gibbonsii*, and *Har. hispanica*, respectively. Importantly, this study describes for the first time in detail the hypotonic disrupture of haloarchaeal cells for facilitated PHA recovery [88].

Antón and colleagues discovered that this strain produces an extracellular polymer, which can be precipitated by cold ethanol, and consists of at least mannose, glucose, and galactose building blocks. Moreover, these authors described that the organism is optimally cultivated in nutrient media containing 200 g/L and more of NaCl, which excludes the overgrowth of culture by foreign organisms [90]. This high stability of cultivation batches of *Hfx. mediterranei* was confirmed later by Hermann-Krauss and colleagues, who ran cultures of this strain for several days without observing any sterilization measures; neither the bioreactor nor the cultivation medium was heat sterilized. According to microscopic control, microbial contamination did not occur [91].

Later, it was revealed that *Hfx. mediterranei* harbors high intracellular amounts of KCl, which demonstrates the mechanistic background of the organism's strategy to cope with high salt loads [92]; this is a typical "salt-in" strategy followed by haloarchaea, which requires enzyme adaptation to maintain their proper conformation and activity near the concentration of salt saturation [28]. The high intracellular osmolarity of *Hfx. mediterranei* opens the door to release PHA granules from biomass by an easy and convenient strategy: in media of drastically lower osmotic pressure, hence, salt concentration below 10%, *Hfx. mediterranei* cells were observed to partially lyse [33]. In highly hypotonic media, such as salt-free water, the high inner-osmotic pressure causes the bursting of *Hfx. mediterranei* cells and liberation of PHA granules into the medium [88]. What remains is the cell debris suspended in the liquid phase, and the PHA granules, which, due to their lower density, are floating to the surface of the liquid phase. This "skimming" of PHA granules, which can be accelerated by the use of dissolved air floatation, allows their convenient harvest by decantation or centrifugation [93].

Compared to other haloarchaeal PHA-producing species reported in the literature, *Hfx. mediterranei* is characterized by higher specific growth rate (μ_{max}) and higher specific PHA production rates (q_{Pmax}), which enhance obtainable space-time yields in the bioprocess. This was first demonstrated in a continuously operated chemostat cultivation process performed at a dilution rate D of 0.12 1/h and a temperature of 38°C. As a result, 3.5 g/L PHA were produced when using 20 g/L glucose as substrate, and 25% of thalassic salts. Exchanging glucose by starch, which was enabled by the high amylase activity of the strain, almost doubled the PHA concentration to 6.5 g/L [94]. In 2003, *Hfx. mediterranei* α-amylase was characterized by Pérez-Pomares, who reveled the salt, pH-value and temperature optima of this enzyme with 3 M NaCl, 7–8, and 50–60°C, and detected that the enzyme occurs in a monomeric form [95].

Remarkably, the characterization of PHA samples produced by *Hfx. mediterranei* revealed that they did not consist of the homopolyester PHB, which is typically produced when using simple carbon sources like carbohydrates but was a poly(3-hydroxybutyrate-*co*-3-hydroxyvalerate) (PHBHV) copolyester. This was the first report on the production of copolyesters from structurally unrelated carbon sources. Classically, biosynthesis of 3-hydroxyvalerate (3HV)-containing copolyesters needs the supplementation of structurally related precursors, e.g. fatty acids with an odd number of carbon atoms (propionic acid, valeric acid, levulinic acid, etc.); application of such precursors adds significantly to the entire copolyester production costs. Most importantly, PHBHV copolyesters are easier to process by melt extrusion or other techniques due to their lower crystallinity and broader window of processibility (the difference between meting temperature and degradation temperature), and are more prone towards biodegradation if compared with the highly crystalline and rather brittle homopolyester PHB [96]. In the context of *Hfx. mediterranei*, the PHBHV copolyester accumulated by this strain displays desirable material features regarding low melting temperature (T_m), low crystallinity (X_c), high molecular mass, and a low dispersity index ($Đ_i$). It took some decades until the biosynthesis of PHBHV from unrelated substrates was also evidenced in additional haloarchaeal species, as detailed later. Han and colleagues revealed that the particularities of *Hfx. mediterranei*'s PHA-biosynthesis metabolism are responsible for this remarkable feature: based on multiple active pathways serving for propionyl-CoA production, the strain accumulates high intracellular concentrations of this compound, which acts as 3HV-precursor; the coupling of propionyl-CoA and acetyl-CoA results in the formation of 3HV, whereas coupling of two acetyl-CoA molecules by the action of the enzyme 3-ketothioase generates 3HB [97].

Meanwhile, a lot of knowledge was generated in the fields of the enzymatic machinery of *Hfx. mediterranei* and particularities of its genome, with special focus on the context of PHA biosynthesis. This encompasses studying PHA synthases active in this strain [98], the nature of phasins, which are important enzymes in PHA granule formation [99], the identification of the genes encoding PHA biosynthesis in *Hfx. mediterranei* [98], the 3HV-precursor supplying pathways [97, 100], and particular enzymes involved in in vitro degradation of native *Hfx. mediterranei* granules [101].

4.3.2.2   Coproduction of PHA and EPS by *Hfx. mediterranei*

In order to get insights into the kinetics of growth and PHA production by this organism, a formal kinetic mathematical model was established by Koller et al. [102]. Further, the context between the production and in vivo degradation of PHA and EPS, which are produced by this strain as a secondary polymeric product of potential industrial use, were studied in detail by the same group of authors; it was particularly shown that intracellular PHA degradation is a rather slow process, even under carbon-limited conditions, which allows a cell harvest even after complete depletion of the carbon source without risking major product loss by intracellular degradation [103]. In this context, the presence of said EPS produced by *Hfx. mediteranei*, a typical feature also of other haloarchaea, was already reported in 1988 by Antòn and colleagues. These researchers noticed that this extracellular polymeric material causes the mucous character of *Hfx. mediterranei* colonies [90]. Chemically, this EPS displays an anionic, sulfated polymer, composed of a regular trisaccharide-repeating unit of one mannose and two 2-acetamido-2-deoxyglucuronic acid monomers, with one sulfate ester bond occurring per trisaccharide unit [79]. This polysaccharide has xanthan-like properties, which makes it interesting as a thickening and gelling agent in food processing. Only recently, production of an EPS with outstanding stabilization performance for water-in-oil emulsions by the haloarchaeon *Hfx. mucosum* was reported by Lopéz-Ortega and colleagues. Within 96 h of cultivation in highly saline (2–3 M NaCl) glucose medium, more than 7 g/L of this EPS was excreted by the cells [104]. Moreover, algal-derived sulfated EPS of compositions similar to those produced by *Hfx. mediterranei* were suggested to find possible application in cosmetics, e.g. as an inhibitor of hyaluronidase, as therapeutic agents due to their anti-allergic, anti-bacterial, antiviral, anti-inflammatory, and anti-tumor activity, and as nutraceuticals due to their antioxidant activity, as comprehensively reviewed by Raposo et al. [105]. In this context, the anti-proliferative effect on gastric cancer cell lines of an EPS excreted by the haloarchaeon *Halorubrum* sp. TBZ112 was described by Hamidi et al. This EPS was a heteropolymer, mainly consisting of mannose, glucosamine, galacturonic acid, arabinose, and glucuronic acid [106].

In a study presented by Cui et al., the effect of different salinities on the carbon flux towards PHA and/or EPS biosynthesis was studied in order to enable shifting the substrate flow towards the desired main product. As a major result, high NaCl concentrations turned out to inhibit EPS formation, but were beneficial for PHA productivity. As the NaCl concentration increased from 75 to 250 g/L, EPS production declined from 370 to 320 mg per g biomass. At this highest investigated NaCl concentration, the intracellular PHA content reached its maximum value of 0.71 g per g biomass; this clearly showed that high salinity significantly activated PHA bio-synthesis, while at the same time inhibiting EPS production. Technologically, these findings open the door to regulate the carbon distribution to the biosynthesis of PHA or EPS by *H. mediterranei* by simply fine-tuning the salinity [107].

4.3.2.3 Optimized Process Parameters for PHA Production by *Hfx. mediterranei*

The effect of using different nitrogen sources (NH_4^+ or NO_3^-) on biomass forma-tion and PHA biosynthesis by *Hfx. mediterranei* was studied by Ferre-Guell and Winterburn [108]. Using NH_4^+ as an inorganic nitrogen source in a nitrogen-rich medium based on glucose and yeast extract and a salinity caused by 156 g/L NaCl, a cell dry mass (CDM) concentration of 10.7 g/L and 4.6 wt.-% PHBHV in CDM were achieved, while using NO_3^-, a CDM of 5.6 g/L, harboring 9.3 wt.-% PHBHV were obtained. Unexpectedly, the applied nitrogen source had an impact on the mono-meric composition of PHBHV. Using NH_4^+, the 3HV fraction in PHBHV was 16.9 mol-%, which was considerably higher than for the NO_3^--supplied cultivation setups (12.5 mol-%). Low carbon-to-nitrogen ratio of 42/1 resulted in the formation of low biomass concentration, but in turn increased the intracellular PHBHV fraction to 6.6 wt.-% in the case of NH_4^+; when using NO_3^-, this effect was less pronounced (9.4 wt.-%). In addition, it was shown that decreasing the carbon-to-nitrogen ratio increases the 3HV fraction in PHBHV, hence, this ratio seems to directly impact the propionyl-CoA generating pathways. As another remarkable finding, integration of 3HV building blocks into growing PHA chains starts with a certain time delay; only after a concentration of 0.45 g/L PHA was produced, 3HV was detected in the polyester [108]. Hence, PHBHV copolyester synthesis by *Hfx. mediterranei* needs further studies in order to become better understood on a mechanistic level; the detailed comprehension of the effect of different nitrogen sources and their optimum concentration will serve for an optimized layout of the production process for halo-archaeal PHBHV copolyesters of pre-defined composition and material properties. In a follow-up experiment, Melanie and colleagues studied the impact of the initial phosphate concentration on PHA production by *Hfx. mediterranei*; these authors reported the accumulation of 0.95 g/L PHBHV with an astonishingly high 3HV frac-tion of 22.36% after one week of cultivation in 500 mL shaking flasks at a salinity of 156 g/L NaCl when using 0.5 g/L KH_2PO_4 as a phosphate source, which was a better outcome than obtained by parallel setups using lower phosphate concentrations (0.25 or 0.00375 g/L KH_2PO_4). Thermal characterization of the polyester revealed data similar to other PHA samples produced by *Hfx. mediterranei* [109].

In addition to the effect of nitrogen and phosphate sources, the effect of tem-perature on biomass formation and PHA accumulation by *Hfx. mediterranei* was studied by Cui and associates [110]. In this study, a mathematical model for the organism's growth and product formation kinetics at low and mesophilic tempera-ture (15, 20, 25, and 35°C) was established, calibrated, and validated. Experimental results elaborated by cultivations of the strain in a molasses-based wastewater-like medium delivered the kinetic coefficients needed to develop the model; 2.5 L stirred and aerated flasks were used as culture vessels. Most importantly, it was demon-strated that *Hfx. mediterranei*-mediated PHA production strongly depends on the cultivation temperature; at 15°C, a volumetric PHA productivity of only 390 mg/(L·h) was achieved, while this value increased to 620 mg/(L·h) at warmer temperature (35°C). In accordance with an estimate by an Arrhenius equation plot, the maximum specific growth rate (μ_{max}), the maximum specific substrate conversion rate (q_{Smax}), and the specific decay rate (k_d) of *Hfx. mediterranei* increased in parallel to increasing temperature. μ_{max} ranged from 0.009 1/h at 15°C to 0.033 1/h at 35°C, q_{Smax} fluctuated between 0.018 g/(g·h) at 15°C and 0.037 g/(g·h) at 35°C, while the value for k_d was calculated between 0.0048 1/h at 15°C and 0.0089 1/h at 35°C. The estimated activation energy for biomass growth, substrate consumption, and biomass decay amounted to 58.31 kJ/mol, 25.59 kJ/mol, and 22.38 kJ/mol, respectively. Generally, the model was of high predictive power for all studied temperature conditions. Moreover, the 3HV fraction in the copolyester was not dependent on the temperature; for all conditions, the PHBHV consisted of 83.3 mol-% 3HB, and 16.7 mol-% 3HV. Even without nitrogen limitation, hence, under nutritionally balanced conditions, a temperature of 35°C significantly increased the PHA accumulation rate and was suggested by the authors as the optimum temperature for *Hfx. mediterranei* when used for PHA production [110]. This finding is of importance because of the contradictory, strongly differing, temperatures used in older literature for cultivations of this organism.

4.3.2.4 Feedstocks Used for PHA Biosynthesis Using *Hfx. mediterranei*

Due to these intriguing features, *Hfx. mediterranei* attracted the interest of many other research groups globally. The quest for different inexpensive raw materials as feedstocks for this strain has been an especially hot topic during the last couple of years. Among these inexpensive feedstocks, numerous carbon-rich waste streams were the subject of investigations, such as lactose-rich whey from the dairy and cheese-making industry [111–113], biodiesel industry-derived crude glycerol phase (CGP) [91], starch processed by extrusion together with enzymes [114, 115], rice bran [116], the ethanol production byproduct stillage [117, 118], wastewater from olive mills [119], vinasse [120], or hydrolyzed macroalgal biomass [121] were successfully tested. Regarding inexpensive feedstocks, *Hfx. mediterranei* is currently considered the most auspicious candidate for industrial-scale PHA production based on surplus whey; in combination with promising productivity, this mainly originates from its high robustness and stability and the convenient downstream processing [111, 112]. In this context, *Hfx. mediterranei* grew well on enzymatically or acidic hydrolyzed whey permeate (about 200 g/L sugars); in bioreactor cultivation setups, high maximum specific growth rates (μ_{max}) of 0.11 1/h were noticed, which is considerably better than reported for other haloarchaeal organisms. The maximum specific production rate (q_P) for PHBHV in these processes reached 0.08 g/(g·h). Further optimization of the cultivation conditions enhanced both volumetric and specific PHBHV productivity to 0.09 g/(L·h) and 0.15 g/(g·h), respectively. The highest values for biomass concentration and PHBHV fractions in biomass were 16.8 g/L and 73 wt.-%, respectively [122]. However, these experiments were characterized by a severe shortcoming: *Hfx. mediterranei* shows a clear preference for glucose conversion when cultivated on hydrolyzed whey, hence, on equimolar mixtures of glucose and galactose; this results in the piling up of galactose in the cultivation broth, causing an unacceptably high biological oxygen demand

(BOD) of the spent fermentation broth, and in the loss of a considerable part of the carbon source, which is not a feasible option from an ecological perspective. Solutions proposed encompassing the separation of galactose post downstreaming for further use (sweetener, nutritional supplement), which is also doubtful from a cost-use point of view. Another solution was offered by the detection of Pais and colleagues, who discovered that the activity of galactose-converting enzymes could be enhanced when fine-tuning the supply with trace elements, which results in a more complete substrate conversion in a medium with 156 g/L NaCl. Surprisingly, these authors isolated PHBHV copolyesters with 3HV fractions below 0.02 mol/mol when using hypo-osmotic shock for cell disintegration [113].

To optimize the quality of whey-based PHA produced by *Hfx. mediterranei*, valeric acid and γ-butyrolactone (GBL) were supplied in cultivation setups containing 200 g/L NaCl as co-substrates in addition to hydrolyzed whey permeate, which acted as main carbon substrate, in order to increase the 3HV fraction in the polyester, and to introduce 4HB as a third building block. Supply of these substrates generated a terpolyester poly(3HB-*co*-21.8%-3HV-*co*-5.1%-4HB) with auspicious material properties in terms of low melting points, high molecular mass, and low crystallinity [112].

Already in the early stages of process development, it became clear that the highly saline byproduct streams occurring in cultivation processes with *Hfx. mediterranei* need reasonable disposal or re-utilization in order to save costs and not to generate any environmental hazards. In this context, it has to be considered that disposing of salt after cell harvest and downstreaming displays a challenge and environmental threat; valid environmental standards prohibit discharging total dissolved solids (TDS) exceeding 2 g/L in wastewater. Hence, both from an economic and a sustainability perspective, it was of interest to study whether saline cell debris remaining after biomass disruption in a hypotonic medium and salt-rich spent fermentation broth can undergo recycling in subsequent cultivation batches. Experiments for using spent fermentation broth and saline cell debris were thus performed, which confirmed the feasibility of recycling spent fermentation broth stemming from PHBHV production by *Hfx. mediterranei* using hydrolyzed whey in follow-up fermentation setups; here, it was demonstrated that a significant fraction of salts needed to prepare fresh saline mineral medium can be replaced when using spent fermentation broth. When replacing up to 29% of yeast extract, which typically acts as costly nitrogen-, phosphate-, and growth factor-containing medium component for optimum *Hfx. mediterranei* cultivations by salt-rich cell debris, which remains after removing PHA granules from *Hfx. mediterranei* biomass by hypotonic cell disruption, growth rates similar to the original cultivation setups were obtained [123]. In analogy, waste stillage stemming from rice-based ethanol production was used for PHA production by *Hfx. mediterranei* by Bhattacharyya and associates; these authors also aimed at closing material recycles and generating environmental benefits by the bioprocess. On a shaking flask scale, 16.4 g/L PHA, a mass fraction of PHA in CDM of about 70 wt.-%, a substrate-to-PHA conversion yield of 0.35 g/g, and a volumetric PHA productivity of 0.17 g/(L·h) were obtained. Similar to the whey-based experiments described above, the PHA produced was identified as a PHBHV copolyester with 15.3 mol-% 3HV. PHA recovery was again accomplished by cell disruption in hypotonic media, delivering PHA granules covered by a membrane composed of proteins, lipids, and (lipo)polysaccharides. This crude product was purified with sodium dodecyl sulfate (SDS) and organic solvents. Environmentally significant, the authors describe a decrease of about 85% of the chemical oxygen demand (COD) and biochemical oxygen demand (BOD) of the feedstock stillage. In addition, the total dissolved solids (TDS) content in the spent fermentation broth was as low as 670 mg/L, which corresponds to only about 30% of the maximum permitted TDS load [117].

In addition to the use of surplus whey as raw material for PHA production by *Hfx. mediterranei*, this strain also biosynthesizes PHA heteropolyester (co- and terpoly-esters) starting from crude glycerol phase (CGP), which is the main side-product of the strongly emerging biodiesel industry. Here, it has to be considered that 1 kg of biodiesel produced generates about 0.1 kg of CGP, which over-saturates the glycerol market. Hermann-Krauss et al. used CGP as feedstock for bioreactor cultivations with 150 g/L NaCl, and obtained a volumetric PHBHV productivity of 0.12 g/(L·h); 75 wt.-% PHBHV were determined in biomass, with the 3HV fraction in PHBHV amounting to 10 mol-%. Thermoanalysis and molecular mass determination of PHBHV copolyesters produced by *Hfx. mediterranei* from CGP in different cultivation batches showed melting points (T_m) in the range of 130 to 140°C, a weight average molecular mass (M_w) between 150 and 253 kDa, and a $Đ_i$ between 2.1 and 2.7. When the 4HB-precursor γ-butyrolactone (GBL) was co-supplied together with CGP as the main carbon source, a PHA terpolyester was accumulated, which was composed of 3HB (83 mol-%), 3HV (12 mol-%), and 4HB (5 mol-%); this material displayed reduced T_m (122 and 137°C, two separated melting endotherms obtained in the thermogram), T_g (2.5°C), and higher M_w values (391 kDa) if compared with the copolyesters (PHBHV) [91]. At that point, it should be emphasized that conversion of glycerol for PHA biosynthesis is not a widespread feature among haloarchaea; it was reported that a range of haloarchaea use glycerol for cell growth and maintenance energy production, but not for accumulation of PHA. From the metabolic point of view, this can be explained by glycerol, after its phosphorylation, being converted exclusively via the tricarboxylic acid cycle (TCA), which results in biomass formation instead of PHA biosynthesis [124].

Extruded rice bran (ERB) and extruded cornstarch (ECS) were supplied by Huang et al. to *Hfx. mediterranei* cultures as carbon-rich feedstocks for PHA production; the cultivations were performed using nutritionally balanced media, which was complete in all nutrients, and contained 234 g/L NaCl. Due to the strain's inability to utilize the native forms of the raw materials, cornstarch, and rice bran directly, they were subjected to extrusion before being applied as substrate. Bioreactor cultivations on a five-liter scale were tested under pH-stat conditions (pH-value was kept constant at 6.9 to 7.1). Mixtures of ERB and ECS in a ratio of 1/8 (g/g) were supplied as the main carbon substrate using a repeated fed-batch feeding protocol. Achieved values for CDM, PHA, and PHA content in biomass amounted to 140 g/L, 77.8 g/L, and 0.56 g/g, respectively. These values were considerably lower when using ECS as the sole substrate (62.6 g/L, 24.2 g/L, and 0.39 g/g, respectively). For the cultivations, highly saline conditions were used, with NaCl concentration amounting to 234 g/L, which gave rise to optimism by the authors to run such setups for extended periods on an industrial scale [116].

Another study carried out by Chen and colleagues reported on the application of cornstarch as a substrate for *Hfx. mediterranei*-based PHA production. In this case, cornstarch was processed by enzymatic reactive extrusion in a single-screw extruder; 1–5 g of the biocatalyst α-amylase was added per 100 g wet mass of cornstarch. This generated extrudate was used by Chen et al. as substrate in a pH-stat fed-batch cultivation with 234 g/L NaCl. Keeping carbon- and nitrogen source levels constant, the feed stream was composed of a mixture of enzymatically ECS and yeast extract in a ratio of 1/1.7 g/g. At the end of the cultivation, the PHA content in biomass amounted to 0.51 g/g. Analogous to the use of other carbon sources for PHA production by *Hfx. mediterranei* such as whey or CGP, a copolyester (PHBHV) was obtained. The product contained 10.4 mol-% 3HV; via DSC analysis, a T_g of −1.2°C, and two separated T_m melting peaks (129.1 and 144.0°C) were determined. For 70 hours of cultivation in a six-liter bioreactor under pH-stat conditions, the highest reported volumetric productivity to date for *Hfx. mediterranei* of about 0.28 g/(L·h) was reported, which corresponds to a PHA (PHBHV copolyester with a 3HV fraction of 4.4 mol-%) concentration of 20 g/L [115].

Green macrolaga *Ulva* sp., an organism typically causing undesired algal bloom in coastal areas, was subjected to hydrolysis by Ghosh and colleagues. The obtained hydrolysate was used as feedstock for *Hfx. mediterranei* cultivations in shaking flasks at 42°C and a pH-value of 7.2. The highest CDM and PHA concentration, achieved when *Haloferax mediterranei* was cultivated in 25% of *Ulva* hydrolysate, amounted to 3.8 and 2.2 g/L, respectively, corresponding to a PHA fraction in CDM of 58%. The isolated PHBHV copolyester contained 8% 3HV building blocks. The authors proposed their outcomes as a new approach to "haloarchaeal sea-agriculture" [121].

4.3.2.5   Techno-Economic Assessment of Whey-Based PHA Production by *Hfx. mediterranei*

Based on data obtained by a 200 L pilot-scale cultivation setup, it was possible to provide a preliminary cost assessment for hydrolyzed whey-based PHBHV production by *Hfx. mediterranei*. This calculation encompassed the benefits of the solvent-free downstream processing technique in a hypotonic medium, the hydrolysis of whey lactose by mineral acids, the waiver of any sterility precautions, the omission of 3HV-related precursor compounds, and importantly, the recycling of saline cell debris and spent fermentation broth; the estimated production price per kg of PHA was reported at less than € 3, which is considerably below previously reported PHA production prices of about € 5–10 per kg. Moreover, this study revealed that the higher value creation from whey when converting it to PHA than, as an established alternative, to whey powder. Apart from the economic consideration, a complete life cycle assessment (LCA) study of this process, using the "sustainable process index" (SPI) as a sustainability indicator, was performed. It was demonstrated that the calculated ecological footprint of whey-based PHA production by haloarchaea potentially outpaces fossil-based plastics, but only if the above-described process side streams are recycled [125]. This is in accordance with more recent considerations by Narodoslawsky and colleagues, who underlined that the ecological footprint of biobased plastics is not intrinsically better than is the case for established plastics, but becomes superior when considering and optimizing the entire cradle-to-grave life cycle of the polymer [126].

4.3.2.6 Microstructure and Structural Fine-Tuning of Hfx. mediterranei PHA Copolyesters

In 2006, Don and associates discovered further particularities of PHBHV produced by *Hfx. mediterranei*. These authors revealed that this biopolyester is not homogenous but composed of different fractions. By using a mixture of chloroform and acetone, it was possible to separate two fractions of *Hfx. mediterranei* PHBHV, with the two fractions having different 3HV contents. The main fraction (about 93 wt.-% of the entire biopolyesters) had a molecular mass of about 570 kDa, and a 3HV fraction of 10.7 mol.-%; the minor fraction of the material revealed a significantly higher 3HV fraction of 12.3 mol.-%, and a considerably lower molecular mass of only 78.2 kDa. This low molecular mass fraction was soluble even in acetone, which is typically described as an "anti-solvent" for short chain length PHA (*scl*-PHA) like PHB or PHBHV. However, both fractions revealed similar values for T_m and T_g, and had a narrow distribution of molecular mass (low $Đ_i$). Differential scanning calorimetry (DSC) demonstrated that at heating rates below 20°C/min result in two overlapping melting peaks in the thermogram, with the relative intensity of the two peaks altering by variation of the heating rate; according to the authors, this phenomenon might originate from a PHA melt-and-recrystallization process [127]. In analogy, it was observed by Koller and colleagues that a low molecular mass fraction (below 200 kDa) of a poly(3HB-*co*-3HV-*co*-4HB) terpolyester produced by *Hfx. mediterranei* is extractable by acetone under reflux conditions in a Soxleth apparatus, whereas the major part (more than 99%) of the PHA was acetone-soluble only under an elevated temperature and pressure above acetone's boiling point. This study also confirmed the formation of intracellular PHA blends by *Hfx. mediterranei* [128]. Only recently, further insights into the microstructure of PHBHV produced by *Hfx. mediterranei* were provided by Han and colleagues; these authors revealed the complex blocky structure of PHBHV (*b*-PHA), which consists of alternating PHB and poly(3-hydroxyvalerate) (PHV) homopolyester blocks linked to PHBHV copolyester blocks of random distribution. This *b*-PHA production can be fine-tuned by the co-feeding of glucose and valerate. *b*-PHA synthesized by *Hfx. mediterranei* features intriguing material characteristics because of its "blocky" structure and the high molar 3HV fraction; low crystallinity and enhanced Young's moduli are examples for these beneficial material features. In addition, this *b*-PHA was also investigated for its potential biomedical application; it was shown that it exhibited better blood platelet adhesion and faster blood clotting in comparison with PHBHV of random distribution; as a consequence, *Hfx. mediterranei*-based *b*-PHA was suggested by the authors as a favorable candidate for use in the medical field [100].

Ferre-Guell and Winterburn developed a fed-batch bioreactor cultivation process for *H. mediterranei* to reproducibly produce PHBHV copolyesters with predefined composition using VFA (butyric and valeric acid) mixtures as substrates. The emulsifier Tween 80 increased the bioavailability of substrates at 37°C. The highest PHBHV mass fraction in CDM amounted to 0.59 g/g and a volumetric productivity of 10.2 mg/(L·h) were obtained using a butyric/valeric acid mix of 56/44.

The produced PHBHV had the aspired 3HV fraction of 0.43 mol/mol. Predefining the PHBHV composition by adaptation of the substrate mix was demonstrated on a shaking flask and bioreactor scale under different cultivation conditions regarding temperature and emulsifier. It was also noted that the uptake of butyric acid is faster during the phase of microbial growth and CDM formation, while consumption rates for butyric and valeric acid were identical during the stationary phase. Importantly, no microbial contamination was detected during the cultivations despite the fact that neither media nor equipment were sterilized prior to use; sampling and feed additions were performed under septic conditions. The negligible differences in PHBHV product quality (molecular mass, thermoanalysis data) from flask cultivations and bioreactor cultivations demonstrated the robustness and scalability of this production technique; a PHBHV copolyester with almost the same material properties can be obtained in a consistent and reliable fashion, which minimizes the well-known risk of uncontrollably and randomly varying material properties, which currently constitutes a stumbling block to large-scale production of tailor-made biopolyesters for defined applications [129].

Another study by Ferre-Guell and Winterburn also aimed at the production of *Hfx. mediterranei* copolyesters of controlled composition and microstructure. In this study, randomly distributed and blocky structured PHBHV copolyesters were produced by supplying cells with different even-numbered (acetic, butyric, hexanoic, octatonic, and decanoic acid) and odd-numbered (propionic, valeric, heptanoic, nonanoic, and undecanoic acid) fatty acids as substrates. Only fatty acids with less than seven carbon atoms were accepted by the strain for growth and PHA accumulation. Using acetic acid, PHBHV with about 10 mol-% 3HV (similar to reports on glucose, glycerol, etc.), almost no 3HV was found in polyesters produced from butyric acid. In contrast, copolyesters with outstandingly high (more than 90 mol-%) 3HV content were obtained when using valeric acid as sole carbon source. In PHBHV from propionic acid, the 3HV content was lower due to a partial conversion of propionyl-CoA to acetyl-CoA of propionyl-CoA towards acetyl-CoA, which acts as 3HB precursor. Investigating the impact of different feeding strategies using butyric acid and valeric acid or defined mixtures thereof, it was shown that sequential feeding of the substrates generated blocky structured PHBHV (alternating PHB and PHV blocks in PHBHV), while co-feeding resulted in random distribution of 3HB and 3HV. Moreover, it was shown that increasing 3HV content in randomly distributed PHBHV results in decreased T_g, hence, in PHA chains of higher mobility in the amorphous phase. Generally, increasing 3HV fractions in random PHBHV resulted in less crystalline/more amorphous polymers. Moreover, 3HV-rich copolyesters had lower melting temperatures, better elasticity, and improved ductility [130].

4.3.3 OTHER HALOARCHAEAL PHA PRODUCERS

4.3.3.1 *Haloarcula [Har.]* sp.

In recent years, a range of haloarcheal PHA producers other than *Hfx. mediterranei* were collected in diverse saline habitats. In 1989, Altekar and Rajagopalan revealed a context between PHA accumulation by haloarchaea (*Hfx. mediterranei*,

Hfx. volcanii, and *Har. marismortui*) and CO_2-fixtion activity of their cell extract due to the presence of ribulose bisphosphate carboxylase (RuBisCo) [131]. In 1999, Nicolaus and colleagues isolated three new organisms from the last pond of a marine saltern near Monastir, Tunisia. These isolates grew well under extremely halophilic conditions of 3.5 M NaCl. When cultivated on minimal media containing diverse carbohydrates, accumulation of PHB homopolyester (identified by ^{1}H- and ^{13}C-NMR) and, in parallel, excretion of an EPS, precipitable by addition of cold ethanol, by all three strains, was observed. Investigating the lipid patterns of the strains, it was revealed that they belong to the genus *Haloarcula [Har.]*. All of them utilized starch as a carbon source, and one of the isolates (T5) showed excellent growth on molasses as inexpensive feedstock. After eight to ten days of cultivation, T5 revealed PHB accumulation of 5 mg per g CDM when cultivated on starch or glucose, and 10 mg per g CDM on molasses. DNA-DNA hybridization tests and other biochemical characterization studies suggested that the organism is a new subspecies of *Haloarcula japonica* [58].

Taran and Amirkhani optimized PHB biosynthesis by *Haloarcula* sp. IRU1, isolated from the hypersaline Iranian Lake Urmia, in shaking flask cultivations using different glucose, nitrogen-, and phosphate-source concentrations in a temperature range between 37 and 55°C. The highest PHB fractions (0.63 g PHB per g CDM) were achieved by *Har.* sp. IRU1 at 42°C when using 2 g/L glucose, 0.2 g/L NH_4Cl, 0.004 g/L KH_2PO_4. Unfortunately, the study only reports on the intracellular PHB fractions, but does not disclose any data for CDM and PHB concentration [132]. Moreover, Taran tested the potential of *Har.* sp. IRU1 to utilize glucose, fructose, sucrose, starch, palmitic acid, and acetate as carbon sources. In a medium containing glucose, a considerable increase in CDM and PHB concentration was observed in comparison to all other carbon sources, with the lowest concentration for CDM and PHB being observed when using palmitic acid or acetate as substrate [133]. Later, the same strain was cultivated on petrochemical wastewater; optimized conditions of 2% wastewater as the carbon source, 0.8% tryptone as the nitrogen source, 0.001% KH_2PO_4 as the phosphate source and a temperature of 47°C revealed the highest PHB fractions in biomass of about 0.47 g/g, as calculated using the Taguchi experimental design method; similar to the preliminary study, data for PHB and biomass concentrations and data for polymer properties are missing in this report [134]. The same author, Taran, cultivated *Har.* sp. IRU1 under axenic conditions in minimal media with crude oil as the sole source of carbon and energy. After five days of cultivation at a temperature of 47°C, 2% crude oil, 0.4% yeast extract, and 0.016% NaH_2PO_4, CDM and PHB production were studied by conventional methods. According to these studies, *Har.* sp. IRU1 constitutes a proficient strain for bioremediation of petrochemically contaminated habitats, combined with PHB biosynthesis [135]. Even textile wastewater was studied as a substrate for the growth of this organism, although without monitoring PHA biosynthesis [136].

The extremely halophilic haloarchaeon *Har. marismortui*, originally described in the 1970s as a PHB producer [57], was later cultivated on vinasse, a byproduct of the molasses-based ethanol production process, which contains non-volatile components like inhibiting phenolic compounds which remain in the fermentation broth after distillative ethanol recovery. On a cultivation medium containing 10% raw vinasse and 200 g/L NaCl, *Har. marismortui* accumulated 0.26 g PHB homopolyester per g CDM at a volumetric productivity of 0.015 g/(L·h). This shaking flask scale process was significantly enhanced by removal of the phenolics using charcoal; on a medium consisting of 100% dephenolized vinasse, 0.30 g PHB per g CDM were obtained by this organism, with the volumetric productivity reaching 0.02 g/(L·h). The polyester was recovered by hypotonic cell disintegration and further purified using NaClO. By means of FT-IR, DSC, UV–VIS spectroscopy, and ^{1}H-NMR, the product was characterized and classified as PHB homopolyester [137].

4.3.3.2 *Halogeometricum [Hgm.]* sp.

In 2013, Salgaonkar and colleagues isolated and screened seven extremely halophilic *Archaea*, which originated from brine and sediments of solar salterns in India. A saline synthetic medium containing 20 wt.-% NaCl turned out to be suitable to cultivate all of these isolates; all seven strains also displayed PHA accumulation. Among these isolates, six were identified as members of the genus *Hfx.* by genotypic and phenotypic tests, they were categorized as laboratory strains TN4, TN5, TN6, TN7, TN10, and BBK2. In addition, the isolate TN9 was identified as the species *Halogeometricum [Hgm.] borinquense*. This new strain, *Hgm. borinquense* TN9, turned out to be most promising among the seven isolates, and therefore was selected to study its kinetics of growth and PHA biosynthesis. It was revealed that the strain already displays maximum rates for PHA production during the logarithmic (exponential) growth phase before the onset of depletion of a growth-essential nutrient, hence, it constitutes a typical "growth-associated PHA-producer", which is in contrast to most PHA production strains for industrial use, which display boosted PHA productivity only under nutritionally balanced conditions. A PHA content in biomass of about 0.14 g/g was reported after five days of cultivation; the PHA was identified as PHB homopolyester [138]. Afterwards, the same team of researchers isolated another haloarchaeal strain from the Indian Marakkanam solar salterns in Tamil Nadu. This new isolate was called *Hgm. borinquense* E3, and, similar to *Hfx. mediterranei* but different from the closely related species *Hgm. borinquense* TN9, produced PHBHV copolyesters when using glucose as the sole carbon source in a highly saline medium. The presence of both 3HB and 3HV building blocks in the copolyester was confirmed by Fourier transform infrared spectroscopy (FT-IR), X-ray diffraction (XRD), and proton nuclear magnetic resonance spectroscopy (^{1}H-NMR). After four days in shaking flask cultivations, the PHBHV content in biomass reached 0.74 g/g, while the 3HV molar fraction in PHBHV amounted to 0.22% [139].

These studies were further intensified by the same research group by cultivating a range of haloarchaeal wild-type strains on hydrolyzed sugarcane bagasse (SCBH), an abundantly available lignocellulosic byproduct from the sugar industry; these experiments aimed at assessing the PHA accumulation potential of these strains on this substrate [140]. According to Nile red staining, *Hgm. borinquense* E3, the strain described above, showed best PHA accumulation potential among the investigated organisms, which were identified as *Haloferax volcanii* BBK2, *Haloarcula japonica* BS2, and *Halococcus salifodinae* BK6. *Hgm. borinquense* E3 formed light pinkish and highly mucous colonies, which is in accordance with the appearance of *Hfx. mediteterranei*, and evidences its potential for EPS and pigment biosynthesis. In shaking flask cultivation setups in a highly saline cultivation medium containing 20% NaCl at 37°C, *Hgm. borinquense* E3 was cultivated on 25 or 50%, respectively, of SCBH for six days. Obtained PHA contents in biomass amounted to 0.50 g/g (25% SCBH) and 0.46 g/g (50% SBCH), respectively, while values for the specific production rate (q_p) were reported with 3.0 mg/(g·h) (25% SCBH) and 2.7 mg/(g·h) (50% SCBH), respectively. The biomass produced in the 25% SCBH setups was oven-dried; subsequently, the PHA was chloroform-extracted and characterized by a spectroscopic crotonic acid assay, DSC, FT-IR, XRD, and ^{1}H-NMR. It turned out that the isolated product was a PHBHV copolyester with a 3HV fraction of 0.133 mol/mol [140]. Follow-up experiments aimed at elucidating the strain's potential of converting starch and starch-based waste materials to PHA. On a shaking flask scale, acid-hydrolyzed cassava waste was used as substrate in parallel with cultivations using pure starch. Again, 200 g/L NaCl was used in the setups, which lasted ten days. In the case of starch, 4.6 g/L PHBHV (13.1 mol-% 3HV in PHBHV) were obtained, while the use of cassava waste resulted in the production of 1.5 g/L PHBHV (19.7% 3HV in PHBHV) [141].

4.3.3.3 *Natrinema [Nnm.]* sp.

In 2015, Danis and associates studied a series of five extremely halophilic archaeal isolates in order to trace the species displaying the highest PHA production capacity among them. Different inexpensive carbon-rich feedstocks were individually examined for their potential application for PHA production, among them cornstarch, whey, sucrose, apple waste, melon waste, and waste from tomato processing. Among these substrates, cornstarch turned out as the most promising feedstock for PHA production. The isolate 1KYS1 performed best regrading PHA productivity among the five investigated halophilic organisms. Using comparative 16S rRNA gene sequence analysis, the close relationship of 1KYS1 to members of the genus *Natrinema [Nnm.]*, and, within this genus, especially to the species *Natrinema pallidum* JCM 8980 was detected. The strain 1KYS1 reached a PHA content in biomass of 0.53 g/g when cultivated on starch as the only carbon source and 25% NaCl. The accumulated PHA, which was identified as PHBHV copolyester, appeared as large and uniform PHA granules when observed via transmission electron microscopy (TEM). The copolyester PHBHV isolated from *Nnm. pallidum* was further blended with low molar mass poly(ethylene glycol) for the preparation of new biocompatible films to be tested in drug liberation experiments. Rifampicin was used as model bioactive compound in these experiments, with best rifampicin delivery efficiency being obtained at 37°C and a pH-value of 7.4, hence, under conditions approaching physiological conditions in the human organism [142].

Natrinema ajinwuensis RM-G10 (synonym *Natrinema altunense* strain RM-G10), an extremely halophilic archaeon, was isolated in 2015 by Mahansaria and coworkers from Indian salt production pans. In 72 h repeated batch cultivations on glucose, *Nnm. ajinwuensis* accumulated about 0.61 g PHA per g biomass at a volumetric PHA productivity of 0.21 g/(L·h). In contrast to glucose, the use of glycerol enabled biomass growth, but did not result in any PHA production. TEM, DSC, gas chromatography coupled to flame ionization detection (GC-FID), gas chromatography coupled to mass spectrometry (GC-MS), thermogravimetry, FT-IR, XRD, and NMR were used to characterize the glucose-based biopolyester, which turned out to be a PHBHV copolyester, and is similar to other haloarchaea described previously in this chapter; the 3HV content in the polyester amounted to 0.14 mol/mol. Two melting endotherms at 143°C and 157.5°C, a T_g of -12.3°C, an onset of decomposition temperature (T_d) of 284°C, and an X_c of 35.45% were reported [124].

4.3.3.4 Halogranum [Hgr.] sp.

Halogranum [Hgr.] amylolyticum TNN58, a strain isolated from samples collected at a marine solar saltern near Lianyungang, PR China, was investigated in 2015 by Zhao and associates, and described as a powerful haloarchaeal producer of PHBHV copolyesters starting from simple unrelated carbon sources without the co-feeding with 3HV-related precursor compound, as it was unambiguously shown by GC-MS and ^{1}H-NMR. By using TEM, high amounts of PHA granules were detected in the microbial cells. Surprisingly high 3HV fractions surmounting 0.2 mol/mol in PHBHV constituted the highest reported 3HV content in PHBHV to date produced by wild-type strains without supplying structurally related precursor compounds. Nitrogen-limited conditions in the production medium turned out to be beneficial for PHBHV production by this organism if compared with nutritionally balanced media, although PHBHV accumulation was partially growth-associated. Among different tested carbon sources like glucose, starch, glycerol, acetate, benzoic acid, lauric acid, butyric acid, or casamino acids, glucose resulted in the best outcomes for biomass and PHA production. Under controlled conditions in a 7.5 L bioreactor fed-batch (four substrate pulses of glucose after 64, 90, 114, and 144 h) setups, PHBHV production by *Hgn. amylolyticum* was studied in more detail. At the end of the cultivation after 188 h, CDM, PHBHV, and volumetric productivity amounted to 29 g/L, 14 g/L, and 0.074 g/(L·h), respectively [143].

4.3.3.5 Halococcus [Hcc.] sp.

In 2010, Legat and colleagues screened 20 haloarchaeal isolates available in different strain collections by different PHA-staining techniques such as Sudan black B, Nile blue A, and Nile red. PHA granules were further visualized via TEM, while ^{1}H-NMR spectroscopy was used to elucidate the monomeric composition of the biopolyesters produced by the individual strains. PHA production was noticed in the strains *Haloarcula hispanica* DSM 4426[T], *Halobacterium noricense* DSM 9758[T], *Halococcus dombrowskii* DSM 14522[T], *Halococcus hamelinensis* JCM 12892[T], *Halococcus morrhuae* DSM 1307[T], *Halococcus qingdaonensis* JCM 13587[T], *Halococcus saccharolyticus* DSM 5350[T], *Halococcus salifodinae* DSM 8989[T], *Haloferax volcanii* DSM 3757[T], *Halorubrum chaoviator* DSM 19316[T], *Halorubrum coriense* DSM 10284T, *Natronococcus occultus* DSM 3396[T], and *Natronobacterium gregoryi* NCMB 2189[T]. Saline (20% NaCl) complex and minimal media from literature were used for cultivation of the organisms. As a major outcome, *Halococcus [Hcc.]* was identified as a new genus with PHA-producing members. While *Hcc. saccharolyticus* produced PHB homopolyester, the other organisms accumulated PHBHV copolyesters without the need of supplying structurally related precursors [144].

4.3.3.6 *Haloterrigena [Htg.] sp.*

Haloterrigena [Htg.] hispanica DSM 18328T, a haloarchaeon isolated from Fuente de Piedra, southern Spain, and first described by Romano and colleagues to require at least 150 g/L NaCl for growth [145], was cultivated by Di Donato et al. in a saline medium (200 g/L NaCl) containing carrot waste or tomato waste as the sole carbon source. This organism thrives best at a high temperature of 50°C and was cultivated for five days in batch bioreactor cultivations and in bioreactor cultivation setups with a dialysis tube. On carrot waste, only about 1.3 mg PHB were produced per g CDM by *Htg. hispanica*, which is comparable to the outcomes using an expensive cultivation medium based on yeast extract and casamino acids as carbon sources [146].

4.3.3.7 *Halorhabdus [Hrd.] sp.*

Strain AX-2T was isolated by Wainø and colleagues from sediment of the Great Salt Lake in Utah, USA. This aerobic isolate showed a high optimal salinity for growth of 27% NaCl, a high temperature optimum of 50°C, and a neutral pH optimum; only a small number of carbohydrates (glucose, xylose, and fructose), but no tested fatty acids supported its growth; it did not grow on complex substrates like yeast extract or peptone. The strain was tested positively for PHB accumulation, although no quantitative data were provided by the authors. 16S rRNA analysis showed that it was a member of the *Halobacteriaceae* family, but with only restricted similarity to other *Halobacteriaceae*. Therefore, the authors proposed introducing the new taxon, *Halorhabdus utahensis* for this species, which now is deposited as strain DSM 12940[T] [147].

Antunes and colleagues isolated the organism *Halorhabdus tiamatea*, a non-pigmented, extremely halophilic archaeon from a deep-sea, hypersaline anoxic basin of the Red Sea. This strain was reported to thrive bets at 270 g/L NaCl, 45°C, pH in the neutral range, uses starch for growth, and, in contrast to *Hrd. utahensis*, prefers microaerophilic environments. Unfortunately, the principal ability for PHB production by this new species was described by the authors merely as a short note [148]. Some years later, the same authors provided the complete description of the *Hrd. tiamatea* genome, showing considerable differences to the genome of *Hrd. utahensis*, and revealing that the organism possesses putative genes for a trehalose synthase and a lactate dehydrogenase [149].

Also, the third known species of the *Hrd.* genus, namely *Halorhabdus rudnickae*, was described to accumulate PHA. Isolated from samples taken from a borehole in the Barycz mining area belonging to the Polish Wieliczka Salt Mine Company, this facultative anaerobic strain is red pigmented and forms non-motile Gram-negative cocci with optimum growth at 20% NaCl, 40°C, and a pH-value of 6.5–7.5. PHA inclusions were detected by electron microscopic observation, but not quantified [150].

4.3.3.8 *Halobiforma [Hbf.] sp.*

Strain 135[T], a carotenoid-rich (red pigmented) aerobic organism isolated by Hezayen and colleagues from the surface of hypersaline soil in Aswan, Egypt, was able to accumulate up to 0.40 g PHB per g CDM when cultivated for eight days on butyric acid, and 0.15 g PHB per g CDM even on complex substrates such as peptone, casamino acids, or yeast extract. This organism requires at least 2.2 M NaCl for growth, and thrives best at 42°C. The authors suggested labeling the isolate as *Halobiforma [Hbf.] haloter-restris* sp. nov. (DSM 13078^T) [31]. For a closely related strain from the *Halobiforma [Hbf.]* genus, *Hbf. lacisalsi* sp. nov., isolated by Xu et al. from a Chinese salt lake and shown to thrive best at 1.7 M NaCl, no PHA accumulation was reported [151].

4.3.3.9 *Haloquadratum [Hqr.] sp.*

Unusual organisms, originally isolated in the Egyptian Sinai Peninsula, were described in 1980 by Walsby, who was originally interested in the highly refractive gas vesicles produced by this organism; this researcher described his isolates as "ultra-thin square bacteria" [152]. Later, Walsh reported that cells of this strain resemble "thin, square or rectangular sheets with sharp corners", and reported their dimensions as being 2–5 µm wide but not even 0.2 µm thick. The outstandingly low thickness of the sheets makes them bulge slightly, with gas vesicles visible along their edges; he also noted that this organism thrives at the edge of water activity. Importantly, Walsh also observed "poly-β-hydroxybutyrate granules in the corners" [153]. The organism was, for a long time, believed to be not cultureable in monoseptic setups; its genome was deciphered in 2006 by Bolhuis et al., who revealed that the strain's genome encodes photoactive retinal proteins of the membrane and S-layer glycoproteins of the cell wall [24]. Burns et al. investigated two closely related novel square-shaped aerobic, extremely halophilic members of the haloarchaea, isolated from saltern crystallizers in Australia and Spain, and classified both of them as the new species *Hqr. walsbyi*. They described that growth of this species occurs at pH 6.0–8.5, 25–45°C, and 14–36% (w/v) NaCl. The extremely halophilic cells lyse immediately in distilled water and a minimum of ~14% (w/v) salts is required for growth. Optimal growth occurs under neutrophilic to alkaliphilic conditions, above 18% salinity. By electron cryo-microscopy, PHA inclusions were reported by the authors, but not studied further [154]. In 2011, the strain was grown aerobically with illumination on a medium containing 195 g/L NaCl and 0.5 g glycerol, 0.1 g yeast extract and 1 g sodium pyruvate as carbon sources; atomic force microscopy (AFM) was used for a detailed study of the cellular morphology; importantly, these AFM studies showed corrugation of the cellular surface due to the presence of PHA granules, which had almost all the same size within a single cell, and were packaged in tight bags. It was assumed that the primary function of PHA granules was to reduce the cytosol volume, thus reducing the cellular energy demand for osmotic homeostasis; hence, they play a pivotal role for the strain to cope with the high salinity [155]. This assumption was later substantiated by Sedlacek and associates, who, when subjecting *Halomonas halophila* towards osmotic up- and down-shock, postulated that practically water-free PHA granules serve as hydrophobic scaffolds that considerably decrease cytoplasm shrinkage during cell dehydration, which substantially impedes plasmolysis [47].

4.4 GRAM-NEGATIVE HALOPHILIC EUBACTERIA AS PHA PRODUCERS

In addition to haloarchaea, a range of eubacteria are also described as halophile PHA producers, which grow well and store considerable amounts of PHA under conditions of elevated salinity. In this context, several halotolerant to halophilic strains were isolated in 2014 by Singh from agricultural sources and cultivated on different solid media containing 2 M NaCl. Solidified Luria Bertani (LB) medium turned out to be the best medium to grow the organisms if compared to trypticase agar or nutrient agar; using the selective medium DSC97, substantial bacterial growth was also detected. After optimizing the cultivation conditions, the bacteria were stained with different methods such as Gram staining, methylene blue, Sudan black, etc., and subjected to biochemical characterization, encompassing the testing of citrate utilization, catalase test, and others. By 16S rRNA analysis, a Gram-positive *Bacillus subtilis* ssp. and a Gram-negative *Pseudomonas* sp. were identified as those strains revealing the highest potential for PHA biosynthesis [156].

4.4.1 *HALOMONAS BOLIVIENSIS* AS HALOPHILIC PHA PRODUCERS

In 2004, Quillaguamán and colleagues described the isolation of two novel moderately halophilic, psychrophilic, and alkaliphilic strains, classified as *Halomonas boliviensis* sp. nov. from soil samples collected around the Laguna Colorada lake, located in the south-west region of Bolivia at 4300 m above sea level; the strains were labeled as LC1 and LC2 and deposited in strain collections as DSM 15516^T=ATCC BAA-759^T, and DSM 15517=ATCC BAA-760, respectively. These strains are members of the broad *Halomonadaceae* family, which belongs to the class of gamma-proteobacteria and, within them, to the order of *Oceanospirales* [157]. Soon after its discovery, *H. boliviensis* LC1 was shown to use maltose and its oligomers, the products of incomplete amylase-mediated starch hydrolysis, for biomass growth and PHB accumulation; up to 0.56 g PHB per g CDM were obtained in shaking flask setups, while the maximum PHB fractions in batch bioreactor cultivations without oxygen limitation amounted to 0.35 g/g. Careful assessment of the impact of oxygen supply on growth and PHB biosynthesis by this strain showed the high dependence of the strain's metabolism on oxygen availability; under aeration conditions mimicking the oxygen availability in shaking flasks, the carbon flux is diverted from biomass growth towards PHB accumulation, with a strict separation between growth and PHB accumulation phase. The best results for PHB production (ca. 25 g/L PHB, 0.49 g/g PHB in CDM) were obtained in bioreactor setups with constant aeration and stirring, but uncontrolled pH-value. The optimum salt concentration to grow this organism was determined between 4 and 5%. This study constitutes the first report on PHA biosynthesis of a member of the *Halomonas* genus, which harbors the prototype strains of moderately halophile PHA producers [158]. PHB production by *H. boliviensis* LC1 was also reported from different carbon sources like glucose, xylose, sucrose, sodium acetate, and butyrate. Excess carbon sources and limited supply of the nitrogen source yeast extract boosted PHB biosynthesis by this organism. When testing different salt concentrations, the authors noticed the formation of smaller sized cells at higher salinity as a consequence of the formation of compatible organic solutes to withstand osmotic stress at the expense of cell growth; in contrast, PHB productivity was not impacted by higher salinity. The highest PHB mass fractions of 0.88 g/g were obtained using combinations of butyric acid and acetate at 45 g/L NaCl and a pH-value of 8.0. In addition, it was noticed that large and uniform PHB granules are produced by this strain, which contributed to a convenient recovery process [37]. Further studies were carried out by the same team of research-ers to optimize the medium composition in order to enhance biomass growth and PHB productivity. Shaking flask setups revealed that the addition of aspartic acid, glycine, glutamine, or glutamate drastically boosted biomass growth of *H. bolivien-sis* LC1. 0.4% NH$_4$Cl, 0.22%, K$_2$HPO$_4$, and repeated supply with glutamate turned out as the best conditions for growth and product formation; after 18 h of cultivation, PHB mass fraction and CDM reached 0.90 g/g and 23 g/L, respectively. Follow-up experiments tested the feeding of these growth components only at the beginning of the cultivation; after 36 h, CDM, PHB fraction in CDM, and volumetric PHB pro-ductivity amounted to 44 g/L, 0.81 g/g, and 1.1 g/(L·h), respectively [159].

Later, the strain *H. boliviensis* LC1 was also cultivated on wheat bran, which was hydrolyzed to a sugar cocktail containing mainly glucose, mannose, xylose, and arabinose by a crude enzyme preparation from *Aspergillus oryzae*. After 30 h cultivation on a 3+1 mixture of glucose and xylose, a PHB concentration and mass fraction in CDM of 1.1 g/L and 0.45 g/g, respectively, were obtained. On wheat bran hydrolysate as substrate, a similar PHB concentration, but a lower PHB content of 0.34 g/g were obtained. In a batch-mode bioreactor cultivation with 18 g/L sugars from hydrolyzed wheat bran, a maximum PHB content of about 0.3 g/g was reached within 20 h. Using 8 g/L butyric acid and 8 g/L sodium acetate (both available from anaerobic digestion of potato waste) and 10 g/L reducing sugars from hydrolyzed wheat brain in a 20 h cultivation batch, PHB concentration and mass fraction were increased to 4 g/L and 0.5 g/g, respectively [160].

4.4.2 *HALOMONAS VENUSTA* AS HALOPHILIC PHA PRODUCER

In 2018, PHA accumulation by the moderately halophilic strain *Halomonas venusta* KT832796 was described by Stanley and colleagues. This new proteobacterium was isolated from samples taken on the Indian coast. After optimization of the feedstock supply and the cultivation parameters, the strain was cultivated in 2 L laboratory bio-reactors under controlled conditions. Evaluating the optimum salinity for PHA pro-duction by this strain was carried out by testing a NaCl concentration range between 1.5 and 15% NaCl; as a result, the lowest concentration of 1.5% delivered best results, thus characterizing this strain as a moderate halophile according to the classifica-tion presented at the beginning of this chapter. Other studies assessed the impact of different nitrogen- (ammonium acetate, -chloride, -citrate, -nitrate, -oxalate, -phos-phate, and -sulfate) and carbon sources (glucose, fructose, sucrose, lactose, and glyc-erol). As a result, ammonium citrate and -sulfate acted as the most suitable nitrogen sources, while glucose turned out to be the most efficient carbon source. Based on shaking flask experiments with glucose and ammonium citrate at a ratio of 20/1, *H. venusta* KT832796 produced 3.52 g/L biomass, with 0.7 g PHA per g biomass. When running fed-batch cultivation bioreactor experiments, switching from ammonium citrate to ammonium sulfate turned out to be more beneficial for overall productivity. For increasing PHA productivity, the authors tested various feeding strategies like the addition of substrate pulses and substrate supply coupled to changing pH-values. The coupling of feeding to changing pH-values increased the PHA concentration to 26 g/L; as a drawback, the main part of the carbon source was used by the strain for production of non-PHA biomass, which explains the rather low PHA content in biomass of only about 0.4 g/g. Supplying a highly concentrated single substrate pulse delivered the highest PHA concentration of 33.4 g/L, and an excellent PHA content in biomass of about 0.88 g/g, which constituted a nine-fold increase if compared to simple batch cultivation setups [161]. A close relative of the representatives of the genus *Halomonas* described above, namely the thalassic organism *Halomonas profundus*, was isolated from a shrimp collected next to a hydrothermal vent and turned also out to be a producer of PHB homopolyester producer from diverse carbon sources such as glucose, glycerol, octanoate, or acetate. When adding propionic acid or valeric acid, PHBHV copolyesters were produced. However, this organism, in contrast to other *Halomonas* sp., grows best at a salinity of 2–3% NaCl, hence in the moderate salinity range. The pH-optimum of this organism is reported at moderate alkalinity of pH-value 8–9 [162].

4.4.3 *HALOMONAS CAMPANIENSIS* SSP. AS A HALOPHILIC PHA PRODUCER

Halomonas campaniensis, a haloalkaliphilic bacterium isolated from an algae-covered mineral pool in the Campania Region, Italy, was described in 2005 by Romano and colleagues. This strain showed optimum growth at 100 g/L NaCl and pH 9, but also grew in the absence of salt. Importantly, this strain accumulated the osmo-protectants glycine-betaine, ectoine, and glutamate as typical organic solutes, and excretes EPS [163]. In 2014, Yue and colleagues carried out the first cultivations with halophile eubacteria on a larger scale to study PHA biosynthesis under reduced sterility precautions. For these experiments, the recombinant halophilic and alkalophilic bacterium *H. campaniensis* LS21 was used, the wild type of this strain was originally isolated from sludge and plant residues collected from a Chinese salt lake. PHA production by *H. campaniensis* LS21 was studied in a continuously operated, non-sterile, open process. The cultivation medium was composed of alkaline seawater instead of fresh water and carbonaceous artificial kitchen waste mainly consisting of carbohydrates, proteins, and lipids in order to save compounds of nutritional significance. PHA acted as the model product to study the feasibility of running industrially relevant long-term cultivation processes in open systems by bacteria [164].

To an increasing extent, genetic engineering techniques are used to optimize PHA biosynthesis in terms of manipulating growth kinetics for fast biomass growth, modifications to expand cell volumes to provide more space for PHA storage, reprogramming the PHA synthesis pathways by resorting to optimized ribosome binding sites and/or PHA operon promoters, rerouting the intracellular metabolic fluxes towards PHA biosynthesis by today's omnipresent gene scissors (CRISPRi), and, as the topic of this chapter, the use of still underexplored host strains like proficient, robust halophiles [165–168].

Genetic engineering was also studied in the case of *H. campaniensis* LS21 to further optimize the process by equipping the strain with additional *phbCAB* genes encoding for the PHB synthesis enzymes. Both the wild type and the genetically engineered strain were cultivated for more than two months in a continuously operated, open process in the above-described artificial medium. A temperature of

37°C, elevated salinity (27 g/L NaCl), and high alkalinity (pH-value 10) turned out to be the optimal conditions for the cultivation of this organism. The engineered *H. campaniensis* LS21 strain reached a PHA content in biomass of about 0.7 g/g under the described optimum conditions, while the wild type organism accumulated not more than 0.26 g PHA per g biomass. Most importantly, during the entire cultivation period of 65 days microbial contamination was neither observed for the wild type, nor for the genetically engineered strain; both cultivation setups maintained their monoseptic character, even though they were operated under open, non-sterile conditions. Extracellular hydrolytic enzymes were excreted by the cells, which enabled the conversion of the complex substrates present in the artificial kitchen waste. The plasmid carrying the *phbCAB* PHA biosynthesis genes was also kept by the recombinant cells during the long cultivation period, which illustrates the possibility of using this recombinant organism in long-running continuous cultivation processes. In combination with its useful accessibility to genetic engineering, *H. campaniensis* LS21 definitely seems an auspicious candidate for cost-, water-, and energy-saving biosynthesis of PHA and other marketable products such as extremozymes (alkali-proof halozymes) starting from inexpensive raw materials [165]. In the context of haloalkaliphiles, Legat and colleagues described another alkaliphilic haloarchaeon, *Natronobacterium gregoryi* NCMB 2189T, as a PHA producer. This strain was cultivated at 20% NaCl and a pH-value of 9.0, and accumulated PHBHV copolyester [144].

A convenient downstream processing technique to directly recover PHA from wet biomass of *H. campaniensis* without any pre-treatment like drying was presented by Strazzullo and associates. These authors added sodium dodecyl sulfate (SDS) directly to ultrasonic-assisted dispersions of PHA-containing humid cell pellets in distilled water; this was followed by shaking, autoclaving, and repeated washing. Regardless of the density of the cell dispersion, the recovered PHA produced by *H. campaniensis* had a purity of more than 95%; the best recovery yields obtained were 0.12 g PHA per g cell wet mass when the microorganism was cultivated in a glucose-based medium or a glucose/propionate-based medium [169].

4.4.4 *HALOMONAS* TD01 (*HALOMONAS BLUEPHAGENESIS*) AND ITS GENETICALLY ENGINEERED STRAINS AS HALOPHILIC PHA PRODUCERS

The halophilic proteobacterium *Halomonas* TD01 was isolated from the Aydingkol salt lake in PR China, and later named *Halomonas bluephagenesis* TD01 [170]. In 2011, the genome of this strain was completely deciphered by Cai and colleagues, which enabled genetic engineering of this organism. In particular, these authors studied the genes involved in PHA and osmolyte production by this strain in silico to identify horizontal gene transfer events and co-evolutionary relationships between individual genes for the production of PHA and factors involved in osmoregulation. As a major outcome, it was shown that *Halomonas* TD01 favors the accumulation of organic solutes than the salt-in strategy for balancing intracellular osmotic pressure [171]. In the same year, Tan and colleagues cultivated this strain in a continuous, open, non-sterile process [170], analogous to the experiments using *H. campaniensis* LS21 described by Yue et al. [164]. Fed-batch cultivations were carried out at 37°C,

5–6% NaCl, a pH-value of 9.0, with glucose as the sole carbon source, and lasted for 56 h. Under these conditions, the strain reached 80 g/L of biomass and 0.8 g PHA per g biomass. Further, this strain was cultivated in a two-stage continuous process, which was also operated under open, non-sterile conditions; the setup consisted of two parallel continuously operated stirred-tank bioreactors with working volumes of 3 and 1 L, respectively, at the start of the process, and was operated for two weeks. A saline medium rich in both glucose and nitrogen sources was used for the first stage; here, an average biomass concentration of 40 g/L and a PHA content of about 0.6 g per g biomass were achieved. The bacterial suspension was continuously transferred from this first into the second bioreactor; here, the concentration of nitrogen source was zero in order to increase PHA productivity. This transfer resulted in a dilution of the biomass concentration; however, a constant PHA content between 0.65 and 0.7 g per g biomass was reached in the second bioreactor. The conversion yield of glucose to PHA in the first bioreactor amounted to 0.20– 0.30 g/g, but, remarkably, exceeded 0.50 g/g in the second stage. During the process, the cultivation volume in the second bioreactor was kept constant at 3 L. Biomass continuously taken from the second bioreactor was harvested via centrifugation; the resulting liquid phase (spent fermentation broth, supernatant) was treated at 50–60°C and pH-value 10, cooled to room temperature, and recycled in feed streams to bioreactors 1 and 2 [170]. This process for recycling the spent supernatant constitutes a continuous version of the reuse of saline waste streams, which was also accomplished in batch mode for spent *Hfx. mediterranei* fermentation broth [123].

   H. bluephagensis TD01 was also genetically modified by knocking out the gene encoding the enzyme 2-methylcitrate synthase, which causes intracellular depletion of propionyl-CoA. When adding the 3HV precursor propionate, this genetic engineering approach almost doubled the yield of propionate-to-3HV conversion and resulted in a duplication of the molar 3HV share in randomly distributed PHBHV copolyesters. Using a medium of defined composition with glucose as the main substrate and 0.5 g/L of the 3HV precursor propionate, 0.7 g PHBHV were obtained per g biomass, together with a molar 3HV fraction in PHBHV of 0.12. Further, the genes encoding for three PHA depolymerases were deleted from the strain's genome in order to increase PHA productivity by avoiding intracellular PHA degradation, which is typically observed during the later phases, particularly in large-scale cultivations. However, the deletion did not cause a considerable enhancement of overall PHA productivity. In 500 L pilot-scale setups supplied with glucose as the sole carbon source, the genetically modified *H. bluephagenesis* TD01 reached a biomass concentration of 112 g/L and 0.7 g PHB per g biomass. After supplementation of propionate, 80 g/L biomass and 0.7 g PHBHV per g biomass were achieved, with the molar 3HV fraction in PHBHV amounting to 0.08. In experimental setups on a shaking flask scale, even 0.92 g PHB per g biomass and drastically increased substrate-to-PHB conversion yields were obtained [172]. Further, the DNA restriction/methylation system of the strain was partially inhibited by Tan et al. in order to increase the genetic stability of this engineered *Halomonas* TD01; moreover, a stable conjugative plasmid called *pSEVA341* was designed for inducing the expression of genes involved in multiple pathways, such as overexpression of the threonine synthesis pathway. Hence, the new engineered strain called *Halomonas* TD08 was characterized by the deletion of 2-methylcitrate synthase and the knockout of three depolymerases; the construct accumulated up to 0.82 g PHBHV per g biomass. Long-term continuous cultivation test runs were suggested by the authors as necessary follow-up studies in order to evaluate the stability and performance of these promising modified halophilic cell factories [173].

In 2017, Tao and colleagues used CRISPRi successfully for the first time in *H. bluephagenesis* TD01 to repress the expression of the *ftsZ* gene encoding the bacterial fission ring (Z-ring) formation protein, which leads to elongated cells of filamentous shape. CRISPRi was employed to regulate the expression of the *prpC* gene encoding 2-methylcitrate synthase in order to trigger the 3HV fraction in PHBHV copolyesters from less than 1 to 13%. In addition, repression of the *gltA* gene encoding citrate synthase directed more acetyl-CoA from the TCA cycle to PHB biosynthesis, which resulted in an increased PHB accumulation by about 8% compared with the parental strain. Only recently, it was shown by Chen and colleagues (2019) how chromosome engineering of the TCA cycle via CRISPRi can enhance PHBHV biosynthesis in *H. bluephagenesis* using glucose as the sole carbon source. In this study, the TCA cycle genes *sdhE* and *icl* encoding the succinate dehydrogenase assembly factor 2 and isocitrate lyase, respectively, were deleted. This resulted in redirection of the metabolite flux towards 3HV biosynthesis. Resulting strain *H. bluephagenesis* TY194 ($\Delta sdhE\Delta icl$), equipped with a P_{porin}-194 promoter, accumulated 17 mol-% 3HV in PHBV. Moreover, gluconate was used to regulate the NADH/NAD$^+$ ratio to improve TCA cycle activity, and thus, to further enhance 3HV synthesis. Additional insertion of genes encoding phosphoenolpyruvate carboxylase and *Vitreoscilla* hemoglobin (increases oxygen availability according to [174]) into the chromosome enhanced the TCA cycle activity; this resulted in 6.3 g/L CDM, 65 wt.-% PHBHV in CDM and 25 mol-% 3HV in PHBHV when the new genetically engineered strains were cultivated on a shaking flask scale using glucose and gluconate as sole carbon sources [168].

Recently, various strategies were examined to produce the copolyester poly(3HB-*co*-4HB) in a cost-effective manner. One of these approaches resorts to *H. bluephagenesis* TD01, which is not a natural producer of 4HB-containing PHA biopolyesters. *H. bluephagenesis* TD01 was genetically modified by inserting the gene *orfZ* encoding *Clostridium kluyveri* 4HB-CoA transferase; this strategy enables the organisms to synthesize poly(3HB-*co*-4HB) when supplied with glucose and the 4HB precursor compound GBL. One m³ bioreactor pilot-scale cultivation runs were carried out with this engineered strain, resulting in the generation of 83 g/L biomass, a poly(3HB-*co*-4HB) fraction in biomass of 61%, and a molar share of 4HB in poly(3HB-*co*-4HB) of 16 mol-%. The produced copolyester revealed an extraordinarily high elongation at break of about 1000%. Also in this case, it was possible to carry out the process under open, non-sterile conditions because of the high salt concentration (60 g/L NaCl) in the cultivation medium, and to achieve similar results both at 1.7 L and 1000 L scales [175]. Further pilot scale studies demonstrated the possibility of using this strain of for poly(3HB-*co*-4HB) production on a 5 m³ production scale under non-sterile, continuous cultivation conditions; for these setups, waste gluconate was used as cheap carbon source. Volumetric productivity for poly(3HB-*co*-14%-4HB) of 1.67 g/(L·h) was reported; within 36 h of cultivation, 100 g/L CDM, and about 60 g/L PHA were obtained [176]. In such modified *H. bluephagenesis* strains harboring the *orfZ* gene encoding 4HB-CoA transferase, gene expression was boosted by selected promoters from a newly constructed promotor library; this library, based on the P_{porin} core region, was developed for the transcription of various different heterologous genes, and was used as a platform for fine-tuned metabolic engineering of *H. bluephagenesis*. The best among the promotor-driven genetic constructs was able to produce a CDM of more than 100 g/L, harboring 80% poly(3-HB-*co*-11 mol-% 4-HB). After 50 h of cultivation under non-sterile fed-batch conditions, a productivity of 1.59 g/ (L·h) was achieved [177]. Further engineering of this strain was accomplished by Ye and colleagues, who succeeded in making it capable of poly(3HB-*co*-4HB) production even from glucose as sole carbon source without the addition of GBL or other 4HB precursor compounds; by this "low cost platform", which was designed by "pathway debugging" and whole genome sequencing and comparative genomic analysis, 25 mol-% 4HB was determined in the poly(3HB-*co*-4HB) copolyester. This was enabled by knocking out the genes encoding succinate semialdehyde dehydrogenase, which catalyzed a reaction competing with 4HB biosynthesis [178]. Additional studies involving this promising bacterium resorted to genome-wide random mutagenesis as a tool to make the organism resistant to inhibiting metabolites like short chain organic acids or ethanol, which are increasingly accumulated during aspired long-term cultivation at high biomass concentration. Such genetic modification of the generated constructs were performed to achieve overexpression of the PHA synthesis genes by overexpression an optimized *phaCAB* operon; in 7 L bioreactor setups, a biomass concentration of 90 g/L CDM and a polyester fraction of 0.79 g PHA per g biomass were obtained by this construct labeled *H. bluephagenesis* TDHCD-R$_3$-8-$_3$, which is superior to values obtained using the wild type strain, where only 81 g/L biomass and 0.10 g/g polyester content in biomass were achieved [179].

Considering the fact that high cell density in aerobic processes, such as the cultivation of *Halomonas* sp., can only be obtained by a sufficient oxygen supply, technological solutions are needed to enhance oxygen availability in cells. However, the use of pure oxygen for aeration instead of compressed air is expensive, and intrinsically bears the risk of explosion. As a novel approach, the *Vitreoscilla* hemoglobin (VHb), which creates complexes with molecular oxygen, can be expressed in cells of PHA-producing strains like *H. bluephagenesis* TD01 and *H. campaniensis* LS21 with the aim of increasing oxygen availability for the cells and consequently enhancing cell growth. As shown by Ouyang and colleagues, bacterial cell membranes impose barriers for oxygen transfer towards VHb when present intracellularly in the cytoplasm. As a solution, the authors succeeded in expressing VHb in the periplasmic space, which turned out to be an excellent compartment for short-term storage of VHb-bound oxygen. In this study, the twin-arginine translocase pathway was used to transport VHb into the periplasm; together with the insertion of promotors inducing the expression of both the VHb encoding gene (*vgb*) and the PHB synthesis operon under low-oxygen conditions, it was possible to double the obtainable biomass concentration of the investigated halophilic PHA producers, which is a prerequisite to boost the volumetric PHA productivity in the later stages of the production process [172].

Surprising findings in the field of PHA biosynthesis by *H. bluephagenesis* were reported by Ling and colleagues, who showed for the first time that this strain uses NADH instead of NADPH as a cofactor for PHB production, which results in enhanced PHA accumulation under oxygen-limited cultivation conditions. To increase the NADH/NAD⁺ ratio under oxygen limitation, the authors blocked an electron transport pathway containing electron transfer flavoprotein subunits α and β encoded by the *etf* operon was to increase NADH supply; this way, 90% PHA in CDM was achieved, which is higher compared with the wild type strain (84 wt.-%). Acetate was supplied together with glucose to balance the redox state and to alleviate inhibition of pyruvate metabolism, resulting in 22% more biomass and 94 wt.-% PHA in CDM. The cellular redox state changes induced by acetate addition increased the 3HV fraction in PHBV from 4 to 8%, and the 4HB fraction in poly(3HB-*co*-4HB) from 8 to 12%, respectively. Hence, systematically modulating the redox potential in *H. bluephagenesis*, which enhances PHA accumulation and allows fine-tuning the monomeric composition of PHA copolymers under oxygen limitation, the reduction of energy consumption and, finally, makes the cultivation system less complicated [180].

In order to facilitate PHA recovery, especially when considering large-scale production process, it is beneficial to generate large PHA granules, which have a positive impact on the PHA recovery via filtration, sedimentation, or centrifugation. In this context, genes encoding for the production of three different phasins, hence, amphiphilic enzymes on the PHA granule surface triggering the number and size of granules in cells, were deleted in *H. bluephagenesis*. Although this approach resulted in a reduced number of granules per cell, and in larger granule sizes, especially when deleting phasin 1 or deleting together phasin 1 and 2, and 1 and 3, respectively, an increase in granule sizes was confined by the cell size. Therefore, the genes encoding for the enzymes that block cell fission rings ("Z-rings") needed for binary cell fission were also overexpressed in phasin-knock out mutants. This way, "giant cells" of *H. bluephagenesis* were created, which harbored PHA granules with sizes up to 10 μm, which is roughly 20 times the size of granules observed in the parental strain. Such "giant cells" with huge granules can be conveniently harvested by low-speed centrifugation. The authors assume that "if the PHA granule sizes are engineered further to be as large as the starch particle sizes, PHA can be purified and recovered as conveniently as starch powder"; which could definitely contribute to making PHA production even more economic [181].

4.4.5 *Halomonas halophila* as a Halophilic PHA Producer

The strain *Halomonas halophila* was recently identified as a potent PHA producer. *Halomonas halophila*, formerly *Deleya halophila*, belongs among moderately halophilic bacteria, which encompass those bacteria which grow optimally at NaCl concentrations between 30 and 150 g/L. *H. halophila* is a Gram-negative, strictly aerobic bacterium capable of utilizing a wide range of carbohydrates. It was originally isolated near Alicante in southeast Spain, and was deposited at the Czechoslovak Collection of Microorganisms, Brno, Czech Republic, as strain CCM 3662T [182] The bacterium is capable of growing on simple and inexpensive mineral media without the requirement of expensive complex nitrogen sources such as yeast extract or peptone. Further, the bacterium is capable of PHB homopolymer production in extensive amounts up to 82% per CDM under elevated salinity of 66 g/L NaCl. It was

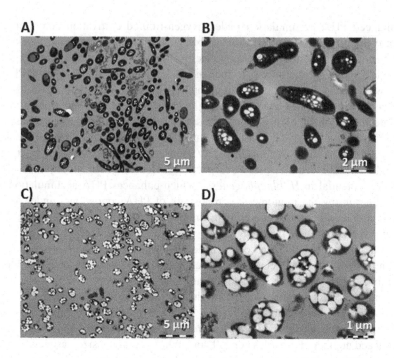

FIGURE 4.3 TEM microphotographs of cells of *H. halophila* cultivated in nitrogen-rich cultivation media (A, B; 4 wt.-% PHA in biomass) and nitrogen limited mineral media (C, D; 70 wt.-% PHA in biomass). PHA granules are well visible as light-refractile inclusion bodies.

observed that salinity of the cultivation media affects the molecular weight of the polymer: an increase in NaCl concentration results in an increase in the molecular weight of PHB. While higher salt concentration up to 100 g/L increased M_w to up to 810 kDa, M_w of about 420 kDa was obtained under moderately halophilic conditions of 20 g/L NaCl. In parallel, Đi of PHA by trend increased with increasing salt concentration. Hence, salinity can be used as a tool to produce polymers with tailored properties. Except lactose, *H. halophila* is able to utilize most common sugars, including glucose, saccharose, galactose, xylose, or mannose. Therefore, it seems to be an auspicious candidate for PHA production from lignocellulose substrates, which are rich in these carbohydrates. Nevertheless, prior to conversion of lignocellulose materials into PHA, polysaccharides must be cleaved to yield fermentable sugars. This process usually includes chemical hydrolysis of hemicellulose by mineral acids and subsequent enzymatic hydrolysis of cellulose. Necessary neutralization of the hydrolysates results in the presence of high NaCl concentration in cultivation media. Therefore, it is wise to employ halophiles for the biotechnological valorization of lignocellulose hydrolysates since they are naturally tolerant to high salt concentrations, and there is no need to remove salts from cultivation media prior to cultivation. In this context, *H. halophila* was successfully tested for PHA production from hydrolyzed inexpensive substrates such as spent coffee grounds, sawdust, or corn stover which indicates its potential for industrial PHB production [12]. Figure 4.3

shows TEM photographs of this organism under different cultivation conditions (salinity: 66 g/L NaCl, carbon source: 20 g/L glucose; A and B: nitrogen-rich conditions, C and D: nitrogen limitation).

More recently, this strain was also cultivated on differently pretreated hydrolyzed spent coffee grounds. Hydrolysis of this abundant food waste generates inhibiting compounds which, together with phenolics constitutively present in coffee grounds, impede growth and PHA biosynthesis by this strain. A viable solution was developed by detoxifying the hydrolysate using styrene-divinylbenzene -based resins as sorbents for the inhibitors after extracting before originally present phenolics. In 72 h shaking flask setups, almost 1 g/L PHB was accumulated by *H. halophila* using this inexpensive feedstock, with polymer fractions in biomass amounting to 0.27 g/g [183].

4.4.6 OTHER *HALOMONAS* SP.

Additional *Halomonas* sp. strains are currently in investigation status for their potential for PHB production under open, non-sterile conditions. In 2006, Romano and colleagues described co-production of osmolytes, PHB, and extracellular polysaccharides by the new salt-pool isolate 18bAG^T, which tolerates op to 200 g/L NaCl and thrives optimally under alkaline conditions of pH 9; therefore, the authors proposed to new species name *Halomonas alkaliphila* sp. nov. for this organism [184].

The moderately halophilic bacterium *Halomonas marina* strain HMA 103, deposited as MTCC 8968, was isolated from solar salterns in Orissa, India, and characterized by Biswas et al. The strain grows best at 100 g/L NaCl and synthesizes PHB during growth. PHB production by this strain was optimized in shaking flask batch cultivation setups using 20 g/L glucose as the carbon source; PHA fraction of 0.59 g per g CDM was obtained after 50 h of cultivation. Optimum PHB biosynthesis was achieved by supplying NH$_4$Cl and yeast extract as nitrogen sources, 0.01% phosphate, 1.5% sulfate together with 100 g/L NaCl. Growing the strain on alkanoic acids (propionic acid, butyric acid, valeric acid, capronic acid) resulted in the production of PHBHV copolyesters. A two-step cultivation approach considerably increased PHBHV accumulation up to 0.8 g PHBHV (12.8 mol-% 3HV) per g CDM in a glucose medium supplemented with 1 g/L valerate as 3HV precursor [185].

Co-production of PHB homopolyester and the fructan-type EPS levan was demonstrated by Tohme and colleagues, who cultivated *Halomonas smyrnensis* under unlimited cultivation conditions starting from simple carbon sources like glucose or sucrose. In the presence of 20 g/L glucose, biomass and PHA concentrations reached their maxima after 120 h of fermentation in shaking flasks with a PHA concentration of 1.34 g/L (46% PHA), while cultures grown on 50 g/L sucrose reached stationary growth phase after 120 h, with the highest PHA accumulation reported after 144 h with PHA concentrations of 0.48 g/L (27% PHA in biomass). For these setups, 137.2 g/L sea salt from the Çamaltı region (Turkey), a pH-value of 7.0, and a temperature of 37°C were used. Together with PHB, 15.3 g/L levan were produced in the presence of sucrose, but no detectable levan could be recovered from cell-free fermentation media of the cultures grown on glucose [186]. In general, the co-production of PHA and EPS is a frequently observed feature among *Halomonadaceae*; besides

PHA and levan co-production by *H. smyrnensis*, parallel PHA and EPS biosynthesis is described for *Halomonas maura, Halomonas ventosa, Halomonas anticariensis, Halomonas alkaliphila, Cobetia marina, Halomonas eurihalina, Halomonas sinaiensis*, and *Halomonas almeriensis* [reviewed by 21].

The marine bacterium *Halomonas* sp. SF2003, originally isolated from the Iroise Sea, was cultivated in 2019 by Lemechko and colleagues on agro-industrial effluents as the sole carbon sources in a medium containing 11 wt.-% sea salts; experiments were carried out in two-stage processes (growth medium, followed by accumulation medium) in two subsequent 5 L bioreactors. In 40 h cultivation, this organism and was able to produce PHB homopolyester at a productivity of 1.3 g/L with a M_n of 342,000 g/mol. When adding valeric acid as 3HV precursor, PHBHV copolyesters with controllable 3HV fractions (0–35 wt.-%) were obtained. Thermal and mechanical characteristics of the copolyesters were investigated in dependence on the 3HV fraction; it was shown that the glass transition temperature T_g, the melting temperature T_m, tensile strength, and the Young's modulus by trend decrease with increasing 3HV content, while elongation at break increases with increasing 3HV content in PHBHV [187]. Only recently, the complete genome sequence of *Halomonas* sp. SF2003 was deciphered by Thomas et al., who, inter alia, reported the presence of typical PHA biosynthesis genes like *phaA, phaB*, and *phaC*, together with a preliminary analysis of their intracellular organization and characteristics. Further, the authors detected the high versatility of this strain to adapt to various salinities and temperatures and its high potential for future industrial applications such as PHA production [188].

In 2019, Hernández-Núñez et al. reported production of PHB by *Halomonas salina*, an organism isolated from a hypersaline microbial mat at the Exportadora de Sal (ESSA) saltworks at Guerrero Negro, Baja California Sur, Mexico. This strain was able to produce isotactic PHB homopolyester when supplied with 10 g/L glucose as the sole carbon source and a total salt concentration of about 25 g/L for 72 h in shaking flasks. About 0.6 g/L PHB, corresponding to about 26 wt.-% in CDM, were obtained by these setups. The polyester was thermally stable up to 225°C, and revealed a melting temperature T_m of 173.6°C. It showed lower crystallinity (39.3%) in comparison with PHB produced by other bacteria and even when compared with PHA copolyesters. The authors suggested this PHB for use in further potential biotechnological applications in which elastic polymers are needed [189].

Considerable PHA production potential was also observed in *Halomonas neptunia* and *Halomonas hydrothermalis* since both strains are capable of converting lipids, such as waste frying oil, into PHA. It was observed that *H. hydrothermalis* reveals a unique morphology of PHA granules since each cell contains a remarkably high number (more than 25) of granules with substantially low diameter; this indicates a high level of expression of phasins, which are responsible for number and size of PHA granules, in *H. hydrothermalis*. Initial concentration of NaCl was identified as a crucial parameter influencing PHA yields as well as molecular mass of the polymer especially in *Halomonas hydrothermalis*, which seems to be a more promising strain than *H. neptunia*. Optimal salinity of the cultivation medium is 40 g/L of NaCl, which provides maximal productivity in flask experiments with

H. hydrothermalis; here, PHB content in biomass reached 62% of CDM, and the PHB titer achieved 2.26 g/L. In addition, *H. hydrothermalis* was capable of efficient biosynthesis of PHBHV copolymer, when valerate was utilized as a precursor, the 3HV fraction in the copolymer reached high values of 50.15 mol.-% [190]. Figure 4.4 shows the different arrangement and morphology of PHA granules in the two different strains by TEM pictures.

Halomonas neptunia   *Halomonas hydrothermalis*

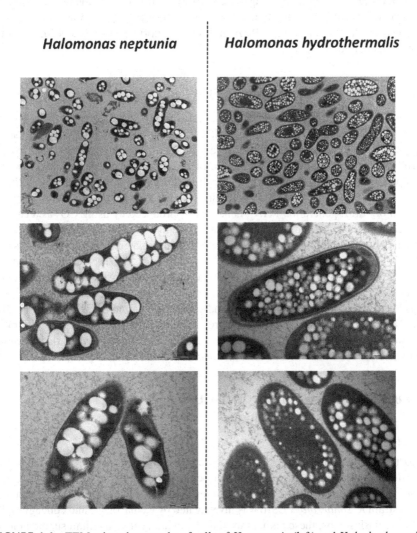

FIGURE 4.4   TEM microphotographs of cells of *H. neptunia* (left) and *H. hydrothermalis* (right) under different magnifications. The different intracellular arrangement and morphology of PHA granules (*H. neptunia*, left row: large granules, uniformly distributed in the cytoplasm; *H. hydrothermalis*, right row: characteristic "pearl necklace" arrangement of granules in the cells). *H. hydrothermalis*: 62 wt.-% PHA in biomass, salinity 40 g/L NaCl; *H. neptunia*: 50 wt.-% PHA in biomass, 60 g/L NaCl. Waste frying oil from a canteen was used as carbon source for both organisms.

In a nutshell, the genus *Halomonas*, which encompasses diverse PHA-producing strains, which can be cultivated at moderately alkaline pH-values and high salt concentrations under open and non-sterile conditions without risking microbial contamination, display an excellent platform for process optimization via the tools of genetic engineering and synthetic biology. This allows better control of bacterial growth and product formation, shifting of the intracellular carbon flux, and modification of cell morphology for facilitating accumulation of high PHA loads in enlarged and elongated cells. It is very likely that such approaches will open up new avenues for efficient production of tailored PHA and other important bioproducts like EPS or special osmolytes [171, 186, 191].

4.4.7 HALOPHILIC PHOTOTROPHS AS PHA PRODUCERS

Among Gram-negative bacteria, cyanobacteria display a distinguished phylogenetic group, characterized by their phototrophic life form and the capability of fixation of molecular nitrogen (diazotropy). Cyanobacteria occur in various aquatic and terrestrial habitats, and besides being described as producers of various bioactive compounds, pigments (phycobilines), extracellular polysaccharides, and others, they are more and more in the spotlight of researchers as cell factories for generation of "third-generation PHA", hence, PHA produced by using CO_2 as a carbon source [192–194]. In 2010, Shrivastav and colleagues studied *Spirulina subsalsa*, a marine photoautotrophic cyanobacterium organism isolated from Indian coast samples. When exposed to an elevated salt concentration of 50 g/L NaCl, this organism revealed increased PHA biosynthesis, which was proposed by the authors as an indicator for PHA's protective role against osmotic up-shock. The produced biopolyester was identified as PHB homopolyester, and was characterized by DSC, TGA, FT-IR, and NMR [195]. Cyanobacteria, which are at the same time halophilic and alkaliphilic, were also described as PHA producers, e.g. for *Spirulina platensis*, the optimum pH-value for PHA anabolism and catabolism amounts to 9 to 11 [196].

4.4.8 *RHODOBACTERACAEA* AS A FAMILY ENCOMPASSING HALOPHILIC PHA PRODUCERS

4.4.8.1 *Rhodobacter* sp.

Apart from cyanobacteria, photosynthetic non-sulfur purple bacteria from the *Rhodobacteracaea* family are also described to produce PHA. In this context, the halotolerant strain *Rhodobacter sphaeroides* U7, a UV mutant strain of the halotolerant strain *R. sphaeroides* ES16 was investigated by Kemavongse and colleagues. In shaking flask setups, these authors used a glutamate-acetate medium supplemented with sodium valerate and 30 g/L NaCl under different cultivation temperatures, with and without the addition of ammonium sulfate. The optimum values to grow this organism turned out to be 0.2 g/L ammonium sulfate and 30°C, which were used for 5 L bioreactor cultivations under both aerobic-dark conditions with different aeration rates and under anaerobic-light (3000 lux) conditions. Different stirrer agitation rates were tested to investigate the impact. Anaerobic-light (3000 lux) cultivation with an aeration rate of 1.0 vvm and a stirrer speed of 200 rpm resulted in the highest PHA concentration of 2.5 g/L, while the highest PHA mass fraction in cells of 0.65 g/g was obtained at an aeration of 0.5 vvm and an agitation of 200 rpm. The product was recovered from biomass and characterized via ^{13}C-NMR, FT-IR, DSC, X-ray diffraction, and intrinsic viscosimetry, and identified as PHBHV copolyester with a 3HV content of 15.2 mol/mol [197].

4.4.8.2 *Yangia* sp.

High salinity often prevails in microbial consortia found at coastal marine locations and in estuarine systems, which, therefore, offer promising environments to discover new halophilic microbes. In this context, the entire genome sequence of a halophilic bacterium isolated from Malaysian Matang Mangrove soil sediments, which was identified as *Yangia* sp. CCB-MM3, was described in 2017 by Lau and associates. Both marine conditions and rivers characterize the Malaysian Mangrove ecosystem and its bionetwork. This was the first entirely decoded gene sequence of a member of *Yangia*, which is a rather new genus of the *Rhodobacteraceae* family. The *Yangia* sp. CCB-MM3 genome consists of two chromosomes and five plasmids, the entire genome has a size of 5.522,061 bp, with the average GC share amounting to 0.65 bp/bp. Genome sequence analysis confirmed the presence of a propionyl-CoA synthesis pathway and a PHA biosynthesis gene cluster in this strain. Its PHA production capability was experimentally evidenced and verified also in vitro. This study might considerably support the understanding of unique metabolic and physiological features of the genus *Yangia* in particular, and generally of the *Rhodobacteraceae* family [198]. Similarly, another representative of the *Yangia* genus isolated from mangrove forest soil, *Yangia* sp. ND199, was shown to produce PHB homopolyester on different substrates like glucose, fructose, glycerol, sugarcane molasses, and corn syrup, although with rather modest biomass concentrations and PHB fractions in cells [199]. Using glycerol as main carbon source and yeast extract in 3 L bioreactor cultivations at 32°C and 45 g/L NaCl, this strain produced a maximum biomass concentration of 5.7 g/L and 0.53 g/g PHA in biomass; in this case, the polyester contained about 3 mol-% 3HV monomers. Switching from yeast extract to glutamate as the nitrogen source, homopolyester PHB was produced [200]. Later, Phong and colleagues tested this organism in enhanced ten-liter batch and fed-batch bioreactor cultivation setups. This time, the salinity of the media amounted to 30 g/L NaCl, fructose was supplied as the main substrate beside peptone and yeast extract. While in batch setups about 8 g/L biomass, less than 4 g/L PHA, a maximum PHA content on biomass of 0.49 g/g, and a volumetric productivity of about 0.12 g/(L·h) were achieved, the fed-batch setups, which maintained a fructose concentration of about 15 g/L by substrate pulse feeds, delivered considerably better results, with about 21 g/L biomass, almost 20 g/L PHA, a maximum PHA content on biomass of 0.45 g/g, and a volumetric productivity of about 0.16 g/(L·h). This process was further enhanced by running a two-stage fed-batch process, where a nitrogen feed was supplied in the first stage (18 h) of the process in parallel to low fructose concentration of only 5 g/L; in this case, the PHA content reached only about 0.1 g/g due to the ample availability of nitrogen source. In the second stage, fructose concentration was increased to about 20 g/L and a nitrogen-free fructose feed was supplied, causing an increase of the intracellular PHA mass fraction to 67.5% after 54 h of cultivation. After 45 h of cultivation, the volumetric PHA productivity reached its highest value of 1 g/(L·h), which displays an outstanding value for halophilic organisms [201].

4.4.8.3 *Paracoccus* sp.

The moderate halophilic strain *Paracoccus* sp. LL1, another member of the *Rhodobacteraceae* family, was isolated from soil and water samples collected from the Lonar crater located in the Indian district Buldhana, Maharashtra, by Sawant and associates, and first described in 2015 as a proficient PHA producer. 16S rRNA sequencing and *phaC* partial gene isolation revealed that the isolate is a new species of the genus *Paracoccus*. In this study, the lignocellulosic waste material corn stover was hydrolyzed to a crude cocktail of hexose and pentose sugars by a cellulase mixture obtained by co-cultivation of the fungi *Trichoderma reesei* and *Aspergillus niger*. In reference cultivations on pure sugars (glucose, xylose, cellobiose), *Paracoccus* sp. LL1 grew significantly better and accumulated considerably more PHA than other isolates tested in this study. The strain accumulated PHA up to 72.4% of its CDM, corresponding to a CDM of about 7 g/L and a PHB concentration of about 4.5 g/L at optimum C/N ratios of 20/1 on artificial mixtures of the three mentioned sugars, and a salinity of 10 g/L NaCl. Changing the initial sugar concentration in shaking flask studies at a fixed C/N ratio (20/1) and a glucose/xylose/cellobiose ratio of 14/2/2, the highest PHA production (almost 20 g/L, about 0.72 g PHA per g CDM) was observed when using 100 g/L of total initial sugar concentration. Using the hydrolyzed corn stover cocktail containing 40 g/L sugars in a 5 L scale bioreactor fermentation operated in batch mode, PHA concentration and PHA fraction in biomass reached 10 g/L and 0.72 g/g within 72 h; with a yield of about 0.25 g/g, the conversion of sugars from the complex substrate mix to PHA was even slightly higher than in the small-scale processes on purified sugars. Based on FT-IR spectroscopy, the product was identified as PHB homopolyester [202]. Later, this moderately halophilic organism successfully converted a wide range of carbon sources, namely glycerol, galactose, fructose, lactose, mannitol, and methanol; in this study performed by Kumar et al., glycerol resulted in the highest PHA contents in biomass. The high osmo-tolerance of the organism was manifested by its high growth at glycerol concentrations of even 100 g/L; the growth and PHA production on a medium containing 5 g/L methanol as the sole carbon source suggested the use of this organism for cultivations on crude glycerol phase from biodiesel production. Further, a pH-optimum of 7.5 was determined on a shaking flask scale. In a batch fermentation in a 5 L bioreactor using a mineral media supplemented with 20 g/L glycerol, *Paracoccus* sp. LL1 produced 3.77 g/L PHB (0.43 g/g in CDM) within 96 h; in parallel, 3.6 mg/L of carotenoids (mainly the highly demanded pigment astaxanthin) were produced in this cultivation setup. This process revealed that PHA accumulation by this strain occurs in a growth-associated way. As a process improvement, a cell retention cultivation system based on a ceramic filter placed inside the bioreactor to permanently supply substrate (50 g/L glycerol in nutrient solution) at a D of 0.07 h^{-1} resulted in a 2.2-fold increase of CDM; 24.2 g/L, 0.39 g PHA per g biomass, and 9.5 g/L PHA were obtained by this strategy after 120 h of cultivation, which means a duplication of the volumetric PHA productivity (0.08 vs 0.04 g/(L·h)). Surprisingly, PHA samples produced in this study turned out to be PHBHV copolyesters with 3HV fractions up to 0.11 mol/mol at the end of the bioreactor fermentation (0.03 mol/mol after 24 h), as shown by ^{1}H-NMR, FT-IR, and GC-FID [202]; this is in contrast to the first report on this organism (see above; [202]). Moreover, shaking flask setups performed in this study showed the molar 3HV fraction in PHA accumulated by this strain is highly dependent on the applied carbon source. At the same time, the production of total carotenoids was also doubled if compared to the simple batch setup [203].

Only recently, Kumar and Kim addressed the fact that vegetable oils and waste cooking oil (WCO) originating from the food and agro-industry are among the cheapest abundantly available carbon sources. In their study, these authors studied the potential of *Paracoccus* sp. LL1 to use vegetable oils and WCO for the parallel production of PHA and carotenoid pigments. Several surfactants were applied to improve the capacity of the organism to utilize the lipids. Using 1 vol.-% WCO as carbon source, Tween-80 as surfactant, and a salinity of 10 g/L, 1.0 g/L poly(3-hydroxybutyrate-*co*-3%-3-hydroxyvalerate) (31 wt.-% in biomass) and 0.89 mg/L of carotenoids were produced in batch bioreactor cultivation setups after 96 h. Importantly, this was the first report demonstrating the potential of a halophilic production strain to convert lipophilic substrates into PHA [204].

4.4.9 *Vibrio* as a Genus Encompassing Halophilic PHA Producers

The halophilic Gram-negative strain *Vibrio proteolyticus*, a member of the *Vibrionaceae* family, was isolated from a marine environment on the Korean peninsula. To determine optimal growth and production conditions for this organism, different salinity, carbon sources, and nitrogen sources were evaluated by Hong and colleagues. It turned out that an M9 minimal medium containing 20 g/L fructose, 3 g/L yeast extract, and 50 g/L NaCl resulted in a PHA fraction of 55% in biomass and 2.7 g/L PHA after 48 h cultivation in shaking flasks, which was considerably higher than when using other sugars (glycerol, glucose, sucrose, or maltose) or organic acids (lactate, acetate, butyrate). Importantly, NaCl concentrations significantly higher than 50 g/L resulted in a drastic decrease in PHA production. The addition of 3 g/L propionate as 3HV precursor resulted in the production of PHBHV copolyesters with a 3HV fraction of 0.158 mol/mol. Moreover, *V. proteolyticus* was also cultured under non-sterile conditions without microbial contamination, which demonstrates again the potential of halophilic PHA producers for implementation on a larger scale [205].

In addition, a new moderately halophilic isolate isolated from fermented shrimp paste in Vietnam, designated as *Salinivibrio* sp. M318, was characterized as a potent PHA producer using fish sauce as nitrogen source and mixtures of waste fish oil and glycerol as carbon sources at NaCl concentration of 30 g/L. In a bioreactor cultivation, the strain produced 10.7 g/L PHB homopolyester with a PHB content of 0.57 g/g in biomass within 48 h. The strain was able to incorporate 4HB or 3HV monomer units into the polymer structure when appropriate precursors of 4HB such as sodium 4-hydroxybutyrate, 1,4-butanediol, or GBL, and 3HV such as valerate, propionate, or heptanoate, were applied. High CDM of 69.1 g/L, a volumetric PHB productivity of 0.46 g/(L·h), and a PHB content of 0.515 g/g of CDM were obtained by the strain after 78 h of bioreactor cultivation in fed-batch culture in a laboratory bioreactor [206].

4.5   GRAM-POSITIVE HALOPHILIC PHA PRODUCERS

4.5.1   PARTICULARITIES OF GRAM-POSITIVE PHA PRODUCTION STRAINS

Studies reporting PHA biosynthesis by Gram-positive microorganisms are scarcely found in the literature in comparison to the vast variety of reports about Gram-negative PHA producers [207]. However, it is often forgotten that the first microorganism ever described as a PHA producer in fact was a Gram-positive strain, namely *Bacillus megaterium*, which was studied by Lemoigne in the 1920s [208]. In the subsequent decades, PHA production by *Bacilli* and other Gram-positives was rather considered a myth by the global scientific community. Only about ten years ago, this situation suddenly changed by detecting some PHA-producing representatives of the genus *Bacillus*. Examples are *Bacillus* sp. JMa5, an organism with outstanding PHA production capacity from molasses, a side stream of sugar industry [209], *Bacillus cereus* CFR06, a bacterium, which accumulates intracellular PHA pools of more than half of its biomass [210], *Bacillus cereus* YB-4 [211], or *Bacillus* sp. IPCB-403, a strain with PHA accumulation exceeding even 0.7 g per g biomass [212]. Finally, PHA production on a larger scale was carried out using the new Gram-positive isolate *Bacillus cereus* SPV [213, 214]. Importantly, in contrast to Gram-negative strains, Gram-positive bacteria do not synthesize unwanted lipopolysaccharides (LPS), a group of inflammatory active endotoxins, in their cell wall. When PHA produced by Gram-negative biomass is recovered, LPS are co-isolated together with PHA, which drastically hampers using PHA from Gram-negative bacteria in vivo, in the medical field, which encompasses the production of surgical sutures, implants, and others [215–217]. Among the genus *Bacillus*, some halotolerant to halophilic representatives are also described, as detailed in the following paragraphs.

4.5.2   BACILLUS MEGATERIUM SSP. AS HALOPHILIC PHA PRODUCERS

Bacillus megaterium uyuni S29, a strain originally isolated from Bolivian water and mud samples from the Uyuni salt lake, is a representative of the rather scarce number of Gram-positive organisms which revealed accumulation of significant PHA quantities in environments of elevated salinity. This organism was first recognized in 2013 as a proficient PHA production strain by Rodríguez-Contreras and coworkers. Using glucose as sole carbon source in shaking flask cultivation experiments revealed accumulation of PHB by this strain. Unexpectedly, this PHB turned out to constitute a polymer blend of different PHB fractions of diverse molecular mass [5]. In defined minimal cultivation media, PHA production capacity of this strain was investigated in more detail under controlled conditions in three-liter bioreactors, showing its high capability for this purpose. In these experiments, a product concentration of 8.5 g/L PHB, a volumetric productivity of 0.25 g/(L·h), and an intracellular PHB content of 0.30 g/g were reported; to date, these are the best results for PHA biosynthesis by a Gram-positive production strain [218]. Later, the adaptation of *B. megaterium* uyuni S29 to fluctuating salinity conditions was studied by evaluating its growth and PHA production kinetics under different NaCl concentrations. The salt concentration range was chosen based on reported salinity conditions to the salinity dominant in different regions in and around the Bolivian salt lake where the organism was isolated. Surprising metabolic flexibility of the strain was monitored regarding its response to altering salinity; for both biomass growth and PHA production, the best results were achieved in media containing 45 g/L NaCl. Even at an extremely high salt concentration of 100 g/L NaCl, *B. megaterium* uyuni S29 grew well, and PHA biosynthesis was detected even at salinity as high as 250 g/L NaCl. In addition, no sporulation was observed by microscopic observations of cells cultivated under conditions enhancing PHA biosynthesis, which displays a considerable advantage regarding the substrate-to-product conversion yields [9]. Typically, spore formation causes the loss of parts of the carbon source, which was the main reason for avoiding the use of Gram-positive strains for PHA production for decades. This fact was also addressed when using *B. cereus* SPV, the organism isolated and studied by Valappil and associates [213, 214, 219]; in this case, it was demonstrated that spore formation can be blocked by decreasing the pH-value of the cultivation medium to a range of 4.5 to 5.8; surprisingly, this low pH-value also enhanced also the yield of PHA biosynthesis [214]. As shown later by Philip et al., these higher yields originate from a lack of PHA degradation at low pH-values [220]. When using the halophilic organism *B. megaterium* uyuni S29, acidification was not needed to suppress sporulation [5, 9, 218].

Recently, Schmid and colleagues used desugarized sugar beet molasses, a highly saline side stream of the sugar industry for PHB production by *B. megaterium* uyuni S29. After 24 h of batch cultivation with 10 g/L NaCl and phosphate-limited conditions, a biomass concentration of up to 16.7 g/L with a PHB fraction of 0.60 g/g and a volumetric productivity of 0.42 g/(L·h) were obtained. The starting concentration of the sugars in the molasse medium were in the region of 30 g/L. Utilization of this inexpensive sugar beet molasses medium resulted in an up to threefold increase in biomass production compared to parallel cultivations in a mineral-based medium. These authors also reported that sporulation did not occur. Moreover, the polymer produced in this study also consisted of two fractions characterized by different molecular mass [221], which is analogous to the previous findings by Rodríguez-Contreras et al. [5].

To sum up, *B. megaterium* uyuni S29 looks highly proficient as a robust PHA production strain able to synthesize PHA biopolyesters of a quality matching the needs for use in the biomedical field; moreover, this strain was also suggested to be used for the biological management of highly saline wastewater [9]. Figure 4.5 shows a light microscopic picture of *B. megaterium* uyuni S29 cells.

Three other isolates from solar salterns in Goa, India, identified as *Bacillus megaterium* ssp. were described by Salgaonkar and colleagues. Among these strains, the organism *Bacillus megaterium* H16 revealed similar PHB homopolyester production capacity both without NaCl (0.40 g/g PHA in CDM) and with 50 g/L NaCl (0.39 g/g PHA in CDM) when using glucose as the sole carbon source; hence, this organism is another representative of halotolerant to moderately halophilic PHA production strains [222].

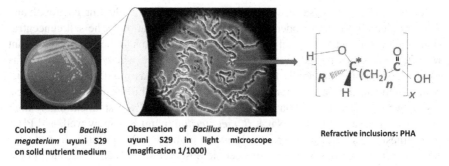

Colonies of *Bacillus megaterium* uyuni S29 on solid nutrient medium

Observation of *Bacillus megaterium* uyuni S29 in light microscope (magification 1/1000)

Refractive inclusions: PHA

FIGURE 4.5 Light microscopic observation of *B. megaterium* uyuni S29 cells; PHA granules visible as refractive inclusion bodies.

4.5.3 *BACILLUS* MG12

Moorkoth and Nampoothiri studied a halophilic mangrove isolate, which, via 16S rRNA sequencing, was identified as a *Bacillus* spp. closely related to the *Bacillus cereus* group and labeled as *Bacillus* MG12. This isolate was able to use a broad range of carbon sources sugars (glucose, fructose, sucrose, and the pentoses xylose and arabinose), acetic acid, propionic acid, octanoic acid, succinic acid, glycerol, and acidically pre-treated sugarcane trash liquor for growth and PHA production. This wild-type strain was able to produce PHBHV copolyesters by both propionate-dependent (addition of the 3HV precursor propionate) and propionate-independent pathways (utilization of threonine). When adding 10 mM propionic acid to 20 g/L of different sugars, the highest PHBHV productivity in 48 h shaking flask experiments was obtained using glucose as the main carbon source (2 g/L), with 48 mol-% 3HV in PHBHV, and a PHBHV mass fraction in CDM of 73%. Investigating PHBHV biosynthesis without propionic acid supplementation, the strain was supplied with acetate, mixture of acetate and threonine, mixture of acetate, cyanocobalamin and threonine, and threonine alone. With acetate as the sole carbon source, PHB homopolyester was produced, while threonine-containing setups delivered PHBHV copolyesters. Acidic pre-treated sugarcane trash liquor from bioethanol production generation, which contained xylose, glucose, and arabinose as major sugars, but also inhibitors like furfural or small organic acids, resulted in a maximum CDM of 2.89 g/L and 0.24 g/L PHBHV with a 3HV fraction of 8 mol-% within 48 h. For all cultivation, a moderately high NaCl concentration of 5 g/L was used, while the authors report this halotolerant strain to endure up to 90 g/L of NaCl [223].

4.6 CONCLUSIONS AND OUTLOOK

Halophiles definitely contain proficient candidates for the production of PHA biopolyesters and other marketable bio-products. As demonstrated for several examples in this chapter, fine-tuning the salinity offers a possibility of shifting the intracellular carbon flux between different end-products such as PHA, extracellular polysaccharides, organic osmolytes, or pigments, and even triggers the properties of PHA in terms of molecular mass and intramolecular architecture. Implementation of halophiles as biotechnological platforms in low energy and low sterility production processes paves new ways in White Biotechnology. It is likely that the combination of cultivation in open bioreactor vessels and tanks, application of inexpensive feedstocks, high-quality PHA production from simple, structurally unrelated substrates, and convenient techniques for PHA recovery from such microorganisms will finally enable economically competitive and ethical biopolyester production. The application of production strains which are halophilic and thermophilic at the same time holds particular promise for energy efficient PHA production processes.

As demonstrated in this chapter, genetic engineering approaches, recently also resorting to CRISPRi tools, have already been successfully tested; this enables the closing of specific metabolic bottlenecks, hence, de facto "dream organisms", which are auspicious candidates for large-scale implementation, have already been designed. Apart from the upscaling of available lab-scale processes described in the chapter, the next R&D efforts should be directed towards further fine-tuning of the microstructure of PHA produced by halophiles in order to meet the requirements of diverse applications in the plastics market. Moreover, innumerable halophiles still are awaiting exploration for biotechnological use; this wealth of natural organisms needs to be tapped!

REFERENCES

1. Koller M. Production of Polyhydroxyalkanoate (PHA) biopolyesters by extremophiles, *MOJ Poly Sci* 2017; 1(2): 1–19.
2. Koller M, Obruca S, Pernicova I, *et al.* Physiological, kinetic and process engineering aspects of Polyhydroxyalkanoate biosynthesis by extremophiles. In: Williams, Harvey, Kelly, Patricia (Eds.). *Polyhydroxyalkanoates: Biosynthesis, Chemical Structures and Applications.* New York: Nova Science Publishers, Inc, 2018, pp. 1–70.
3. Hsiao LJ, Lee MC, Chuang PJ, *et al.* The production of poly(3-hydroxybutyrate) by thermophilic *Caldimonas manganoxidans* from glycerol. *J Polym Res* 2018; 25(4): 85.
4. Mani K, Salgaonkar BB, Bragança JM. Culturable halophilic archaea at the initial and crystallization stages of salt production in a natural solar saltern of Goa, India. *Aquatic Biosystems* 2012; 8(1): 15.
5. Rodriguez-Contreras A, Koller M, Miranda de Sousa Dias M, *et al.* Novel Poly[(*R*)-3-hydroxybutyrate]-producing bacterium isolated from a Bolivian hypersaline lake. *Food Technol Biotechnol* 2013; 51(1): 123–130.
6. Kavitha G, Rengasamy R, Inbakandan D. Polyhydroxybutyrate production from marine source and its application. *Int J Biol Macromol* 2018; 11: 102–108.
7. Melanie S, Winterburn J. Production of Poly(3-hydroxybutyrate-*co*-3-hydroxyvalerate) by marine archaea *Haloferax mediterranei* DSM 1411 with yeast extract as nutrient source. *Squalen Bull Mar Fish Postharvest Biotechnol* 2017; 12(3): 117–125.
8. Borowitzka MA, Borowitzka LJ, Kessly D. Effects of salinity increase on carotenoid accumulation in the green alga *Dunaliella salina*. *J Appl Phycol* 1990; 2(2): 111–119.
9. Rodriguez-Contreras A, Koller M, Braunegg G, *et al.* Poly[(*R*)-3-hydroxybutyrate] production under different salinity conditions by a novel *Bacillus megaterium* strain. *New Biotechnol* 2016; 33: 73–77.
10. Yin J, Chen JC, Wu Q, Chen GQ. Halophiles, coming stars for industrial biotechnology. *Biotechnol Adv* 2015; 33: 1433–1442.
11. Wen Q, Ji Y, Hao Y, *et al.* Effect of sodium chloride on polyhydroxyalkanoate production from food waste fermentation leachate under different organic loading rate. *Biores Technol* 2018; 267: 133–140.

12. Kucera D, Pernicová I, Kovalcik A, Koller M, *et al.* Characterization of the promising poly (3-hydroxybutyrate) producing halophilic bacterium *Halomonas halophila. Biores Technol* 2018; 256: 552–556.
13. Obruca S, Benesova P, Petrik S, *et al.* Production of polyhydroxyalkanoates using hydrolysate of spent coffee grounds. *Proc Biochem* 2014; 49(9): 1409–1414.
14. Obruca S, Benesova P, Marsalek L, Marova I. Use of lignocellulosic materials for PHA production. *Chem Biochem Eng Q* 2015; 29(2): 135–144.
15. Oren A. Microbial life at high salt concentrations: Phylogenetic and metabolic diversity. *Saline Syst* 2008; 4(2): 1–13.
16. Torregrosa-Crespo J, Galiana CP, Martínez-Espinosa RM. Biocompounds from haloarchaea and their uses in biotechnology. In: Sghaier H, Najjari A, Ghedira K (eds.), *Archaea-New Biocatalysts, Novel Pharmaceuticals and Various Biotechnological Applications.* Rijeka, Croatia: InTech, 2017.
17. Marquez MC, Ventosa A, Ruiz-Berraquero F. A taxonomic study of heterotrophic halophilic and non-halophilic bacteria from a solar saltern. *Microbiology* 1987; 133(1): 45–56.
18. Mozumder MSI, Garcia-Gonzalez L, De Wever H, Volcke EIP. Effect of sodium accumulation on heterotrophic growth and polyhydroxybutyrate (PHB) production by *Cupriavidus necator. Biores Technol* 2015; 191: 213–218.
19. Obruca S, Marova I, Svoboda Z, Mikulikova R. Use of controlled exogenous stress for improvement of poly(3-hydroxybutyrate) production in *Cupriavidus necator. Folia Microbiol* 2010; 55(1): 17–22.
20. Passanha P, Kedia G, Dinsdale RM, *et al.* The use of NaCl addition for the improvement of polyhydroxyalkanoate production by *Cupriavidus necator. Biores Technol* 2014; 163: 287–294.
21. Quillaguamán J, Guzmán H, Van-Thuoc D, Hatti-Kaul R. Synthesis and production of polyhydroxyalkanoates by halophiles: Current potential and future prospects. *Appl Microbiol Biotechnol* 2010; 85(6): 1687–1696.
22. Mirete S, Mora-Ruiz MR, Lamprecht-Grandío M, *et al.* Salt resistance genes revealed by functional metagenomics from brines and moderate-salinity rhizosphere within a hypersaline environment. *Front Microbiol* 2015; 6: 1121.
23. Kennedy SP, Ng WV, Salzberg SL, *et al.* Understanding the adaptation of *Halobacterium* species NRC-1 to its extreme environment through computational analysis of its genome sequence. *Genome Res* 2001; 11: 1641–1650.
24. Bolhuis H, Palm P, Wende A, *et al.* The genome of the square archaeon *Haloquadratum walsbyi*: Life at the limits of water activity. *BMC Genomics* 2006; 7: 169–180.
25. Ng WV, Kennedy SP, Mahairas GG, *et al.* Genome sequence of *Halobacterium* species NRC-1. *P Nat Acad Sci USA* 2000; 97(22): 12176–12181.
26. Baliga NS, Bonneau R, Facciotti MT, *et al.* Genome sequence of *Haloarcula marismortui*: A halophilic archaeon from the Dead Sea. *Genome Res* 2004; 14(11): 2221–2234.
27. Danson MJ, Hough DW. Structure, function and stability of enzymes from the Archaea. *Trends Microbiol* 1998; 6(8): 307–314.
28. Oren A. Life at high salt concentrations. In: Dworkin, M., Falkow, S., Rosenberg, E., Schleifer, K. H., Stackebrandt, E. (Eds.). *The Prokaryotes, a Handbook on the Biology of Bacteria: Ecophysiology and Biochemistry.* New York: Springer, 2006. pp. 263–282.
29. Bonete MJ, Pire C, Llorca FI, *et al.* Glucose dehydrogenase from the halophilic Archaeon *Haloferax mediterranei*: Enzyme purification, characterisation and N-terminal sequence. *FEBS Lett* 1996; 383(3): 227–229.
30. Britton KL, Baker PJ, Fisher M, *et al.* Analysis of protein solvent interactions in glucose dehydrogenase from the extreme halophile *Haloferax mediterranei. P Nat Acad Sci USA* 2006; 103(13): 4846–4851.

31. Hezayen FF, Tindall BJ, Steinbüchel A, Rehm BHA. Characterization of a novel halophilic archaeon, *Halobiforma haloterrestris* gen. nov., sp. nov., and transfer of *Natronobacterium nitratireducens* to *Halobiforma nitratireducens* comb. nov. *Int J Syst Evol Microbiol* 2002; 52(6): 2271–2280.
32. Hezayen FF, Gutiérrez MC, Steinbüchel A, *et al. Halopiger aswanensis* sp. nov., a polymer-producing and extremely halophilic archaeon isolated from hypersaline soil. *Int J Syst Evol Microbiol* 2010; 60(3): 633–637.
34. Chen CW, Hsu SH, Lin MT, Hsu YH. Mass production of C50 carotenoids by *Haloferax mediterranei* in using extruded rice bran and starch under optimal conductivity of brined medium. *Bioproc Biosyst Eng* 2015; 38(12): 2361–2367.
35. Wood J, Bremer, E, Csonka LN, *et al.* Osmosensing and osmoregulatory compatible solute accumulation by bacteria. *Comp Biochem Phys A* 2001; 130(3): 437–460.
36. Sauer T, Galinski EA. Bacterial milking: A novel bioprocess for production of compatible solutes. *Biotechnol Bioeng* 1998; 57(3): 306–313.
37. Quillaguaman J, Delgado O, Mattiasson B, *et al.* Poly(β-hydroxybutyrate) production by a moderate halophile, *Halomonas boliviensis* LC1. *Enzyme Microb Tech* 2006; 38(1–2): 148–154.
38. Mothes G, Schubert T, Harms H, *et al.* Biotechnological coproduction of compatible solutes and polyhydroxyalkanoates using the genus *Halomonas*. *Eng Life Sci* 2008; 8(6): 658–662.
39. Cristea A, Floare C, Borodi G, *et al.* Physical and chemical detection of Polyhydroxybutyrate from *Halomonas elongata* strain 2FF. *New Front Chem* 2017; 26(2): S4_OP3-S4_OP3.
40. Guzmán H, Van-Thuoc D, Martí J., *et al.* A process for the production of ectoine and poly(3-hydroxybutyrate) by *Halomonas boliviensis*. *Appl Microbiol Biotechnol* 2009; 84(6): 1069–1077.
41. Soto G, Setten L, Lisi C, *et al.* Hydroxybutyrate prevents protein aggregation in the halotolerant bacterium *Pseudomonas* sp. CT13 under abiotic stress. *Extremophiles* 2012; 16(3): 455–462.
42. Obruca S, Sedlacek P, Mravec F, *et al.* Evaluation of 3-hydroxybutyrate as an enzyme-protective agent against heating and oxidative damage and its potential role in stress response of poly(3-hydroxybutyrate) accumulating cells. *Appl Microbiol Biotechnol* 2016; 100(3): 1365–1376.
43. Obruca S, Sedlacek P, Mravec F, *et al.* The presence of PHB granules in cytoplasm protects non-halophilic bacterial cells against the harmful impact of hypertonic environments. *New Biotechnol* 2017; 39(A): 68–80.
44. Ginzburg M, Sachs L, Ginzburg, BZ. Ion Metabolism in a Halobacterium: I. Influence of age of culture on intracellular concentrations. *J Gen Physiol* 1970; 55(2): 187–207.
45. Oren A, Ginzburg M, Ginzburg BZ, *et al. Haloarcula marismortui* (Volcani) sp. nov., nom. rev., an extremely halophilic bacterium from the Dead Sea. *Int J Syst Evol Microbiol* 1990; 40(2): 209–210.
46. Desmarais D, Jablonski PE, Fedarko NS, *et al.* 2-Sulfotrehalose, a novel osmolyte in haloalkaliphilic archaea. *J Bacteriol* 1997; 179(10): 3146–3153.
47. Sedlacek P, Slaninova E, Koller M, *et al.* PHA granules help bacterial cells to preserve cell integrity when exposed to sudden osmotic imbalances. *New Biotechnol* 2019; 49: 129–136.
48. Koller M, Rodríguez-Contreras A. Techniques for tracing PHA-producing organisms and for qualitative and quantitative analysis of intra-and extracellular PHA. *Eng Life Sci* 2015; 15(6): 558–581.
49. Spiekermann P, Rehm BH, Kalscheuer R, *et al.* A sensitive, viable-colony staining method using Nile red for direct screening of bacteria that accumulate polyhydroxyalkanoic acids and other lipid storage compounds. *Arch Microbiol* 1999; 171(2): 73–80.

50. Fathima N, Krishnaswamy V. Isolation and Screening of PHB producing halotolerant bacterial strains from a saline environment. *J Biotechnol Biomed Sci* 2017; 1(1): 34.

51. Mahansaria R, Choudhur JD, Mukherjee J. Polymerase chain reaction-based screening method applicable universally to environmental haloarchaea and halobacteria for identifying polyhydroxyalkanoate producers among them. *Extremophiles* 2015; 19(5): 1041–1054.

52. Han J, Hou J, Liu H, *et al.* Wide distribution among halophilic archaea of a novel polyhydroxyalkanoate synthase subtype with homology to bacterial type III synthases. *Appl Environ Microbiol* 2010; 76(23): 7811–7819.

53. García-Torreiro M, López-Abelairas M, Lu-Chau TA, *et al.* Application of flow cytometry for monitoring the production of poly(3-hydroxybutyrate) by *Halomonas boliviensis*. *Biotechnol Progr* 2017; 33(2): 276–284.

54. Lorantfy B, Seyer B, Herwig C. Stoichiometric and kinetic analysis of extreme halophilic Archaea on various substrates in a corrosion resistant bioreactor. *New Biotechnol* 2014; 31(1): 80–89.

55. Hezayen FF, Rehm BHA, Eberhardt R, *et al.* Polymer production by two newly isolated extremely halophilic archaea: Application of a novel corrosion-resistant bioreactor. *Appl Microbiol Biotechnol* 2000; 54(3): 319–325.

56. Mahler N, Tschirren S, Pflügl S, Herwig C. Optimized bioreactor setup for scale-up studies of extreme halophilic cultures. *Biochem Eng J* 2018; 130: 39–46.

57. Kirk RG, Ginzburg M. Ultrastructure of two species of halobacterium. *J Ultrastruct Res* 1972; 41(1–2): 80–94.

58. Nicolaus B, Lama L, Esposito E, *et al. Haloarcula* spp able to biosynthesize exo-and endopolymers. *J Ind Microbiol Biotechnol* 1999; 23(6): 489–496.

59. Soppa J. From genomes to function: Haloarchaea as model organisms. *Microbiology* 2006; 152(3): 585–590.

60. Singh A, Singh AK. Haloarchaea: Worth exploring for their biotechnological potential. *Biotechnol Lett* 2017; 39(12): 1793–1800.

61. Tokunaga M, Tokunaga H, Ishibashi M, *et al.* Halophilic enzymes: Negative charges determine halophilicity. *Seikagaku, J Jap Biochem Soc* 2009; 81(5): 401–406.

62. Karan R, Capes MD, DasSarma S. Function and biotechnology of extremophilic enzymes in low water activity. *Aquatic Biosystems* 2012; 8(1): 4.

63. Delgado-García M, Valdivia-Urdiales B, Aguilar-González CN, *et al.* Halophilic hydrolases as a new tool for the biotechnological industries. *J Sci Food Agr* 2012; 92(13): 2575–2580.

64. Lozier RH, Bogomolni RA, Stoeckenius W. Bacteriorhodopsin: A light-driven proton pump in *Halobacterium halobium*. *Biophys J* 1975; 15(9): 955.

65. Vsevolodov NN, Druzhko AB, Djukova TV. Actual possibilities of bacteriorhodopsin application in optoelectronics. In: Hong FT (ed.), *Molecular Electronics*. Boston, MA: Springer, 1989, pp. 381–384.

66. Henderson R. The purple membrane from *Halobacterium halobium*. *Annu Rev Biophys Bio* 1977; 6(1): 87–109.

67. Henderson R, Baldwin JM, Downing KH, *et al.* Structure of purple membrane from *Halobacterium halobium*: Recording, measurement and evaluation of electron micrographs at 3.5 Å resolution. *Ultramicroscopy* 1986; 19(2): 147–178.

68. Englert C, Wanner G, Pfeifer F. Functional analysis of the gas vesicle gene cluster of the halophilic archaeon *Haloferax mediterranei* defines the vac-region boundary and suggests a regulatory role for the gvpD gene or its product. *Mol Microbiol* 1992; 6(23): 3543–3550.

69. Pfeifer F. Haloarchaea and the formation of gas vesicles. *Life* 2015; 5(1): 385–402.

70. Sleytr UB, Schuster B, Egelseer EM, Pum D. S-layers: Principles and applications. *FEMS Microbiol Rev* 2014; 38(5): 823–864.

71. Torregrosa-Crespo J, Rodrigo-Baños M, Pire C, *et al.* Innovative applications of halo-archaea to waste water treatments and biotechnological uses of the biomass produced. *New Biotechnol* 2016; 33(3): 421.
72. Bonete MJ, Bautista V, Esclapez J, *et al.* New uses of haloarchaeal species in bioreme-diation processes. In: Shiomi, N. (Ed.). *Advances in Bioremediation of Wastewater and Polluted Soil.* Rijeka, Croatia: InTech, 2015.
73. Bonfá MR, Grossman MJ, Mellado E, *et al.* Biodegradation of aromatic hydrocar-bons by Haloarchaea and their use for the reduction of the chemical oxygen demand of hypersaline petroleum produced water. *Chemosphere* 2011; 84(11): 1671–1676.
74. Martínez-Espinosa RM, Richardson DJ, Bonete MJ. Characterisation of chlorate reduc-tion in the haloarchaeon *Haloferax mediterranei. BBA-Gen Subjects* 2015; 1850(4): 587–594.
75. Al-Mailem DM, Eliyas M, Khanafer M, *et al.* Biofilms constructed for the removal of hydrocarbon pollutants from hypersaline liquids. *Extremophiles* 2015; 19(1): 189–196.
76. Al-Mailem DM, Eliyas M, Radwan SS. Enhanced haloarchaeal oil removal in hyper-saline environments via organic nitrogen fertilization and illumination. *Extremophiles* 2012; 16(5): 751–758.
77. Koller M. Polyhydroxyalkanoate biosynthesis at the edge of water activity - haloar-chaea as biopolyester factories. *Bioengineering* 2019; 6(2): 34.
78. Poli A, Di Donato P, Abbamondi GR, *et al.* Synthesis, production, and biotechnological applications of exopolysaccharides and polyhydroxyalkanoates by archaea. *Archaea* 2011; 2011: Article ID 693253.
79. Parolis H, Parolis LA, Boán IF, *et al.* The structure of the exopolysaccharide pro-duced by the halophilic Archaeon *Haloferax mediterranei* strain R4 (ATCC 33500). *Carbohyd Res* 1996; 295: 147–156.
80. Cheung J, Danna KJ, O'Connor EM, *et al.* Isolation, sequence, and expression of the gene encoding halocin H4, a bacteriocin from the halophilic archaeon *Haloferax medi-terranei* R4. *J Bacteriol* 1997; 179(2): 548–551.
81. Kamekura M, Seno Y, Dyall-Smith M. Halolysin R4, a serine proteinase from the halo-philic archaeon *Haloferax mediterranei*; gene cloning, expression and structural stud-ies. *BBA-Protein Struct M* 1996; 1294(2): 159–167.
82. Rodrigo-Baños M, Garbayo I, Vílchez C, *et al.* Carotenoids from haloarchaea and their potential in biotechnology. *Marine Drugs* 2015; 13(9): 5508–5532.
83. Ashwini R, Vijayanand S, Hemapriya J. Photonic potential of haloarchaeal pig-ment bacteriorhodopsin for future electronics: A review. *Curr Microbiol* 2017; 74(8): 996–1002.
84. Salgaonkar BB, Das D, Bragança JM. Resistance of extremely halophilic archaea to zinc and zinc oxide nanoparticles. *Appl Nanosci* 2016; 6(2): 251–258.
85. Han J, Lu Q, Zhou L, *et al.* Molecular characterization of the *phaEC<sub>Hm* genes, required for biosynthesis of poly(3-hydroxybutyrate) in the extremely halophilic archaeon *Haloarcula marismortui. Appl Environ Microbiol* 2007; 73(19): 6058–6065.
86. Ding JY, Chiang PW, Hong MJ, *et al.* Complete genome sequence of the extremely halophilic archaeon *Haloarcula hispanica* strain N601. *Genome Announc* 2014; 2(2): e00178–e00114.
87. Rodriguez-Valera F, Juez G, Kushner DJ. *Halobacterium mediterranei* spec, nov., a new carbohydrate-utilizing extreme halophile. *Syst Appl Microbiol* 1983; 4(3): 369–381.
88. Fernandez-Castillo R, Rodriguez-Valera F, Gonzalez-Ramos J, *et al.* Accumulation of poly(β-hydroxybutyrate) by halobacteria. *Appl Environ Microbiol* 1986; 51(1): 214–216.
89. Torreblanca M, Rodriguez-Valera F, Juez G, *et al.* Classification of non-alkaliphilic halobacteria based on numerical taxonomy and polar lipid composition, and descrip-tion of *Haloarcula* gen. nov. and *Haloferax* gen. nov. *Syst Appl Microbiol* 1986; 8(1–2): 89–99.

90. Antón J, Meseguer I, Rodriguez-Valera F. Production of an extracellular polysaccharide by *Haloferax mediterranei*. *Appl Environ Microbiol* 1988; 54(10): 2381–2386.
91. Hermann-Krauss C, Koller M, Muhr A, *et al*. Archaeal production of polyhydroxyalkanoate (PHA) co-and terpolyesters from biodiesel industry-derived by-products. *Archaea* 2013; 2013, Article ID 129268.
92. Rodriguez-Valera F, Lillo J. Halobacteria as producers of polyhydroxyalkanoates. *FEMS Microbiol Lett* 1992; 103(2-4): 181–186.
93. Koller M, Niebelschütz H, Braunegg G. Strategies for recovery and purification of poly[(R)-3-hydroxyalkanoates] (PHA) biopolyesters from surrounding biomass. *Eng Life Sci* 2013; 13: 549–562.
94. Lillo JG, Rodriguez-Valera F. Effects of culture conditions on poly (β-hydroxybutyric acid) production by *Haloferax mediterranei*. *Appl Environ Microbiol* 1990; 56(8): 2517–2521.
95. Pérez-Pomares F, Bautista V, Ferrer J, *et al*. α-Amylase activity from the halophilic archaeon *Haloferax mediterranei*. *Extremophiles* 2003; 7(4): 299–306.
96. Braunegg G, Lefebvre G, Genser KF. Polyhydroxyalkanoates, biopolyesters from renewable resources: Physiological and engineering aspects. *J Biotechnol* 1998; 65: 127–161.
97. Han J, Hou J, Zhang F, *et al*. Multiple propionyl coenzyme A-supplying pathways for production of the bioplastic poly(3-hydroxybutyrate-*co*-3-hydroxyvalerate) in *Haloferax mediterranei*. *Appl Environ Microbiol* 2013; 79(9): 2922–2931.
98. Lu Q, Han J, Zhou L, *et al*. Genetic and biochemical characterization of the poly(3-hydroxybutyrate-*co*-3-hydroxyvalerate) synthase in *Haloferax mediterranei*. *J Bacteriol* 2008; 190(12): 4173–4180.
99. Cai S, Cai L, Liu H, *et al*. Identification of the haloarchaeal phasin (PhaP) that functions in polyhydroxyalkanoate accumulation and granule formation in *Haloferax mediterranei*. *Appl Environ Microbiol* 2012; 78(6): 1946–1952.
100. Han J, Wu LP, Hou J, *et al*. Biosynthesis, characterization, and hemostasis potential of tailor-made poly(3-hydroxybutyrate-*co*-3-hydroxyvalerate) produced by *Haloferax mediterranei*. *Biomacromolecules* 2015; 16(2): 578–588.
101. Liu G, Hou J, Cai S, *et al*. A patatin-like protein associated with the polyhydroxyalkanoate (PHA) granules of *Haloferax mediterranei* acts as an efficient depolymerase in the degradation of native PHA. *Appl Environ Microbiol* 2015; 81(9): 3029–3038.
102. Koller M, Horvat P, Hesse P, *et al*. Assessment of formal and low structured kinetic modeling of polyhydroxyalkanoate synthesis from complex substrates. *Bioproc Biosyst Eng* 2006; 29(5–6): 367–377.
103. Koller M, Chiellini E, Braunegg G. Study on the production and re-use of poly(3-hydroxybutyrate-*co*-3-hydroxyvalerate) and extracellular polysaccharide by the archaeon Haloferax mediterranei strain DSM 1411. *Chem Biochem Eng Q* 2015; 29(2): 87–98.
104. López-Ortega MA, Rodríguez-Hernández AI, Camacho-Ruíz RM, *et al*. Physicochemical characterization and emulsifying properties of a novel exopolysaccharide produced by haloarchaeon *Haloferax mucosum*. *Int J Biol Macromol* 2020; 142: 152–162.
105. Raposo MFDJ, de Morais RMSC, Bernardo de Morais AMM. Bioactivity and applications of sulphated polysaccharides from marine microalgae. *Marine Drugs* 2013; 11(1): 233–252.
106. Hamidi M, Mirzaei R, Delattre C, *et al*. Characterization of a new exopolysaccharide produced by *Halorubrum* sp. TBZ112 and evaluation of its anti-proliferative effect on gastric cancer cells. *3 Biotech* 2019; 9(1): 1.
107. Cui YW, Gong XY, Shi YP, Wang ZD. Salinity effect on production of PHA and EPS by *Haloferax mediterranei*. *RSC Adv* 2017; 7(84): 53587–53595.

108. Ferre-Guell A, Winterburn J Production of the copolymer poly (3-hydroxybutyrate-*co*-3-hydroxyvalerate) with varied composition using different nitrogen sources with *Haloferax mediterranei. Extremophiles* 2017; 21(6): 1037–1047.

109. Melanie S, Winterburn JB, Devianto H. Production of biopolymer Polyhydroxyalkanoates (PHA) by extreme halophilic marine archaea *Haloferax mediterranei* in medium with varying phosphorus concentration. *J Eng Technol Sci* 2018; 50(2): 255–271.

110. Cui YW, Zhang HY, Ji SY, Wang ZW. Kinetic analysis of the temperature effect on polyhydroxyalkanoate production by *Haloferax mediterranei* in synthetic molasses wastewater. *J Polym Environ* 2017; 25(2): 277–285.

111. Koller M, Hesse P, Bona R, *et al.* Potential of various archae-and eubacterial strains as industrial polyhydroxyalkanoate producers from whey. *Macromol Biosci* 2007; 7(2): 218–226.

112. Koller M, Hesse P, Bona R, *et al.* Biosynthesis of high quality polyhydroxyalkanoate co-and terpolyesters for potential medical application by the archaeon *Haloferax mediterranei. Macromol Symp* 2077; 253(1): 33–39.

113. Pais J, Serafim LS, Freitas F, *et al.* Conversion of cheese whey into poly(3-hydroxybutyrate-*co*-3-hydroxyvalerate) by *Haloferax mediterranei. New Biotechnol* 2016; 33(1): 224–230.

114. Chen CW, Hsu SH, Lin MT, *et al.* Mass production of C50 carotenoids by *Haloferax mediterranei* in using extruded rice bran and starch under optimal conductivity of brined medium. *Bioproc Biosyst Eng* 2015; 38(12): 2361–2367.

115. Chen CW, Don TM, Yen HF. Enzymatic extruded starch as a carbon source for the production of poly(3-hydroxybutyrate-*co*-3-hydroxyvalerate) by *Haloferax mediterranei. Proc Biochem* 2006; 41(11): 2289–2296.

116. Huang TY, Duan KJ, Huang SY, *et al.* Production of polyhydroxyalkanoates from inexpensive extruded rice bran and starch by *Haloferax mediterranei. J Ind Microbiol Biotechnol* 2006; 33(8): 701–706.

117. Bhattacharyya A, Saha J, Haldar S, *et al.* Production of poly-3-(hydroxybutyrate-*co*-hydroxyvalerate) by *Haloferax mediterranei* using rice-based ethanol stillage with simultaneous recovery and re-use of medium salts. *Extremophiles* 2014; 18(2): 463–470.

118. Bhattacharyya A, Jana K, Haldar S, *et al.* Integration of poly-3-(hydroxybutyrate-*co*-hydroxyvalerate) production by *Haloferax mediterranei* through utilization of stillage from rice-based ethanol manufacture in India and its techno-economic analysis. *World J Microbiol Biotechnol* 2015; 31(5): 717–727.

119. Alsafadi D, Al-Mashaqbeh O. A one-stage cultivation process for the production of poly-3-(hydroxybutyrate-*co*-hydroxyvalerate) from olive mill wastewater by *Haloferax mediterranei. New Biotechnol* 2017; 34: 47–53.

120. Bhattacharyya A, Pramanik A, Maji SK, *et al.* Utilization of vinasse for production of poly-3-(hydroxybutyrate-*co*-hydroxyvalerate) by *Haloferax mediterranei. AMB Express* 2012; 2(1): 34.

121. Ghosh S, Gnaim R, Greiserman S, *et al.* Macroalgal biomass subcritical hydrolysates for the production of polyhydroxyalkanoate (PHA) by *Haloferax mediterranei. Biores Technol* 2019; 271: 166–173.

122. Koller M, Atlić A, Gonzalez-Garcia Y, *et al.* Polyhydroxyalkanoate (PHA) biosynthesis from whey lactose. *Macromol Symp* 2008; 27: 287–292.

123. Koller M. Recycling of waste streams of the biotechnological poly(hydroxyalkanoate) production by *Haloferax mediterranei* on whey. *Int J Polym Sci* 2015; 2015: Article ID 370164.

124. Mahansaria R, Dhara A, Saha A, *et al.* Production enhancement and characterization of the polyhydroxyalkanoate produced by *Natrinema ajinwuensis* (as synonym)≡ *Natrinema altunense* strain RM-G10. *Int J Biol Macromol* 2018; 107: 1480–1490.

125. Koller M, Sandholzer D, Salerno A, *et al.* Biopolymer from industrial residues: Life cycle assessment of poly(hydroxyalkanoates) from whey. *Resour Conserv Recy* 2013; 73: (2013) 64–71.
126. Narodoslawsky M, Shazad K, Kollmann R, *et al.* LCA of PHA Production–Identifying the ecological potential of bio-plastic. *Chem Biochem Eng Q* 2015; 29(2): 299–305.
127. Don TM, Chen CW, Chan TH. Preparation and characterization of poly(hydroxyalkanoate) from the fermentation of *Haloferax mediterranei. J Biomat Sci-Polym E* 2006; 17(12): 1425–1438.
128. Koller M, Bona R, Chiellini E, Braunegg G. Extraction of short-chain-length poly-[(*R*)-hydroxyalkanoates] (*scl*-PHA) by the "anti-solvent" acetone under elevated temperature and pressure. *Biotechnol Lett* 2013; 35(7): 1023–1028.
129. Ferre-Guell A, Winterburn J. Increased production of polyhydroxyalkanoates with controllable composition and consistent material properties by fed-batch fermentation. *Biochem Eng J* 2019; 141: 35–42.
130. Ferre-Guell A, Winterburn J. Biosynthesis and characterization of polyhydroxyalkanoates with controlled composition and microstructure. *Biomacromolecules* 2018: 19(3); 996–1005.
131. Altekar W, Rajagopalan R. Ribulose bisphosphate carboxylase activity in halophilic Archaebacteria. *Arch Microbiol* 1990; 153(2): 169–174.
132. Taran M, Amirkhani H. Strategies of poly(3-hydroxybutyrate) synthesis by *Haloarcula* sp. IRU1 utilizing glucose as carbon source: Optimization of culture conditions by Taguchi methodology. *Int J Biol Macromol* 2010; 47(5): 632–634.
133. Taran M. Synthesis of poly(3-hydroxybutyrate) from different carbon sources by *Haloarcula* sp. IRU1. *Polym-Plast Technol* 2011; 50(5): 530–532.
134. Taran M. Utilization of petrochemical wastewater for the production of poly (3-hydroxybutyrate) by *Haloarcula* sp. IRU1. *J Hazard Mater* 2011; 188(1–3): 26–28.
135. Taran M. Poly (3-hydroxybutyrate) production from crude oil by *Haloarcula* sp. IRU1: Optimization of culture conditions by Taguchi method. *Pet Sci Technol* 2011; 29(12): 1264–1269.
136. Taran M, Sharifi M, Bagheri S. Utilization of textile wastewater as carbon source by newly isolated *Haloarcula* sp. IRU1: Optimization of conditions by Taguchi methodology. *Clean Technol Envir* 2011; 13(3): 535–538.
137. Pramanik A, Mitra A, Arumugam M, *et al.* Utilization of vinasse for the production of polyhydroxybutyrate by *Haloarcula marismortui. Folia Microbiol* 2012; 57(1): 71–79.
138. Salgaonkar BB, Mani K, Bragança JM. Accumulation of polyhydroxyalkanoates by halophilic archaea isolated from traditional solar salterns of India. *Extremophiles* 2013; 17(5): 787–795.
139. Salgaonkar BB, Bragança JM. Biosynthesis of poly(3-hydroxybutyrate-*co*-3-hydroxyvalerate) by *Halogeometricum borinquense* strain E3. *Int J Biol Macromol* 2015; 78: 339–346.
140. Salgaonkar BB, Bragança JM. Utilization of sugarcane bagasse by *Halogeometricum borinquense* strain E3 for biosynthesis of poly(3-hydroxybutyrate-*co*-3-hydroxyvalerate). *Bioengineering* 2017; 4(2): 50.
141. Salgaonkar BB, Mani K, Bragança JM. Sustainable bioconversion of cassava waste to poly(3-hydroxybutyrate-*co*-3-hydroxyvalerate) by *Halogeometricum borinquense* strain E3. *J Polym Environ* 2019; 27(2): 299–308.
142. Danis O, Ogan A, Tatlican P, *et al.* Preparation of poly(3-hydroxybutyrate-*co*-hydroxyvalerate) films from halophilic archaea and their potential use in drug delivery. *Extremophiles* 2015; 19(2): 515–524.
143. Zhao YX, Rao ZM, Xue YF, *et al.* Poly(3-hydroxybutyrate-*co*-3-hydroxyvalerate) production by Haloarchaeon *Halogranum amylolyticum. Appl Microbiol Biotechnol* 2015; 99(18): 7639–7649.

144. Legat A, Gruber C, Zangger K, *et al.* Identification of polyhydroxyalkanoates in *Halococcus* and other haloarchaeal species. *Appl Microbiol Biotechnol* 2010; 87(3): 1119–1127.

145. Romano I, Poli A, Finore I, *et al. Haloterrigena hispanica* sp. nov., an extremely halophilic archaeon from Fuente de Piedra, southern Spain. *Int J Syst Evol Microbiol* 2007; 57(7): 1499–1503.

146. Di Donato P, Fiorentino G, Anzelmo G, *et al.* Re-use of vegetable wastes as cheap substrates for extremophile biomass production. *Waste Biomass Valorization* 2011; 2(2): 103–111.

147. Wainø M, Tindall BJ, Ingvorsen K. *Halorhabdus utahensis* gen. nov., sp. nov., an aerobic, extremely halophilic member of the Archaea from Great Salt Lake, Utah. *Int J Syst Evol Microbiol* 2000; 50(1): 183–190.

148. Antunes A, Taborda M, Huber R, *et al. Halorhabdus tiamatea* sp. nov., a non-pigmented, extremely halophilic archaeon from a deep-sea, hypersaline anoxic basin of the Red Sea, and emended description of the genus *Halorhabdus. Int J Syst Evol Microbiol* 2008; 58(1): 215–220.

149. Antunes A, Alam I, Bajic VB, Stingl U. Genome sequence of *Halorhabdus tiamatea*, the first archaeon isolated from a deep-sea anoxic brine lake. *J Bacteriol* 2011; 193(17): 4553–4554.

150. Albuquerque L, Kowalewicz-Kulbat M, Drzewiecka D, *et al. Halorhabdus rudnickae* sp. nov., a halophilic archaeon isolated from a salt mine borehole in Poland. *Syst Appl Microbiol* 2016; 39(2): 100–105.

151. Xu XW, Wu M, Zhou PJ, Liu SJ. *Halobiforma lacisalsi* sp. nov., isolated from a salt lake in China. *Int J Syst Evol Microbiol* 2005; 55(5): 1949–1952.

152. Walsby AE. A square bacterium. *Nature* 1980: 283(5742): 69.

153. Walsby AE. Archaea with square cells. *Trends Microbiol* 2005; 13(5): 193–195.

154. Burns DG, Janssen PH, Itoh T, *et al. Haloquadratum walsbyi* gen. nov., sp. nov., the square haloarchaeon of Walsby, isolated from saltern crystallizers in Australia and Spain. *Int J Syst Evol Microbiol* 2007; 57(2): 387–392.

155. Saponetti MS, Bobba F, Salerno G, *et al.* Morphological and structural aspects of the extremely halophilic archaeon *Haloquadratum walsbyi. PLoS One* 2011; 6(4): e18653.

156. Singh R. Isolation and characterization of efficient poly-β-hydroxybutyrate (PHB) synthesizing bacteria from agricultural and industrial land. *Int J Curr Microbiol Appl Sci* 2014; 3(6): 304–308.

157. Quillaguamán J, Hatti-Kaul R, Mattiasson B, *et al. Halomonas boliviensis* sp. nov., an alkalitolerant, moderate halophile isolated from soil around a Bolivian hypersaline lake. *Int J Syst Evol Microbiol* 2004; 54(3): 721–725.

158. Quillaguaman J, Hashim S, Bento F, *et al.* Poly(β-hydroxybutyrate) production by a moderate halophile, *Halomonas boliviensis* LC1 using starch hydrolysate as substrate. *J Appl Microbiol* 2005; 99(1): 151–157.

159. Quillaguamán J, Doan-Van T, Guzmán H, *et al.* Poly(3-hydroxybutyrate) production by *Halomonas boliviensis* in fed-batch culture. *Appl Microbiol Biotechnol* 2008; 78(2): 227–232.

160. Van-Thuoc D, Quillaguaman J, Mamo G, *et al.* Utilization of agricultural residues for poly (3-hydroxybutyrate) production by *Halomonas boliviensis* LC1. *J Appl Microbiol* 2008; 104(2): 420–428.

161. Stanley A, Kumar HP, Mutturi S, *et al.* Fed-batch strategies for production of PHA using a native isolate of *Halomonas venusta* KT832796 strain. *Appl Biochem Biotechnol* 2018; 184(3): 935–952.

162. Simon-Colin C, Raguénès G, Cozien J, *et al. Halomonas profundus* sp. nov., a new PHA-producing bacterium isolated from a deep-sea hydrothermal vent shrimp. *J Appl Microbiol* 2008; 104(5): 1425–1432.

163. Romano I, Giordano A, Lama L, *et al. Halomonas campaniensis* sp. nov., a haloalkaliphilic bacterium isolated from a mineral pool of Campania Region, Italy. *Syst Appl Microbiol* 2005; 28(7): 610–618.
164. Yue H, Ling C, Yang T, *et al.* A seawater-based open and continuous process for polyhydroxyalkanoates production by recombinant *Halomonas campaniensis* LS21 grown in mixed substrates. *Biotechnol Biofuels* 2014; 7(1): 108.
165. Chen GQ, Jiang XR. Engineering microorganisms for improving polyhydroxyalkanoate biosynthesis. *Curr Opin Biotech* 2018; 53: 20–25.
166. Li D, Lv L, Chen JC, Chen GQ. Controlling microbial PHB synthesis via CRISPRi. *Appl Microbiol Biotechnol* 2017; 101(14): 5861–5867.
167. Lv L, Ren YL, Chen JC, *et al.* Application of CRISPRi for prokaryotic metabolic engineering involving multiple genes, a case study: controllable P(3HB-*co*-4HB) biosynthesis. *Metab Eng* 2015; 29: 160–168.
168. Tao W, Lv L, Chen GQ. Engineering *Halomonas* species TD01 for enhanced polyhydroxyalkanoates synthesis via CRISPRi. *Microb Cell Fact* 2017; 16(48): 1–10.
169. Strazzullo G, Gambacorta A, Vella FM, *et al.* Chemical-physical characterization of polyhydroxyalkanoates recovered by means of a simplified method from cultures of *Halomonas campaniensis. World J Microbiol Biotechnol* 2008; 24(8): 1513–1519.
170. Tan D, Xue YS, Aibaidula G, Chen GQ. Unsterile and continuous production of polyhydroxybutyrate by *Halomonas* TD01. *Bioresource Technol* 2011; 102: 8130–8136.
171. Cai L, Tan D, Aibaidula G, *et al.* Comparative genomics study of polyhydroxyalkanoates (PHA) and ectoine relevant genes from *Halomonas* sp. TD01 revealed extensive horizontal gene transfer events and co-evolutionary relationships. *Microb Cell Fact* 2011; 10(1): 88.
172. Fu XZ, Tan D, Aibaidula G, *et al.* Development of *Halomonas* TD01 as a host for open production of chemicals. *Metab Eng* 2014; 23: 78–91.
173. Tan D, Wu Q, Chen JC, et al. Engineering halomonas TD01 for the low-cost production of polyhydroxyalkanoates. *Metab Eng* 2014; 26: 34–47.
174. Ouyang P, Wang H, Hajnal I, et al. Increasing oxygen availability for improving poly(3-hydroxybutyrate) production by *Halomonas. Metab Eng* 2018; 45: 20–31.
175. Chen X, Yin J, Ye J, *et al.* Engineering *Halomonas bluephagenesis* TD01 for non-sterile production of poly(3-hydroxybutyrate-*co*-4-hydroxybutyrate). *Bioresour Technol* 2017; 244: 534–541.
176. Ye J, Huang W, Wang D, *et al.* Pilot Scale-up of poly(3-hydroxybutyrate-*co*-4-hydroxybutyrate) production by *Halomonas bluephagenesis* via cell growth adapted optimization process. *Biotechnol J* 2018; 13: 1800074.
177. Shen R, Yin J, Ye JW, *et al.* Promoter engineering for enhanced P(3HB-*co*-4HB) production by *Halomonas bluephagenesis. ACS Synth Biol* 2018; 7(8): 1897–1906.
178. Ye J, Hu D, Che X, *et al.* Engineering of *Halomonas bluephagenesis* for low cost production of poly(3-hydroxybutyrate-*co*-4-hydroxybutyrate) from glucose. *Metab Eng* 2018; 47: 143–152.
179. Ren Y, Ling C, Hajnal I, *et al.* Construction of *Halomonas bluephagenesis* capable of high cell density growth for efficient PHA production. *Appl Microbiol Biotechnol* 2018; 102: 4499–4510.
180. Ling C, Qiao GQ, Shuai BW, *et al.* Engineering NADH/NAD+ ratio in *Halomonas bluephagenesis* for enhanced production of polyhydroxyalkanoates (PHA). *Metab Eng* 2018; 49: 275–286.
181. Shen R, Ning ZY, Lan YX, *et al.* Manipulation of polyhydroxyalkanoate granular sizes in *Halomonas bluephagenesis. Metab Eng* 2019; 54: 117–126.
182. Quesada E, Ventosa A, Ruiz-Berraquero F, *et al. Deleya halophila*, a new species of moderately halophilic bacteria. *Int J Syst Bacteriol* 1984; 34: 287–292.

183. Kovalcik A, Kucera D, Matouskova P, *et al.* Influence of removal of microbial inhibitors on PHA production from spent coffee grounds employing *Halomonas halophile. J Environ Chem Eng* 2018; 6(2): 3495–3501.

184. Romano I, Lama L, Nicolaus B, *et al. Halomonas alkaliphila* sp. nov., a novel halotolerant alkaliphilic bacterium isolated from a salt pool in Campania (Italy). *J Gen Appl Microbiol* 2006; 52(6): 339–348.

185. Biswas A, Patra A, Paul A. Production of poly-3-hydroxyalkanoic acids by a moderately halophilic bacterium, *Halomonas marina* HMA 103 isolated from solar saltern of Orissa, India. *Acta Microbiol Imm H* 2009; 56(2): 125–143.

186. Tohme S, Hacıosmanoğlu GG, Eroğlu MS, *et al. Halomonas smyrnensis* as a cell factory for co-production of PHB and levan. *Int J Biol Macromol* 2018; 118: 1238–1246.

187. Lemechko P, Le Fellic M, Bruzaud S. Production of poly(3-hydroxybutyrate-*co*-3-hydroxyvalerate) using agro-industrial effluents with tunable proportion of 3-hydroxyvalerate monomer units. *Int J Biol Macromol* 2019; 128: 429–434.

188. Thomas T, Elain A, Bazire A, *et al.* Complete genome sequence of the halophilic PHA-producing bacterium *Halomonas* sp. SF2003: Insights into its biotechnological potential. *World J Microbiol Biotechnol* 2019; 35(3): 50.

189. Hernández-Núñez E, Martínez-Gutiérrez CA, López-Cortés A, *et al.* Physico-chemical characterization of poly(3-Hydroxybutyrate) produced by *Halomonas salina*, isolated from a hypersaline microbial mat. *J Polym Environ* 2019; 27(5): 1105–1111.

190. Pernicova I, Kucera D, Nebesarova J, *et al.* Production of polyhydroxyalkanoates on waste frying oil employing selected *Halomonas* strains. *Bioresource Technol* 2019; 122028.

191. Chen X, Yu L, Qiao G, Chen GQ. Reprogramming *Halomonas* for industrial production of chemicals. *J Ind Microbiol Biotechnol* 2018; 45(7): 545–554.

192. Troschl C, Meixner K, Drosg B. Cyanobacterial PHA production—Review of recent advances and a summary of three years' working experience running a pilot plant. *Bioengineering* 2017; 4(2): 26.

193. Drosg B, Fritz I, Gattermayr F, *et al.* Photo-autotrophic production of poly (hydroxyalkanoates) in cyanobacteria. *Chem Biochem Eng Q* 2015; 29(2): 145–156.

194. Koller M, Marsalek L. Cyanobacterial polyhydroxyalkanoate production: Status quo and quo vadis? *Curr Biotechnol* 2015; 4(4): 464–480.

195. Shrivastav A, Mishra SK, Mishra S. Polyhydroxyalkanoate (PHA) synthesis by *Spirulina subsalsa* from Gujarat coast of India. *Int J Biol Macromol* 2010; 46(2): 255–260.

196. Jau MH, Yew SP, Toh PS. Biosynthesis and mobilization of poly(3-hydroxybutyrate) [P(3HB)] by *Spirulina platensis. Int J Biol Macromol* 2005; 36(3): 144–151.

197. Kemavongse K, Prasertsan P, Upaichit A, *et al.* Poly-β-hydroxyalkanoate production by halotolerant *Rhodobacter sphaeroides* U7. *World J Microbiol Biotechnol* 2008; 24(10): 2073–2085.

198. Lau NS, Sam KK, Amirul AAA. Genome features of moderately halophilic polyhydroxyalkanoate-producing *Yangia* sp. CCB-MM3. *Stand Genomic Sci* 2017; 12(1): 12.

199. Van-Thuoc D, Huu-Phong T, Thi-Binh N, *et al.* Polyester production by halophilic and halotolerant bacterial strains obtained from mangrove soil samples located in Northern Vietnam. *MicrobiologyOpen* 2012; 1(4): 395–406.

200. Van-Thuoc D, Huu-Phong T, Minh-Khuong D, *et al.* Poly(3-hydroxybutyrate-*co*-3-hydroxyvalerate) production by a moderate halophile *Yangia* sp. ND199 using glycerol as a carbon source. *Appl Biochem Biotechnol* 2015; 175(6): 3120–3132.

201. Phong TH, Khuong DM, Van Hop D, *et al.* Different fructose feeding strategies for poly(3-hydroxybutyrtae) production by *Yangia* sp. ND199. *Vietnam J Sci Technol* 2017; 55(2): 195.

202. Sawant SS, Salunke BK, Kim BS. Degradation of corn stover by fungal cellulase cocktail for production of polyhydroxyalkanoates by moderate halophile *Paracoccus* sp. LL1. *Bioresource Technol* 2015; 194: 247–255.

203. Kumar P, Jun HB, Kim BS. Co-production of polyhydroxyalkanoates and carotenoids through bioconversion of glycerol by *Paracoccus* sp. strain LL1. *Int J Biol Macromol* 2017; 107(B): 2552–2558.

204. Kumar P, Kim BS. *Paracoccus* sp. strain LL1 as a single cell factory for the conversion of waste cooking oil to polyhydroxyalkanoates and carotenoids. *Appl Food Biotechnol* 2019; 6(1): 53–60.

205. Hong JW, Song HS, Moon YM, *et al.* Polyhydroxybutyrate production in halophilic marine bacteria *Vibrio proteolyticus* isolated from the Korean peninsula. *Bioproc Biosyst Eng* 2019; 42(4): 603–610.

206. Van Thuoc D, Ngoc My D, Thi Loan T, Sudesh, K. Utilization of waste fish oil and glycerol as carbon sources for polyhydroxyalkanoate (PHA) production by *Salinivibrio* sp. M318. *Int J Biol Macromol.* 2019; 141: 885–892.

207. Valappil SP, Boccaccini AR, Bucke C, Roy I. Polyhydroxyalkanoates in Gram-positive bacteria: Insights from the genera *Bacillus* and *Streptomyces*. *Anton Leeuw* 2007; 91(1): 1–17.

208. Lemoigne M. Produits de deshydration et de polymerisation de l'acide β-oxybutyrique. *Bull Soc Chim Biol* 1926; 8: 770–782.

209. Wu Q, Huang H, Hu G, *et al.* Production of poly-3-hydroxybutyrate by *Bacillus* sp. JMa5 cultivated in molasses media. *Anton Leeuw* 2001; 80(2): 111–118.

210. Halami PM. Production of polyhydroxyalkanoate from starch by the native isolate *Bacillus cereus* CFR06. *World J Microbiol Biotechnol* 2008; 24(6): 805–812.

211. Mizuno K, Ohta A, Hyakutake M, *et al.* Isolation of polyhydroxyalkanoate-producing bacteria from a polluted soil and characterization of the isolated strain *Bacillus cereus* YB-4. *Polym Degrad Stab* 2010; 95(8): 1335–1339.

212. Dave H, Ramakrishna C, Desai JD. Production of polyhydroxybutyrate by petrochemical activated sludge and *Bacillus* sp. IPCB-403. *Indian J Exp Biol* 1996; 34(3): 216–219.

213. Valappil SP, Misra SK, Boccaccini AR, *et al.* Large-scale production and efficient recovery of PHB with desirable material properties, from the newly characterised *Bacillus cereus* SPV. *J Biotechnol* 2007; 132(3): 251–258.

214. Valappil SP, Rai R, Bucke C, Roy I. Polyhydroxyalkanoate biosynthesis in *Bacillus cereus* SPV under varied limiting conditions and an insight into the biosynthetic genes involved. *J Appl Microbiol* 2008; 104(6): 1624–1635.

215. Koller, M. Biodegradable and biocompatible polyhydroxy-alkanoates (PHA): Auspicious microbial macromolecules for pharmaceutical and therapeutic applications. *Molecules* 2018; 23(2): 362.

216. Luef KP, Stelzer F, Wiesbrock F. Poly(hydroxy alkanoate)s in medical applications. *Chem Biochem Eng Q* 2015; 29(2): 287–297.

217. Zinn M, Witholt B, Egli T. Occurrence, synthesis and medical application of bacterial polyhydroxyalkanoate. *Adv Drug Deliver Rev* 2001; 53(1): 5–21.

218. Rodríguez-Contreras A, Koller M, Miranda de Sousa Dias M, *et al.* High production of poly(3-hydroxybutyrate) from a wild *Bacillus megaterium* Bolivian strain. *J Appl Microbiol* 2013; 114(5): 1378–1387.

219. Valappil SP, Peiris D, Langley GJ, *et al.* Polyhydroxyalkanoate (PHA) biosynthesis from structurally unrelated carbon sources by a newly characterized *Bacillus* spp. *J Biotechnol* 2007; 127(3): 475–487.

220. Philip S, Sengupta S, Keshavarz T, Roy I. Effect of impeller speed and pH on the production of poly(3-hydroxybutyrate) using *Bacillus cereu*s SPV. *Biomacromolecules* 2009; 10(4): 691–699.

221. Schmid MT, Song H, Raschbauer M, *et al.* Utilization of desugarized sugar beet molasses for the production of poly(3-hydroxybutyrate) by halophilic *Bacillus megaterium* uyuni S29. *Proc Biochem.* 2019; 85: 9–15.
222. Salgaonkar BB, Mani K, Braganca JM. Characterization of polyhydroxyalkanoates accumulated by a moderately halophilic salt pan isolate *Bacillus megaterium* strain H16. *J Appl Microbiol* 2013; 114(5): 1347–1356.
223. Moorkoth D, Nampoothiri KM. Production and characterization of poly(3-hydroxy butyrate-*co*-3-hydroxyvalerate) (PHBV) by a novel halotolerant mangrove isolate. *Bioresource Technol* 2016; 201: 253–260.

Part III

Bioengineering

5 Role of Different Bioreactor Types and Feeding Regimes in Polyhydroxyalkanoate Production

Geeta Gahlawat, Sujata Sinha and Guneet Kaur

CONTENTS

5.1 INTRODUCTION

Polyhydroxyalkanoates (PHA) are considered to be the future of bioplastics, which have similar processing and material properties as conventional plastics (e.g. propylene), and can undergo processing extrusion, injection and molding [1]. These can be easily broken or decomposed by soil microbes and are understood to be completely biodegradable in nature. Industrially synthesized PHA have the potential to have a wide range of applications such as (a) pharmaceuticals: controlled release and in drug delivery systems; (b) biofuel: methyl ester of 3-hydroxybutyrate and other

3-hydroxyalkanoates are used as biofuel; (c) medicine: absorbable sutures, pin, film and staples, plates of bones, implants, grafts in tissue engineering, (d) disposables like food trays, diapers, razors, utensils, cosmetics packaging, glasses, carpet, compostable lids, etc. (e) chromatography: in chromatography columns they may be used as stationary phase; (f) agriculture: in regulated discharge of pesticides, herbicides, fertilizers and plant growth regulators [2].

Some of the reported commercial PHA are Biopol, Nodax, Degra pol and Biogreen. Biopol and Nodax are copolymers. For example, Biopol is poly(3-hydroxybutyrate-co-3-hydroxyvalerate), while Nodax is a copolymer of poly(3-hydroxybutyrate) (P(3HB)) with a comparatively small quantity of medium chain length monomer with side groups greater or equal than three carbon units. Degrapol consists of two blocks of polymer and is polyester-urethane in nature. Biogreen is a polymer of P(3HB) which has been produced by methanol and is a product of Mitsubishi Gas Chemicals [3].

PHA can be classified into three major groups depending upon the number of carbon atoms present in the functional group. Monomers of short chain hydroxyalkanoic acid (scl-PHA) consist of carbon chains of three to five carbon atoms; medium chain hydroxyalkanoic acids (mcl-PHA) consist of six to 14 carbon atoms in the chain and monomers of large chain hydroxyalkanoic acids (lcl-PHA) consist of more than 15 carbon atoms. All types of bacteria, i.e., Gram-negative and Gram-positive, produce and accumulate PHA intracellularly, when there is excess carbon but limitations of other nutrients like nitrogen, phosphorus and oxygen. Whenever these limited nutrients are supplied, these storage molecules are degraded and used for growth as a carbon source.

In the current scenario, the high cost of industrial PHA production compared to petroleum-based polymers is a major challenge in the field of extensive PHA production and commercialization. Optimization of various steps in biochemical technology for sustainable and cost-effective PHA production has been summarized in Figure 5.1.

The most important factors contributing to costs are expensive raw materials and chemicals as a source of organic matter, but these can be offset by an excellent alternative to plastic replacement. Efforts are being made to use economically feasible and more efficient carbon sources such as hemicellulose, whey, agricultural waste and molasses for industrial production. Cyanobacterial PHA production using more sustainable carbon dioxide (CO_2) has gained importance recently, which also provides a means to address the problem of global warming by allowing carbon capture via cyanobacteria [4]. Other important factors are continuous oxygen supply, equipment depreciation and the high demand for energy and chemicals used for downstream processing [5]. A number of alternatives have been proposed at laboratory scale, one of which includes the use of industrial waste and byproducts (side streams). These include agricultural feedstock, oils from waste plants and wastewater [6]. However, at pilot- and industrial-scale production of PHA, new technological advancements are required for the use of waste streams as raw material. Three crucial factors like the cost of substrate, downstream processing and process development govern successful large-scale production of PHA. Design and implementation

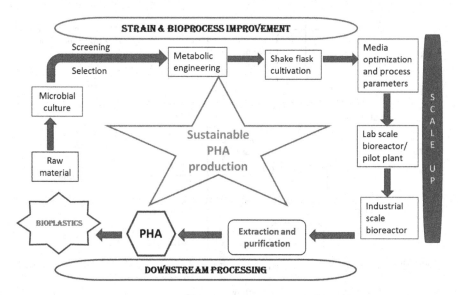

FIGURE 5.1 Various steps in biochemical technology for sustainable and cost-effective PHA production and its process optimization.

of efficient bioprocess strategies lead to high concentration, yield and productivity of PHA. Inexpensive substrates like agro- and industrial waste facilitate further reduction in cost and benefit large scale production. Various bioprocess strategies for large scale production and their novel copolymers have been reported [7]. Batch, fed-batch and continuous fermentation strategies have been reported for large scale industrial production of PHA (Figure 5.2). A batch process is considered to be the simplest, most commonly used and primary investigation for any bioprocess and is a closed system where substrate is added to the system at the start of the process and product is removed at the end of the process. Fed-batch mode is specially operated to achieve high cell density where substrate is also added during cultivation in addition at the starting time and operated as a batch where product is removed at the end of fermentation. P(3HB) has been found to be the most investigated biopolymer by fed-batch cultivation, and unconventional substrates like biodiesel have also been reported in some studies [7]. Batch and fed-batch modes of operation are generally compared, and batch operation is preferred over simplicity of operation. However, in terms of improving product concentration, yield and productivity, fed-batch is preferred which is also easy to design and implement a simple feeding strategy. In addition, looking further for inexpensive products can also improve the economy of operation and reduce the cost of the product which is important in the case of PHA production at large scale. The best results are obtained in terms of higher productivity in the case of continuous fermentation, especially if culture has a high specific growth rate [8–9]. In such cases, reactors can be operated at a high dilution rate without the problem of wash out, and lead to high productivity and product concentration. Very few reports of continuous cultivation for PHA production have

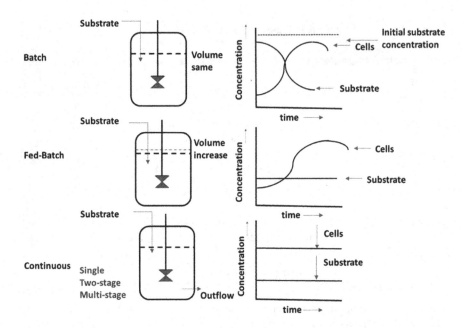

FIGURE 5.2 Various cultivation modes of batch, fed-batch and continuous fermentation for PHA production.

been reported as the chances of microbial contamination are higher and thus it may lead to higher financial losses. However, it has been suggested that by maintaining a proper cultivation environment and robust strains in terms of continuous and stable operation of the bioreactor for longer periods, contamination problems can be overcome. Three-stage continuous cultivation at elevated pressure has been reported using *Pseudomonas putida* KT2440 for the production of *mcl*-PHA. This set-up included batch cultivation on C8 carbon sources, fed-batch on C8/C11:1 and final continuous cultivation on C8/C11:1 at a dilution rate of 0.15 h^{-1}. Yield and volumetric productivity increased with this feeding strategy. The cost of maintaining a sterile environment in continuous cultivation settings can pay off in terms of low production costs of PHA.

Another interesting strategy is a two-stage fermentation strategy which can be used for all the above modes of cultivation and involves physical separation of two stages of PHA production. In the first bioreactor, microbial growth takes place while product formation is achieved in the second bioreactor and it has an edge in terms of product synthesis over the traditional batch mode. Maintenance of different media and environmental conditions are also allowed in these two stages, for example, a high growth rate in the first stage and a high rate of product accumulation in the second stage. *Aeromonas hydrophila* has been used in a two-stage batch fermentation for improving P(3HB) production. When high biomass density was achieved in the first stage of fermentation, cells were transferred to the second stage and were grown in nitrogen-limiting conditions for maximum P(3HB) accumulation [10]. A similar two-stage approach has also been reported for production of poly(3-hydroxy-4-methylvalerate) by relieving the toxicity of isocaproic acid. A cell yield of 39% of cell dry mass (CDM) was achieved by this method [11].

In this chapter, work related mainly to microbial (bacterial) production is reported. In recent times, microalgae are also being reported for PHA production. Algae-based production can be achieved at comparatively lower cost as their growth/ nutritional requirements are minimal and they can also use CO_2 and light as the main energy sources due to their photoautotrophic nature. Moreover, this approach helps in reducing the greenhouse effect as CO_2 released by industry can be consumed by PHA producers in this process. However, a metabolic pathway for synthesis of PHA in algae has not been totally elucidated and their extraction and culture conditions need to be studied in detail as these influence the properties of PHA.

5.2 PROCESS OPTIMIZATION FOR PHA PRODUCTION

PHA are typical intracellular storage (reserve) materials and products of secondary metabolism. Similar to other secondary metabolites, PHA production by most bacterial producers occurs under limited availability of one or more essential nutrients. In other words, it is boosted under conditions that limit the growth and cell biomass production. Typically, the kinetics of cell growth and PHA production consists of two distinguishable phases. First is the nutrient-rich phase during which active biomass proliferation is observed and the concentration of biomass increases in accordance with the kinetics of an autocatalytic process until a limited concentration of an essential nutrient, e.g., nitrogen or phosphate remains in the fermentation broth. Thereafter, the second phase of cultivation is initiated which is characterized by an almost constant biomass concentration which starts to accumulate increasingly high intracellular PHA according to the zero-order reaction kinetics [12]. The increase in PHA concentration in the second phase continues until the carbon source is depleted or due to the increased steric hindrance offered by the presence of high cell biomass. This is called 'non-growth associated PHA production' [13]. However, this is not the only mode of PHA production mechanism and some PHA producers accumulate PHA under nutritionally balanced conditions, which is called 'growth-associated PHA production'. Yet another category of producers synthesizes PHA using the two-phase mode but with no nutrient limitation. In addition, there is another class of producers such as *Pseudomonas* 2F, which are called PHA hyperproducers and demonstrate extremely high PHA production rates following a period of starvation and subsequent re-feeding of carbon source [14]. The existence of different modes of PHA production in different microorganisms implies that a thorough understanding of PHA bioprocess kinetics is crucial for PHA fermentation optimization. This provides essential information on the design of nutrient feeding regimes based on biomass growth and PHA accumulation profiles. Regardless of the degree of growth association of PHA accumulation, it is apparent that a high cell density is directly related to PHA accumulation and obtaining a high PHA volumetric productivity [7]. This indeed is the most crucial factor for the successful development of an industrial process and thus both biomass and PHA formation need to be optimized.

With respect to process optimization, the technical factors which influence PHA production include pH, dissolved oxygen (DO) and temperature. Additionally, strain selection and/or engineering, bioreactor design, nutrient feeding, substrate selection and downstream processing are other important aspects. As discussed above, high PHA accumulation in most production processes requires the maintenance of excess carbon and limited availability of another key nutrient (nitrogen/phosphate) in the fermentation medium. On the other hand, the adoption of dual nutrient limitation for both carbon and nitrogen has also been reported [15]. An additional concern is the possible inhibition caused by certain types of substrates, e.g., volatile fatty acids (VFA), waste glycerol, etc., which affect the biomass growth and product formation rates. Knowledge about limiting and/or inhibitory nutrient concentrations and their relationship with growth and production rates is obtained from process kinetics analysis and mathematical models. Mathematical models are invaluable tools in bioprocess engineering which facilitate process understanding and informed optimization of process conditions in a minimum time [16–17]. Usually the models are empirical in nature and adequately describe offline the kinetics of biomass growth, substrate consumption and product formation. In addition to process understanding, these can be used as tools to simulate and identify the consequence of different substrate feeding concentrations and strategies on product accumulation. The best strategy can then be selected from several tested strategies (offline) and experimentally implemented. Thus, the model obviates the need for extensive experimentation, and trial and error approaches allow the identification of metabolic bottlenecks; it is therefore particularly useful for industrial implementation of a bioprocess.

Numerous models have been developed in literature for PHA processes. Some only describe the PHA production process involved mathematically while some others have been successfully used to design and optimize the nutrient feeding for high PHA production in fed-batch and/or continuous cultivations [16, 18]. The approach has also been extended to the production of PHA copolymers which require a more complicated feeding of multiple key substrates for the synthesis of two or more monomers [19]. This is a useful approach to allow the realization of potential applications of PHA copolymers. Overall, mathematical models involving low structured/kinetic, high structured, i.e. metabolic, cybernetic or hybrid-type models, have the potential to understand metabolic fluxes, identify critical bottlenecks and accelerate up-scaling of laboratory experiments to pilot and industrial scales.

Another important aspect of process optimization is the design and use of different bioreactor types and nutrient feeding strategies. Their selection is aimed at achieving optimal cell density and PHA accumulation under conditions of optimal DO availability, mass transfer and substrate availability [20]. Continuous stirred-tank reactors (CSTR) for submerged PHA formation are the most commonly used bioreactors mainly because of the homogenous environment and substrate conditions, and ease of design and operation. These have been extensively used for PHA production in batch, fed-batch and continuous fermentations. The provision of impellers along with an aeration system provides uniform mixing conditions inside the reactor. Other reactor configurations include the air-lift reactor and bubble column reactors [21].

These reactors are mechanically simpler with no rotating shaft or impellers and this allows a reduction in construction or operation cost and reduced risks of contamination. Their high relevance for PHA production is due to the provision of a definite fluid flow pattern which results in high oxygen and heat transfer, lower mixing time and consequently a higher cell density which in turn allows a higher PHA accumulation and productivity [22].

In recent times, the possible use of inexpensive or waste raw materials as substrates for PHA production has been increasingly considered. While this helps to decrease the production cost and increase the profitability, another advantage is the alternative route for the management of waste materials which are otherwise discarded, and their conversion from waste to value-addition. This also adds to the merit of a bio-based process and is being developed in view of sustainable development goals and circular economy perspectives. To this end, the valorization of industrial byproducts and/or side (waste) streams, e.g., agricultural feedstock, waste plant oils or wastewater are good carbon sources for PHA production. In fact, PHA yields of 0.6–0.8 g/g have been reported for fatty acids as compared to lower yields of 0.3–0.4 g/g reported for sugars [6].

However, to ensure a successful build-up of industrial production based on waste streams, relevant testing at pilot scale after the development of a laboratory-scale process is necessary. Most of the reports on laboratory scale PHA processes are not carried further to higher scales. A few examples for implementation of such processes at pilot scales of 100 L for *mcl*-PHA [23], 30 L for P(3HB-*co*-4HB) [24] and 1200 L for P(3HB-*co*-3HV) [25] are available in the literature. It is worthwhile to note that the use of urban waste streams, e.g., wastewater, waste activated sludge, etc. offers the opportunity to integrate PHA production into urban settings (wastewater treatment plants) which allows a constant supply of carbon source without any extra cost of waste collection. This helps to solve the major problem for scale-up of the PHA production process and undoubtedly warrants more research efforts to facilitate the successful development of an industrial PHA process.

5.3 REACTOR OPERATING STRATEGIES FOR PHA PRODUCTION

Different strategies related to bioreactor operations have been used for optimizing the PHA production. CSTR (one-stage, two-stage and multi-stage), bubble column, airlift and other types have special design and operational attributes, and these are discussed for PHA bioproduction in this section. Each type of reactor has its advantages and disadvantages and has been developed to cater for special purposes like types of PHA, yield and productivity.

5.3.1 CONTINUOUS STIRRED TANK REACTORS (CSTRs)

Stirred tank reactors have been used for centuries for submerged fermentation and are still widely used for bioprocessing and in the chemical industry due to the simplicity of the procedure. CSTRs consist of a cylindrical vessel with central driven shaft supporting one or more impellers which depend on bioreactor size and aspect ratio, and are the simplest reactor type. These can be of batch, continuous and fed-batch types. The batch type is closed system and considered to be simplest for primary investigation in bioprocessing technology, however, it is limited by product concentration and productivity due to starvation of culture at the end of the bioprocess. Fed-batch is preferred to achieve high cell densities and in turn leads to high production of metabolites. CSTR has the advantage of maintaining a constant nutrient environment and allows the study of the effect of nutrient-limiting conditions on growth and productivity in a reliable and real time manner. Establishing single-, two- and multi-stage continuous (chemostat) processes in stirred tank reactors has been found to be feasible for the production of PHA.

Some successful examples of single-stage systems of continuous PHA productions are available [26]. Use of extremophiles for continuous cultivation has been found to be feasible for lowering the risk of microbial contamination. Continuous production in a one-stage set up was possible with the highly halophilic archaeon *Haloferax mediterranei*, a member of the class of haloarchaea which was able to grow a minimum of 1.5 M NaCl for growth and was able to tolerate up to 5 M concentration. The organism was already able to accumulate considerable amounts of PHA during exponential growth which led to their successful growth in a continuous one-stage set up. Robustness and stability of the system were tested by maintaining only minimal sterility precautions and at D = 0.12 h^{-1} the system stayed monoseptic for 90 days [27]. The continuous system was compared with that of batch and fed-batch, and glucose was used as carbon source with two different concentrations of phosphate. Five different dilution rates (D = 0.02–1.0 h^{-1}) were tried with two phosphate concentrations. The best operating concentrations were found to be at D = 0.02 h^{-1} and a specific productivity of 0.014 g/(g·h) was achieved. However, the exact composition of the produced PHA was not reported due to the simple analytical method used in this study. Optimal conditions for both cell growth and PHA accumulation for some organisms cannot be maintained in a single-stage continuous system; therefore, two-stage systems are preferred over a single stage. In another study, single-stage continuous production was not found to be suitable for *C. necator* where PHA accumulation was not found to be growth associated or can be said to be partly growth associated. In a single-stage continuous process, the growth rate was found to be the maximum but PHA production was not related to growth. Also, incomplete conversion of carbon source occurred due to the need for higher carbon to nitrogen ratio (C/N) in the feed stream. It doubled the cost of substrate expense as carbon cannot be recycled and treatment of spent fermentation broth due to higher biochemical oxygen demand (BOD) adds to overall process cost. For such cases, two-stage processes consisting of serially attached CSTR have been recommended. These have been found to be superior to a single-stage chemostat in terms of PHA production [28]. Biomass produced continuously under balanced nutrient supply is transferred to a second vessel where continuous production of PHA from carbon source takes place under a growth-limiting nitrogen source. Both the stages of PHA production can be optimized separately in this process [29]. However, one-stage continuous processes work well in conditions where high productivity is seen in parallel with microbial growth [30]. In this case too, the addition of a second reactor led to an increase in productivity by providing the cells the required time of exposure, or residence time, for complete conversion of substrate. The first two-stage chemostat was described using *A. lata* on sucrose and propanoic acid for P(3HB-*co*-3HV) production. Propionic acid was completely converted in the first stage, but sucrose was present at a low concentration, and 48% biomass concentration was produced under these conditions. Biomass was transferred continuously to the second reactor so leftover sucrose was continuously utilized and 58% of copolymer in biomass was achieved. Propionic acid concentration of more than 8.5 g/L in the feed triggered an inhibition of sucrose uptake. In one of the studies, it has been emphasized that *mcl*-PHA production in two stages is far superior to that of a single-stage set-up due to entirely different kinetics related to biomass growth and PHA accumulation when octanoate is used as carbon source. Accumulation of *mcl*-PHA is maximized when cells grow at specific growth rate (μ) of 0.21 h^{-1} compared to single-stage chemostat cultures where maximum μ was observed at 0.48 h^{-1} [31]. Multi-stage systems are difficult to handle as they are complicated and pose a contamination problem due to the large number of connecting devices between the vessels. They resemble a continuous plug flow tubular reactor (CPFTR) as they provide different cultivation conditions in each reactor. The number of reactors (minimum of five) in a series is considered to be the best in terms of product quality. It could be explained by the fact that cells are in a uniform physiological state due to narrow residence time distribution and this has led to the recent development of a five-stage bioreactor for PHA biosynthesis with high productivity. Such multi-stage systems are discussed in detail in the section on bioreactor cascade (Section 5.3.4).

5.3.2 BUBBLE COLUMN REACTORS

Bubble columns are a type of loop bioreactor in which aeration is supplied from the bottom for optimum gas exchange. These have been designed specifically to avoid mechanical agitation, which otherwise creates shear stress on culture growth. These types of reactors are not very popular for PHA production, but their simple design without any mechanical agitators allows low construction and operational costs [32]. Considering their low production cost and relatively lower risk of contamination, they have been studied by various researchers to check their feasibility for PHA production. One such report investigated the biosynthesis of P(3HB) by *Anabaena solitaria* in a flat panel bubble column photo-bioreactor [33]. A photo-bioreactor (PBR) was continuously fed from the bottom with carbon substrate by bubbling air enriched with CO_2 at a rate of 2.5 L/h. *A. solitaria* exhibited very low P(3HB) accumulation capacity in a PBR system, and P(3HB) concentrations of around 7 mg/L was obtained after the 12th day of cultivation at light intensity and temperature of 500 $\mu mol/(m^2.s)$ and 28°C, respectively.

In 2012, *Methylocyctis hirsuta* was used for the first time for evaluating the synthesis of P(3HB) from methane gas in a bubble column bioreactor [34]. A full factorial statistical model was designed to study the effect of CH_4 to air ratio and the amount of nitrogen on P(3HB) accumulation in a bubble bioreactor. P(3HB) synthesis by *M. hirsuta* was observed to be growth associated and both factors showed a significant effect on the P(3HB) synthesis potential, resulting in a total P(3HB) concentration of 1.41 g/L and content of 42.5 wt.-% of CDM. Recently, the same methanotrophic culture was studied for P(3HB) production from methane using high cell density cultivation [35]. The effect of different carbon sources such as ethanol, methanol, acetate and glucose was tested in a bubble column bioreactor. The cultivation was carried out in two phases; in growth phase, cells were allowed to grow under different concentrations of carbon substrates to check their effect on methane uptake by methanotrophic culture. During P(3HB) synthesis phase, the feeding of different carbon sources was started along with nitrogen, magnesium or phosphorus limitation. Results suggested that *M. hirsuta* had the potential to grow on tested substrates and limitation of magnesium and/or phosphorus played an important role in achieving high cell densities in bubble bioreactors. The best results were achieved for a ratio of 1:1 of methanol:ethanol and it showed accumulation of 8 g/L of cell biomass with polymer content of 73.4% of CDM at optimized conditions. More recently, a new isolated strain of methanotroph, *Microbacterium* sp. was also investigated for the P(3HB) synthesis from methane gas in a bubble column bioreactor [32]. Four different media were tested for cell growth and P(3HB) accumulation by isolated strain. After selection of the best medium, a Taguchi design was used for evaluating the effect of five process parameters on P(3HB) production. The effect of five process variables, mainly nitrogen source, Na_2HPO_4 content, methane/air ratio, inoculum age and pH-value was investigated for P(3HB) accumulation inside a bubble column bioreactor. The results revealed that pH was the most important parameter under selected conditions for P(3HB) accumulation. However, P(3HB) accumulation was still very low (25% of CDM) under optimized conditions.

5.3.3 AIRLIFT REACTORS

Airlift reactors (ALRs) are a special type of bubble column reactor (pneumatic reactor) which have net-draft tube without any mechanical components such as agitators. They have a simple design with no rotating impellers which ultimately results in low operation and maintenance costs along with a lower risk of contamination. ALRs are particularly used because of the several advantages provided by them such as simple construction, better aseptic control, a definite fluid flow pattern resulting in high oxygen-mass transfer, shorter mixing time and less power consumption due to the absence of mechanical agitators. ALR was studied for the first time for the production of P(3HB) using a two-stage culture of *Alcaligenes eutrophus* [36]. Initial experiments on P(3HB) production in ALR resulted in low polymer concentrations and productivity in a chemostat culture due to the low oxygen mass transfer coefficient (k_La) inside the bioreactor. Later, a surface-active agent such as carboxymethylcellulose (CMC) was added to increase the oxygen-mass transfer rates. After the supplementation of 0.05 wt.-% CMC inside the reactor, k_La was increased to 375 h^{-1} from 250 h^{-1}, and P(3HB) concentration, productivity and content were enhanced to 56.4 g/L, 1.02 g/(L·h) and 81.4 wt.-%, respectively [36].

Subsequently, a new strategy was developed for the production of biodegradable PHA from food scraps using an air-bubbling bioreactor [37]. The food waste was initially digested in an anaerobic bioreactor resulting in four organic acids which were then transferred into an ALR for the production of PHA by *Ralstonia eutropha*. Feeding of butyric acid, acetic acid and propionic acids resulted in production of a homopolymer, P(3HB). A maximum cell density of 11.3 g/L and polymer content of 60.2% of CDM was obtained when a silicone rubber membrane was used as net-draft tube. On the other hand, the mass transfer rate of acids and PHA accumulation was increased when a dialysis membrane was used. The cell biomass and its PHA content reached to 22 g/L and 72% of CDM, respectively. The synthesized polymer was a copolyester of poly(3HB-*co*-3HV) with 2.8 mol-% of 3HV monomer unit.

PHA production has also been investigated using different configurations of ALR such as a concentric draught tube and outer aeration and internal settling (OAIS), and their performance has been compared with that of a CSTR [20–21]. In the case of *R. eutropha* DSM 545, ALR exhibited better performance than CSTR in terms of P(3HB) accumulation along with the advantage of low energy consumption [20]. The effect of different aeration rates from 12 to 50 L/min was investigated on P(3HB) accumulation by *R. eutropha*. The P(3HB) production rate and yield were increased as the aeration rate or superficial gas velocity was increased during the accumulation phase. The highest P(3HB) production rate of 0.6 g/(L·h) and content of 50% was obtained at 30 L/min in ALR, while in the CSTR, P(3HB) production rate was 0.82 g/(L·h) with 50% polymer content. However, the latter configuration requires a higher power consumption than the airlift reactor, indicating the suitability of ALR for economical production of PHA.

In another study, a different configuration of ALR with a cell settling arrangement was used [21]. Batch experiments with ALR with in situ cell retention were carried out using *A. australica* DSM 1124. ALR demonstrated a significant improvement in biomass synthesis and PHA accumulation in comparison to CSTR. It demonstrated an accumulation of 10.76 g/L biomass and 7.81 g/L P(3HB) with an overall P(3HB) yield of 0.32 g/g sucrose, while CSTR yielded a biomass and P(3HB) concentration of 8.31 and 5.45 g/L, respectively. The feasibility of ALR for high cell density cultivation was evaluated using a fed-batch fermentation strategy [38]. During two-stage fed-batch cultivation of *Burkholderia sacchari* IPT 189, a stepwise addition of sucrose was done to promote cell growth, and nitrogen limitation was applied to increase the P(3HB) synthesis rate. This approach yielded a significantly high cell density of 150 g/L in ALR and induced the P(3HB) accumulation up to 42% of CDM. Results showed the P(3HB) productivity of 1.7 g/(L·h) and P(3HB) yield on sucrose was 0.22 g/g. This report suggested that ALR could be an appropriate choice for cost-effective production of P(3HB) at high cell density.

5.3.4 BIOREACTOR CASCADE

In a bioreactor cascade system, three or more than three CSTR are attached in series, and it is also known as multi-stage chemostat. The potential of a multi-stage process is based on the fact that different controlled nutrient conditions can be applied along the reactor cascade [39]. Multi-stage processes are more important than single-stage or two-stage continuous cultivations for maintaining optimal conditions for both cell growth and PHA synthesis, and for achieving high productivity along with superior quality PHA. They are especially useful for non-growth associated PHA production or partial-growth associated PHA production [40]. Apart from high productivities, the multistep cascade provides the advantage of maintaining the exact process conditions at each step, by supplying nutrients (substrates and precursors) and parameters like temperature, pH and oxygen supply. This is a promising strategy to synthesize novel types of tailor-made biopolymers for special applications, e.g. block-polymers ('*b*-PHA') consisting of soft and hard units.

This approach was first studied by Atlíc and group using a five-stage bioreactor cascade, wherein the first CSTR was used for cell biomass synthesis under continuous feeding of nutrients and the other four CSTRs were used for PHA production under nutrient limitation (Figure 5.3; [40]). The CSTRs attached in series, and involved in active PHA synthesis, mimicked the characteristics of a tubular plug flow reactor (TPFR), which was recommended two decades ago to be highly appropriate for synthesis of intracellular products [41]. The first reactor was dedicated to balanced bacterial growth under nutritional balanced conditions which led to a catalytically active biomass in high concentration. Fermentation broth was continuously passed from subsequent bioreactors which were operated under limiting nitrogen conditions (Figure 5.3).

A regular supply of carbon source was ensured in each of the vessels and microorganism *C. necator* DSM 545 was used as microbial strain. The aim of this cascade reactor was to improve the productivity and polymer content. Volumetric and specific productivity of 1.85 g/L and 0.1 g/(L·h) PHA was achieved, having excellent polymer properties. The cascade was operated under steady state for more than 200 h. At steady state, the residual biomass, P(3HB) productivity and content were 19 g/L, 1.97 g/(L·h) and 77% of CDM, respectively (from the last bioreactor). Thus, a multi-stage continuous high cell density culture can surpass fed-batch, one-stage and

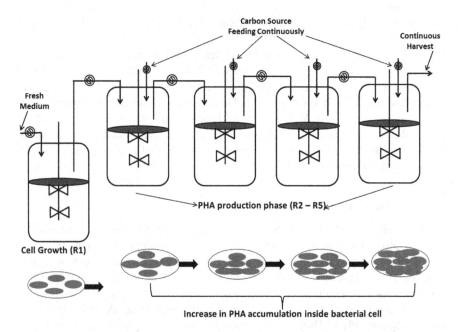

FIGURE 5.3 Schematic representation of different stages of cultivation in a bioreactor cascade (adapted from Koller (2018) [12] with permission from MDPI).

two-stage continuous cultivation systems in terms of polymer productivity and final product concentrations. Optimization of the entire process also reduced the glucose content in the outlet stream resulting in a lower biochemical oxygen demand of the spent fermentation broth. It was concluded that a multi-stage cascade of reactors after optimization can lead to an economically viable process. Another advantage was that due to the flexibility of the system, the final product can be fine-tuned, e.g. the polymer composition can be changed if a copolymer is added to each vessel or a block copolymer of P(3HB) and P(3HV/4HB) can be made by adding alternate hard and soft segments.

After performing several experimental studies, these multi-stage bioreactor cascade systems were investigated in detail for PHA production using a mathematical modeling tool [39, 42]. In one study, a multi-stage kinetic model was developed for mixed-type synthesis of P(3HB) under nitrogen limitation in order to obtain further insights about the operation of such cascade systems [39]. The main aim of the study was to develop a dynamic mathematical model as a predictive tool for process optimization. PHA production was observed to be partially growth-associated under nitrogen limitation, therefore the Luedeking-Piret's model of partial growth associated product formation was adopted as a working model. Various model predictions suggested that the P(3HB) production rates can be increased from 2.13 to 9.96 g/(L·h) if different dilution rates and feed concentrations are used at each stage. Here, the first reactor (R1) in the cascade system was maintained under nutrient balanced conditions for continuous production of biomass. Reactor R2 was modeled as a process controlled by glucose and nitrogen, whereas reactors R3 to R5 were maintained under excess glucose and limiting nitrogen concentration for PHA biosynthesis. The total biomass of 164 g/L and P(3HB) concentration of 123 g/L was obtained under these optimized conditions.

Further in-depth metabolic flux analysis of *C. necator* cultivated in a bioreactor cascade for PHA production was demonstrated by Lopar and associates, who designed a highly structured metabolic model for continuous P(3HB) synthesis process [42]. The metabolic state of bacteria in each reactor cascade (R1 to R5) was studied by two-dimensional yield space analysis and elementary flux modes for P(3HB) synthesis. The metabolic model developed exhibited good correlation with the experimental results for biomass and P(3HB) yields at all stages of the cascade system. Although a number of CSTR reports are available in the literature with advantages of continuous harvesting and tweaking of nutrient concentration according to requirements, it is overshadowed by the challenges associated with maintaining a continuously sterile environment and chances of strain mutability.

More recently, a new technique of specific growth rate and specific production rate estimation was established based upon footprint area analysis of pictures obtained by electron microscopy (Figure 5.4; [43–44]. This group developed a formal kinetic and structured kinetic model, accompanied by footprint area analysis of binary imaged cells in a multi-stage bioreactor cascade. A continuous five-stage bioreactor cascade mimicked the process features of TPFR and was developed for high-throughput PHA production by *C. necator* at high volumetric and specific PHA productivity of 2.31 g/(L·h) and 0.105 g/(g·h), respectively. Formal kinetic modeling optimized the fermentation process in terms of reactor volumes, dilution rate,

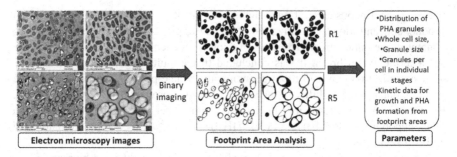

FIGURE 5.4 The footprint area analysis of Transmission Electron Microscopy images of *C. necator* DSM 545 culture (*left group of four pictures*) during first stage (R1) and fifth stage (R5) of cultivation in a bioreactor cascade (adapted from Vadija et al. (2016) [43] with permission from Springer).

and substrate concentration, whereas high structured metabolic model was based on elementary flux analysis reflecting the metabolic states of the cells at different prospective scenarios (Figure 5.4). Footprint area analysis helped in understanding the morphological changes of the cells and PHA granules under fluctuating environmental conditions. This should facilitate the design of the cascade on a larger scale continuous cultivation, e.g. regarding the optimum number of stages.

5.4 NUTRIENT FEEDING REGIMES FOR PHA PRODUCTION

5.4.1 BATCH CULTIVATION

The batch method is the simplest and most extensive method of any bioprocess and is considered to be a closed system where substrate is added to the bioreactor at the beginning of cultivation and product is removed at the end. No substrate is added during cultivation and no cells are withdrawn from the reactor. It is especially useful for studying the influence of various process parameters, bioconversion of new carbon substrates and operating conditions in a bioprocess. Very few reports of uses of refined sugar are available in the literature for PHB production, while mostly inexpensive carbon sources like agro-industrial wastes including cane molasses, rice straw hydrolysate, grass biomass hydrolysate, plant oils, etc. have been reported for batch PHA production [7]. The batch mode of feeding for enhanced PHA synthesis in two-stage fermentation is well established in bioprocess technology. It involves physical separation of two phases like growth in one bioreactor and PHA formation in a second bioreactor. It leads to enhanced product synthesis over a single batch reactor. It involves different growth conditions and media in two reactors with each suited for particular conditions, i.e. high growth rate and product accumulation. This strategy was tried for *Aeromonas hydrophila* for batch production of P(3HB) [10]. Once maximum growth was achieved in the first reactor, PHA was produced under nitrogen-limiting conditions to accumulate cell biomass of 16.8 g/L with a polymer concentration and content of 10.4 g/L and 62.1% of CDM, respectively, in the second reactor.

A strategy was developed to convert crude glycerol and activated sludge for production of PHA through aerobic dynamic feeding [45]. A PHA accumulating mixed microbial culture in a sequencing batch reactor (SBR) was able to accumulate both P(3HB) and 3HV monomers in the molar ratio of 60:40. A maximum PHA content of 80% was achieved as a biomass dry weight with a production yield of 0.7 mg C PHA/mg C. This method was the classical case of converting waste to value-added products. Copolymers showed better mechanical strength than their homopolymer counterparts. The maximum PHA productivity achieved was 193–236 mg/(L·h) and biomass concentration increased from 0.7 to 2 g/L when the organic loading rate (OLR) was increased from 360–1000 mgC/(L.d). Direct utilization of crude glycerol without pre-treatment reduced the cost significantly.

Sequential batch reactors (SBR) have been used for enrichment of PHA accumulating microbes under feast-famine regimes and non-sterile conditions [46]. Two enrichment approaches were used, one with nitrogen availability during the complete cycle and the other with nitrogen limitation during the feast phase. Crude glycerol, a byproduct of the biodiesel industry was converted to volatile fatty acids and 1,3-propanediol using CSTR with a focus on 1,3-propanediol. It was then converted to PHA using aerobic dynamic feeding strategy. No PHA from 1,3-propanediol was observed when nitrogen was present during the entire cycle or when it was in limiting condition. A different enrichment strategy when limited nitrogen was present in the feast phase triggered the accumulation response but nitrogen was still supplied in famine phase. A PHA yield of 0.99 ± 0.07 C_{mol} / C_{mol}S (0.84 g COD PHA/g chemical oxygen demand (COD) S), PHA content of 76 ± 3.1 g PHA per 100 g total suspended solids (TSS) and $99 \pm 2.1\%$ recovery of 1,3-propanediol was obtained. An ultimate yield of 0.19 g COD PHA per gram of input COD was achieved. This was a novel enrichment strategy for targeting specific biotransformation from byproducts of fermentation to high value products such as PHA.

In a recent study, P(3HB) production and recovery were performed from agro-industrial residues [47]. Hydrolysates from hemp hurd biomass and NH_4Cl were used as carbon and nitrogen sources respectively for the production of P(3HB) using *Ralstonia eutropha*. A maximum hydrolysis yield of 72.4% was achieved with total sugar hydrolysate (glucose/xylose) of 53.0 g/L. The preferred metabolism of glucose over xylose was observed in this study and under optimum conditions, a total of 56.3% of CDM could be accumulated with total PHA production of 13.4 g/L and a productivity of 0.16 g/(L·h). No adverse effect of the nitrogen source was observed on PHA biosynthesis but showed an adverse impact on the growth of cells. The average molecular weight of recovered plastic was 150–270 KDa with polydispersity index (M_w/M_n) in the range of 2.12 [47].

Currently, there is a need to develop improved cultivation strategies which could help in cultivation under non-limiting and non-inhibitory substrate concentrations and subsequently enhance the total yield and productivity of the PHA accumulation during the cultivation. Fermentation processes with high cell densities are highly desirable as they favor enhanced PHA accumulation, especially with significant reduction in the culture volume, reduction in production costs, equipment cost and in wastewater production. Continuous, fed-batch and multi-stage cultivations are the three important nutrient feeding strategies for achieving high cell density cultivation of microorganisms.

5.4.2 FED-BATCH CULTIVATION

For PHA production, fed-batch cultivations have been extensively studied with the aim of achieving high cell density; these cultures yield high PHA productivity and require low investment cost [48]. The fed-batch cultures are initiated as batch cultivation and after some time intermittent or continuous addition of nutrients is done to maintain optimum substrate concentration inside the bioreactor. The nutrient feed solution should be balanced enough with respect to substrate concentration and rate of feeding primarily to maintain the growth of the microorganisms at a specific growth rate and simultaneously eliminate the generation of inhibitory byproducts. Fed-batch cultures are usually carried out by the control of nutrient feeding with respect to dissolved oxygen content [49], culture pH values [50–51] or availability of residual carbon sources [51–52]. Reports are available in literature wherein mathematical models have also been used intelligently to selectively feed the fresh nutrients such as carbon and nitrogen during fed-batch fermentation [16, 48, 53].

A simple fed-batch cultivation strategy was designed for growth-associated P(3HB) production on sucrose using *Alcaligenes latus* ATCC 29714 [50]. This cultivation strategy was based on the feeding of sucrose and ammonia solution by the pH-stat method and the addition of nutrient solution was done to maintain a constant pH value. This approach yielded a total biomass concentration of 35.4 g/L with a maximum P(3HB) productivity and content of 0.99 g/(L·h) and 51%, respectively. During nutrient feeding, it is extremely important to maintain carbon concentration at an optimum level inside the bioreactor [52]. It was established earlier that pulse addition of carbon should not be attempted after carbon exhaustion, but rather it should be initiated when carbon concentration is close to half its velocity coefficient (K_s). Thus, the feeding rate of carbon source during fed-batch cultivation is crucial for the biopolymer accumulation. This phenomenon was further explored in detail by designing different feeding strategies such as pH-stat fed-batch and variants of pulse feeding to maintain glucose concentration at optimum levels to increase PHA accumulation [51]. The pH-stat based feeding showed an accumulation of PHA up to 26 g/L by *Halomonas venusta*, but most of the carbon was fluxed towards biomass formation instead of PHA synthesis. The highest PHA concentration of 33.4 g/L and PHA content of 88.12% of CDM was obtained from single pulse and high residual glucose concentration feeding strategy. In total, there was an 8.6-fold increase in PHA concentration by adopting this pulse-feeding strategy, as compared to batch mode.

In another report, two different nutrient feeding strategies such as exponential feeding and DO-stat mode feeding were designed for the production of P(3HB) by *C. necator* DSM 428 using waste cooking oil [49]. The bioreactor cultivation was initiated in batch mode for 17–20 h and after that fed-batch cultivation was started by feeding used cooking oil. The exponential feeding strategy was designed with the aim of maintaining the specific cell growth rate recorded during the batch cultivation (0.14 ± 0.02 h^{-1}) for the entire duration during the fed-batch to achieve high cell density. In the case of DO-stat mode, the cooking oil feeding rate was automatically controlled on the basis of dissolved oxygen concentration inside the bioreactor under a constant stirring speed of 500 rpm. A total cell biomass of 21.3 g/L and P(3HB) content of 84.0% was achieved with exponential feeding. This strategy yielded a volumetric productivity of 4.5 g/(L·d). However, the highest P(3HB) productivity of 12.6 g/(L·d) was achieved with the DO-stat based feeding. During DO-stat mode, NH_4OH was replaced with NaOH for controlling pH and imposing nitrogen limitation, which ultimately resulted in a high polymer concentration of 19.8 g/L within 37 h.

Mathematical models can also serve as a powerful tool to study the systemic behavior of culture [53] and design of adequate nutrient feeding strategies [48] for enhanced biomass and P(3HB) accumulation. In a recent study, a mathematical model was developed for improving the production of P(3HB) by *Azohydromonas australica* [48]. Two dynamic fed-batch cultivation strategies such as pseudo-steady state with respect to sucrose and pseudo-steady state with respect to nitrogen along with carbon substrate feeding were designed using the model developed. These nutrient feeding strategies were designed by performing various offline simulations on the computer as guided by the model. One such strategy of nitrogen limitation during pseudo-steady state fed-batch cultivation resulted in a significantly high biomass concentration of 39.17 g/L and P(3HB) concentration of 29.64 g/L in 50 h of cultivation. This nutrient feeding strategy resulted in high P(3HB) productivity of 0.6 g/(L·h) and P(3HB) content of 75% of CDM. Therefore, it can be hypothesized that mathematical modeling of microorganisms and its subsequent computer simulation under different cultivation conditions are effective tools in biochemical engineering which can lead to better understanding and optimization of the microbial processes with minimum experiments.

5.4.3 CONTINUOUS CULTIVATION

In continuous mode, fresh media are continuously fed into the bioreactor, and a small amount of the culture medium is withdrawn from the other end at the same dilution rate [12]. The parameters (cell, substrate and product concentrations) inside the bioreactor are maintained at steady state. The continuous cultures are generally referred to as chemostat because of static chemical environment inside the reactor. The continuous fermentation processes are of great commercial importance due their high productivities, particularly with the cultures having high maximum specific growth rates [54]. However, it must be noted that the adoption of continuous cultivation for PHA production is a big challenge. Most of the bacteria generally synthesize PHA in a non-growth-associated manner, and therefore its synthesis cannot be optimized during growth phase inside a reactor maintained at steady state [55]. One solution to avoid this issue is to perform a two-stage cultivation in which growth phase and PHA synthesis phase are separated from each other. In the first stage, growth is promoted and in the second stage, conditions favorable for PHA accumulation (i.e. high carbon and low nitrogen or phosphorous), can be applied.

One of the earliest studies on continuous cultivation for P(3HB) and poly(3HB-*co*-3HV) production was carried out on *A. eutrophus* and *A. latus* [28]. In a single-stage continuous cultivation, *A. eutrophus* DSM 545 (or *C. necator* DSM 545) accumulated P(3HB) concentration up to 33% of CDM when grown on glucose under nitrogen limitation. Poly(3HB-*co*-3HV) copolymer was produced by *A. latus* ATCC 29714 in a two-stage continuous cultivation when valeric or propionic acids were added to the feed medium containing sucrose as carbon source. In the first stage of chemostat, feeding of propionic acid at 5 g/L yielded 43 wt.-% poly(3HB-*co*-3HV) copolymer with a final 3HV content of 20 mol-%. In this case, shifting the bioreactor to a second stage demonstrated a complete consumption of sucrose and poly(3HB-*co*-3HV) content increased to 58% of CDM with a 3HV content of 11 mol-%.

A similar strategy of two-stage continuous cultivation was designed for enhanced P(3HB) production by *C. necator* WSH3 [56]. The first stage was used for the culture biomass growth, and the second stage featured P(3HB) production under nitrogen limitation. A significantly high cell density of 50 g/L and P(3HB) concentrations of 30.5 g/L were obtained at 0.075 h^{-1} dilution rate. Moreover, this strategy resulted in a high P(3HB) productivity of 1.23 g/(L·h). Later, a non-sterile, two-stage continuous cultivation process was developed for P(3HB) production by halophilic strain, *Halomonas* TD01 to reduce the production cost [57]. The fermentation was initially performed in bioreactor I for two weeks. Thereafter, the contents of bioreactor I were transferred to bioreactor II for P(3HB) production under nitrogen limitation. In the final stage (second stage), a total biomass concentration of 20 g/L was obtained with a maximum P(3HB) content of 65% of CDM. Halophilic strains provide various advantages, such as growth on high salt containing seawater and the possibility of open, unsterile and continuous fermentation. These characteristics will significantly help in the reduction of PHA production cost and minimization of the fresh water requirement [58]. Moreover, distilled water can be used for the lysis of halophilic bacterial cells, thus helping to reduce downstream processing costs.

The dual nutrient limitation feeding regimes during continuous cultivation could help in the synthesis of tailor-made PHA, provided the carbon source is totally consumed by the end of fermentation [59]. This group demonstrated the continuous production of PHA in a one-stage chemostat culture of *Pseudomonas putida* ATCC 29147. Dual-nutrient (carbon and nitrogen) limiting conditions were applied at a dilution rate of 0.1 h^{-1}. Different mixtures of 5-phenylvalerate, octanoate and 10-decanoate were fed to the culture which resulted in the formation of five new tailor-made copolyesters with varying compositions of aromatic side chains. It was therefore concluded by the group that the steady state in chemostat culture conditions provides a suitable strategy for the production of tailor-made *mcl*-PHA.

5.5   INDUSTRIAL PRODUCTION OF PHA – TECHNO-ECONOMIC CHALLENGES AND THEIR SOLUTIONS

The market opportunity for PHA is critically dependent on its availability which is governed by economics and the performance of industrial-scale PHA production. To this end, the suitability of industrial production is evaluated on the basis of factors such as the safety and stability of microorganisms, utilizable range and cost of raw material (substrates), rate of microbial growth and PHA accumulation, yield on substrate, achievable cell densities and PHA content and extractability of intracellular PHA [60]. With respect to process development for high cell growth and product formation, once a bioprocess has been established successfully in laboratory-scale experiments, it is then carried out in bioreactors of increasing scale, with a normal scale-up ratio of 1:10. The final process optimization occurs in the pilot plant scale with 50 to 300 L fermenter volumes since the operational conditions and hydrodynamic and mixing parameters are very similar to those in industrial-scale fermenters. Such a rigorous scale-up in optimization and development leads to a successful industrial production process [61]. However, except for a very few examples of such a sequential process development, scale-up to successive scales is not usually observed in PHA literature reports. In fact, a high percentage of reports are based only on shake flask studies, and follow-up to large bioreactor scales even at laboratory level is generally not performed.

Effective scale-up of bioprocesses including PHA is dependent on the proper and careful application of one of the following scale-up criteria, i.e. constant specific power unit (P/V), constant volumetric mass transfer coefficient (k_La), constant impeller tip speed of agitator or shear and constant dissolved oxygen (DO) concentration. It is known that the specific scale-up criterion could result in different process conditions on a particular scale [62]. For PHA production, the influence of oxygen gas-liquid mass transport is perhaps the most significant factor for scale-up. Biopolymer synthesis generally occurs under high aerobic conditions and usually a polymer with higher molecular weight is produced under non-limited conditions. However, the aqueous fermentation medium containing salts and organic substances has a high viscosity and shows non-Newtonian behavior. Under these conditions, oxygen is an important nutrient which influences microbial growth, maintenance and product formation and scarcity of oxygen negatively affects the process performance. Thus, DO becomes a limiting nutrient in high oxygen demand processes, i.e. under conditions of fast-growing microorganisms, high biomass and polymer production or when rheological properties of broth offer a high resistance to mass transfer (under polymer production) [63]. The supply of oxygen from gas stream to culture broth is important and, consequently, the accurate estimation of the oxygen transfer rate (OTR) at different scales and different operational conditions has a relevant role for selection, design and scale-up of bioreactors.

OTR can be described as proportional to the concentration gradient or the volumetric mass transfer coefficient (k_La). The maximum value of k_La is limited due to the low solubility of most gases associated with aerobic fermentation (such as PHA), notably oxygen. Thus, the maximum mass transfer rate from gas to liquid in a bioreactor can be estimated by the product $k_La.C^*$, where C^* is the saturation concentration in the liquid phase. The solubility of oxygen in aqueous phase solutions under 1 atm pressure of air is extremely low (~10 ppm near ambient temperature), thus resulting in a low availability to producing cells. In particularly high biomass conditions during PHA fermentation in suspension cultures, the OTR diminishes significantly due to the increasing viscosity and the non-Newtonian characteristic of the fermentation broth [64]. As a result, the availability of oxygen to cells is not only dependent on its concentration in broth but also on its transfer to cells. Thus, optimization of process conditions for high k_La and their efficient translation to a larger scale is the crucial requirement and indeed a challenge for successful industrial scale-up.

The increased production of PHA means the emergence of new PHA applications when increasing amounts of PHA are available on the market for exploration. This in turn means that in the future, large scale PHA applications will be dependent on its production cost [7]. This urgently necessitates the development of low-cost PHA production technology, including the use of low-cost substrates, transition from discontinuous batch and fed-batch cultivation to continuous and even non-sterile processes based on mixed cultures and/or mixed substrates, and lower use of solvents and/or labor for their downstream processing.

The early industrial implementation of PHA production was fueled by the anticipated burgeoning of petroleum prices from the 1970s, which was considered an opportunity to promote the market for bio-based plastics in place of conventional petroleum-based ones [65–66]. The most relevant example of this ideology was Imperial Chemical Industries (ICI), UK, which has produced PHA on a large scale since the 1980s. ICI sold its polymer under the trade name of Biopol which was essentially an *scl*-PHA copolymer poly(3HB-*co*-3HV) produced by *Ralstonia eutropha* and used for packaging applications. The high robustness and production capacity of ICI process allowing a high cell density of 100 g/L in just 72 h fermentation time made Biopol a huge success. However, the company made a huge mistake in anticipation of sky-rocketing petroleum prices which actually did not increase to that extent and thus jeopardized the growth of Biopol over traditional petroleum-based plastics. Globally, over 24 companies had engaged in PHA research and development as well as production, but some of them stopped their PHA activities, mainly in the 1990s, owing to the plummeting oil prices. Another opportunity for PHA came in the early 2000s when the oil prices increased to US$ 100 per barrel in 2003 [13]. However, this did not continue for too long and PHA production has struggled to witness a constantly increasing trend ever since. In recent times, the greatest opportunity for PHA comes from the growing understanding and acceptance that petroleum is still an exhaustible resource and petroleum-based products are in opposition to the current renewability, green chemistry and sustainable development goals. Since PHA production and applications are related to environmental protection, reduced CO_2 emissions, green chemistry approaches and overall sustainable development, it explains the consistent ongoing effort by companies to keep improving the process technology to produce PHA and aim to make it competitive with petroleum-based plastic.

In addition to the improvements required in PHA production technology, another significant realization is the fact that both low- and high-value added applications are required to be developed to widen the value of PHA. For low-value applications, the major market remains environment-friendly packaging applications. Amidst the increasing global concern over conventional plastic packaging, the option of a 'bioplastic' could be viewed more positively and holds a greater promise in recent times than ever before. Not surprisingly, even with increased appreciation as a low-value 'bio/green' packaging product, large amounts of PHA are needed to fulfill this purpose. On the other hand, for high-value applications, such as bio-implant materials, tissue engineering materials and smart materials, relatively lesser PHA quantities are needed while much higher revenues can be generated. This provides the economic advantage to industrial PHA processes. In fact, this approach was adopted by ICI following a decline in profitability due to decreasing petroleum prices. ICI was reluctant to let the PHA project die since Biopol exhibited very interesting properties. Thus,

ICI set up a separate entity called Marlborough Biopolymers in 1983 from which various other companies and patents focusing on diverse applications of the original Biopol were created [7, 67]. Newer applications, other than packaging, such as medical implants, raw materials for other products, drug delivery, blending with other polymers, e.g., Ecoflex, were developed by the producers. This allowed the profitable production of PHA at large scale. The classic example of Biopol demonstrates the requirement to develop diverse applications of PHA other than the traditional large-volume, low-value ones in order to promote the large-scale production and increase the economic competitiveness of PHA polymers over conventional plastics. Table 5.1 lists the major vendors of PHA in recent years, the PHA produced and their respective production capacities.

The increasing interest and impetus from government and institutional organizations also help to promote the development of industrial processes. The research efforts in PHA production and applications create new directions for potential industrial implementation in the future. Research efforts in PHA production have been extensively carried out for decades and significant leads have been obtained as discussed in detail in Sections 3 and 4 of this chapter. Funding support has greatly facilitated the development of new production technologies, exploration of unconventional raw materials and new market applications for PHA. Some of the recent European projects based on PHA production and/or applications are listed in Table 5.2. Along with academic partners, the active role of small and medium enterprises

TABLE 5.1
Summary of Major PHA Vendors and Their Production Capacity

Company Name	Main Product	Main Substrate	Applications	Production Capacity [tons/year]
Bio-On, Italy	PHB, PHBV	Beet sugar	Medical, bioremediation	10,000
Danimer Scientific, USA	mcl-PHA (Nodax™ PHA)	Cold pressed canola oil	Additives, aqueous coatings, hot-melt adhesives, film resins	3500
BioMatera, Canada	PHA resins	Renewable raw materials	Food packaging, agricultural products, inks, cosmetics, biomedical	–
TianAn Biologic Materials Co., China	PHBV	Dextrose from corn	–	2000
Tianjin GreenBio Materials Co., China	P(3HB)	Sucrose	Pure resin, films, injection moldings, foam	10,000
Yield10 Bioscience (Metabolix, Inc.), USA	P(3HB), P(4HB)	Corn	Industrial, packaging and personal care products	50,000

TABLE 5.2

Recent Examples of European Projects Based On PHA Production and/or Applications

Project Name	Year	Project Coordinator	Substrate Used
A novel and efficient method for the production of polyhydroxyalkanoate polymer-based packaging from olive oil wastewater (OLI-PHA)	2012–2015	Iris Technology Solutions, Sociedad Limitada, Spain	Olive oil wastewater
Biopolymers from Syngas Fermentation (SYNPOL)	2012–2016	CSIC (Consejo Superior de InvestigacionesCientíficas)	Syngas
Novel technology to boost the European Bioeconomy: reducing the production costs of PHA biopolymer and expanding its applications as 100% compostable food packaging bioplastic (EUROPHA)	2013–2016	Federacion De CooperativasAgrariasde Murcia S Coop, Spain	Agri-food industry wastes
Exploitation of oily wastes for the simultaneous production of polyhydroxyalkanoates (PHA) and rhamnolipids (RLs) (SIMPHASRLS)	2014–2016	Faculdade De Ciencias E TecnologiadaUniversidade Nova De Lisboa, Portugal	Wastes generated from oil-processing plants and industries (waste vegetable oils such as corn oil and olive oil)
Reinforced Bioresorbable Biomaterials for Therapeutic Drug Eluting Stents (ReBioStent)	2014–2017	Lucideon Limited, UK, and University of Westminster, London	Synthetic medium components
Biowaste derived volatile fatty acid platform for biopolymers, bioactive compounds and chemical building blocks (Volatile)	2016–2020	Tecnalia – Tecnalia Research & Innovation, Spain	VFA from municipal waste and sludgy biowaste from food industry and wastewater treatment plants
High Performance Polyhydroxyalkanoates Based Packaging to Minimize Food Waste (YPACK)	2017–2020	Agencia Estatal Consejo Superior Deinvestigaciones Cientificas, Spain	Fruit waste
Full recyclable food package with enhanced gas barrier properties and new functionalities using high-performance coatings (REFUCOAT)	2017–2020	AIMPLAS – Asociacion de Investigacion de Materiales Plasticos y Conexas, Spain	–

(Continued)

TABLE 5.2 (CONTINUED)

Recent Examples of European Projects Based On PHA Production and/or Applications

Project Name	Year	Project Coordinator	Substrate Used
RESources from URbanBIo-waSte (ResUrbis)	2017–2020	CIABC – Research Centre for Protection of Environment and Cultural Heritage (Head), University of Rome La Sapienza	Urban bio-waste (organic fraction of municipal solid waste, excess sludge from urban wastewater treatment, garden and parks waste, selected waste from food-processing)
Advanced Filtration TEchnologies for the Recovery and Later conversIon of relevant Fractions from wastewater (AFTERLIFE)	2017–2021	Optimizaciónorientada a la sostenibilidad SL, Spain	VFA extracted from wastewater

(SMEs) is also seen in these projects. It is believed that such a cooperation between industry and academia could facilitate a faster realization of potential ideas and concepts towards the development of a profitable PHA process.

5.6   CONCLUSIONS AND OUTLOOK

It is a fact that PHA research has witnessed a great deal of progress in recent decades. As discussed in detail in this chapter, the innovation in bioreactor designs and operating strategies and the implementation of newer nutrient feeding strategies have been instrumental in spearheading this research. In addition, a major contribution has also come from the utilization of waste streams such as industrial/urban wastewater, solid industrial and agricultural wastes, etc. The fact that higher PHA yields can be obtained from fatty acids which are present in several waste streams, over those obtained from sugars is indeed very promising and provides a strong impetus for further efforts in this direction. Furthermore, the effective conversion of gaseous waste such as CO_2 into PHA is another interesting route to PHA production, both from an economic and environmental perspective. The production of a high-value product such as PHA along with the possibility of greenhouse gas mitigation is a win-win situation.

Regarding the use of waste streams, the future lies in the successful demonstration of efficient fermenter-based productions and particularly (at least) at pilot scale. The results from some successful pilot-scale run by Follonier et al. (2015), Huong et al. (2017) [23–24], etc. are already impressive and provide confidence in support of this scheme for PHA production. The possible integration of PHA production technology with the existing urban infrastructure in the coming years will pave the way for advanced biorefineries which are conceived to be the future biochemical industries as a replacement for petrochemical-based industries. To this end, in addition to the economic evaluation of PHA production technology, the product use for various applications and the overall environmental footprint of the process also demand consideration. The consumer perception and market for waste-based PHA need to be evaluated. In addition, the requirements for energy and water for processes including cultivation, filtration/centrifugation, product extraction and processing to final (finished) product need to be carefully assessed. Additionally, the methods for disposal of process wastewater are also a concern. The key question to be answered in the future is whether the above burdens are outweighed by the environmental credits associated with waste-based and biotechnologically produced high-value products such as PHA. Thus, both technological and environmental efficiency of the PHA production schemes will govern their success in the future.

ACKNOWLEDGMENT

Geeta Gahlawat thanks the University Grants Commission (UGC), Government of India, for providing the Dr D. S. Kothari Post-Doctoral Fellowship for the execution of the project on PHA. Sujata Sinha is thankful to the Department of Science and Technology, Government of India, for providing the grant in the form of the Women Scientist Scheme (WOS-A) (Grant no-SR/WOS-A/LS-1004/2015).

REFERENCES

1. Costa, S. S., Miranda, A. L., de Morais, M. G., et al. Microalgae as source of polyhydroxyalkanoates (PHAs) - A review. Int J Biol Macromol 2019; 131: 536–547.
2. Singh, A. K., Sharma, L., Mallick, N., et al. Progress and challenges in producing polyhydroxyalkanoate biopolymers from cyanobacteria. J Appl Phycol 2017; 29(3): 1213–1232.
3. Anjum, A., Zuber, M., Zia, K. M., et al. Microbial production of polyhydroxyalkanoates (PHAs) and its copolymers: A review of recent advancements. Int J Biol Macromol 2016; 89: 161–174.
4. Troschl, C., Meixner, K., Drosg, B., Cyanobacterial PHA production-review of recent advances and a summary of three years working experience running a pilot plant. Bioengineering 2017; 4(26). doi:10.3390/bioengineering4020026.
5. Kamravamanesh, D., Lackner, M., Herwig, C., Bioprocess engineering aspects of sustainable polyhydroxyalkanoate production in cyanobacteria. Bioengineering (Basel) 2018; 5(4):111. doi: 10.3390/bioengineering5040111
6. Rodriguez-Perez, S., Searrano, A., Pantion, A. A., et al. Challenges of scaling-up PHA production from waste stream: A review. J Environ Manage 2018; 205: 215–230.
7. Kaur, G., Roy, I., Strategies for large-scale production of polyhydroxyalkanoates. Chem Biochem Eng Q 2015; 29(2): 157–172.
8. Kaur, G., Srivastava, A. K., Chand, S., Advances in biotechnological production of 1,3-propanediol. Biochem Eng J 2012; 64: 106–118.
9. Ienczak, J. L., Schmidell, W., Aragão, G. M., High cell density culture strategies for polyhydroxyalkanoate production: A review. J Ind Microbiol Biotechnol 2013; 40(3–4): 275–286.

10. Chen, B. Y., Hung, J. Y., Shiau, T. J., *et al.* Exploring two-stage fermentation strategy of polyhydroxyalkanoate production using *Aeromonas hydrophila. Biochem Eng J* 2013; 78: 80–84.
11. Kulkarni, S. O., Kanekar, P. P., Nilegaonkar, S. S., *et al.* Production and characterization of a biodegradable poly (hydroxybutyrate-*co*-hydroxyvalerate) (PHB-co-PHV) copolymer by moderately haloalkalitolerant *Halomonas campisalis* MCM B-1027 isolated from Lonar Lake, India. *Bioresour Technol* 2010; 101(24): 9765–9771.
12. Koller, M., A review on established and emerging fermentation schemes for microbial production of 3 polyhydroxyalkanoate (PHA) biopolyesters. *Fermentation* 2018; 4: 30.
13. Chen, G. Q. Industrial production of PHA. In: Chen GQ, Ed., *Plastics from Bacteria: Natural Functions and Applications.* Berlin, Heidelberg: Springer, 2010; pp. 121–132.
15. Zinn, M., Witholt, B., Egli, T., Dual nutrient limited growth: Models, experimental observations, and applications. *J Biotechnol* 2004; 113: 263–279.
16. Gahlawat, G., Srivastava, A. K., Development of a mathematical model for the growth associated Polyhydroxybutyrate fermentation by *Azohydromonas australica* and its use for the design of fed-batch cultivation strategies. *Bioresour Technol* 2013; 137: 98–105.
17. Novak, M., Koller, M., Braunegg, M., *et al.* Mathematical modelling as a tool for optimized PHA production. *Chem Biochem Eng Q* 2015; 29: 183–220.
18. Penloglou, G., Vasileiadou, A., Chatzidoukas, C., *et al.* Model-based intensification of a fed-batch microbial process for the maximization of polyhydroxybutyrate (PHB) production rate. *Bioprocess Biosyst Eng* 2017; 40: 1247–1260.
19. Špoljarić, I. V., Lopar, M., Koller, M., *et al.* Mathematical modeling of poly[(R)-3-hydroxyalkanoate] synthesis by *Cupriavidus necator* DSM 545 on substrates stemming from biodiesel production. *Bioresour Technol* 2013; 133: 482–494.
20. Tavares, L. Z., da Silva, E. S., da Cruz Pradella, J. G., Production of poly(3-hydroxybutyrate) in an airlift bioreactor by *Ralstonia eutropha. Biochem Eng* 2004; 18: 21–31.
21. Gahlawat, G., Srivastava, A. K., Enhanced production of poly(3-hydroxybutyrate) in a novel airlift reactor with in situ cell retention using *Azohydromonas australica. J Ind Microbiol Biotechnol* 2012; 39:1377–1384.
22. Chisti, Y., Moo-young, M., Prediction of liquid circulation velocity in airlift reactors with biological media. *J Chem Tech Biotechnol* 1998; 42: 211–219.
23. Follonier, S., Riesen, R., Zinn, M., Pilot-scale production of functionalized mcl-PHA from grape pomace 1094 supplemented with fatty acids. *Chem Biochem Eng Q* 2015; 29: 113–121.
24. Huong, K. H., Azuraini, M. J., Aziz, N. A., *et al.* Pilot scale production of poly(3-hydroxybutyrate-*co*-4-hydroxybutyrate) biopolymers with high molecular weight and elastomeric properties. *J Bioscience Bioeng* 2017: 124: 76–83.
25. Werker, A., Bengtsson, S., Korving, L., *et al.* Consistent production of high quality PHA using activated sludge harvested from full scale municipal wastewater treatment – PHARIO. *Water Sci Technol* 2018; 78: 2256–2269.
26. Koller, M., Muhr, A., Continuous production mode as a viable process-engineering tool for efficient poly(hydroxyalkanoate) (PHA) bio-production. *Chem Biochem Eng Q* 2014; 28: 65–77.
27. Lillo, J. G., Rodriguez-Valera, F., Effects of culture conditions on poly (β-hydroxybutyric acid) production by *Haloferax mediterranei. Appl Environ Microbiol* 1990; 56(8): 2517–2521.
28. Ramsay, B. A., Lomaliza, K., Chavarie, C., *et al.* Production of poly-(beta-hydroxybutyric-co-beta-hydroxyvaleric) acids. *Appl Environ Microbiol* 1990; 56(7): 2093–2098.
29. Braunegg, G., Rona, R., Schellauf, F., *et al.* Polyhydroxyalkanoates (PHAs) sustainable biopolyester production. *Polymery* 2002; 7/8: 479–484.
30. Moser, A., *Bioprocess Technology: Kinetics and Reactors.* Springer-Verlag, New York, 1988.

31. Jung, K., Hazenberg, W., Prieto, M., *et al.* Two-stage continuous process development for the production of medium-chain-length poly (3-hydroxyalkanoates). *Biotechnol Bioeng* 2001; 72(1): 19–24.

32. Khosravi-Darani, K., Yazdian, F., Babapour, F., *et al.* Poly (3-hydroxybutyrate) Production from natural gas by a methanotroph native bacterium in a bubble column bioreactor. *Chem Biochem Eng Q* 2019; 33(1): 69–77.

33. Roberts, M., Screening and selection of a cyanobacteria for production of polyhydroxybutyrate in a closed photobioreactor. Ph.D. Thesis, University of Adelaide, Adelaide, Australia, 2009.

34. Rahnama, F., Vasheghani-Farahani, E., Yazdian, F., *et al.* PHB production by *Methylocystis hirsuta* from natural gas in a bubble column and a vertical loop bioreactor. *Biochem Eng J* 2012; 65: 51–56.

35. Ghoddosi, F., Golzar, H., Yazdian, F., *et al.* Effect of carbon sources for PHB production in bubble column bioreactor: Emphasis on improvement of methane uptake. *J Environ Chem Eng* 2019; 7(2): 102978.

36. Taga, N., Tanaka, K., Ishizaki, A., Effects of rheological change by addition of carboxymethylcellulose in culture media of an air-lift fermentor on poly-D-3-hydroxybutyric acid productivity in autotrophic culture of hydrogen-oxidizing bacterium, *Alcaligenes eutrophus*. *Biotechnol Bioeng* 1997; 53(5): 529–533.

37. Du, G., Yu, J., Green technology for conversion of food scraps to biodegradable thermoplastic polyhydroxyalkanoates. *Environ Sci Technol* 2002; 36(24): 5511–5516.

38. Da Cruz Pradella, J. G., Ienczak, J. L., Delgado, C. R., *et al.* Carbon source pulsed feeding to attain high yield and high productivity in poly(3-hydroxybutyrate) (PHB) production from soybean oil using *Cupriavidus necator*. *Biotechnol Lett* 2012; 34: 1003–1007.

39. Horvat, P., Špoljarić, I. V., Lopar, M., *et al.* Mathematical modelling and process optimization of a continuous 5-stage bioreactor cascade for production of poly [-(R)-3-hydroxybutyrate] by *Cupriavidus necator*. *Bioprocess Biosyst Eng* 2013; 36(9): 1235–1250.

40. Atlić, A., Koller, M., Scherzer, D., *et al.* Continuous production of poly([R]-3-hydroxybutyrate) by *Cupriavidus necator* in a multistage bioreactor cascade. *Appl Microbiol Biotechnol* 2011; 91(2): 295–304.

41. Braunegg, G., Lefebvre, G., Renner, G., *et al.* Kinetics as a tool for polyhydroxyalkanoate production optimization. *Can J Microbiol* 1995; 41: 239–248.

42. Lopar, M., Vrana Špoljarić, I. V., Atlić, A., *et al.* Five-step continuous production of PHB analyzed by elementary flux, modes, yield space analysis and high structured metabolic model. *Biochem Eng J* 2013; 79: 57–70.

43. Vadlja, D., Koller, M., Novak, M., *et al.* Footprint area analysis of binary imaged *Cupriavidus necator* cells to study PHB production at balanced, transient, and limited growth conditions in a cascade process. *Appl Microbiol Biotechnol* 2016; 100(23): 10065–10080.

44. Koller, M., Vadlja, D., Braunegg, G., *et al.* Formal and high-structured kinetic process modelling and footprint area analysis of binary imaged cells: Tools to understand and optimize multistage-continuous PHA biosynthesis. *EuroBiotech J* 2017; 1(3): 1–9.

45. Fauzi, A. H. M., Chua, A. S. M., Yoon, L. W., *et al.* Enrichment of PHA-accumulators for sustainable PHA production from crude glycerol. *Process Safety Environ Protect* 2019; 122: 200–208.

46. Burniol-Figols, A., Varrone, C., Le, S. B., *et al.* Combined polyhydroxyalkanoates (PHA) and 1, 3-propanediol production from crude glycerol: Selective conversion of volatile fatty acids into PHA by mixed microbial consortia. *Water Res* 2018; 136: 180–191.

47. Khattab, M. M., Dahman, Y., Production and recovery of poly-3-hydroxybutyrate bio-plastics using agro-industrial residues of hemp hurd biomass. *Bioprocess Biosys Eng* 2019; 42: 1115–1127.
48. Gahlawat, G., Srivastava, A. K., Model-based nutrient feeding strategies for the increased production of Polyhydroxybutyrate (PHB) by *Alcaligenes latus*. *Appl Biochem Biotechnol* 2017; 183: 530–542.
49. Cruz, M. V., Gouveia, A. R., Dionísio, M., *et al.* A process engineering approach to improve production of P (3HB) by *Cupriavidus necator* from used cooking oil. *Int J Polym Sci* 2019; 2019: 1–7.
50. Grothe, E., Chisti, Y., Poly (β-hydroxybutyric acid) thermoplastic production by *Alcaligenes latus*: Behavior of fed-batch cultures. *Bioprocess Eng* 2000; 22: 441–449.
51. Stanley, A., Kumar, H. P., Mutturi, S., *et al.* Fed-batch strategies for production of PHA using a native isolate of *Halomonas venusta* KT832796 strain. *Appl Biochem Biotechnol* 2018; 184(3): 935–952.
52. Ienczak, J. L., Quines, L. K., De Melo, A. A., *et al.* High cell density strategy for poly (3-hydroxybutyrate) production by *Cupriavidus necator*. *Braz J Chem Eng* 2011; 28(4): 585–596.
53. Porras, M. A., Ramos, F. D., Diaz, M. S., *et al.* Modeling the bioconversion of starch to P(HB-*co*-HV) optimized by experimental design using *Bacillus megaterium* BBST4 strain. *Environ Technol* 2019; 40(9): 1185–1202.
54. Blunt, W., Levin, D., Cicek, N., Bioreactor operating strategies for improved polyhydroxyalkanoate (PHA) productivity. *Polymers* 2018; 10(11): 1197.
55. Koller, M., Muhr, A., Continuous production mode as a viable process-engineering tool for efficient poly(hydroxyalkanoate) (PHA) bio-production. *Chem Biochem Eng Q* 2014; 28(1): 65–77.
56. Du, G., Chen, J., Yu, J., *et al.* Continuous production of poly-3-hydroxybutyrate by Ralstonia eutropha in a two-stage culture system. *J Biotechnol* 2001; 88(1): 59–65.
57. Tan, D., Xue, Y. S., Aibaidula, G., *et al.* Unsterile and continuous production of polyhydroxybutyrate by *Halomonas* TD01. *Bioresour Technol* 2011; 102(17): 8130–8136.
58. Kourmentza, C., Plácido, J., Venetsaneas, N., *et al.* Recent advances and challenges towards sustainable polyhydroxyalkanoate (PHA) production. *Bioeng* 2017; 4(2): 55.
59. Hartmann, R., Hany, R., Geiger, T., *et al.* Tailored biosynthesis of olefinic medium-chain-length poly [(R)-3-hydroxyalkanoates] in *Pseudomonas putida* GPo1 with improved thermal properties. *Macromolecules* 2004; 37(18): 6780–6785.
60. Garcia-Ochoa, F., Gomez, E., Theoretical prediction of gas-liquid mass transfer coefficient, specific area and hold-up in sparged stirred tanks. *Chem Eng Sci* 2004; 59: 2489–2501.
61. Papapostolou, A., Karasavvas, E., Chatzidoukas, C., Oxygen mass transfer limitations set the performance boundaries of microbial PHA production processes – A model-based problem investigation supporting scale-up studies. *Biochem Eng J* 2019; 148: 224–238.
62. Bandaiphet, C., Prasertsan, P., Effect of aeration and agitation rates and scaleup on oxygen transfer coefficient, kLa in exopolysaccharide production from *Enterobacter cloacae* WD7. *Carbohydr Polym* 2006; 66: 216–228.
63. Galaction, A. I., Cascaval, D., Oniscu, C., *et al.* Prediction of oxygen mass transfer coefficients in stirred bioreactors for bacteria, yeasts and fungus broths. *Biochem Eng J* 2004; 20: 85–94.
64. Badino, A. C., Facciotti, M. C. R., Schmidell, W., Volumetric oxygen transfer coefficients ($k_L a$) in batch cultivations involving non-Newtonian broths. *Biochem Eng J* 2001; 8: 111–119.

65. Chen, G. Q., A microbial polyhydroxyalkanoates (PHA) based bio-and materials industry. *Chem Soc Rev* 2009; 38: 2434–2446.
66. Chanprateep, S., Current trends in biodegradable polyhydroxyalkanoates. *J Bioscience Bioeng* 2010; 110: 621–632.
67. Feder, B. J., Technology; 'Bugs' that make plastics (1985). http://www.nytimes.com/19 85/05/03/business/technology--bugs-that-make-plastics.html.

6 Recovery of Polyhydroxyalkanoates from Microbial Biomass

Maria R. Kosseva and Edy Rusbandi

CONTENTS

6.1   INTRODUCTION

In spite of worldwide efforts committed to biopolymer research, polyhydroxyalkano-ates (PHA) are still not competitive with petrochemical plastics mainly because of production costs, and to a certain extent due to their material properties. One of the factors limiting implementation of the biopolymers is their higher price compared to conventional polymers. The recovery and purification of biopolymers are known to contribute significantly to the overall PHA biomanufacturing costs. Intracellular accu-mulation of PHA and a relatively low content of product can result in a high recovery cost. The extraction step represents around 50% or more of the total cost of the PHA production process [1]. Despite intensive investigations carried out in the area of PHA biosynthesis, research on downstream processing of PHA is relatively limited [2, 3].

The aim of the current review is to analyze trends in development of an efficient downstream processing technology for recovery and purification of PHA. We will consider advances in development of the PHA recovery methods, which can increase yield and purity of the products but reduce the cost and environmental impact of large-scale production.

6.2   PHA RECOVERY METHODS

Several studies on isolation and purification of PHA were published in the last decade; their final aim was to develop competitive processes for industrial implementation [1, 4–12, 22]. Based on the research, numerous factors have to be taken into account when selecting the PHA recovery method such as the microbial producer, starting material, type and composition of biopolymers, product purity requirements, impact on the PHA properties, PHA load in biomass [9], cost and environmental consid-erations. Figure 6.1 shows the technological steps involved in the PHA recovery including various extraction methods that have been reported.

Overall, the recovery methods from biomass can be classified as direct chemical extraction of PHA (involving the use of solvents, supercritical fluids or aqueous two-phase systems); chemical digestion, mechanical cell rupture methods (using high pres-sure homogenizers, or ultrasound, or milling), and biological alternatives to chemical digestion, which apply enzymes and living organisms for direct dissolution of biomass.

Broadly, there are two schemes for recovering PHA from the reaction medium post-fermentation, i.e. dissolving biomass to separate PHA granules with strong oxi-dants such as acids, or surfactants, and extracting PHA directly from the biomass using suitable solvents. The role of solvent is to change the permeability of cell mem-branes and then to dissolve the polymer inside the cells. A pretreatment step could be used to achieve better recovery, and a purification step – to obtain higher purity of the biopolymer [2]. Independent of the applied extraction solvent, microbial PHA-rich biomass has to be subjected to a drying step prior to the polymer extraction. This can either be done via lyophilization of biomass, or by thermal treatment. Water resi-dues in biomass would obstruct the efficiency of the extraction process, thus result-ing in a hydrolytic shortage of PHA molar mass [9, 10]. Notably, the extraction and/or purification method may positively or negatively influence the characteristics of the biopolymer obtained [1].

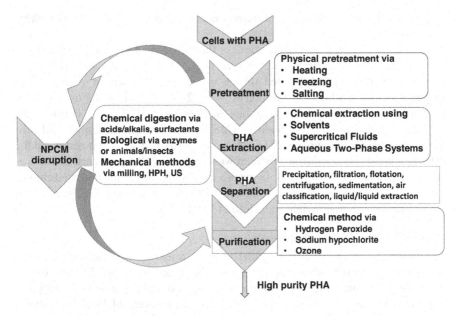

FIGURE 6.1 The technological steps for PHA recovery and purification.

A combination of various recovery techniques have been employed to further improve purity and/or recovery of PHA [11–13].

6.2.1 CHEMICAL METHODS

6.2.1.1 Solvent Extraction from Biomass

Direct extraction is the common recovery technique adopted in the production of PHA, it is also the best-established method at both the research and commercial scale. This method consists of immersing the PHA-containing biomass in a suitable solvent, or a solvent mixture. The extracted PHA are then recovered by addition of a precipitant solvent, where the PHA is collected as a crystal. Typical solvents used are halogen-containing solvents such as chloroform and methylene chloride. There is, however, a considerable environmental concern when such solvents are employed due to their toxicity and adverse environmental impact, particularly when they are used for mass production of PHA.

6.2.1.1.1 Application of Halogenated Solvents

At room temperature, typical halogenated extraction solvents like chloroform, or, to a minor extent, dichloromethane (methylene chloride), polychlorinated ethane (1,2-dichloroethane, 1,1,2-trichloroethane, 1,1,2,2-tetrachloroethane) perform well for extraction of short-chain length (*scl-*) and medium-chain-length (*mcl-*)PHA regarding extraction yields and purity [9]. After PHA extraction, the polyester's solubility is drastically minimized by adding a "PHA anti-solvent", typically low-molecular alcohols (ethanol or methanol), hexane, ether, acetone or water. This results in the precipitation of highly pure PHA. Ether, hexane and acetone can only be applied for precipitation of *scl*-PHA, as most representatives of *mcl*-PHA are soluble in these compounds [9].

Regardless of undesirable applications of the halogenated solvents and the harmful character of chloroform, recently Kumar et al. [14] developed a method for *mcl*-PHA extraction using halogenated solvent including chloroform at a temperature ranging from 40 to 80°C for a duration of 24 to 120 hours. The optimum conditions for extractions were 76.4°C and a duration of about 5 days (120 hours). In this study, municipal secondary wastewater sludge (MSWS) containing mixed microbial consortia was used as the source of PHA. Moreover, Ntaikou et al. [15] investigated the PHA production obtained from acidified waste glycerol (AWG) via enriched microbial consortium derived from soil. In the process, the extraction method used chloroform as the extracting medium, carried out at 70°C for about one hour. The reported average molecular mass weights are close to 1.8×10^6 Da.

6.2.1.1.2 Non-Halogenated Solvents

To develop sustainability, it is essential to focus research on extraction processes based on easily recyclable solvents that do not display any biohazard. For this reason, other non-halogenic extracting solvents are proposed such as methanol, propanol, etc. Kinoshita and coworkers [16, 17] reported the recovery of P(3HB-*co*-3HHx) from dried or wet biomass using isobutanol/toluene as the extracting agent at high temperature – 100°C. This study demonstrates that gradual addition of precipitant solvent, such as heptane, to the extracting solvent produces PHA crystals with fluidity, which were capable of being brushed away. The purity and yield of PHA recovered using this method are 99 and 97%, respectively. The same researchers [18] later discovered that water can be used in substitute of heptane, rendering lower recovery cost. Kinoshita and coworkers [17] reported that water content in the feed stream prior to extraction has to be carefully controlled. Small water content of 5% or less in biomass, results in a less significant decrease of weight average molecular mass (M_w) during the recovery process. Excessive water content can be reduced, preferably by drying and heating. The equipment for drying by heating used for this purpose ranges from spray drier, heating furnace, microwave heating equipment to vacuum drier. A moderate temperature, typically 50°C, is used for drying to prevent polymer degradation. This study also revealed that drying time has little impact on the recovery rate and purity of PHA. Kinoshita and coworkers [18] also reported a process to deliberately lower the M_w of the biopolymers, hence enhancing the filterability of the fluid. This process efficiently produces PHA with good operability. The dried biomass has to be heated in an oven at 50°C for 120 hours, after which it is subjected to extraction at the same temperature for 120 hours. During this period, the mixture has to be stirred smoothly. A precipitant solvent is then added to yield solid biopolymers. The purity and recovery of this method are reasonably high, i.e. 99% and up to 95%, respectively. Technically, the major disadvantage of this approach is the long time required to carry out heating pretreatment and extraction. Besides, the use of solvents may destroy the natural morphology of PHA granules, which is important for the production of strong fibers.

Currently, Rosengart et al. [19] reported usage of non-chlorinated industrial solvents such as anisole, cyclohexanone and phenetole for extraction of P(3HB) (Table 6.1).

TABLE 6.1

Common PHA Isolation Methods and Techniques Reported in Literature (Modified from [8])

Isolation Method	Chemicals Conditions	Microorganisms	Polymers	Purity and Yield [%]	Ref.
Digestion method	Surfactant Na dodecyl sulphate	*Ralstonia eutropha*	P(3HB)	Purity >95 Yield >90	[63]
Chemical digestion	Surfactant: ammonium laurate (a switchable anionic surfactant) with pre-treatment using NaOCl	Mixed microbial consortia	PHB	Recovery 74 ± 8	[36]
	Surfactant: Linear alkylbenzene sulfonic acid (LAS) at 80°C and pH 5.0	*Cupriavidus necator*	P(3HB-*co*-3HV)	Purity 90 Yield 84	[30]
	Sodium Hypochlorite NaClO, chloroform/ethanol	*Enterobacter cloacae* SU-1	P[(R3)HA]	Yield 94	[64]
	NaClO, chloroform (multiphase) for one hour at 37°C	*Bacillus* sp. BPPI-14 and *Bacillus* sp. BPPI-19.	PHA	PHA production of 49.46 ± 2.79% by *Bacillus* sp. BPPI-14 and 45.86 ± 2.17% by *Bacillus* sp. BPPI-19	[32]
	SDS, LAS-99, ES702 AOS-04, Brij_58, NaOH	*Ralstonia eutropha* *Escherichia coli*	PHA	Yield 99 Purity 90	[65]
	0.5N NaOH 37°C 5 hours H_2SO_4, bleaching NaClO 40°C	*Cupriavidus necator H16*	P(3HB)	Purity >90 Yield ~80 Purity 98 Yield 70–82	[4]

(Continued)

TABLE 6.1 (CONTINUED)

Common PHA Isolation Methods and Techniques Reported in Literature (Modified from [8])

Isolation Method	Chemicals Conditions	Microorganisms	Polymers	Purity and Yield [%]	Ref.
	0.1M H_2SO_4 at 100°C Base (pH = 10), NaClO	Cell slurry of Gram-negative bacteria	P(3HB) (3HV) (4HV)	Yield 95 Purity 97 M_w 500 kDa	[33, 34]
	Alkaline digestion	Cupriavidus necator	P(3HB-co-3HHx)		[66]
Enzymatic digestion	Alcalase, SDS, EDTA	Pseudomonas putida	PHA	Purity 92.6 Yield 90	[51]
	Benzonase Alcalase lysozyme flavourzyme microfluidizer	P. putida	Biopolymers	-	[67]
	Lytic enzymes from Microbispora sp inactivated by heat	Sinorhizobium meliloti	PHA	Yield 94 Purity 92	[54]
	Bromelain pancreatin	Ralstonia eutropha DSM545	P(3HB)	Purity 89 Purity 90	[51]
	Lysozyme/methanol	Cupriavidus necator	P(3HB-co-3HV)	Yield 75 Purity 41	[42]
	Crude enzyme Aspergillus oryzae	Cupriavidus necator	PHA	Yield 98 Purity 96.7	[68]
Solvent extraction	Chloroform methanol	Halomonas campisalis MCM B-1027	PHA	Purity 82 Yield 36	[2]
	Ethanol Acetone	Azotobacter vinelandii OPNA	P(3HB)	Purity 95 Yield 85	[21]

(Continued)

TABLE 6.1 (CONTINUED)
Common PHA Isolation Methods and Techniques Reported in Literature (Modified from [8])

Isolation Method	Chemicals Conditions	Microorganisms	Polymers	Purity and Yield [%]	Ref.
	Chloroform hexane	Cupriavidus necator ATCC17699	P(3HB-co-3HV)	Yield up to 70	[69]
	Chloroform hexane (Optimum condition 76.4°C and duration about 5 days)	Mixed microbial consortia obtained from municipal secondary wastewater sludge	Predominantly mcl-PHA (58%)	0.605 g PHA Recovery	[14]
	Chloroform	Bacillus cereus SPV	P(3HB)	Purity 92 Yield 31	[70]
	Chloroform	Mixed culture	PHA	Yield 77	[71]
	Chloroform extraction complimented with liquid-liquid phase separation	Fresh Ralstonia solanacearum	PHA	Recovery 98	[1]
	Chloroform	Enriched microbial consortium	P(3HB) and P(3HB-co-3HV) or P (3HB-co-HHx)	Mw 1.8×10^6 Da	[15]
	Isobutanol, toluene 50–130°C Heptane 90°C 100°C	Ralstonia eutropha	PHPB 3HHx	Purity 99 Yield 97 P99, Y95	[17] [18]
	NaClO+CH$_2$Cl$_2$ (1:1) CH$_2$Cl$_2$, EtOH	Cupriavidus necator H16	P(3HB)	Purity >98 Yield 90 P >99(Y90)	[4]

(Continued)

TABLE 6.1 (CONTINUED)
Common PHA Isolation Methods and Techniques Reported in Literature (Modified from [8])

Isolation Method	Chemicals Conditions	Microorganisms	Polymers	Purity and Yield [%]	Ref.
	Anisole Cyclohexanone 120–130°C	*Burkholderia sacchari*	P(3HB)	Yield 97 Yield 93	[19]
		Alcaligenes eutrophus	PHA	Purity 98	[72]
	Isoamyl propionate Propyl butyrate Isoamyl valerate Isoamyl Isoamylate (Isovalerate) (3-methyl-1-butanoate of 3-methyl-1-butanol)				
	Propylene carbonate Butyl acetate Isoamyl alcohol Ethyl acetate (at 120°C after 30 min)	*Cupriavidus necator*	P(3HB-*co*-3HV)	Purity 95 Yield 90	[30]
	Dimethyl carbonate (DMC) at 90°C for about 2 hours	Mixed microbial consortia (MMC)	PHA	Purity 54 ± 5 Yield 92 ± 4	[31]
	Acetone at 120°C 7 bar for 20 min	*Haloferax mediterranei* DSM 1411	*Scl*-PHA	Purity 98.4 Yield 96.8	[46]
	Acetone at room temperature	*Pseudomonas putida* GPo1	P(3HO acid)	Yield 94	[67]
	Cyclohexanone or γ-butyrolactone (heated at 120°C for 3 min)	*Cupriavidus necator* H16	PHA	Yield 95	[20]
	1,2-Propylene carbonate	*Cupriavidus necator* DSM 545	P(3HB)	Purity 84 Yield 95	[69]

(Continued)

TABLE 6.1 (CONTINUED)
Common PHA Isolation Methods and Techniques Reported in Literature (Modified from [8])

Isolation Method	Chemicals Conditions	Microorganisms	Polymers	Purity and Yield [%]	Ref.
	Thermo-separating-based Aqueous Two-Phase Extraction (ATPE)	*Cupriavidus necator*	PHA	Yield 72.2	[26, 27]
Solvent gelation	Acetone, methyl ethyl ketone, methyl isobutyl ketone and alcohols, e.g. Toluene, butanol	Mixed culture	PHBV	Purity >90 Yield ~81	[23]
Supercritical fluids	CO_2 100 min, 200 atm, 40°C, 0.2 ml CH_3OH	*Ralstonia eutropha*	P(βHB)	Yield 89	[72]
	CO_2 20 min, 200 bar, 30°C, 0.1 vol.-% toluene 0.4 wt.-% NaOH	Cell slurry	P(R-HB)	Purity 99	[39, 40]
Air classification		*Escherichia coli* *C. necator*	PHA	P 97 Y 90 P 95, Y 85	[43]
Extraction via Ionic Liquids	At 20–130°C From 5 to 24 hours	*Alcaligenes eutrophus*	*scl-co-mcl*-PHA	Yield < 50	[28]
	At 60°C for 24 hours using 1-Ethyl-3-methylimidazolium diethyl phosphate, $[C_2mim][(C_2)_2OPO_3]$	*Halomonas hydrothermalis*	PHA	Yield 60	[29]
Mechanical disruption	SDS-high pressure homogenization	*Methylobacterium* sp V49	P(3HB)	Purity 95 Yield 98	[74]

(Continued)

TABLE 6.1 (CONTINUED)
Common PHA Isolation Methods and Techniques Reported in Literature (Modified from [8])

Isolation Method	Chemicals Conditions	Microorganisms	Polymers	Purity and Yield [%]	Ref.
	Ultrasound in ATPS with poly(ethylene glycol) and phosphate	Bacillus flexus	PHA	Purity 92 Yield 20	[75]
	Pressure cell disruption (at 0.25 to 2.25 kbar) HPH	Cupriavidus necator DSM 545 Delftia acidovorans DSM 39	P(3HB) P(3HB)	~84.6% Dry cell mass ~15.4% Dcm	[76]
	High pressure (90 MPa) and solvent extraction using propylene carbonate Ultrasound (frequency of 10 kHz) and solvent extraction using propylene carbonate	C. necator	PHB	Capacity 97.8% Capacity 92.2%	[11]
	French Press Trypsin/ chloroform In trypsin at 37°C for 1 h Freeze–dried overnight and boiled in chloroform for 6 h, polymer dried at 55°C	Yeast Arxula adeninivorans	PHB-V	52.1% of CDM	[48]
Cell fragility	Alkaline hydrolysis	Bacillus flexus grown on inorganic medium	PHA	Yield >86	[37]
Biological Recovery	Living organisms: mealworms, small cockroaches, big cockroaches, crickets and Sprague Dawley rats	C. necator H16	PHA	Purity 94	[59]
Heat Treatment	Heat treatment (at 95°C for 45 min) followed by solvent extraction using propylene carbonate	C. necator	PHB	Capacity 92.1%	[11]
Microwave-Assisted Solvent Extraction	Combined treatment with EDTA and microwave	Bacillus aryabhattai PHB10	PHB	Purity 97.2	[12]

Non-chlorinated solvents, in particular, cyclohexanone and ɣ-butyrolactone were used for extracting PHA produced by the bacterial strain *Cupriavidus necator* H16 cultivated on vegetable oil as a sole carbon source as demonstrated by Gouzhan et al. [20] recently. When cyclohexanone was used as the extraction solvent at 120°C for three minutes, 95% of the PHB was recovered from the cells with a similar purity as that extracted using chloroform.

Garcia et al. [21] demonstrated the production and recovery process of P(3HB) synthesized using *Azotobacter vinelandii* OPNA strain, using a fed-batch fermentation, in which the polymer was precipitated with ethanol and washed with acetone, reaching a purity close to 95% and a yield of 85%. The P(3HB) obtained was of ultra-high molecular mass, with values of 5693 ± 615 kDa. This study explored the separation method of P(3HB) using "soft" solvents such as ethanol performed at a temperature close to its boiling point (77°C) to break the cell membranes and solubilize solvent-related compounds (proteins, carbohydrates, etc.) to allow the separation of the P(3HB) granules.

Lately, Perez-Rivero and co-authors [22] have summarized the main halogen-free solvents used in the last decade for PHA extraction. Those alternative options were described and patented by international companies such as Agroferm, Procter & Gamble, Monsanto, Metabolix and Kaneka.

6.2.1.1.3 Solvent Gelation

Werker et al. [23] patented methods for recovering PHA from a mixed culture biomass produced in a biological wastewater treatment process. Compared with the previous treatments these methods could account for the quality of the polymer obtained with regard to their anticipated final applications. It permitted for controlling the M_w of the polymer and for real time fine tuning the extraction time from batch to batch. Sustaining strict criteria for the polymer quality, while minimizing process steps and optimizing for the solvent use and operating economy, they contributed to a more economical process with robust and consistent PHA recovery methods. PHA-rich solvent gelation was exploited by controlling PHA-rich solvent gel formation. Using the same pilot system, a PHBV-rich biomass was produced, granulated to particle size between 0.71 and 2 mm, and contained to a greater extent, copolymer blends with 3-hydroxyvalerate, which had a temperature range of 63 to 175°C. With batch-to-batch variation, the extraction conditions of time and/or temperature for the PHA recovery were altered systematically. PHA-poor solvents were generally found within the groups consisting of aliphatic alcohols, ketones, esters and/or aromatic hydrocarbons. The solvent dissolved non-PHA components of the PHA-rich biomass such that upon cooling, the PHA-rich solution forms a gel and the non-PHA components of the biomass remained in solution. Thus, dissolved non-PHA biomass components can be separated with the solvent from the extracted gelled PHA. Desirable PHA-poor solvents were those capable of dissolving at least 30 g PHB/L of solvent at temperatures in the range of about 100 to 160°C, and further formed a solvent-gel when cooled to temperatures above 60°C. The PHA-poor solvents reported in the literature, include ketones such as acetone, methyl ethyl ketone and methyl isobutyl ketone, alcohols such as propanol, butanol, pentanol and isomers thereof, toluene and propylene carbonate. Additionally, Werker et al. [23]

achieved good extractions of PHB with 2-butanol within a maximum temperatures range between 120–160°C. Werker et al. [24] also defined the suitable recovery processes – those that can accommodate a variety of biomass sources, biomass PHA contents and PHA types, and were readily scalable at a starting capacity in the order of 1000 ton PHA per year.

6.2.1.2   Aqueous Two-Phase Systems

An aqueous two-phase system (ATPS) is a co-existing two immiscible phases that form when two structurally different polymers or an inorganic salt and one polymer are mixed in water beyond critical concentration [25]. Comparing with the traditional solvent extraction, the ATPS for PHA recovery has several advantages and essential characteristics that can be useful for industrial application. One of the main advantages comprises high water content (70–90 wt.-%) of the ATPS, which provides a beneficial environment for separation of sensitive biomaterials. Additionally, the materials that form different phases/layers of ATPS are harmless and environmentally friendly compared to conventional solvent extraction methods. Other advantages are the ability for high capacity processing, leading to reduced purification steps, and easy scalability of the purification step. So ATPS is a potential solution for the highly efficient large-scale separation technology with a short processing time. Leong et al. [25], examined the effect of pH and salts addition in *Ralstonia eutropha* H16 cultures during the extractive fermentation, using an ATPS system. The main principle of extractive fermentation with ATPS is promoting the partitioning of the target product in one phase, while cell and substrates accumulated in the other phase, which helps to retrieve the components. The optimum result obtained in this study was a PHA concentration of 0.139 g/L and recovery yield of 55–65% using ATPS of poly(ethylene glycol) (PEG) 8000/sodium sulphate adjusted to pH 6 plus addition of 0.5 M NaCl.

6.2.1.2.1   Thermo-Separating ATPE Method

A thermo-separating aqueous two-phase extraction (ATPE) has been proposed for ecofriendly and cost-effective PHA primary recovery and purification [26]. ATPE is a technique that utilizes the preferable partitioning of bioproducts in the system formed due to the immiscibility of polymer/polymer and salting-out of polymer by sulfate or phosphate salts. The most important ATPS parameters (type and concentration of thermo-separating polymer, salt addition, feedstock load and separating temperature) were optimized in order to achieve high PHA retrieval from the bacterial lysate. The recovery of PHA obtained from a crude feedstock of *Cupriavidus necator* was carried out under the condition of 14 wt.-% of EOPO 3900 concentration (ethylene oxide – propylene oxide), 14 wt.-% of ammonium sulfate concentration and pH 6. The yield and purification factor up to 72.2% and 1.61-fold, respectively, can be achieved with the thermo-separating copolymers, which can be recycled and reused twice. The best conditions for PHA extraction (recovery yield of 94.8% and purification factor of 1.42) were reached under the following conditions: 20 wt.-% EOPO, 10 mM NaCl addition, and a separating temperature of 60°C with a crude feedstock limit of 37.5 wt.-% [27]. An additional benefit of this process is the ability to recycle and reuse EOPO 3900 at least twice, achieving a satisfying yield and purification factor [26].

Leong et al. [27] further investigated the economic and environmental performance of the PHA production process using ATPE as the recovery method. The analysis revealed that the application of thermo-separating ATPE results in the reduction of the chemical consumption, downstream equipment cost and wastewater treatment cost. Further, it is also stated that thermo-separating ATPE is an economically viable and technically feasible PHA primary purification strategy. Hence, moving forward, the above strategy provides alternative options to academia and industry for the effective recovery of PHA.

6.2.1.3 Usage of the Ionic Liquids

Methods for using ionic liquids to extract and separate a biopolymer from a biomass containing the PHA were disclosed [28]. The invention also relates to methods for dissolving the PHA biopolymer in the ionic liquid. Gentle heating, agitation, sonication, pressure and radiation energy (e.g. microwave, infrared), may be applied to accelerate the dissolution of the biopolymer, thus reducing the processing time (Table 6.1). However, the cost of separation processes carried out in the ionic liquids is relatively high.

The use of ionic liquids (ILs) as the extracting medium was also reported by Dubey et al. [29]. In this study, 1-Ethyl-3-methylimidazolium diethyl phosphate, $[C_2mim]$ $[(C_2)_2OPO_3]$ was used to extract the PHA from both wet and dried *Halomonas hydrothermalis* biomass using glycerol as the carbon source. The extraction was undertaken at about 60°C for 24 hours. Unlike organic solvents, the IL was able to break up the bacterial cell wall easily (believed via electrostatic forces) and removes the PHA in a single pot without any discharge into the environment. Following the extraction, excess ethanol was added to isolate the biopolymer and remove the IL completely. The separated IL was reused to evaluate its recyclability. It was shown that no chemical degradation of the structure of IL was observed after two cycles.

6.2.1.4 Microwave-Assisted Extraction

Microwave-assisted extraction involves rapid heating of the sample/solvent mixture in closed vessels aiming at a more productive extraction reducing time and solvent consumption simultaneously. A study to investigate the effect of a combined treatment with ethylenediaminetetraacetic acid (EDTA) and microwaves was recently demonstrated by Pillai et al. [12] to recover PHB synthesized with *Bacillus aryabhattai* PHB10, which were cloned in *E. coli*. The combined extraction technique was used as an alternative to sodium hypochlorite cell lysis extraction. It is shown that the combined method attained 97.21% purity and 2.9-fold improvement in molecular weight and better polydispersity index (PDI) compared to the sodium hypochlorite cell lysis extraction. In the EDTA microwave-assisted cell lysis, the cell suspensions were exposed to microwaves at maximum power of 700 W for about ten minutes.

6.2.1.5 Recyclable Solvents and Renewable Materials

The use of recyclable solvents instead of pure solvents was demonstrated by Gahlawat et al. [30]. In their study, the recovery of poly(3-hydroxybutyrate-*co*-3-hydroxyvalerate) copolymer from *Cupriavidus necator* was undertaken using recyclable solvents, i.e. propylene carbonate, butyl acetate, isoamyl alcohol and ethyl acetate. The recovery efficiency of 90% and polymer purity of 95% achieved at 120°C after 30 minutes were reported.

One of the recent developments in this area is directed towards the use of renewable starting material such as waste. An example of this is demonstrated by Samori et al. [31], where PHA were produced from anaerobically digested sewage sludge. The extraction of the PHA was undertaken using dimethyl carbonate (DMC) at an elevated temperature of 90°C for about two hours. The PHA can then be recovered upon solvent evaporation and drying.

6.2.2   CHEMICAL CELL MASS DISSOLUTION

Dissolving the non-PHA cell mass (NPCM) and conserving PHA granules intact are the primary objectives in cell mass dissolution, which can be achieved using chemicals (acids/ alkalis) or an alternative enzymatic treatment [4, 9]. In the early studies, strong oxidation agents, such as sodium hypochlorite and sodium hydroxide, were used to dissolve NPCM. However, if the concentration of the oxidation agent is not well controlled, upon dissolution, not only NPCM dissolves but so does the PHA produced, leading to low recovery and also degradation of PHA. This nonselective method was subsequently replaced by a selective dissolution whereby only the NPCM is digested. For selective digestion, there have been a number of digesting agents used such as anionic surfactants and acids.

Mohammed et al. [32] modified the NPCM dissolution method by using a solvent mixture consisting of NaOCl and chloroform to extract PHA from *Bacillus* sp. BPPI-14 and *Bacillus* sp. BPPI-19 isolated from plastic waste landfill. The digestion and extraction were carried out at 37°C for about one hour. Up to $49.5 \pm 2.79\%$ of PHA concentration was reported in this study using glucose as the carbon source.

For efficient polymer recovery, the increase of the intracellular polymer contents as well as appropriate size of the PHA granules is of high importance. The NPCM has also to be converted in a sustainable, value-adding way. Research in this direction is devoted to the anaerobic digestion of NPCM in biogas plants, or to the chemical or enzymatic hydrolysis of NPCM to a rich carbon and nitrogen sources for subsequent microbial cultivations. As an alternative, NCPM can be applied in agriculture as "green fertilizer" [9].

6.2.2.1   Digestion by Acids

Yu [33, 34] reported the recovery of PHA using acid digestion. This study shows that the concentration of acid has an important role in the degradation of PHA and in determining the mechanical strength of PHA. Using a higher acid concentration of c. 0.1 M H_2SO_4 tends to lower the molecular weight of the biopolymer produced, but with the desired mechanical strength. Nevertheless, pH adjustment after the acid digestion is necessary before further treatment process.

6.2.2.2   Digestion by Surfactants

The use of surfactants to dissolve the NPCM is an effective way to extract the polymer with a good purity while maintaining its molecular weight [35]. The problems of this method are the high surfactant to biomass ratio needed and the difficulty of recovering the surfactant, which then generates a large quantity of wastewater if not implemented [2]. Mannina et al. [10] proposed an advanced protocol to recover PHA from mixed microbial cultures (MMCs) via destruction of NPCM in aqueous phase using ammonium laurate, a switchable anionic surfactant (SAS). SASs are effectively and simply recoverable by using CO_2 as pH-trigger [36]. They can be directly and reversibly converted to the least soluble form in the reaction medium, so that they can be removed from the liquid phase and recovered to be reused afterward [35]. In this strategy, sodium hypochlorite was nonetheless used as a pretreatment agent to break the cell walls. During the accumulation step, lyophilization has been applied to freeze-dry the PHA containing biomass before the extraction step. About 74 ± 8% of PHB recovered was reported using this method. The efficacy of the digestion using an anionic surfactant, linear alkylbenzene sulfonic acid (LAS) was reported by Gahlawat et al. [30]. In their study, the results revealed a maximum yield of 84% and purity of 90% at 80°C and pH of 5.0. Nevertheless, the use of surfactant in PHA extraction is still considered as an expensive and non-ecofriendly method.

6.2.2.3   Digestion by Alkaline and Surfactant

The saponification reaction between sodium hydroxide and the lipid layer in the bacterial cell wall destabilizes the membrane and increases its permeability. Sodium and potassium hydroxides serve as mild digestion agents, which can replace use of harsh chemicals during the PHA recovery [22].

Fernández-Dacosta et al. [5] evaluated three downstream processing (DSP) paths for the recovery of intracellular P(3HB) from mixed culture of the wastewater derived from a paper mill or the food industry. The production capacity of P(3HB) from this stock was at 1.5 kilotons per annum (kt/a) based on 6.8 kt COD (chemical oxygen demand)/per annum available from wastewater, the final product purity was 99.9 wt-%. One of the treatment techniques described was based on chemical treatment with alkali and surfactant. Digestion of the cell material was conducted with NaOH (0.2 M) and SDS (0.2% wt./vol.), and a residence time of 1 h in a separation reactor [35]; 95% of the intracellular P(3HB) was released [35] and recovered.

The economic evaluation revealed that the alkali/surfactant trial was the most cost-competitive among the last three options reported by Fernández-Dacosta et al. [5]. The total production cost was € 1.40/kg P(3HB). In all three options the cost of DSP was larger than that for the fermentation stage with contributions of 70% for the alkali treatment, and higher for the other options. Current industrial PHA production typically uses sugars, which can account for 23% of the total production costs [8]. Usage of wastewater as a feedstock for the mixed-culture bacterial fermentation can explain the lower costs found in this study, since sterilization costs were negligible.

The life cycle assessment (LCA) proved: the alkali treatment was the most favorable with production cost of 1.40 €/kg PHB, GWP of 2.4 kg CO2-eq/kg PHB and NREU of 106 MJ/kg PHB. These results were similar to those reported for sugar-based PHA production (2.0 kg CO2-eq/kg PHA) and analogous to fossil-based plastics (2.15 kg CO2-eq/kg PET) [8]. However, the thermal stability of the final product obtained by digestion with chemicals is lower, while in a solvent-based DSP, the quality of the P(3HB) is comparable to that of a commercial polymer [35].

6.2.3 Cell Fragility

The cell wall strength of some Gram-negative and Gram-positive microorganisms could be compromised by modifying the composition of the growth medium. Cell fragility can also occur due to the accumulation of PHA in the cells. This mechanism was applied to the strain of *Bacillus flexus* [37]. The cells grown in the medium containing inorganic salts experienced absence of diaminopimelic acid and decreased concentrations of other amino acids. PHA recovery in the range of 86–100% was obtained using hot chloroform or mild alkaline hydrolysis with cells cultivated in inorganic medium. Although, to obtain high microbial growth with elevated PHA content it was necessary to balance the cell wall softening with cell wall integrity [3].

PHA recovery can also be carried out with processes involving supercritical fluid extraction shown in Table 6.1.

6.2.4 Supercritical Fluid Extraction

Substances above their critical pressure and critical temperature exhibit a useful intermediate behavior. They have the diffusivity properties of a gas and the solvation power of a liquid. For this reason, supercritical fluids (SCFs) can diffuse through solids and dissolve materials [22]. SCFs [38, 39] have attracted attention for biopolymer recovery due to their availability, nontoxicity, harmlessness, non-flammability and low reactivity. This emerging method relies on cell disruption via supercritical carbon dioxide (sCO_2). The temperature and pressure values (31°C and 74 bar) of sCO_2 allow working in mild conditions. Thus, it has been used to extract PHA from biomass with purities ranging from 86 to 99%. The downsides of this method are the capital and maintenance costs of the equipment.

While the PHA extraction methods using solvents and chemical digestion are considered as the most widely used, most of the studies applied toxic and highly volatile chemicals. To mitigate its impact on the environment, physical and biological methods are regarded as a positive alternative for PHA recovery. However, the challenge of using these methods is, among others, lower polymer recovery. As such, in general, it would be preferable to use a combination of a chemical method and a physical or biological method. Further, moving forward, the extraction step ideally requires reducing product losses, obtaining a product of high purity and preserved physico-chemical characteristics, obtaining a product that fits the term sustainable, reducing process costs and developing a method applied on an industrial scale. Some of the important advantages of the solvent extraction (shown in Table 6.2) are the abilities to remove endotoxins not needed in medical applications and to cause negligible degradation to the polymers [9].

6.3 MECHANICAL METHODS

6.3.1 Mechanical Disruption

Among mechanical methods, bead milling and high-pressure homogenization dominate cell disruption on the industrial scale. It is favored mainly because of the little damage it causes to the products and environment. The disadvantages of the above

TABLE 6.2

Advantages and Disadvantages of Various PHA Recovery Methods (modified from [3, 22])

Recovery methods	Benefits	Downsides
Solvent extraction	High purity, high yield High molecular weight Useful for medical applications because of endotoxin removal Cause limited degradation to the polymer	Toxic for human health and environment High capital and operation costs Lengthy process Native order of polymer chains might be disrupted
Chemical digestion	Lower toxicity for human health Low capital cost	Can affect polymer quality Costly wastewater treatment hard to reuse
Selective dissolution of NPCM by acids	Low operating cost Higher recovery yield	Requires strict control of process parameters Severe reduction in molecular weight if no process control
Supercritical fluids	Simple and Rapid Non-toxic Environmentally friendly	Problems with extracting polar analytes and natural samples High cost of the equipment
Enzymatic digestion	Mild operation conditions Good recovery with good quality	Complex process High cost of enzymes
Biological recovery	High purity No chemicals involved Valorization of NPCM	Slow process Problem with reproducibility
Aqueous two-phase system (ATPS)	High yield and a relatively high capacity Short processing time Low material cost Low energy consumption Good scalability	Problematic robustness and reproducibility Mechanism is under research
Mechanical methods Bead milling	Efficient at low cell concentrations Easily scalable No chemicals used Less contamination	Pre-treatment stage involved Several numbers of passes required Long processing time
High pressure homogenization	No chemicals used Less contamination Scale-up potential High yield without pre-treatment	Severe micronization of PHA granules Depends on microbial and process parameters High heat generation
Ultrasound-assisted extraction	High extraction yield Low cost Easy scalable No pollution	Used in combination with other methods

method are high capital investment cost, long processing time and difficulty in scaling up [40]. Used with glass beads in a container holding the cell suspension, bead milling causes cell disruption by the spheres colliding with the biomass, promoting a transfer of energy and consequently cellular disruption. Some interferences must be taken into account, such as the size of the spheres, speed of agitation and cellular concentration. The efficiency of the mill depends on the residence time, shear forces, type of microorganism, concentration of the cell suspension and speed of agitation [1]. The main problems related to bead mills are a temperature rise with an increase in the bead volume, poor scale-up and a high probability of contamination [41].

Mechanical disruption of NPCM can also be combined with subsequent treatment using surfactants (e.g., anionic SDS) or chemicals [2]. Despite the high polymer purity achieved by chemical methods, the environmental impact of such downstream separation methods is high due to the large volumes of solvents and surfactants required.

Gutt et al. [42] applied the analysis of variance (ANOVA) models to compare and optimize P(3HBHV) extraction protocols. They used a mill containing 300 metal beads at 500 rpm for 2 min, where 1 g of the biomass (C. necator) was dispersed in 10 ml of buffer solution. After the procedure, the samples were centrifuged for 30 min at 10,000×g, obtaining a 74% accumulation result at 46% purity. The authors also performed ball milling in combination with other methods, where 1–4 g biomass was used in buffer solutions ranging from pH 2 to 7.5 under stirring at 500 rpm for 2–5 min. After the SDS addition at 1.6–18.4% concentration and centrifugation for 60 min at 10,000×g, the best purity achieved was 74% at pH 7.5 [1].

6.3.2   Ultrasound-Assisted Extraction

The most practical way to reduce the cost of PHA recovery is to obtain a high amount of PHA within a short duration of extraction time using a minimum volume of mild solvents [7]. For this reason, efficient mass transfer in the extraction system is required, which can be achieved via sonication. This ultrasound treatment could significantly improve the release of soluble compounds by disrupting the cell walls, enhancing the diffusion and facilitating the solvent access to the intracellular compartments, an advantage related to the cavitation phenomenon of ultrasound. Ishaks et al. [7] carried out mcl-PHA extraction from bacterial biomass suspended in a solvent/marginal non-solvent mixture and irradiated by ultrasonic waves. The effects of sonication volumetric energy dissipation, time, type of marginal non-solvent and solvent/marginal non-solvent ratio on mcl-PHA extraction efficiency were investigated. The use of heptane as a marginal non-solvent in an ultrasound-assisted system led to improved PHA extraction yield. The marginal non-solvent is a substance, which is a non-solvent by itself, but when mixed in an appropriate ratio with a solvent, it does not weaken its solvating power and becomes capable of dissolving solute [43]. It can also act as a precipitant at high concentration. Under optimal conditions, slight effects of ultrasound irradiation on post-extraction thermal properties, molecular weights and chemical structure of mcl-PHA were observed by the same researchers [7].

The use of ultrasound in the PHA recovery was conducted by Pavan et al. [11]. In this study, an economic assessment of the PHB production process by C. necator was conducted. Four different extraction methods were compared, i.e. a) ultrasonication (with frequency of 10 kHz) followed by solvent extraction; b) heat treatment (at 95°C for 45 min) followed by solvent extraction; c) high pressure (90 MPa) followed by solvent extraction; and d) the use of solvent extraction alone. In each of the extraction methods, propylene carbonate was used as solvent or extracting medium. The study revealed that the purification without pretreatment, i.e. approach d) consumes more resources and results in the highest production cost. It is demonstrated that method b) provides the lowest total capital investment. Further, it is interesting to note that methods b) and c) incurred the same production cost.

PHA recovery based on mechanical disruption was patented by Tianan Biological Material Co. Bead milling and sonication processes were used for breaking the cells in the fermentation broth. The pH was adjusted to be alkaline before or after surfactant and coagulant agents were added. The final product was separated after one hour by centrifugation or vacuum suction filtration. It was claimed that the invented process has low cost, high extraction yield and no pollution. Those attributes make it an easily scalable method [44].

6.3.3 HIGH-PRESSURE HOMOGENIZATION (HPH)

The high-pressure homogenization processes are based on cell disruption through the passage of the biomass through a valve or narrow orifice, followed by depressurizing and a large increase in flow velocity, with high shear stress and consequent cavitation causing the deformation or the rupture of cells that are in suspension [1]. Homogenizers can vary in design and contain a high number of solids, up to 50% of the feed. Heat generation is also high at 1.5°C/69 bar [45]. The homogenizer can be applied to large-scale production.

Koller et al. [46] extracted scl-PHA from acetone under high temperature and high pressure. In the process, 21 g of biomass were used in 700 mL of acetone at a temperature of 120°C for 20 min under continuous stirring at a pressure of 7 bar. The molar mass presented by the method and proposed by the authors was 2.0×10^5 Da and polydispersity amounted to 2.55.

6.3.4 FRENCH PRESS

In a French press the cell suspension is drawn through a valve into a pump cylinder. Then it is forced under pressure of up to 1500 bar, through a narrow annular orifice and discharge valve, where the pressure drops to atmospheric. Cell disruption is achieved due to the sudden drop in pressure upon discharge, causing the cells to rupture. The French press is a small-scale method, which is often applied to disintegrate yeast cells [46].

Murugan et al. [47] performed the passage of the washed and resuspended *Cupriavidus necator* H16 cells with Tris-HCl (pH 7.2) for three cycles in the French press at a pressure of 69 bar, resulting in polymer almost 100% pure.

Biernacki et al. [48] increased the synthesis of PHB-V in the non-conventional yeast *Arxula adeninivorans* by stabilization of polymer accumulation via genetic modification and optimization of culture conditions. A poly(3-hydroxy)butyrate copolymer produced was isolated by solvent extraction method. Yeast was harvested, washed twice with water and processed in a French Press. 30 mg/L^{-1} of trypsin was added to the disrupted cell suspension and incubated at 37°C for 1 h. The resulting cell debris was freeze-dried overnight, and 5 g of lyophilized material was boiled for 6 h in 100 mL of chloroform. This resulted in 71.9% polymer recovery and 99.2% purity, having a number average mass of 8.6×10^5 Da and weight average of 1.7×10^5 Da molar mass and a polydispersity of 2. The extracted material consisted of 73.1%wt of PHB and 26.1%wt of PHV [48].

6.3.5   AIR CLASSIFICATION

This technique consists of the separation of finely ground solid particles based on their size or weight. The resulting finer fraction, with a high concentration of the product of interest, is later recovered with 85–95 % yield and 85–90% purity, using physical methods such as filtration or centrifugation [22]. Van Hee and colleagues [49] carried out study on the mechanism of flotation as a separating mechanism of *mcl*-PHA granules from the cell debris of *P. putida*. They used a flotation device with an enzyme treated broth, near the iso-electric point of bacterial debris and inclusion bodies, and water injection. The samples recovered at the bottom and the top were lyophilised. The PHA purity of 86% was achieved.

6.4   BIOLOGICAL RECOVERY METHODS

6.4.1   ENZYMATIC DIGESTION

Proteolytic enzymes are highly active in hydrolysis reactions, dissolving biomass with little PHA degradation [41, 49–51]. Due to their high selectivity, proteolytic enzymes are good candidates for PHA recovery, their other advantages are low energy requirements, biological specificity, mild operating conditions and low capital investment [52]. In the industrial production of Biopol®, Holmes and Lim [41] implemented a method applying proteolytic enzymes and surfactant (sodium dodecyl sulfate, SDS) treatment for PHB extraction. The combined action of proteolytic enzymes with phospholipase enhanced cell wall disruption compared to the separate action achieved when individual enzymes (e.g., Alcalase or Lecitase) were used [52]. Yasotha et al. [51] employed Alcalase for enzymatic digestion combined with surfactant (SDS and ethylene diamine tetra-acetic acid) treatments achieving 90% recovery and 92.6% purity of PHA. A high PHA recovery yield (94%) and purity (92%) was reported via PHA separation from 5 g/L of thermally inactivated cell mass of *Sinorhizobium meliloti* using a process involving the application of lytic enzymes from *Microbispora* sp. [53] (Table 6.1).

Cost competitive production of enzymes could be achieved through solid state fermentation (SSF) of fungal strains using a variety of agro-industrial waste and byproduct streams [54, 55]. Fungal strains of *Aspergillus oryzae* produce various enzymes including protease, phosphatase, pectinase and lipase [56, 57]. Kachrimanidou et al. [58] used a crude enzyme produced via SSF performed on sunflower meal, a biodiesel byproduct, with a fungal strain *A. oryzae* producing mainly proteases. The enzymatic recovery of intracellular P(3HB-*co*-3HV) granules was achieved with high recovery yields (98%) and purities (96.7%) under uncontrolled pH value conditions and 48°C.

6.4.1.1 Heat Pretreatment

A heat treatment often precedes the enzymatic reaction to deactivate nucleic acids and PHB depolymerase, which inhibit the recovery of intracellular biopolymers [2]. Kapritchkoff et al. [50] employed a heat treatment step prior to the assessment of various enzymes, including trypsin, bromelain and lysozyme, for PHB extraction from *Ralstonia eutropha* DSM 545 cells. Bromelain yielded 88.8% of PHB purity, though high purity (90%) was also obtained with pancreatin after 8 h reaction. The potential valorization of hydrolyzed cell mass was also proposed as a potential approach for the reutilization of nutrients released during enzymatic lysis [52]. All the above studies demonstrated good potential for industrial and bacterial enzymes to hydrolyze NPCM.

6.4.2 PHA Recovery by Insects

The new biological methods are based on the ability of insects and animals to consume lyophilized cells of the *Cupriavidus necator* H16 containing polymer granules. Murugan et al. [47] showed that the larvae of mealworm beetles, *Tenebrio molitor*, can recover PHA granules from the cells by intake. The cells containing 54 wt-% P(3HB-*co*-25 mol-% 3HHx) were used, which were harvested and freeze-dried according to standard procedures. After purification with water, detergent and heat, almost pure PHA were yielded. The molecular weight and dispersion of the product molecules compared with the chloroform extracted PHA was not changed. Nevertheless, the feasibility of the biological recovery process at the industrial scale has to be proven.

Another example of biological recovery of PHA involving the use of insects under alkaline conditions and uncontrolled pH (treatment done in the water) was proposed by Ong et al. [57]. Up to 94% of purity of the PHA with M_w of about 240 kDa was reported. In the biological recovery, PHA granules were excreted by mealworms, small cockroaches, big cockroaches and crickets after eating freeze-dried cells containing PHA. The PHA in the study was synthesized by bacterial strains *C. necator* H16 cells, which is previously reported to be nontoxic and beneficial to animals as it promotes physiological benefits to digestive tracts and supports intestinal microbial growth.

6.4.3 P(3HB) Recovery by Other Animals

An additional biological method for recovery of P(3HB) was developed by Kunasundari et al. [59]. Laboratory rats consumed freeze-dried cells of *Cupriavidus necator* H16, which were cultivated using palm oil containing 39 wt-% P(3HB) as the sole diet source. The test animals excreted pellets containing 82–97 wt-% P(3HB). The remaining impurities were removed by washing the pellets with water and/or low concentrations of detergent, which resulted in P(3HB) granules of a high purity.

The molecular masses and thermal properties of P(3HB) obtained by this method were comparable with P(3HB) extracted from bacterial cells using chloroform. This technique eliminated the need for solvents and hazardous chemicals, thus resulting in ecofriendly P(3HB) polymer with potential applications in agriculture. Ong et al. [57] likewise reported the recovery of PHA using Sprague Dawley rats.

6.4.4 CELL LYSIS INDUCED VIA GENETIC MANIPULATIONS

A number of bacterial and yeast strains have been designed to secrete particular fermentation products, which can naturally accumulate in the cytoplasm. This can reduce the costs associated with DSP procedures required for the recuperation of intracellular metabolites [22]. A recent breakthrough for producing PHA via the fermentation route was published by Sabirova and coworkers [60, 61]. This study demonstrates that it is possible to produce PHA extracellularly using genetically engineered microorganisms to avoid costly multiple recovery steps involving hazardous solvents. This technique relies on modifying gene encoding enzymes interfering with the production of PHA to allow PHA-producing microorganisms to deposit the biopolymers in extracellular medium. Ongoing efforts are focusing on improving the selectivity and yield of the polymers. Hence, maximizing the cell culture density and PHA concentration is highly desirable to optimize the carbon, chemical and energy efficiency.

6.5 PHYSICAL PURIFICATION METHODS

When the action of enzymes or chelating agents is combined with a hydrogen peroxide treatment of PHA, it can form purification methods [2]. Horowitz and Brennan [62] applied ozone as a method of purification substituting hydrogen peroxide treatment of PHA with ozone, which was delivered as 2 and 5% of an oxygen stream to the biomass or solution. The beneficial effects of bleaching, deodorization and solubilization of impurities were observed. Thus, the latest method was able to replace the hydrogen peroxide treatment, which disadvantages are high operating temperatures (80–100°C), an instability of peroxides, and a decrease of the polymer M_w [62].

A summary of downstream processes for the recovery of PHA is shown in Table 6.1.

All the above methods described have certain limitations either from the economic point of view, based on ecological considerations, safety aspects, low PHA recovery yield and purity, or because of scaling difficulties, which are crucial for industry [8, 77]. Advantages and disadvantages of various PHA recovery methods are presented in Table 6.2.

The methods evaluated vary in their cost requirement, isolation yield, product purity/ quality, and applicability to the biological PHA production system. Based on the circumstances, one has to decide if the carefully chosen technology matches the requirements for a final application of the isolated biopolyester [9]. The selection of a suitable combination of methods depends upon different factors [1]. Thus, before considering any DSP method, the polymer quality and property requirements should be specified. Mechanical and physical properties, such as molecular weight, will be affected by the type of extraction method. The PHA natural form (amorphous or crystalline) might change with certain treatments [3]. Depending on the targeted purity and endotoxin level allowed, an additional purification step might be required.

Current developments aiming to optimize PHA recovery methods involve the application of harmless solvents, supercritical fluids, ionic liquids or a combination between chemical and mechanical or biological techniques. PHA granules of surprising purity can be excreted by some animals and insects, which selectively digest the non-PHA fraction if they are fed with PHA-rich bacterial biomass [77]. Within the innovative separation methods, ATPE performs as a cost-effective and scalable system, which can serve as a first step in the purification process. As already stated, the combination of operations must be carefully evaluated based on the final product specifications but, finally, life cycle assessment should determine whether the selected technology can act as a sustainable solution [22].

In spite of the rapid development of techniques for recovery and purification of PHA, major problems remain, e.g. toxicity of the chemicals used, environmental impact, sustainability and substantial amounts of heat and power input. The challenge for industrial PHA production lies in downstream processing, because it requires substantial amounts of process energy and also auxiliary materials (e.g. solvents, detergents) resulting in considerable indirect energy use [78].

6.6 CONCLUSIONS AND OUTLOOK

This review evaluates the variety of methods and techniques which have been developed for the recovery and purification of PHA granules from microbial cell biomass during the past decade. Development of effective recovery methods is essential for the overall economics of PHA production. The choice of efficient extraction and purification methods is based on several criteria including costs and ability to maintain the original M_w while not compromising the degree of purity for various applications to be achieved in an environmentally friendly and economically feasible manner. Simultaneously with PHA recovery techniques, fermentation strategies should also be considered. The final intended application for PHA will dictate the degree of purity of the PHA granules. The major limitations of downstream processing: toxicity of the chemicals used, environmental impact, sustainability, and substantial amounts of heat and power input and their optimization remain to be solved for the industrial applications of these versatile biopolymers.

REFERENCES

1. Macagnan, K.L., M.I. Alves, A.S. Moreira. Approaches for enhancing extraction of bacterial polyhydroxyalkanoates for industrial applications. In: V.C. Kalia, Ed., *Biotechnological Applications of Polyhydroxyalkanoates*, Springer Nature Singapore Pte Ltd., 2019; pp. 389–408.
2. Jacquel, N., C.W. Lo, Y.H. Wei, H.S. Wu, S.S. Wang. Isolation and purification of bacterial poly(3-hydroxyalkanoates). *Biochem Eng J* 2008; 39: 15–27.
3. Kunasundari, B., K. Sudesh. Isolation and recovery of microbial polyhydroxyalkanoates. *eXPESS Polymer Lett* 2011; 5: 620–634.

4. López-Abelairas, M., M. García-Torreiro, T. Lú-Chau, J.M. Lema, A. Steinbüchel. Comparison of several methods for the separation of poly(3-hydroxybutyrate) from *Cupriavidus necator* H16 cultures. *Biochem Eng J* 2015; 93: 250–259.

5. Fernández-Dacosta, C., J.A. Posada, R. Kleerebezem, M.C. Cuellar, A. Ramirez. Microbial community-based polyhydroxyalkanoates (PHAs) production from wastewater: techno-economic analysis and ex-ante environmental assessment. *Bioresour. Technol.* 2015; 185: 368–377.

6. Hajnal, I., X. Chen, G.-Q. Chen. A novel cell autolysis system for cost-competitive downstream processing. *Appl Microbiol Biotechnol* 2016; 100 (21): 9103–9110.

7. Ishak, K.A., M.S.M. Annuar, T. Heidelberg, A.M. Gumel. Ultrasound-assisted rapid extraction of bacterial intracellular medium-chain-Length poly(3-Hydroxyalkanoates) (mcl-PHAs) in medium mixture of solvent/marginal nonsolvent. *Arab J Sci Eng* 2016; 41: 33–44.

8. Kosseva, M.R., E. Rusbandi. Trends in the biomanufacture of polyhydroxyalkanoates with focus on downstream processing. *Int J Biol Macromol* 2018; 107 (Pt A): 762–778.

9. Koller, M., H. Niebelschütz, G. Braunegg. Strategies for recovery and purification of poly[(R)-3-hydroxyalkanoates] (PHA) biopolyesters from surrounding biomass. *Eng Life Sci* 2013; 13: 549–562.

10. Mannina, G., D. Presti, G. Montiel-Jarillo, M.E.S. Ojeda. Bioplastic recovery from wastewater: a new protocol for polyhydroxyalkanoates (PHA) extraction from mixed microbial cultures. *Bioresour Technol* 2019; 282: 361–369.

11. Pavan, F.A., T.L. Junqueira, M.D.B. Watanabe, A. Bonomi, L.K. Quines, W. Schmidell, G.M.F. de Aragao. Economic analysis of polyhydroxybutyrate production by *Cupriavidus necator* using different routes for product recovery. *Biochem Eng J* 2019; 146: 97–104.

12. Pillai, A.B., A.J. Kumar, H. Kumarapillai. Enhanced production of poly(3hydroxybutyrate) in recombinant *Escherichia coli* and EDTA–microwave-assisted cell lysis for polymer recovery. *AMB Expr* 2018; 8: 142–157.

13. Yu, L., H. de Alwis Weerasekera, M. Forattini Lemos Igreja, V. Sankar, M.J. Williamson, S.S. Soman, K. Chow. US Patent 2019/0203237 A1, 2019.

14. Kumar, M., P. Ghosh, K. Khosla, I.S. Thakura. Recovery of polyhydroxyalkanoates from municipal secondary wastewater sludge. *Bioresour Technol* 2018; 255: 111–115.

15. Ntaikou, I., I. Koumelis, C. Tsitsilianis, J. Parthenios, G. Lyberatos. Comparison of yields and properties of microbial polyhydroxyalkanoates generated from waste glycerol-based substrates. *Int J Biol Macromol* 2018; 112: 273–283.

16. Kinoshita, K., F. Osakada, Y. Ueda, K. Narasimhan, A.C. Cearley, K. Yee, I. Noda. US Patent 20,050,222,373, 2005.

17. Kinoshita, K., F. Osakada, Y. Ueda, K. Narasimhan, A.C. Cearley, K. Yee, I. Noda. US Patent 20,050,239,998, 2005.

18. Kinoshita, K., Y. Yanagida, F. Osakada, Y. Ueda. Method of producing polyhydroxyalkanoate. US Patent 20,060,105,440, 2006.

19. Rosengart, A., M.T. Cesário, M.C.M.D. de Almeida, R.S. Raposo, A. Espert, E.D. de Apodaca, M.M.R. da Fonseca. Efficient P(3HB) extraction from *Burkholderia sacchari* cells using non-chlorinated solvents. *Biochem Eng J* 2015; 103: 39–46.

20. Guozhan, J., B. Johnston, D.E. Townrow, I. Radecka, M. Koller, P. Chaber, G. Adamus, M. Kowalczuk. Biomass extraction using non-chlorinated solvents for biocompatibility improvement of polyhydroxyalkanoates. *Polymers* 2018; 10: 731–744.

21. Garcia, A., D. Pérez, M. Castro, V. Urtuvia, T. Castillo, A. Díaz-Barrera, G. Espín, C. Peña. Production and recovery of poly-3-hydroxybutyrate (P(3HB)) of ultra-high molecular weight using fed-batch cultures of *Azotobacter vinelandii* OPNA strain. *J Chem Technol Biotechnol* 2019; 94: 1853–1860.

22. Perez-Rivero, C., J.P. Lopez-Gomez, I. Roy. A sustainable approach for the downstream processing of bacterial polyhydroxyalkanoates: state-of-the-art and latest developments. *Biochem Eng J* 2019; 150: 107283.
23. Werker, A.G., P.S.T. Johansson, P.O.G. Magnusson. Process for the extraction of polyhydroxyalkanoates from biomass. US Patent 20150368393, 2015.
24. Werker, A.G., P.S.T. Johansson, P.O.G. Magnusson. WO 2014/125422 A1, WIPO/PCT, 2014.
25. Leong, Y.K., F.E. Koroh, P.L. Show, J.C.W. Lan, H.S. Loh. Optimisation of extractive bioconversion for green polymer via aqueous two-phase system. *Chem Eng Trans* 2015; 45: 1495–1500.
26. Leong, Y.K., P.L. Show, J. C-W. Lan, H-S. Loh, Y.J. Yap, T.C. Ling. Extraction and purification of Polyhydroxyalkanoates (PHAs): application of Thermoseparating aqueous two-phase extraction. *J Polym Res* 2017; 24: 158–167.
27. Leong, Y.K., P.L. Show, J. C-W. Lan, H-S. Loh, H.L. Lam, T.C. Ling. Economic and environmental analysis of PHAs production process. *Clean Techn Environ Policy* 2017; 19: 1941–1953.
28. Hecht, S.E., R.L. Niehoff, K., Narasimhan, C.W. Neil, P.A. Forshey, D.V. Phan, A.D.M. Brooker, K.H. Combs. Extracting biopolymers from a biomass using ionic liquids. US Patent 07763715, 2010.
29. Dubey, S., P. Bharmoria, P.S. Gehlot, V. Agrawal, A. Kumar, S. Mishra. 1-Ethyl-3-methylimidazolium diethylphosphate based extraction of Bioplastic "Polyhydroxyalkanoates" from bacteria: green and sustainable approach. *ACS Sustainable Chem Eng*2018; 6: 766–773.
30. Gahlawat, G., S. Kumar Sonib. Study on sustainable recovery and extraction of polyhydroxyalkanoates (PHAs) produced by *Cupriavidus necator* using waste glycerol for medical applications. *Chem Biochem Eng Q* 2019; 33: 99–110.
31. Samorì, C., A. Kiwan, C. Torri, R. Conti, P. Galletti, Emilio Tagliavini, Polyhydroxyalkanoates and crotonic acid from anaerobically digested sewage sludge. *ACS Sustainable Chem Eng* 2019; 7: 10266–10273.
32. Mohammed, S., A.N. Panda, L. Ray. An investigation for recovery of polyhydroxyalkanoates (PHA) from Bacillus sp. BPPI-14 and *Bacillus* sp. BPPI-19 isolated from plastic waste landfill. *Int J Biol Macromol* 2019; 134: 1085–1096.
33. Yu, J. Recovery and purification of polyhydroxyalkanoates. US Patent 20,080,220,505, 2008.
34. Yu, J. Recovery and purification of polyhydroxyalkanoates. US Patent 7,514,525, 2009.
35. Jiang, Y., G. Mikova, R. Kleerebezem, L.A.M. van der Wielen, M.C. Cuellar. Feasibility study of an alkaline-based chemical treatment for the purification of polyhydroxybutyrate produced by a mixed enriched culture. *AMB Express* 2015; 5(1): 5–18.
36. Samorì, C., M. Basaglia, S. Casella, L. Favaro, P. Galletti, L. Giorgini, D. Marchi, L. Mazzocchetti, C. Torri, E. Tagliavini. Dimethyl carbonate and switchable anionic surfactants: two effective tools for the extraction of polyhydroxyalkanoates from microbial biomass. *Green Chem* 2015; 17 (2): 1047–1056.
37. Divyashree, M.S., T.R. Shamala, Extractability of polyhydroxyalkanoate synthesized by *Bacillus flexus* cultivated in organic and inorganic nutrient media. *Indian J Microbiol* 2010; 50: 63–69.
38. Darani, K.K., M. Reza Mozafari, Supercritical fluids technology in bioprocess industries: a review. *J Biochem Technol* 2010; 2: 144–152.
39. Khosravi-Darani, K. Research activities on supercritical fluid science in food biotechnology. *Crit Rev Food Sci Nutr* 2010; 50: 479–488.
40. Tamer, I.M., M. Moo-Young, Y. Chisti. Disruption of *Alcaligenes latus* for recovery of poly(-hydroxybutyric Acid): comparison of high-pressure homogenization, bead milling, and chemically induced lysis. *Ind Eng Chem Res* 1998; 37: 1807–1814.

41. Holmes, P.A., G.B. Lim. Separation process. US Patent 4 910 145, 1990.
42. Gutt, B., K. Kehl, Q. Ren, L.F. Boesel. Using ANOVA models to compare and optimize extraction protocols of P(3HB)HV from *Cupriavidus necator. Ind Eng Chem Res* 2016; 55(39): 10355–10365.
43. Noda, I. Process for recovering polyhydroxyalkanotes using air classification. US Patent 5,849,854, 1998.
44. Chen, X. Method for separating, extracting and purifying poly-B hydroxyalkanoates directly from bacterial fermentation broth. US Patent 7,582,456 B2, 2003.
45. Geciova, J., D. Bury, P. Jelen. Methods for disruption of microbial cells for potential use in the dairy industry- a review. *Int Dairy J* 2002; 12: 541–553.
46. Koller, M., R. Bona, E. Chiellini, G. Braunegg. Extraction of short-chain-length poly-[(R)-hydroxyalkanoates] (*scl*-PHA) by the "anti-solvent" acetone under elevated temperature and pressure. *Biotechnol Lett* 2013; 35: 1023–1028.
47. Murugan, P., L. Han, C.Y. Gan, F.H.J. Maurer, K. Sudesh. A new biological recovery approach for PHA using mealworm, *Tenebrio molitor. J Biotechnol* 2016; 239: 98–105.
48. Biernacki, M., M. Marzec, T. Roick, R. Pätz, K. Baronian, R. Bode. Enhancement of poly(3-hydroxybutyrate-co-3-hydroxyvalerate) accumulation in *Arxula adeninivorans* by stabilization of production. *Microb Cell Factories* 2017; 16: 144.
49. van Hee, P., A.C.M.R. Elumbaring, R.G.J.M. van der Lans, L.A.M. Van der Wielen. Selective recovery of polyhydroxyalkanoate inclusion bodies from fermentation broth by dissolved-air flotation. *J Colloid Interface Sci* 2006; 297: 595–606.
50. Kapritchkoff, F.M., A.D. Viotti, R.C.P. Alli, M. Zuccolo, J.G.C. Pradella, A.E. Maiorano, E.A. Miranda, A.Bonomi. Enzymatic recovery of polyhydroxybutyrate produced by *Ralstonia eutropha. J Biotechnol* 2006; 122: 453–462.
51. Yasotha, K., M.K. Aroua, K.B. Ramachandran, I.K.P. Tan. Recovery of medium chain-length polyhydroxyalkanoates (PHAs) through enzymatic digestion treatments and ultrafiltration. *Biochem Eng J* 2006; 30: 260–268.
52. Harrison, S.T.L. Bacterial cell disruption: a key unit operation in the recovery of intracellular products. *Biotechnol Adv* 1991; 9: 217–240.
53. Lakshman, K., T.R. Shamala. Extraction of polyhydroxyalkanoate from *Sinorhizobium meliloti* cells using *Microbispora* sp. culture and its enzymes. *Enzym Microb Technol* 2006; 39: 1471–1475.
54. Melikoglu, M., C.S.K. Lin, C. Webb. Stepwise optimisation of enzyme production in solid state fermentation of waste bread pieces. *Food Bioprod Process* 2013; 91: 638–646.
55. Diaz, A.B., O. Alvarado, I. de Ory, I. Caro, A. Blandino. Valorization of grape pomace and orange peels: improved production of hydrolytic enzymes for the clarification of orange juice. *Food Bioprod Process* 2013; 91(4): 580–586.
56. Heerd, D., S. Yegin, C. Tari, M. Fernandez-Lahore. Pectinase enzyme-complex production by *Aspergillus* spp. in solid-state fermentation: a comparative study. *Food Bioprod Process* 2012; 90: 102–110.
57. Ong, S.Y., Z-L. Idris, S. Pyary, K. Sudesh. A novel biological recovery approach for PHA employing selective digestion of bacterial biomass in animals. *Appl Microbiol Biotechnol* 2018; 102: 2117–2127.
58. Kachrimanidou, V., N. Kopsahelis, A.Chatzifragkou, S.Papanikolaou, S.Yanniotis, I. Kookos, A.A. Koutinas. Utilisation of by-products from sunflower-based biodiesel production processes for the production of fermentation feedstock. *Waste Biomass Valor* 2013; 4: 529–537.
59. Kunasundari, B., C.R. Arza, F.H.J. Maurer, V. Murugaiyah, G. Kaur, K. Sud. Biological recovery and properties of poly(3-hydroxybutyrate) from *Cupriavidus necator* H16. *Sep Purif Technol* 2017; 172: 1–6.

60. Sabirova, J., P. Golyshin, M. Ferrer, H. Lunsdorf, W.-r. Abraham, K. Timmis. Extracellular polyhydroxyalkanoates produced by genetically engineered microorganisms. US Patent 20 110 183 388, 2011.
61. Sabirova, J., P. Golyshin, M. Ferrer, H. Lunsdorf, W.-r. Abraham, K. Timmis. Extracellular polyhydroxyalkanoates produced by genetically engineered microorganisms. US Patent 8 623 632, 2014.
62. Horowitz, D., E.B. Brennan. Methods for separation and purification of biopolymers. US Patent 20 010 006 802, 2001.
63. Kim, M., K.S. Cho, H.W. Ryu, E.G. Lee, Y.K. Chang. Recovery of poly(3-hydroxybutyrate) from high cell density culture of *Ralstonia eutropha* by direct addition of sodium dodecyl sulfate. *Biotechnol Lett* 2003; 25: 55–59.
64. Posada, J.A., J.M. Naranjo, J.A. López, J.C. Higuita, C.A. Cardona. Design and analysis of poly-3-hydroxybutyrate production processes from crude glycerol. *Process Biochem* 2011; 46: 310–317.
65. Yang, Y-H, C. Brigham, L. Willis, C.K. Rha, A. Sinskey. Improved detergent-based recovery of polyhydroxyalkanoates (PHAs). *Biotechnol Lett* 2011. doi:10.1007/s10529-010-0513-4.
66. Anis, S.N.S, M.I. Nurhezreen, K. Sudesh, A.A. Amirul. Enhanced recovery and purification of P(3HB−co−3HHx) from recombinant *Cupriavidus necator* using alkaline digestion method. *Appl Biochem Biotechnol* 2012; 167(3): 524–535.
67. Horowitz, D., E.B. Brennan. Methods for separation and purification of biopolymers. EP Patent 1 070 135 B1, 2001.
68. Kachrimanidou, V, N. Kopsahelis, A. Vlysidis, S. Papanikolaou, J.K. Kookos, B.M. Martínez, M.C.E. Rondán, A.A. Koutinas. Downstream separation of poly(hydroxyalkanoates) using crude enzyme consortia produced via solid state fermentation integrated in a biorefinery concept. *Food Bioprod Process* 2016; 100: 323–334.
69. Flores-Sánchez, A, del Rocío López-Cuellar, M, Pérez-Guevara, F, U.F. López, J.M. Martín-Bufájer, B. Vergara-Porras. Synthesis of Poly-(R-hydroxyalkanoates) by *Cupriavidus necator* ATCC 17699 using Mexican avocado (*Persea americana*) oil as a carbon source. *Hindawi Int J Polym Sci* 2017; 2017: 10. Article ID 6942950.
70. Valappil, S.P., S.K. Misra, A.R. Boccaccini, T. Keshavarz, C. Bucke, I. Roy. Large-scale production and efficient recovery of PHB with desirable material properties, from the newly characterised *Bacillus cereus* SPV. *J Biotechnol* 2007; 132: 251–258.
71. Albuquerque, M.G.E., V. Martino, E. Pollet, L. Avérous, M.A.M. Reis. Mixed culture polyhydroxyalkanoate (PHA) production from volatile fatty acid (VFA)-rich streams: effect of substrate composition and feeding regime on PHA productivity, composition and properties. *J Biotechnol* 2011; 151: 66–76.
72. Mantellato, P.E., N.A.S. Durão. Process for extracting and recovering polyhydroxyalkanoates from cellular biomass. U.S Pat. 20080193987 A1, 2008.
73. Hejazi, P., E. Vasheghani-Farahani, Y. Yamini. Supercritical fluid disruption of *Ralstonia eutropha* for poly(beta-hydroxybutyrate) recovery. *Biotechnol Progr* 2003; 19: 1519–1523.
74. Ghatnekar, M.S., J.S. Pai, M. Ganesh. Production and recovery of poly-3-hydroxybutyrate from *Methylobacterium* sp V49. *J Chem Technol Biotechnol* 2002; 77: 444–448.
75. Divyashree, M.S., T.R. Shamala, N.K. Rastogi. Isolation of polyhydroxyalkanoate from hydrolyzed cells of *Bacillus flexus* using aqueous two-phase system containing poly(ethylene glycol) and phosphate. *Biotechnol Bioprocess Eng* 2009; 14: 482–489.
76. Gamero, J.E.R., L. Favaro, V. Pizzocchero, G. Lomolino, M. Basaglia, S. Casella. Nuclease expression in efficient polyhydroxyalkanoates-producing bacteria could yield cost reduction during downstream processing. *Bioresour Technol* 2018; 261: 176–181.

77. Koller, M., G. Braunegg. Advanced approaches to produce polyhydroxyalkanoate (PHA) biopolyesters in a sustainable and economic fashion. *EuroBiotech J* 2018; 2(2): 89–103.

78. Chen, G.Q., M.K. Patel, Plastics derived from biological sources: present and future: a technical and environmental review. *Chem Rev* 2012; 112: 2082–2099.

Part IV

Mixed Microbial Culture
Approaches

7 Polyhydroxyalkanoates by Mixed Microbial Cultures

The Journey So Far and Challenges Ahead

Luísa Seuanes Serafim, Joana Pereira, and Paulo Costa Lemos

CONTENTS

7.1   THE JOURNEY SO FAR

In nature, the production of polyhydroxyalkanoates (PHA) is a characteristic of diverse taxonomic groups. Despite the well-known function as internal carbon and energy sources, other roles can be attributed to PHA, namely as electron donors or compatible solutes, among others [1]. Most of these functions are related to the survival of microorganisms under stress conditions that result from fluctuations in nutrient availability or adverse environmental conditions. The majority of the PHA-storing microorganisms were isolated or identified under such conditions, including *Cupriavidus necator*, isolated from freshwater sludge [2]. Since wastewater treatment plants were designed to collect and treat waste streams with a composition that can change with time, the activated sludge populations can cope with fluctuations in nutrient availability. Consequently, it is expected that PHA-storing organisms would thrive in these environments.

Activated sludge populations have different relationships, that include symbiosis and predation, and as a result, the isolation of bacteria with specific characteristics is quite difficult. It is believed that only around 1% of microorganisms were already isolated and identified successfully [3]. Molecular techniques would help to characterize and identify the remaining 99%, but it is assumed that most of those will be hardly isolated and cultivated [4]. Regarding the high number of known genera being able to produce PHA, surely, among the non-isolated or -cultivated organisms a considerable amount of PHA-producing strains could exist [5].

7.2   DEFINITION OF MMCS

Mixed microbial cultures (MMCs) are a microbial population of undefined composition and composed of many different types of species. MMCs are not always easy to understand by those used to single-species processes and are often mistaken with co-cultures of two or three different microbial strains. Diverse microbial populations are almost as old as life on Earth. However, single-species processes are demanded by industry since, under strict sterilized conditions, they are much easier to control. In every ecosystem on Earth, dense and highly diverse bacterial populations can be found. These populations have a proper balance maintained by multiple interactions between community members, implying chemical interactions as metabolite transfer, communication or antibiosis, as well as interactions at the cellular level, including cell aggregation, biofilm formation, and predation [6]. The ecological relationships can be classified as cooperative or non-cooperative, not only between prokaryotic organisms but also with eukaryotic ones [7]. Anaerobic digestion processes, microbial mats, deep-sea vents, phototrophic consortia, acidophilic communities, microflora of the intestine or the oral cavity, and communities of the rhizo- and phyllosphere are well-known examples for stable natural MMCs [6].

These ecosystems can also develop in human-built structures like wastewater treatment plants (WWTP), where activated sludge (AS) populations thrive. AS have become the most studied type of MMCs, since most of the processes developed using microbial consortia are related to water treatment, pollution control, and bioenergy production. The use of MMCs has become an accepted research trend of applied microbiology or bioprocess engineering [7].

Despite the ubiquity of PHA production capacity in nature, the amount usually produced by MMCs is low and far from the amount desired by the industrial processes. Also, microorganisms without this ability can divert the carbon source for other functions, lowering the yield and the productivity of the process. The optimization of the process demands the enrichment of PHA producers and the elimination of those without this ability. The imposition of a selective pressure allows for the enrichment of the MMCs in the targeted microorganisms [8]. It is known that variations on nutritional availability (e.g. carbon or oxygen) are a selective pressure under which the accumulation of PHA granules constitutes a competitive advantage for bacteria over those without that ability.

The establishment of MMCs with a certain function, as in PHA production, also depends on the type of ecological relationships developed among the different species present. These relationships, together with the selective pressure applied, play a key role in the optimization of the process [9]. For this reason, not only the microbial composition of the MMCs should be analyzed but also the interactions between the different species.

The model of circular economy under the scope of the concept of biorefinery is a route for the use of MMCs. This model comprises the complete use of raw materials, independently from their complexity. In nature, complex substrates are metabolized in a concerted action by the different members of MMCs [6]. The use of waste to produce different bioproducts depends on the existence of microorganisms able to utilize a broader variety of carbon sources. The microbial diversity of MMCs allows for a higher probability of converting different carbon sources [7].

7.3   A LITTLE BIT OF HISTORY

The existence of PHA inclusions inside microorganisms has been known since the 1920s but only in the 1960s–70s their presence was detected and identified in AS from a WWTP. In the middle of the 1960s, the detection of the biopolyester poly(3-hydroxybutyrate) (P3(HB)) in AS cells was recognized [10], and later identified after extraction with chloroform [11]. Its function as endogenous substrate was determined and the possibility of some role in floc formation was discarded [12]. In 1974 the presence of the copolymer of poly(3-hydroxybutyrate-co-3-hydroxyvalerate) (P(3HB/3HV)) was detected for the first time in bacteria, also after extraction with chloroform from AS samples [13].

The development of biological nutrient removal processes in the late 1980s and early 1990s showed that PHA have a key role. This role is essential in enhanced biological phosphorus removal (EBPR). By supplying the external carbon source in the absence of an electron sink, the production of PHA is triggered because bacteria are unable to grow. The stored PHA will be used when an electron sink is available again, after the exhaustion of the external substrate [14]. The main microbial groups present in EBPR systems were able to store PHA, even those not participating in phosphorus accumulation [15]. For this reason, EBPR was the foundation of some of the first processes of PHA production by MMCs.

Also in the 1980s, the bulking problem in aerobic tanks of WWTPs was a concern for researchers. By alternating periods of an excess of external carbon source with depletion, floc-forming bacteria were favored over filamentous ones in AS systems. Floc-forming bacteria were more efficient in storing the external substrate as PHA [16]. This mechanism was already described and is a consequence of the absence of an external carbon source that results in the dropping of levels of enzymes and RNA required for growth [17]. The sudden increase in the external substrate can trigger two sets of metabolic pathways that compete for the carbon source. One is growth that requires the synthesis of a large variety of enzymes and high amounts of RNA, resulting in a complex physiological adaptation with a slow response. PHA-storing bacteria have a faster response by activating metabolic pathways that allow the production of internal reserves, which require lower amounts of enzymes and RNA [17]. This process was first designated as aerobic "feast and famine" and was proposed for the production of PHA by MMCs [16]. This process was later known as aerobic dynamic feeding (ADF), and widely used in the subsequent years since it led to the highest PHA storage contents and yields, becoming the most popular [5, 18]. The process itself does not allow for a high amount of PHA to be produced. Several researchers observed that keeping the MMCs under feast and famine conditions can result in a stable population for a long time but does not maximize the amount of PHA produced [19]. Attempts to increase the amount of PHA produced during the enrichment step resulted in unstable MMCs. Then, the maximization should be performed in a subsequent step, designated as accumulation, using the MMCs obtained in the enrichment step, resulting in the so-called two-step process [18].

The development of processes using MMCs is believed to contribute to a decrease in PHA production costs and this includes the use of waste or surplus-based feedstock as a carbon source. Also, the use of these complex raw materials is a demand for the circular economy concept. This was a problem when the first ADF systems were developed since MMCs need a carbon source enriched in short-chain organic acids (SCOAs) to store PHA. The presence of sugars, which often occurs, usually results in glucose-based polymers [20]. To enrich the carbon source in SCOAs, a preliminary step of acidification of raw materials using anaerobic MMCs is required. In 2004, a three-step process based on the ADF concept was proposed for the use of complex feedstock as substrate [21]. This process starts with the acidification of organic components of the feedstock, mostly sugars, to obtain SCOAs, followed by the enrichment of the MMCs in PHA-storing microorganisms. Finally, the MMCs obtained are used in PHA accumulation assays [21]. The present accepted strategies and the use of waste as substrate are discussed in more detail in Section 2.

At the beginning of the twenty-first century the storage contents and productivities achieved with MMCs were very close to the values obtained with pure or recombinant cultures. After an efficient enrichment step that allowed obtaining an almost pure culture of *Plasticicumulans acidivorans* from MMCs, using a chemically defined substrate, acetate, a content of 89% was obtained [22]. Using fermented cheese whey, selected MMCs were able to accumulate the highest PHA content so far reported with a complex substrate, 81.4% of cell dry mass [23].

In the 2010s, the first pilot plants started to operate, mainly due to the funding from the European Union. Some were installed in municipal WWTPs [24, 25], others installed in industrial facilities such as paper factories [26] or candy factories [27].

7.4 WHAT DO WE KNOW ABOUT PHA BY MMCS?

7.4.1 MMCs CAN USE CHEAP AND RENEWABLE RAW MATERIALS

The use of inexpensive substrates as agroforestry residues, industrial byproducts, and industrial and urban waste can decrease the operational costs. Several complex substrates, namely olive mill wastewater [28], paper mill wastewater [29, 30], pulp and paper mill byproducts [31–33], palm oil mill effluents [34], pyrolysis byproduct and crude glycerol [35–37], and sugar cane molasses [38] were successfully tested in PHA production. Due to the acidification step that already converted many complex substrates to SCOAs and the diversity of microbial compositions, MMCs are more versatile than pure or recombinant cultures for the production of PHA.

7.4.2 MMCs RESULT IN LOW OPERATING AND EQUIPMENT COSTS

In MMCs processes, the dominance of PHA-storing organisms results from the operational conditions imposed. Processes using pure or recombinant cultures deeply depend on tight sterile conditions to avoid contamination by non-PHA-producing bacteria. The establishment of true feast and famine conditions usually only allows for the survival of PHA-storing organisms. Most of the non-storing bacteria usually called "side populations", are eliminated during the famine period, while PHA-producing organisms will use their internal reserves as carbon and energy sources. However, since for many raw materials not all the carbon is converted into SCOAs in the acidification step, it can be consumed by the side population during the famine phase [20, 39]. The remaining side population, even in a very low amount, contributes negatively to the process and should be eliminated. Several strategies were proposed, including biomass settling and withdrawal of medium exhausted from SCOAs before establishing the famine phase [40].

7.4.3 HIGH STORAGE YIELDS, SPECIFIC PRODUCTIVITIES, AND CONTENTS BUT LOW CELL YIELDS WITH MMCs

As already mentioned, storage yields, specific productivities, and contents observed in processes using MMCs were close to those usually obtained with pure and recombinant organisms. Although slightly lower, this is not an issue since the cost savings associated with the MMCs process can compensate and this process also contributes to waste management and valorization that have become critical nowadays. However, low biomass concentration is still a weakness that can be appointed to MMCs processes, due to its strong impact on the amount of polymer produced and on the downstream processing. One of the maximum amounts of biomass ever recorded, 11.8 g/L, allowed the doubling of volumetric productivity, but the value was still far from pure cultures [41].

7.4.4 TAILORING THE POLYMER COMPOSITION IS POSSIBLE

In MMCs processes, the tailored synthesis of PHA with controlled functionality and properties is possible by adjusting the concentration and composition of the carbon sources. This can be achieved by feeding the MMCs with the right mixture of SCOAs. SCOAs with an even number of carbons promote the formation of 3HB and those with an odd number result in 3HV [42, 43]. However, the variety of monomers that can be obtained by MMCs is limited when compared with pure cultures. Only in a few works with MMCs, monomers with more than six carbons were obtained [44].

7.4.5 THROUGH HIGH MICROBIAL POPULATION DIVERSITY, MMCs PRODUCE POLYMERS WITH HOMOGENEOUS COMPOSITION AND CHARACTERISTICS

Despite the wide variety of bacteria already identified in MMCs processes (as reviewed earlier [45, 46]), some of them are already used in pure culture systems, such as members of the genera *Pseudomonas*, *Alcaligenes*, *Paracoccus*, or *Bacillus*. The diversity of populations, instead of being a drawback of the process, since it could signify too much variability on the polymer produced, is usually related to the robustness of the systems to cope with a feedstock that suffers periodic variations in its composition [47, 48]. Moreover, several researchers have already shown that the diversity of genera in MMCs does not result in variability of PHA that bear the same characteristics as those that are achieved with pure cultures. This was confirmed after analyzing molecular weights, polydispersities, melting, and glass transition temperatures, melting enthalpies or crystallinity [42, 49]. Nevertheless, pure and recombinant cultures still allow for better control of PHA characteristics. The microbial characterization of PHA-producing systems based in MMCs processes, as well as the main metabolic pathways involved, are discussed in more detail in Section 7.6.

7.5 PRESENTLY ACCEPTED STRATEGIES

PHA production by MMCs is based on the selection of a microbial population with a high capacity to accumulate intracellular reserves by imposing transient conditions. The key aspect of the process is the selection of conditions that lead to a stable population in terms of PHA storage efficiency. Heterogenous populations composed of PHA and non-PHA accumulators may be more robust to changes in the system but have a negative impact on productivity owing to lower average PHA cell content and increased complexity in downstream processing [45]. Due to the necessity of using complex substrates and the characteristics of the microbial communities used, the enrichment step was complemented with a preliminary step of acidification for the production of SCOAs and a last step for maximization of the PHA produced.

7.5.1 THREE-STEP PROCESS

To obtain an effective PHA accumulation process by MMCs it is essential to have an efficient culture selection and high amounts of accumulated biopolymer that should be based on adequate reactor operational strategies and substrate composition.

A three-step process from PHA production using complex wastes was proposed [21], where a physical separation of the different steps to impose optimal conditions instead of compromising ones occurred. The process includes: (1) acidogenic fermentation of the carbon source to produce mixtures of SCOAs that will serve as precursors of PHA biosynthesis; (2) culture selection, by imposing high selective pressure for microorganisms with PHA storage ability; and finally, (3) PHA production stage where the selected microorganisms accumulate PHA at maximum capacity [21]. This process becomes particularly useful when working with real substrates and wastes that, due to their complexity, have to be converted into simpler and readily biodegradable carbon sources before being transformed into PHA. In the following sections, each step of this process and the respective key parameters will be discussed. Although trying to discuss each parameter separately, it must be kept in mind that their influence is highly dependent, and as such their synergetic interactions need to be considered. Figure 7.1 gives an overview of the three-step process as well as highlighting the key parameters in each step of the process.

7.5.1.1 Acidogenic Fermentation

Most agro-industrial feedstocks are rich in carbohydrates and other compounds that cannot be directly converted into PHA, or that are preferably transformed into other products (ex. glycogen), which makes them less suitable substrates for PHA production by MMCs [50, 51]. This obstacle can be overcome by fermenting the residues into SCOAs since they are, from a metabolic perspective, more energetically advantageous substrates than sugars. Their transformation by β-oxidation generates more chemical energy than the oxidation of a molar equivalent of glucose [52]. The sugar to SCOAs conversion can occur under acidogenic conditions, a stage of the anaerobic digestion (AnD) process. AnD is a strategy, commonly used for the treatment of industrial and urban waste/wastewater, where complex organic compounds are fermented into intermediate products that will be later converted into methane

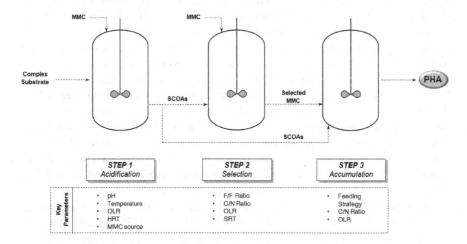

FIGURE 7.1 Schematic representation of the three-step process and the key parameters to be considered in each step.

and carbon dioxide. The process is composed of four sequential stages – hydrolysis, acidogenesis (also known as acidogenic fermentation, AF), acetogenesis, and methanogenesis – conducted by distinct groups of organisms. In hydrolysis, complex carbon sources, like polysaccharides, lipids, and proteins, are fragmented into simpler monomers, by hydrolytic bacteria. Under acidogenic fermentation conditions, bacteria promote the fermentation of the monomers mostly into SCOAs, ethanol, carbon dioxide, and hydrogen. In acetogenesis, SCOAs are converted into acetate, hydrogen, and carbon dioxide by acetogenic bacteria. Finally, methanogenic organisms convert the products of acetogenesis into carbon dioxide and methane [53–55]. If the goal is to obtain SCOAs, it is essential to eliminate methanogenesis by manipulating parameters like pH, temperature, hydraulic retention time (HRT), sludge retention time (SRT), and organic loading rate (OLR) [56].

It is known that the monomeric composition of PHA is related to their mechanical and thermal properties, and consequentially, determines the type of application. Variations in the type and proportions of SCOAs fed to the MMCs can lead to a great variety of monomer compositions and influence the final product, which is another interesting advantage of using SCOAs. By manipulating the AF process in order to obtain different SCOAs profiles, residues can be transformed into tailor-made biopolymers [42, 51, 57].

Like any other biological process, AF is constrained by the operating conditions imposed on the reactor and their optimization is crucial for the success of AF. A literature review can help to understand how MMCs under AF conditions react to variations of each operational parameter [58, 59]. However, studies showing how the manipulation of acidogenic conditions can lead to different monomer compositions for PHA production are still scarce.

One of the key parameters of AF is the pH-value that highly influences the type of SCOAs produced as it affects the metabolic pathways used and the toxicity of substrates in the environment [55]. In AnD, methanogens have a neutral optimum pH and the methane-forming process is inhibited when the pH is higher than 8 or lower than 6. In contrast, acidogens are more adaptable to pH, being inhibited at extreme conditions, under 3 and over 12 [60]. Between 3 and 12, for AF conditions, the ideal pH seems to be strongly correlated to the type of carbon source. The influence of pH in the composition of SCOAs from waste for PHA production was previously studied [58, 61, 62]. These works showed that by increasing the pH, the production of certain SCOAs could be promoted. The production of different types of SCOAs depends on the substrate used and, consequently, influences the PHA's monomeric composition [58, 61, 62]. Furthermore, it was demonstrated that different groups of microorganisms become dominant when different substrates are used or different values of pH imposed, which had a direct outcome on the types of SCOA produced [63]. Still, AF under alkaline conditions seems to be advantageous since it helps to increase the hydrolysis rate and inhibits methanogenic activity [64–67].

The temperature in AF processes can be divided into three ranges: psychrophilic, 0–20°C, mesophilic, 20–42°C, and thermophilic, 42–75°C [55]. Lowering the temperature usually lowers the biological activity and increases the conversion time of organic matter. Thermophilic conditions can lead to an increase in SCOAs concentration [68, 69], as it impacts the MMCs growth rate, the enzymatic activity, and the hydrolysis rate [70], being also associated with higher energy costs. Thereby, meso-philic conditions are usually the most suitable since they are a good compromise between productivity and economic viability of the process [69]. Temperature has an effect, not only on acid production efficiency, but also on the composition of SCOAs. It was reported that butyric acid was the main product when reactors were operated at 55°C whereas acetate and propionate were the main products at 35°C [71].

The influence of OLR in SCOAs production is not quite understood in literature and has an inconsistent influence on the SCOAs profile. Nonetheless, it can be gen-erally assumed that the yield of AF increases with the increase of OLR until the value where OLR becomes inhibitory, and the SCOAs concentration drops drasti-cally. This inhibition could result from the increase of the medium viscosity, reduc-ing mass and heat transfers and, consequently, substrate conversion [72]. The type of SCOAs produced seems to be more correlated to the type of waste used than to the OLR value. An increase in acetate and valerate content and a decrease in propionate and butyrate with the increase of OLR was observed, using food waste [71]. On the other hand, when using synthetic dairy wastewater, an increase of OLR led to an increase in propionate and to the decrease of acetate [73].

Complex substrates are not easily biodegradable and require more time for the culture to adapt and convert the substrate efficiently. Longer hydraulic retention times (HRT) are generally more suited for AF processes. The longer the contact time between the refractory organic matter and the hydrolytic acidogens, the higher is the hydrolysis efficiency [74]. Several authors already have reported this trend [59, 75, 76]. The problem of imposing longer retention times is the necessity of larger reac-tors, increasing the cost of the process [56] and the enhancement of the proliferation of methanogenic communities [69]. Besides influencing the process efficiency, HRT can also impact the SCOAs profile. It was shown that increasing the HRT in the acid-ification step led to an increase of propionate in the VFA profile, and consequently improved the 3HV content in the copolymer P(HB-co-HV) [58]. Studying anaerobic fermentation of food waste, it was reported that the increase of HRT led to the pro-duction of caproate, decreasing the concentration of the butyrate and acetate [77].

The choice of reactor configuration must take into account all the operational parameters desired, costs of operation, and area occupied. Generally, acidogenic fermentation is operated in a continuous mode to acclimatize the MMCs to the feedstock. However, to minimize the risks of shock loading and washout problems, biofilm systems were chosen as an alternative to suspended growth reactors [58, 78].

Apart from the operational conditions, the type of MMCs used can also affect the process. Regularly, MMCs from engineering-designed reactors are used to treat the waste streams anaerobically [69]. Nonetheless, other sources have also been reported like sediments [79] and rumen [80]. Benefits when using aerobic MMCs were also reported, since acidogenic bacteria are facultative anaerobes and methanogens are strict anaerobes, which help to inhibit methane formation [81]. A maximum SCOA production of 7 g/L was achieved from a byproduct of the pulp industry, composed mainly of acetate (59%), butyrate (19%) and propionate (17%) [81]. Contrarily, it was found that with an anaerobic inoculum, the SCOAs yield was almost doubled when compared with an aerobic one, obtaining butyric (70%) and acetic acid (17%) as the main products when using food waste as substrate [82].

7.5.2 CULTURE SELECTION

This was the first step to be developed and its main goal is obtaining MMCs enriched in microorganisms with a high PHA storage capacity to guarantee a good accumulation performance in the last step. The strategy is based on the elimination of organisms with low or no storage ability that would have a negative impact on the process productivity and downstream processing, increasing extraction costs [39, 45]. However, MMCs with high PHA content were described to be unstable and sensitive to variations in operating parameters and feedstock composition [33]. Therefore, the selection step should aim to achieve a compromise between maximizing the cell PHA content and assure the MMCs stability [18]. A selection process that results in MMCs with a diverse population of PHA-accumulating organisms are more robust and adaptable [83]. Besides, culture selection must ensure good results in the accumulation stage. Research has shown that higher PHA contents during enrichment usually involve lower biomass productivities, which has a negative impact on production costs [84].

Over the years, several types of enrichment strategies have been studied, with different advantages; either related to process costs or increased productivities. Nonetheless, two types of enrichment dominated the field – anaerobic/aerobic selection and aerobic dynamic feeding.

7.5.2.1 Anaerobic/Aerobic System (AN/AE)

EBPR systems were created in WWTPs to remove phosphorus from wastewater from domestic and agro-industrial activities [85]. Processes that use alternating anaerobic-aerobic conditions are referred to as anaerobic-aerobic (AN/AE) or under limited oxygen availability as anaerobic-anoxic (AN/AO) [86]. The first stage of the process occurs under strict anaerobic conditions in the absence of external electron acceptors. In the presence of an external organic carbon source, polyphosphate accumulating organisms (PAOs) and glycogen accumulating organisms (GAOs) compete for the carbon source and accumulate PHA. In the following aerobic/anoxic stage, the presence of an external electron acceptor (oxygen or nitrate/nitrite) allows for the consumption of the internally stored PHA for growth/maintenance. Nonetheless, PHA formation in the anaerobic stage is dependent on the cycling of polyphosphate and/or glycogen, whose hydrolysis will provide the ATP and reducing equivalents needed for substrate catabolism, being later regenerated in the second phase. This dependence on the glycogen cycle limits PHA production capacity, with maximum PHA contents reported around 50% for cultures selected under AN/AE conditions [47, 87].

7.5.2.2 Aerobic Dynamic Feeding System

In this process, PHA were accumulated aerobically, without depending on glycogen and phosphate reserves. This process consists of submitting the MMCs to alternated periods of external substrate availability (feast) and unavailability (famine) resulting in an unbalanced growth. During the famine phase, the lack of carbon sources led to a decrease in intracellular components essential for cell growth (RNA and enzymes), which will limit maximum growth rates in the feast phase. During this period the excess carbon fed to the MMCs is then accumulated as PHA [16]. Even though this process has high energy costs due to aeration requirements compared to AN/AE, the high storage capacities, above 80%, makes it the most studied and developed selection process in recent years [88].

ADF is the most commonly used methodology for biomass selection in a three-step production process. In ADF systems, the substrate is used almost entirely for the accumulation of intracellular reserves (around 70% of the substrate). Nonetheless, exposure time to the substrate should be long enough to allow for physiological adaptation of microorganisms [89]. Again, parameters such as pH, temperature, HRT, SRT, and OLR, among others, are the key to successfully steer the selection [83].

The operational parameter than can be more easily related to the specific growth rate of the MMCs is sludge retention time (SRT). High SRTs allow for the enrichment of slow-growing bacteria, and consequently a lower fraction of the substrate is used for growth, which could result in a higher storage yield [90]. MMCs selected with an SRT of three days had a better PHA production capacity that one selected with an SRT of ten days [91]. Something similar was also reported, where an SRT of five days led to better results than an SRT of ten [92]. On the contrary, it was observed that faster-growing organisms accumulated less PHA [93, 94]. The yield of P(3HB) from acetate under excess nutrients was constant and independent of the specific growth rate for SRTs higher than two days, according to [90]. Lower SRTs resulted in a decrease in PHB storage yield and productivity. However, some studies successfully selected cultures with good storage capacity using SRTs of one day [51, 90, 95]. Nonetheless, PHA accumulation capacity must not be the only parameter taken into consideration when selecting the adequate SRT. In fact, under short SRT the sludge yield is higher, an advantage for the accumulation step.

Feast and famine ratios (F/F) reflect the relationship between the duration of the feast and the famine phase. This ratio can be manipulated through changes in the OLR or influent substrate concentration. The famine phase must be long enough to ensure the internal growth limitation necessary to induce PHA production, otherwise the culture will be better fitted to grow instead of accumulating polymer during the feast phase. Previous studies showed that in most cases F/F lower than 0.2 are associated with high storage response, whilst for ratios higher than 0.5, the growth response was more significant. Intermediate values are usually associated with unstable systems [83, 95, 96]. OLR not only affects the F/F but also the kinetics of substrate consumption and polymer production that increase for higher substrate concentrations. Even though low OLRs led to lower F/F and prevent substrate inhibition, they can also limit the selection of a PHA-producing culture with high PHA production rates [83]. A compromising strategy could be the operation under increasing OLR. An increase in the applied OLR from 1.0 to 7.1 kg COD/(m^3·day), led to a parallel increase in volumetric biomass productivity and PHA storage capacity [97]. A similar result was reported where the OLR increase led to an increase in biomass concentration from 1.5 to 3.4 g/L and of volumetric productivity from 0.10 to 0.17 g PHA /(L·h) [31].

Selection of PHA-accumulating organisms under ADF conditions happens due to internal growth limitations, as explained before, and the need for nutrient limitation is not mandatory, although it can also play a role in the process. Strategies using both excess and limiting conditions of nutrients were successfully reported [83]. Limiting conditions can prevent substrate uptake towards growth and promote PHA production. On the other hand, excess nutrients in the famine phase will allow for a higher growth yield for the PHA accumulators [18, 61]. As biomass yield is also an influencing parameter of the overall volumetric productivity, a compromise must be reached. Some authors reported better results when nitrogen was present during the feast phase [19, 51] while others observed no competitive advantage [98]. Another study, using complex feedstock as substrate, stated that ammonia was required during both feast and famine, allowing PHA-accumulating organisms to outcompete the non-accumulators [61]. Recently a strategy where nitrogen was limited in the feast phase and added during the famine phase was proposed. Some authors already reported higher PHA contents using this type of system [99–101].

PHA-storing microorganisms' selection and accumulation are also influenced by pH and temperature. However, the influence of these parameters on the MMCs' selection stage is not very clear since they are highly dependent on the microbial community. In spite of the effect of temperature in PHA production, studies of its effect on the selection of MMCs are scarce and do not lead to according conclusions. Studies considered MMCs enrichment with temperatures ranging between 15 and 35°C, where no obvious trend was reported [102–104]. The enrichment at different temperatures (20 and 36°C) was studied and, although no significant difference was observed in terms of storage capacity during this phase, the authors reported an impact on the MMCs microbial composition [105].

Most of the selection processes described in the literature occurred at pH values between 7 and 8. When pH 7 and 8 were tested it was found that acclimatization under both conditions resulted in similar PHA production capability [91]. Several other authors also reported ideal values in this range. When pH selection at 7.5 was compared with a lack of pH control, the first had better storage capacity than the latter [106]. On the other hand, twice the biomass concentration was reported when the enrichment was conducted at pH 8, compared to operation without control, and with no loss in the specific PHA storage efficiency [41]. When the impact of pH on PHA monomeric composition was studied it was observed that at higher pH values more 3HV was produced over 3HB [107].

7.5.2.3 Accumulation Step

In the third and last step of the process, the MMCs obtained in the selection reactor were exploited to obtain high accumulations of PHA. The accumulation step, thus, aims the PHA accumulation to the point of saturation in microbial cells and its efficiency strongly depends on the storage capacity of the selected MMCs. The excess sludge produced during the selection step and the SCOAs produced during the acidification are used in this step as microbial and carbon sources, respectively. Accumulation is usually performed under nutrient-limiting conditions to steer the substrate towards storage instead of growth. Finally, the PHA-rich biomass flows to the downstream processing for extraction and purification of PHA.

Parameters that affect accumulation are highly dependent on other factors like the type of selection strategy and feedstock used. Still, a general overview of the impact of pH, temperature, feeding strategy, and type of limitation are described below.

Studies on the effect of pH on PHA accumulation report that higher productivities are obtained when the same or slightly higher pH values are used in both selection and accumulation stages. Still, previous research shows that PHA accumulation was significantly inhibited at pH <7 [91, 108]. This phenomenon can be explained by the undissociated acids (e.g. acetic acid) in the substrate under low pH conditions [109]. The influence of pH ranged 4.5–10.5 in batch accumulations was studied and the best results were obtained at pH 8 [28]. Another important aspect to keep in mind is that big adjustments in the substrate pH can lead to high processing costs.

The influence of temperature is also very dependent on the conditions used during selection, the microbial community, and other operational parameters. Optimal values reported can be very distinct, maximum production at 30°C was reported by some authors while others had the strongest accumulation at 20 and 10°C [102, 110, 111]. On the other hand, accumulation at different temperatures (15, 20, 25, and 30°C) was tested and similar values of PHA content were obtained for all experiments [104]. Similarly, experiments were conducted at 20, 28, and 36°C and it was observed that, while accumulation at 36°C had the worst performance, similar results could be obtained for 20 and 28°C [105].

Another important condition is the feeding strategy. Traditionally, the accumulation step is performed in batch mode. However, this may not be the best approach since high substrate concentrations can cause inhibition and limit PHA productivity [19, 61]. Pulsed fed-batch was proposed as an alternative [19, 36] but can come with problems when working with substrates of low OLR due to the rapid increase of the working volume in the accumulation reactor [88]. Another disadvantage of the pulsed fed-batch feeding strategy is the variation in productivity rates throughout the process, with high values in the beginning as the substrate is being consumed followed by a decrease over time. Continuous feeding processes can overcome this problem because, even though their productivities are lower than those obtained in the fed-batch, they are constant during the operation. The pH value can be used as an indicator for this addition as it increases with SCOAs consumption [22, 30, 43]. Another possibility is to add substrate based on changes on online monitoring of the DO, because in conditions of constant mixing, and constant aeration, changes in DO could be related to changes in the biomass respiration rate [112].

Nonetheless, the most important aspect of the PHA accumulation step is the operation under nutrient-limiting conditions to avoid a growth response and keep the carbon flow towards PHA production. Most studies limited N and P levels and reported improvement of specific storage rates and productivities [28, 88, 98]. Nonetheless, other studies reported little influence of the nitrogen concentration on the accumulation step [30, 95, 113]. These facts show how nutrient limitation is closely related to the conditions of the selection step, particularly on the composition of the enriched culture and the type of substrate fed [88]. Limitation of DO during accumulation was also studied and it was observed that a higher fraction of substrate was accumulated as PHA under microaerophilic conditions, the overall accumulation capacity was not affected by DO [114]. Limiting DO conditions resulted in a period three times longer to reach the same PHA content. Accumulation at low DO, instead of nutrient limitation, can be an advantage when working with nutrient-rich substrates and the low aeration requirements can decrease operating costs.

7.5.2.4 Other Production Strategies

Although the three-step process with ADF or AN/AE as selection steps are the main strategies used to promote PHA production by MMCs, other approaches have been proposed. Those included modifications or adaptations of the processes above or even resort to specific types of MMCs.

In 1998, Satoh and colleagues proposed a modification of the AN/AE system, the microaerophilic-aerobic system. In this process, the anaerobic conditions are replaced with microaerophilic conditions that authors optimized to be around 0.51% of the chemical oxygen demand (COD) fed to the system. By providing limited amounts of oxygen, the assimilative activity, like the production of proteins and glycogen, is suppressed and the excess carbon is shifted to PHA production, a less energy-demanding process. With this strategy, the authors were able to increase the PHA accumulation to 62% of cell dry weight [115]. The production of PHA using palm oil mill effluent under used nutrient limitation and microaerophilic conditions was proposed to minimize biomass growth and to favor accumulation [34]. However, although in the initial stages the PHA content increased rapidly, oxygen limitation resulted in lower production rates at a later stage, suggesting that this type of limitation would be more suited for the accumulation step [34].

The use of continuous systems instead of using an SBR under ADF conditions was also tested [38]. Feast and famine phases conducted in separate reactors in continuous mode were also studied. Using two continuous stirring tank reactors (CSTR) in sequence, the substrate was fed to the first reactor (feast reactor), and the resulting effluent passing by overflow to the second reactor (famine reactor). The study reported a stable PHA content of 23% in cell dry mass (CDM) during the selection phase and a maximum of 62% in CDM in the accumulation test. Although the results did not show an improvement when compared to the conventional ADF system, it showed that enrichment under continuous mode was possible [38].

An extended cultivation strategy was introduced, similar to what is already applied in pure culture production to obtain high-density biomass that seemed to improve PHA production [116]. This strategy includes a fourth step, between selection and accumulation with the objective to rapidly amplify the biomass in a short period by keeping the dominance of PHA producers. To achieve this, the carbon source and nutrients were fed separately to the MMCs with a step increase of dosages. Using synthetic substrate and a three-step production process with and without extended cultivation, the authors observed that this new strategy increased cell density to 17.22 g/L compared to 7.42 g/L achieved with the conventional process. Volumetric productivity was also better, 1.22 g/(L·h) compared to 0.46 g/(L·h), showing that extended cultivation can be a promising choice for industrial processes [116].

Recently some studies have been focused on PHA production using specific types of MMCs. Since PHA production is also possible using photosynthetic organisms, the use of these types of cultures in illuminated SBR under ADF conditions was tested [117–119]. In the feast phase, PHA are produced using light as an energy source. In the famine phase, PHA are consumed using the oxygen produced by respiration as an electron acceptor and eliminating the need for aeration. In such conditions, a maximum of 20% CDM of PHA was achieved [119]. Since glycogen was also accumulated during the famine phase, alternation with a dark feast phase was also proposed. Higher PHA accumulations were achieved (30%) and this type of system had lower illumination needs [119]. The best results using photosynthetic mixed cultures of 60% CDM were reported using an SBR in constant feast phase and illumination [117]. Even though this strategy saves costs in terms of aeration requirements, it still needs a lot of energy to keep the system illuminated and so its financial impact compared to the conventional methods of production still needs to be addressed.

The use of halophilic bacteria for PHA production has some advantages, like the use of seawater or high salinity industrial wastewater as a medium, and lower costs of downstream processing since cell lysis can be performed using distilled water. Accumulations of 60% using a halophilic community enriched under ADF conditions using glucose and acetic acid as carbon source were reported [120]. Another study used a mixture of organic acids with a concentration of 0.8 g/L Na^+ as substrate during PHA accumulation of previously enriched MMCs and reached a maximum PHA content of 53% CDM. Increasing the concentration of NaCl, however, led to a decrease in accumulation capacity and had an impact in terms of monomer composition and mechanical properties of the final polymer [121].

7.5.3 THE USE OF WASTE

One of the main advantages of using both MMCs and the three-step process is that they allow for the use of complex feedstocks as a carbon source, together with a higher tolerance to inhibitors and process variations. Industrial wastes and byproducts are among the most interesting substrates that can be used for PHA production and their valorization is mandatory under the concept of circular economy. These wastes have a high organic load and low value, representing a disposal problem for the industry. Their integration into a PHA production process would not only valorize a residue but also fit into a biorefinery concept. This can be an advantage to companies as it offers a range of multiple products resulting in a more flexible, and potentially sustainable, industry [122]. Several authors have already explored this possibility and many types of waste were successfully converted into PHA. Table 7.1 summarizes some of those works.

Food waste or wastewater from the food industry, like whey and sugar molasses, have a compositional matrix that varies broadly based on source and type that it is characterized by its high energy potential [123, 124]. This potential can be used for PHA production, as has been shown in several studies (Table 7.1). The highest PHA accumulation (81.4%) was reported using fermented cheese whey and MMCs selected under ADF conditions [23]. The main problem when using food waste is related to storage and transportation as it easily degrades [69].

Urban waste and municipal wastewater are also good carbon sources for PHA production. An increase in the PHA accumulation from 29 to 78% (CDM) was reported when the substrate for MMCs enrichment was changed from leachate to a synthetic SCOAs mixture resembling the composition of the leachate. The switch resulted in high cellular PHA content in batch accumulation with the leachate, increasing productivity and making the overall process more economically attractive [40].

Some authors chose to study PHA production from residues from the wood processing industry. Works report on wastewater from wood mills and paper mills, as

TABLE 7.1

Overview of PHA Production Processes Using MMCs and Waste as Substrate

Waste Source	Production Strategy	PHA [wt.-%]	Composition [% 3HB:3HV]	Organisms Selected [Molecular Method]	Ref.
Whey	Three-step process: Acidification, ADF Selection, Batch Accumulation	44%	99:1	NR	[112]
	Three-step process: Acidification, ADF Selection, Fed-batch Accumulation	81%	60:40	NR	[23]
	Three-step process: Acidification, ADF Selection, Fed-batch Accumulation	65%	81:19	NR	[126]
	Three-step process: Acidification, ADF Selection, Batch Accumulation	33%	87:13	*Xanthomonadaceae, Paracoccus, Lampropedia, Comamonadaceae* (NGS)	[99]
Sugarcane molasses	Three-step process: Acidification, ADF Selection, Fed-batch Accumulation	56%	48:52	NR	[126]
	AN/AO Selection; Fed-batch Accumulation	37%	56:43	NR	[127]
FOOD WASTE	Three-step process: Acidification, ADF Selection, Batch Accumulation	72%	NR	*Azoarcus, Thauera* (FISH)	[38]
	Three-step process: Acidification, ADF Selection, Batch Accumulation	56%	85:15	NR	[43]
	Three-step process: Acidification, ADF Selection, Batch Accumulation	58%	NR	*Azoarcus, Thauera, Paracoccus* (FISH)	[41]

(Continued)

TABLE 7.1 (CONTINUED)
Overview of PHA Production Processes Using MMCs and Waste as Substrate

Waste Source	Production Strategy	PHA [wt.-%]	Composition [% 3HB:3HV]	Organisms Selected [Molecular Method]	Ref.
Food waste	Three-step process: Acidification, ADF Selection, Batch Accumulation	24%	22:78	NR	[128]
	Anoxic-Aerobic Selection (VSMBR)	2%	84:8	NR	[129]
	AN/AO Selection, Batch Accumulation	51%	65:35	NR	[87]
	Three-step process: Acidification, ADF Selection, Batch Accumulation	61%	NR	Paracoccus, Hydrogenophaga, Thauera (NGS)	[130]
Brewery wastewater	Three-step process: Acidification, ADF Selection, Batch Accumulation	39%	NR	NR	[131]
Food processing wastewater	ADF Selection, Batch Accumulation	61%	100:0	NR	[132]
Jowar grain-based distillery spent wash	ADF Selection, Batch Accumulation	42%	100:0	NR	[132]
Rice grain-based distillery spent wash	ADF Selection, Batch Accumulation	40%	100:0	NR	[132]
Olive oil mill pomace	Acidification, ADF Selection	39%	NR	NR	[133]
Olive oil mill wastewater	Three-step process: Acidification, ADF Selection, Batch Accumulation	11%	78:22	Thauera, Acinetobacter (FISH)	[134]
	Three-step process: Acidification, ADF Selection, Batch Accumulation	20%	NR	Lampropedia hyalina, Candidatus Meganema perideroedes (DGGE)	[135]
Tomato wastewater	ADF Selection, Batch Accumulation	20%	NR	NR	[136]

FOOD WASTE

(Continued)

TABLE 7.1 (CONTINUED)
Overview of PHA Production Processes Using MMCs and Waste as Substrate

Waste Source	Production Strategy	PHA [wt.-%]	Composition [% 3HB:3HV]	Organisms Selected [Molecular Method]	Ref.
Beet wash process water	AN/AO Selection, Batch Accumulation	60%	75:15	*Azospirillum* sp., *Azoarcus* sp., *Ralstonia solanacearum* (NGS)	[137]
Sugar cane wastewater	Three-step process: Acidification, ADF Selection, Batch Accumulation, Batch Accumulation	61%	65:35	*Pseudomonas* (T-RFLP)	[92]
Sludge	ADF Selection, Batch Accumulation	57%	88:12	NR	[138]
	Three-step process: Acidification, ADF Selection, Fed-batch Accumulation	38%	74:26	*Actinobacteria, Rhodococcus* (DGGE)	[139]
	Three-step process: Acidification, ADF Selection, Batch Accumulation	37%	69:31	*Thauera, Saprospiraceae* (NGS)	[140]
Primary sludge and waste activated sludge	Three-step process: Acidification, ADF Selection, Batch Accumulation	66%	52:48	*Thauera, Zoogloea* (NGS)	[141]
Sludge hydrolysate	Three-step process: Acidification, ADF Selection, Batch Accumulation	52%	59:41	*Delftia, Lysinibacillus, Petrimonas Rhodobacter* (NGS)	[142]
Municipal Wastewater	AN/AO Selection, Batch Accumulation	53%	50:50	NR	[47]
	Three-step process: Acidification, ADF Selection, Batch Accumulation	11%	60:40	NR	[143]
Municipal solid waste leachate	Three-step process: Acidification, ADF Selection, Batch Accumulation	78%	100:0	*P. acidivorans* (FISH)	[40]

URBAN WASTE

(Continued)

TABLE 7.1 (CONTINUED)

Overview of PHA Production Processes Using MMCs and Waste as Substrate

Waste Source	Production Strategy	PHA [wt.-%]	Composition [% 3HB:3HV]	Organisms Selected [Molecular Method]	Ref.
OFMSW	Three-step process: Acidification, ADF Selection, Fed-batch Accumulation	52%	92:08	Proteobacteria (mostly ß); Cytophaga/Flexibacter/Bacteroidetes (FISH) Hydrogenophaga spp.; Acidovorax spp. (NGS)	[144]
HSSL	ADF Selection, Batch Accumulation	67%	100:0	Alpha, Beta and Gammaproteobacteria (FISH)	[33]
Kraft cellulose mill wastewater	Acidification, ADF Selection	22%	68:32	Acidovorax, Comamonas (NGS)	[31]
	Batch Accumulation	30%	NR	NR	[145]
Paper mill wastewater	Three-step process: Acidification, AN/AO Selection, Batch Accumulation	42%	50:30	Defluviicoccus vanus, Candidatus Competibacter phosphatis (FISH)	[146]
	Three-step process: Acidification, ADF Selection, Batch Accumulation	48%	39:61	NR	[147]
	Three-step process: Acidification, ADF Selection, Fed-batch Accumulation	77%	86:14	P. acidivorans, Thauera (DGGE)	[30]
Wood hydrolysate	Three-step process: Pre-treatment, ADF Selection, Batch Accumulation	28%	100:0	Actinobacteria (Microbacterium), ß-Proteobacteria (Comamonas, Acidovorax, Achromobacter), α-Proteobacteria (Paracoccus) (NGS)	[148]

INDUSTRIAL WASTE

(Continued)

TABLE 7.1 (CONTINUED)

Overview of PHA Production Processes Using MMCs and Waste as Substrate

Waste Source	Production Strategy	PHA [wt.-%]	Composition [% 3HB:3HV]	Organisms Selected [Molecular Method]	Ref.
Wood mill effluent	Three-step process: Acidification, ADF Selection, Batch Accumulation	29%	80:20	NR	[149]
	Three-step process: Acidification, ADF Selection, Batch Accumulation	35%	NR	NR	[150]
Paperboard mill wastewater	Three-step process: Acidification, ADF Selection, Batch Accumulation	59%	92:8	Proteobacteria (NGS)	[151]
Bio-oil	Three-step process: Acidification, ADF Selection, Batch Accumulation	17%	73:23	NR	[125]
Crude glycerol	ADF Selection, Batch Accumulation	59%	100:0	NR	[37]
	Three-step process: Acidification, ADF Selection, Batch Accumulation	76%	84:16	*Thauera, Amaricoccus* (NGS)	[100]
	ADF Selection, Batch Accumulation	80%	60:40	α- and β- (dominant), γ-Proteobacteria (FISH)	[152]
Raw dairy manure	Three-step process: Acidification, ADF Selection, Batch Accumulation	71%	27:73	*Meganema, Zoogloea, Thauera* (NGS)	[153]

INDUSTRIAL WASTE

NR – Not Reported; FISH – Fluorescence in situ hybridization; NGS – Next generation sequencing; DGGE – Denaturing gradient gel electrophoresis; T-RFLP – Terminal restriction length polymorphism.

well as hardwood spent sulfite liquor (HSSL), a byproduct of the pulp industry (Table 7.1). These wastes can have high concentrations of microbial inhibitors resulting from the degradation of lignocellulosic material [33]. The best result reported using these types of effluents used a three-step production process and resulted in an accumulation of 77% (CDM) of P(3HB/3HV) within five hours [30].

Other industrial wastes like bio-oil from fast-pyrolysis of chicken beds and crude glycerol from biodiesel production were also tested. In both cases, the accumulation of other polymers besides PHA limited the yield of accumulation, with 17% CDM for bio-oil and 76% for crude glycerol [100, 125].

As Table 7.1 shows, the three-step production process is the go-to for most substrates and the accumulation capacities reported indicate that there is much potential to combine and integrate PHA production in several industries.

7.6 MICROORGANISMS AND METABOLISM

There is a direct relationship between microorganisms and their metabolism. Both the operational conditions and the type of substrate of the process will determine the microbiota present. As a consequence, the best metabolic fitted organisms will survive in such systems, determining the monomeric composition of the polymer, and at the end, their applications.

7.6.1 MICROORGANISMS INVOLVED

The microbial population present in an MMCs system for PHA production is diverse by its nature. This is one of the most important aspects in these processes so it can provide more efficient utilization of different wastes and cope with fluctuations observed during process operation while remaining efficient in PHA production. Several factors determine the balance between organisms present in MMCs processes. Most of the knowledge came again from experiments with pure cultures. The substrate, and in most cases the relative proportion of SCOAs, selected for a specific type of bacteria as *Azoarcus*, *Thauera* and *Ammaricoccus* sp. Sugars and fatty acids will select for an organism such as *Pseudomonas*. Operational conditions such as pH, temperature, electron acceptor, and its concentration (oxygen, nitrate/nitrate) have their impact. Engineering conditions like HRT, SRT (reflecting the growth rate of the organism), and OLR are also variables which impact on microbial selection.

The techniques used to characterize the microbial communities have evolved with the evolution of the microbial ecology area. While initially most situations recurred to the utilization of culturing methods such as isolation of the organism able to survive by plating in solid medium (agar) and micromanipulation/cell sorting, it became apparent that most diversity was lost in this way. That was the case when determining the responsibility for biological phosphorus removal in wastewater treatment plants since those organisms are non-culturable. With the rise of the development of molecular biology techniques, a multitude of new approaches using polymerase chain reaction (PCR) or non-PCR related methods has surged. The most well-known non-PCR dependent method able to detect an organism directly in samples is fluorescence *in situ* hybridization (FISH). While having the advantage of being of simple utilization and quantitative, with different degrees of determination, from phyla to sub-type, its utilization depends on some knowledge of the available organisms. Known available probes/sequences are easy to use but the construction of new ones is not straightforward.

Most of the PCR-based methods are qualitative. Among, them denaturing gradient gel electrophoresis (DGGE) and temperature gradient gel electrophoresis (TGGE) provide profiling of the available community. Upon total DNA extraction of the sample and amplification with specific primer samples are applied to acrylamide gel being DNA denatured by the action of the concentration the denaturing agent or by the temperature applied. At specific levels, depending on the guanine-cytosine (GC) content of the amplicon, samples stop migrating in the gel, presenting themselves as bands. Their presence/absence can be utilized to follow the MMCs composition/evolution along time and correlate their existence with operational parameters, by applying statistical methods. The bands can also be excised for later sequencing, although sometimes with low accuracy due to the small dimension of the band (number of base pairs). The same type of information can be obtained from terminal restriction fragment length polymorphism (T-RFLP), but this time the patterns are obtained after digestion of the amplified sample with different restriction enzymes and its application of acrylamide gels.

The emergence of next generation sequencing (NGS) completely changes the determination of microbial communities' composition, in particular with the lowering of their prices in recent years. Again, this is a PCR-amplification-base method but allows determining the full community structure, being semi-quantitative. Presently this is the technique most used to evaluate the composition/diversity of microbial communities. Another helpful PCR-base technique is quantitative PCR (qPCR), and as the name states, it is quantitative. Its accuracy is based on the quality of the amplified sample's DNA, the existence of specific primers/probes among others. When used in conjunction with NGS it provides a more precise determination of the microbial community structure.

Some examples of the utilization of these techniques in the microbial characterization of MMCs communities for PHA production with complex substrates are presented in Table 7.1. A more extensive review about this matter can be obtained elsewhere [5, 45]

7.6.2 Most Common Metabolic Pathways

Due to the high diversity of PHA-accumulating organisms the number of known metabolic pathways involved in the process has increased in recent years (Figure 7.2). A few studies regarding the metabolism of PHA production were conducted with MMCs. Those were conducted using bacteria present in activated sludge systems for biological phosphorus removal. The two most important types were PAOs and GAOs that thrive in the system competing for carbon sources. While PAOs need to keep a balance in between three storage polymers, PHA, glycogen, and polyphosphate (PolyP), for GAOs just PHA and glycogen are involved. The metabolism of PHA accumulation is triggered by the uncoupling of carbon and oxygen metabolism. Under anaerobic conditions external carbon substrates (SCOAs) are stored as PHA,

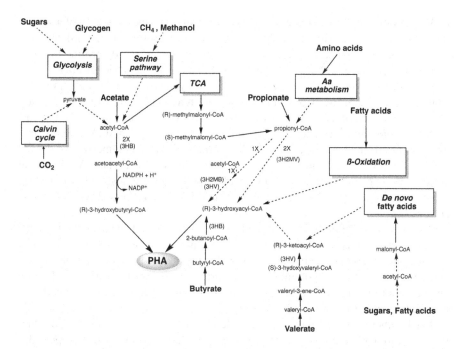

FIGURE 7.2 Some of the important metabolic pathways for production of PHA with MMCs from different substrates (adapted from [5] and [155]).

being the energy needs for the process coming from the hydrolysis of PolyP for PAOs and glycogen for GAOs. For both organisms, glycogen also acts as a precursor for some of the monomers synthesized. Under aerobic conditions the stored PHA are used for growth in the presence of a nitrogen source, for maintenance and replenishing the PolyP pool in PAOs and glycogen for PAOs and GAOs [14,51,154]. Those studies were performed using acetate or propionate as carbon sources. [155]

Most of the studies involving metabolism of PHA production were performed with pure cultures. As shown for acetate and propionate with PAOs and GAOs the metabolism was similar to pure cultures and the same is expected for other substrates. The best-studied pathway for P(3HB) production is the one used by *Cupriavidus necator.* When using acetate, this compound is activated to acetyl-CoA at the expense of ATP. Two units of this compound are joined to form acetoacetyl-CoA by the action of ketothiolase. Upon the activity of a reductase, using NADPH, acetoacetyl-CoA is transformed into R-3-hydroxybutyryl-CoA. Lastly the PHA synthase appends one more monomeric unit into the growing polymeric chain. Other SCOAs, as propionate, butyrate, and valerate, can also be utilized for the production of PHA's short chain length (*scl*) monomers. Sugars and glycogen through glycolysis produce pyruvate that is a precursor of acetyl-CoA. Propionate is an important substrate as it provides the propionyl moiety present in different types of monomers, alone (3-hydroxy-2-methylvalerate, 3H2MV) or when combined with acetate (3HV; 3-hydroxy-2-methylbutyrate, 3H2MB). Butyrate and valerate can be incorporated directly via their corresponding hydroxyacylalkanoates or broken down to acetyl and/or propionyl-CoA using the ß-oxidation pathway. The incorporation of a second monomer improves greatly most of the properties for a useful polymer utilization.

The type of polymerase present in the microorganism will determine the type of polymer produced, being either of the scl or medium-chain length (mcl) type. *Cupriavidus nector* represents the class I PHA polymerase, with a single unit, able to use 3-, 4-, and $5HA_{scl}$-CoA with 3 to 5 carbon atoms. Also, with a single unit, class II PHA polymerase is able to use $3HA_{mcl}$-CoA with more than 5 carbon atoms, being *Pseudomonas* sp. one of the examples. Both class III and IV polymerases are composed of two sub-units. *Allochromatium vinosum* polymerize 3-, 4- and $5HA_{mcl}$-CoA (class III) and *Bacillus megaterium* $3HA_{scl}$-CoA (class IV). In recent years there is evidence that class IV of PHA polymerase is also able to produce medium-chain length PHA (*B. megaterium* DSM 509, *Bacillus* sp.US163, *Bacillus* sp. US177).

The production of *mcl*-PHA derived from the utilization of sugars and fatty acids by the microorganism. While for most of the metabolic pathways used for PHA production the monomer reflects the original substrate, with the conversion of fatty acids via ß-oxidation a shortening of the carbon chain can be achieved while an elongation of the carbon chain can be attained through *de novo* fatty acid synthesis, using either sugars or fatty acids.

Lately other substrates have been regarded as carbon substrates for PHA production. Amino acids are one of them, with a degradation metabolism that provides propionyl-CoA. Single carbon substrates have also gained attention, in particular the possibility of producing PHA from CO_2 through the Calvin cycle. Another greenhouse gas that can be used for PHA production is methane using the serine pathway for acetyl-CoA production. The same pathway is used for methanol incorporation into PHA. Hydrocarbons can be converted by terminal oxidation into the corresponding fatty acids progressing to ß-oxidation [156]. Concerning aromatic compounds, they can be added as a direct precursor for incorporation or internal production of aromatic PHA without addition of those aromatic precursors [157].

7.7 CHALLENGES AHEAD

Despite the huge number of works published and projects funded, PHA produced by MMCs are far from commercialization. In recent years, several issues were addressed but others remained without answer. Three main challenges were identified, one related to the substrates used, and two related to the post-production process, namely downstream processing and polymer characterization.

7.7.1 COMPLEX SUBSTRATES

Notwithstanding the considerable number of types of waste already tested, there are still many to be utilized. The reuse and recycling of waste are mandatory especially after the planned abolition of landfills in the European Union (EU) by 2030. To achieve this objective, the EU defined a target of 70% of waste to be integrated into the value chain [158]. Lignocellulosic biomass (LCB) waste is produced in very high amounts and is considered a good candidate to be used as feedstocks for energy, chemicals, and materials according to the biorefinery concept [159]. However, the complexity of the chemical and structural composition of this type of waste is a serious obstacle to obtain value-added compounds, as PHA [160].

Not many LCB residues were tested for PHA production from MMCs. Most of the works used wastewater from pulp and paper mills and only a few deal with LCB waste. Among these are the works that used bio-oil that resulted from the fast pyrolysis of LCB waste [36, 125] and hardwood spent sulfite liquor [31–33]. Moreover, a biorefinery based on the three-step process with additional production of biohydrogen was proposed but the raw material used was recombinant giant cane modified for easier carbohydrate recovery [161]. Due to the increasing interest in using LCB residues for the biotechnological production of fuels, chemicals, and materials under the concept of biorefinery, PHA production by MMCs has an opportunity to be included in order to convert residues that result from previous steps.

Nevertheless, the production of PHA by MMCs is being developed to establish biorefineries in WWTP, since the initial AF step of the process can work not only for the production of SCOAs but also for the production of biogas [162]. This work describes a pilot plant fed with the organic fraction of municipal solid waste (OFMSW) and biological sludge for the production of PHA and biogas working in a WWTP in Treviso, Italy.

7.7.2 DOWNSTREAM PROCESSING

The downstream processing has been a drawback of PHA production by MMCs since the development of the first processes. In the early years of the process, the recovery of PHA from biomass was not an issue because chlorinated solvents worked well, by showing high yields and not damaging polymer chains. However, the environmental restrictions imposed on the industrial use of halogenated solvents required the search for alternatives. This issue is mostly solved for pure and recombinant cultures since this type of biomass has a more sensitive cell wall. The cell wall becomes fragile when attaining high PHA storage contents and can be disrupted by the use of mild acids or basis. The methods that are applied to pure cultures allow for a high yield of recovery of PHA with good quality [163]. However, with MMCs, the problem is more intricate since the cell wall and floc structure are much more complex and harder to disrupt than pure and recombinant cultures [164].

To replace chlorinated solvents, some methods using dimethylcarbonate (DMC), 2-butanol, NaClO, or simultaneous use of NaOH and surfactant were tested. In the first case, the yield of recovery was rather low, 63%, but provided a polymer with good quality [164], while in the other situations, a higher yield was obtained but a polymer with reduced molecular weight was obtained, which is undesirable from the processing point of view [165].

An efficient method to extract PHA from MMCs was not found and also there is still no hint as to how raw materials can influence the polymer quality. With the presence of contaminants in the substrates, there is doubt as to whether the same methodology can be applied for the different microbial compositions selected or how to process the residues that result from this step. There are still several problems unsolved that need to be answered in the future.

7.7.3 POLYMER CHARACTERIZATION

A complete characterization of PHA produced is crucial for their market acceptance but the characterization is not easy. This acceptance depends not only on the favorable characteristics of PHA but also on their consistency over time. While only a few milligrams are required for the determination of monomeric composition, molecular weight, or thermal characterization, mechanical properties characterization demands a higher amount of polymer, which is not easy to achieve at laboratory scale.

The information found in the literature is still very scarce. If only a few works did the characterization of PHA obtained with pure substrates, even fewer works did it using real substrates. PHA samples obtained using molasses as a carbon source [43], as well as samples produced by MMCs obtained in a pilot plant but fed with acetic and propionic acids were analyzed [166]. Finally, PHA obtained with cheese whey and olive oil mill wastewater were not only characterized but also the impact of the presence of process residues on the polymer was assessed [167].

Without a systematic characterization of the polymer produced, no concrete applications for PHA from MMCs can be found. Due to the sample's origin, in the case of production using MMCs from WWTPs with urban or industrial waste, it is expected that applications cannot be the same as those for PHA obtained by pure cultures.

7.8 CONCLUSIONS AND OUTLOOK

The development of PHA production processes using MMCs and real waste has been a very interesting journey, but it is far from reaching its destination. Since the earlier findings related to the identification of microbial reserves found in AS systems, a lot of research has been done that confirmed the real possibility of implementation of the process. The three-step process is an accepted methodology that can be applied not only on WWTPs but also in industries to allow for waste management. The concepts of biorefinery and circular economy also contributed to the establishment of the process. However, more research is required so to reach the commercialization level.

ACKNOWLEDGMENTS

This work was developed within the scope of the project CICECO-Aveiro Institute of Materials (Ref. UIDB/50011/2020 and UIDP/50011/2020) and LAQV-Associate Laboratory for Green Chemistry (Ref. UIDB/QUI/50006/2020), financed by national funds through the FCT/MCTES and when appropriate co-financed by FEDER under the PT2020 Partnership Agreement. Paulo C. Lemos acknowledges the support by FCT/MCTES for contract IF/01054/2014/CP1224/CT0005. Joana Pereira thanks FCT/MCTES for her PhD grant, SFRH/BD/130003/2017.

REFERENCES

1. Obruca S, Sedlacek P, Koller M, Kucera D, Pernicova I. Involvement of polyhydroxy-alkanoates in stress resistance of microbial cells: Biotechnological consequences and applications. *Biotechnol Adv.* 2018;36(3):856–70.

2. Koller M, Gasser I, Schmid F, Berg G. Linking ecology with economy: Insights into polyhydroxyalkanoate-producing microorganisms. *Eng Life Sci.* 2011;3:222–37.
3. Wintermute EH, Silver PA. Dynamics in the mixed microbial concourse. *Genes Dev.* 2010;24:2603–14.
4. Amann R, Ludwig W, Schleifer K. Phylogenetic identification and in situ detection of individual microbial cells without cultivation. *Microbiol Rev.* 1995;59(1):143–69.
5. Serafim LS, Queirós D, Rossetti S, Lemos PC. Biopolymer production by mixed microbial cultures: Integrating remediation with valorization. In: Koller M, ed. *Recent Advances in Biotechnology – Microbial Biopolyester Production, Performance and Processing – Microbiology, Feedstocks, and Metabolism*, Bentham Science Publishers, Sharjah, 2016. p. 26–264.
6. Jagmann N, Philipp B. Reprint of Design of synthetic microbial communities for bio-technological production processes. *J Biotechnol.* 2014;192:293–301.
7. Ghosh S, Chowdhury R, Bhattacharya P. Mixed consortia in bioprocesses: Role of microbial interactions. *Appl Microbiol Biotechnol.* 2016;100:4283–95.
8. Eng A, Borenstein E. Microbial community design: Methods, applications, and opportunities. *Curr Opin Biotechnol.* 2019;58:117–28.
9. Koch C, Muller S, Harms H, Harnisch F. Microbiomes in bioenergy production: From analysis to management. *Curr Opin Biotechnol.* 2014;27:65–72.
10. Crabtree K, Boyle W, McCoy E, Rohlich GA. A Mechanism of floc formation by zoogloea ramigera. *Water Environ Fed.* 1966;38(12):1968–80.
11. Wallen LL, Davis EN. Biopolymers of activated sludge. *Environ Sci Technol.* 1972;6:161–4.
12. Deinema MH. Bacterial flocculation and production of poly-b-hydroxybutyrate. *Appl Microbiol.* 1972;24(6):857–8.
13. Wallen LL, Rohwedder WK. Poly-b-hydroxyalkanoate from activated sludge. *Environ Sci Technol.* 1974;8(6):576–9.
14. Pereira H, Lemos PC, Reis MAM, Crespo JPSG, Carrondo MJT, Santos H. Model for carbon metabolism in biological phosphorus removal processes based on *in vivo* 13C-NMR labelling experiments. *Water Res.* 1996;30(9):2128–38.
15. Levantesi C, Serafim LS, Crocetti GR, Lemos PC, Rossetti S, Blackall L, Reis MAM, Tandoi V. Analysis of the microbial community structure and function of a laboratory scale enhanced biological phosphorus removal reactor. *Environ Microbiol.* 2002;4(10):559–69.
16. Majone M, Massanisso P, Carucci A, Lindrea K, Tandoi V. Influence of storage on kinetic selection to control aerobic filamentous bulking. *Water Sci Technol.* 1996;34(5–6):223–32.
17. Daigger GT, Grady CPL. An Assessment of the role of physiological adaptation in the transient response of bacterial cultures. *Biotechnol Bioeng.* 1982;XXIV:1427–44.
18. Serafim LS, Lemos PC, Albuquerque MGE, Reis MAM. Strategies for PHA production by mixed cultures and renewable waste materials. *Appl Microbiol Biotechnol.* 2008;81:615–28.
19. Serafim LS, Lemos PC, Oliveira R, Reis MAM. Optimization of polyhydroxybutyrate production by mixed cultures submitted to aerobic dynamic feeding conditions. *Biotechnol Bioeng.* 2004;87(2):145–60.
20. Queirós D, Lemos PC, Rossetti S, Serafim LS. Unveiling PHA-storing populations using molecular methods. *Appl Microbiol Biotechnol.* 2015;99(24):10433–46.
21. Dionisi D, Majone M, Papa V, Beccari M. Biodegradable polymers from organic acids by using activated sludge enriched by aerobic periodic feeding. *Biotechnol Bioeng.* 2004;85(6):569–79.
22. Johnson K, Jiang Y, Kleerebezem R, Muyzer G, Van Loosdrecht MCM. Enrichment of a mixed bacterial culture with a high polyhydroxyalkanoate storage capacity. *Biomacromolecules.* 2009;10(4):670–6.

23. Colombo B, Pepè T, Reis M, Scaglia B, Adani F. Polyhydroxyalkanoates (PHAs) production from fermented cheese whey by using a mixed microbial culture. *Bioresour Technol.* 2016;218:692–9.

24. Valentino F, Moretto G, Lorini L, Bolzonella D, Pavan P. Pilot-scale polyhydroxyalkanoate production from combined treatment of organic fraction of municipal solid waste and sewage sludge. *Ind Eng Chem Res.* 2019;58:12149–58.

25. Bengtsson S, Karlsson A, Alexandersson T, Quadri L, Hjort M, Johansson P, Morgan-Sagastume F, Anterrieu S, Arcos-Hernandez M, Karabegovic L, Magnusson P, Werker A. A process for polyhydroxyalkanoate (PHA) production from municipal wastewater treatment with biological carbon and nitrogen removal demonstrated at pilot-scale. *N Biotechnol.* 2017;35:42–53.

26. Tamis J, Mulders M, Dijkman H, Rozendal R, Van Loosdrecht MCM, Kleerebezem R. Pilot-scale polyhydroxyalkanoate production from paper mill wastewater: Process characteristics and identification of bottlenecks for full-scale implementation. *J Environ Eng.* 2018;144(10):1–9.

27. Tamis J, Lužkov K, Jiang Y, Van Loosdrecht MCM, Kleerebezem R. Enrichment of plasticicumulans acidivorans at pilot-scale for PHA production on industrial wastewater. *J Biotechnol.* 2014;192(Part A):161–9.

28. Dionisi D, Beccari M, Di Gregorio S, Majone M, Papini MP, Vallini G. Storage of biodegradable polymers by an enriched microbial community in a sequencing batch reactor operated at high organic load rate. *J Chem Technol Biotechnol.* 2005;80(11): 1306–18.

29. Bengtsson S, Werker A, Christensson M, Welander T. Production of polyhydroxyalkanoates by activated sludge treating a paper mill wastewater. *Bioresour Technol.* 2008;99(3):509–16.

30. Jiang Y, Marang L, Tamis J, Van Loosdrecht MCM, Dijkman H, Kleerebezem R. Waste to resource: Converting paper mill wastewater to bioplastic. *Water Res.* 2012;46(17):5517–30.

31. Pereira J, Queirós D, Lemos PC, Rossetti S, Serafim LS. Enrichment of a mixed microbial culture of PHA-storing microorganisms by using fermented hardwood spent sulfite liquor. *N Biotechnol.* 2020;56:79–86.

32. Queirós D, Fonseca A, Lemos PC, Serafim L. Long-term operation of a two-stage polyhydroxyalkanoates production process from hardwood sulphite spent liquor. *J Chem Technol Biotechnol.* 2015;91:2480–7.

33. Queirós D, Rossetti S, Serafim LS. PHA production by mixed cultures: A way to valorize wastes from pulp industry. *Bioresour Technol.* 2014;157:197–205.

34. Din M, Mohanadoss P, Ujang Z, Van Loosdrecht M, Yunus S, Chelliapan S, Zambare V, Olsson G. Development of Bio-PORec system for polyhydroxyalkanoates (PHA) production and its storage in mixed cultures of palm oil mill effluent (POME). *Bioresour Technol.* 2012;124:208–16.

35. Moita R, Lemos PC. Biopolymers production from mixed cultures and pyrolysis by-products. *J Biotechnol.* 2012;157(4):578–83.

36. Moita R, Freches A, Lemos PC. Crude glycerol as feedstock for polyhydroxyalkanoates production by mixed microbial cultures. *Water Res.* 2014;58:9–20.

37. Freches A, Lemos PC. Microbial selection strategies for polyhydroxyalkanoates production from crude glycerol: Effect of OLR and cycle length. *N Biotechnol.* 2017;39:22–8.

38. Albuquerque MGE, Concas S, Bengtsson S, Reis MAM. Mixed culture polyhydroxyalkanoates production from sugar molasses: The use of a 2-stage CSTR system for culture selection. *Bioresour Technol.* 2010;101(18):7112–22.

39. Marang L, Jiang Y, Van Loosdrecht MCM, Kleerebezem R. Impact of non-storing biomass on PHA production: An enrichment culture on acetate and methanol. *Int J Biol Macromol.* 2014;71:74–80.

40. Korkakaki E, Van Loosdrecht MCM, Kleerebezem R. Survival of the fastest: Selective removal of the side population for enhanced PHA production in a mixed substrate enrichment. *Bioresour Technol.* 2016;216:1022–9.

41. Oehmen A, Pinto FV, Silva V, Albuquerque MGE, Reis MAM. The impact of pH control on the volumetric productivity of mixed culture PHA production from fermented molasses. *Eng Life Sci.* 2014;14(2):143–52.

42. Serafim S, Lemos PC, Torres C, Reis MAM, Ramos AM. The influence of process parameters on the characteristics of polyhydroxyalkanoates produced by mixed cultures. *Macromol Biosci.* 2008;8(4):355–66.

43. Albuquerque MGE, Martino V, Pollet E, Avérous L, Reis MAM. Mixed culture polyhydroxyalkanoate (PHA) production from volatile fatty acid (VFA)-rich streams: Effect of substrate composition and feeding regime on PHA productivity, composition and properties. *J Biotechnol.* 2011;151(1):66–76.

44. Pisco AR, Bengtsson S, Werker A, Reis MAM, Lemos PC. Community structure evolution and enrichment of glycogen-accumulating organisms producing polyhydroxyalkanoates from fermented molasses. *Appl Environ Microbiol.* 2009;75(14):4676–86.

45. Queirós D, Lemos PC, Rossetti S, Serafim LS. Unveiling PHA-storing populations using molecular methods. *Appl Microbiol Biotechnol.* 2015;99(24):10433–46.

46. Morgan-Sagastume F. Characterisation of open, mixed microbial cultures for polyhydroxyalkanoate (PHA) production. *Rev Environ Sci Biotechnol.* 2016;15(4):593–625.

47. Coats ER, Loge FJ, Wolcott MP, Englund K, McDonald AG. Synthesis of polyhydroxyalkanoates in municipal wastewater treatment. *Water Environ Res.* 2007;79(12):2396–403.

48. Carvalho G, Oehmen A, Albuquerque MGE, Reis MAM. The relationship between mixed microbial culture composition and PHA production performance from fermented molasses. *N Biotechnol.* 2014;31(4):257–63.

49. Laycock B, Halley P, Pratt S, Werker A, Lant P. The chemomechanical properties of microbial polyhydroxyalkanoates. *Prog Polym Sci.* 2013;38(3–4):536–83.

50. Carta F, Beun JJ, Van Loosdrecht MCM, Heijnen JJ. Simultaneous storage and degradation of PHB and glycogen in activated sludge cultures. *Water Res.* 2001;35(11):2693–701.

51. Lemos PC, Serafim LS, Reis MAM. Synthesis of polyhydroxyalkanoates from different short-chain fatty acids by mixed cultures submitted to aerobic dynamic feeding. *J Biotechnol.* 2006;122(2):226–38.

52. Laycock B, Halley P, Pratt S, Werker A, Lant P. The chemomechanical properties of microbial polyhydroxyalkanoates. *Prog Polym Sci.* 2013;38(2):536–83.

53. Demirel B, Yenigün O. Two-phase anaerobic digestion processes: A review. *J Chem Technol Biotechnol.* 2002;77(7):743–55.

54. Singhania RR, Patel AK, Christophe G, Fontanille P, Larroche C. Biological upgrading of volatile fatty acids, key intermediates for the valorization of biowaste through dark anaerobic fermentation. *Bioresour Technol.* 2013;145:166–74.

55. Yuan Y, Hu X, Chen H, Zhou Y, Zhou Y, Wang D. Advances in enhanced volatile fatty acid production from anaerobic fermentation of waste activated sludge. *Sci Total Environ.* 2019;694:133741.

56. Lee WS, Chua ASM, Yeoh HK, Ngoh GC. A review of the production and applications of waste-derived volatile fatty acids. *Chem Eng J.* 2014;235:83–99.

57. Cerrone F, Choudhari SK, Davis R, Cysneiros D, O'Flaherty V, Duane G, Casey E, Guzik MW, Kenny ST, Babu RP, O'Connor K. Medium chain length polyhydroxyalkanoate (mcl-PHA) production from volatile fatty acids derived from the anaerobic digestion of grass. *Appl Microbiol Biotechnol.* 2014;98(2):611–20.

58. Bengtsson S, Hallquist J, Werker A, Welander T. Acidogenic fermentation of industrial wastewaters: Effects of chemostat retention time and pH on volatile fatty acids production. *Biochem Eng J.* 2008;40(3):492–9.

59. Jankowska E, Chwiałkowska J, Stodolny M, Oleskowicz-Popiel P. Effect of pH and retention time on volatile fatty acids production during mixed culture fermentation. *Bioresour Technol.* 2015;190:274–80.

60. Huang YT, Chen PL, Semblante GU, You SJ. Detection of polyhydroxyalkanoate-accumulating bacteria from domestic wastewater treatment plant using highly sensitive PCR primers. *J Microbiol Biotechnol.* 2012;22(8):1141–7.

61. Albuquerque MGE, Eiroa M, Torres C, Nunes BR, Reis MAM. Strategies for the development of a side stream process for polyhydroxyalkanoate (PHA) production from sugar cane molasses. *J Biotechnol.* 2007;130(4):411–21.

62. Gouveia AR, Freitas EB, Galinha CF, Carvalho G, Duque AF, Reis MAM. Dynamic change of pH in acidogenic fermentation of cheese whey towards polyhydroxyalkanoates production: Impact on performance and microbial population. *N Biotechnol.* 2017;37:108–16.

63. Temudo M, Kleerebezem R, Van Loosdrecht MCM. Influence of the pH on (Open) mixed culture fermentation of glucose: A chemostat study. *Biotechnol Bioeng.* 2007;98(1):69–79.

64. Chen Y, Jiang S, Yuan H, Zhou Q, Gu G. Hydrolysis and acidification of waste activated sludge at different pHs. *Water Res.* 2007;41:683–9.

65. Yu G, He P, Shao L, He P. Toward understanding the mechanism of improving the production of volatile fatty acids from activated sludge at pH 10. *Water Res.* 2008;42(18):4637–44.

66. Cabrera F, Serrano A, Torres A, Rodriguez-Gutierrez G, Jeison D, Fermoso FG. The accumulation of volatile fatty acids and phenols through a pH-controlled fermentation of olive mill solid waste. *Sci Total Environ.* 2019;657:1501–7.

67. Zhao J, Wang D, Liu Y, Hao H, Guo W, Yang Q. Novel stepwise pH control strategy to improve short chain fatty acid production from sludge anaerobic fermentation. *Bioresour Technol.* 2018;249(October 2017):431–8.

68. Hao J, Wang H. Volatile fatty acids productions by mesophilic and thermophilic sludge fermentation: Biological responses to fermentation temperature. *Bioresour Technol.* 2015;175:367–73.

69. Atasoy M, Owusu-Agyeman I, Plaza E, Cetecioglu Z. Bio-based volatile fatty acid production and recovery from waste streams: Current status and future challenges. *Bioresour Technol.* 2018;268:773–86.

70. Zhou M, Yan B, Wong JWC, Zhang Y. Enhanced volatile fatty acids production from anaerobic fermentation of food waste: A mini-review focusing on acidogenic metabolic pathways. *Bioresour Technol.* 2018;248:68–78.

71. Jiang J, Zhang Y, Li K, Wang Q, Gong C, Li M. Volatile fatty acids production from food waste: Effects of pH, temperature, and organic loading rate. *Bioresour Technol.* 2013;143:525–30.

72. Strazzera G, Battista F, Garcia NH, Frison N, Bolzonella D. Volatile fatty acids production from food wastes for biorefinery platforms: A review. *J Environ Manage.* 2018;226:278–88.

73. Yu HQ, Fang HHP. Acidogenesis of dairy wastewater at various pH levels. *Water Sci Technol.* 2002;45(10):201–6.

74. Ersahin ME, Gimenez JB, Ozgun H, Tao Y, Spanjers H, Van Lier JB. Gas-lift anaerobic dynamic membrane bioreactors for high strength synthetic wastewater treatment: Effect of biogas sparging velocity and HRT on treatment performance. *Chem Eng J.* 2016;305:46–53.

75. Cavinato C, Da Ros C, Pavan P, Bolzonella D. Influence of temperature and hydraulic retention on the production of volatile fatty acids during anaerobic fermentation of cow manure and maize silage. Bioresour Technol. 2017;223:59–64.

76. Lim S. Anaerobic organic acid production of food waste in once-a-day feeding and drawing-off bioreactor. *Bioresour Technol.* 2008;99(16):7866–74.

77. Bolaji IO, Dionisi D. Acidogenic fermentation of vegetable and salad waste for chemicals production: Effect of pH buffer and retention time. *J Environ Chem Eng.* 2017;5(6):5933–43.

78. Morgan-Sagastume F, Valentino F, Hjort M, Cirne D, Karabegovic L, Gerardin F, Johansson P, Karlsson A, Magnusson P, Alexandersson T, Bengtsson S, Majone M, Werker A. Polyhydroxyalkanoate (PHA) production from sludge and municipal waste-water treatment. *Water Sci Technol.* 2014;69(1):177–84.

79. Fu Z, Holtzapple MT. Consolidated bioprocessing of sugarcane bagasse and chicken manure to ammonium carboxylates by a mixed culture of marine microorganisms. *Bioresour Technol.* 2010;101(8):2825–36.

80. Blasig JD, Holtzapple MT, Dale BE, Engler CR, Byers FM. Volatile fatty acid fermentation of AFEX-treated bagasse and newspaper by rumen microorganisms. *Resour Conserv Recycl.* 1992;7:95–114.

81. Queirós D, Sousa R, Pereira SR, Serafim LS. Valorization of a pulp industry by-product through the production of short-chain organic acids. *Fermentation.* 2017;3(20):1–11.

82. Wang K, Yin J, Shen D, Li N. Anaerobic digestion of food waste for volatile fatty acids (VFAs) production with different types of inoculum: Effect of pH. *Bioresour Technol.* 2014;161:395–401.

83. Reis MAM, Albuquerque MGE, Villano M, Majone M. Mixed culture processes for polyhydroxyalkanoate production from agro-industrial surplus/wastes as feedstocks. In: Mittal V, editor. *Renewable Polymers: Synthesis, Processing and Technology,* Wiley, Hoboken, 2011. p. 670–83.

84. Cabrera F, Torres A, Campos JL, Jeison D. Effect of operational conditions on the behaviour and associated costs of mixed microbial cultures for PHA production. *Polymers (Basel).* 2019;11(2):191.

85. Ramasahayam SK, Guzman L, Gunawan G, Viswanathan T. A Comprehensive review of phosphorus removal technologies and processes. *J Macromol Sci.* 2014;51(6): 538–45.

86. Mino T. Microbial selection of polyphosphate-accumulating bacteria in activated sludge wastewater treatment processes for enhanced biological phosphate removal. *Biochem.* 2000;65(3):341–8.

87. Rhu DH, Lee WH, Kim JY, Choi E. Polyhydroxyalkanoate (PHA) production from waste. *Water Sci Technol.* 2003;48(8):221–8.

88. Kourmentza C, Plácido J, Venetsaneas N, Burniol-Figols A, Varrone C, Gavala HN, Reis MAM. Recent advances and challenges towards sustainable polyhydroxyalkanoate (PHA) production. *Bioengineering.* 2017;4(2):55.

89. Majone M, Dircks K, Beun. Aerobic storage under dynamic conditions in activated sludge processes. the state of the art. *Water Sci Technol.* 1999;39(1):61–73.

90. Beun JJ, Dircks K, Van Loosdrecht MCM, Heijnen JJ. Poly-b-hydroxybutyrate metabolism in dynamically fed mixed microbial cultures. *Water Res.* 2002;36(5):1167–80.

91. Chua ASM, Takabatake H, Satoh H, Mino T. Production of polyhydroxyalkanoates (PHA) by activated sludge treating municipal wastewater: Effect of pH, sludge retention time (SRT), and acetate concentration in influent. *Water Res.* 2003;37(15):3602–11.

92. Chen Z, Huang L, Wen Q, Zhang H, Guo Z. Effects of sludge retention time, carbon and initial biomass concentrations on selection process: From activated sludge to poly-hydroxyalkanoate accumulating cultures. *J Environ Sci.* 2017;52:76–84.

93. Lemos PC, Levantesi C, Serafim LS, Rossetti S, Reis MAM. Microbial characterisation of polyhydroxyalkanoates storing populations selected under different operating conditions using a cell-sorting RT-PCR approach. *Appl Microbiol Biotechnol.* 2008;78:351–60.

94. Van Loosdrecht MCM, Pot MA, Heijnen JJ. Importance of bacterial storage polymers in bioprocesses. *Water Sci Technol.* 1997;35(1):41–7.

95. Dionisi D, Majone M, Vallini G, Di Gregorio S, Beccari M. Effect of the applied organic load rate on biodegradable polymer production by mixed microbial cultures in a sequencing batch reactor. *Biotechnol Bioeng.* 2006;93(1):76–88.

96. Albuquerque MGE, Torres CAV, Reis MAM. Polyhydroxyalkanoate (PHA) production by a mixed microbial culture using sugar molasses: Effect of the influent substrate concentration on culture selection. *Water Res.* 2010;44(11):3419–33.

97. Oliveira H, Yasmine M, Niz K. Effects of the organic loading rate on polyhydroxyalkanoate production from sugarcane stillage by mixed microbial cultures. *Appl Biochem Biotechnol.* 2019;189(4):1039–1055.

98. Johnson K, Kleerebezem R, Van Loosdrecht MCM. Influence of the C/N ratio on the performance of polyhydroxybutyrate (PHB) producing sequencing batch reactors at short SRTs. *Water Res.* 2010;44(7):2141–52.

99. Oliveira CSS, Silva CE, Carvalho G, Reis MAM. Strategies for efficiently selecting PHA producing mixed microbial cultures using complex feedstocks: Feast and famine regime and uncoupled carbon and nitrogen availabilities. *N Biotechnol.* 2017;37:69–79.

100. Burniol-Figols A, Varrone C, Le Balzer S, Daugaard AE, Skiadas IV, Gavala HN. Combined polyhydroxyalkanoates (PHA) and 1,3-propanediol production from crude glycerol: Selective conversion of volatile fatty acids into PHA by mixed microbial consortia. *Water Res.* 2018;136:180–91.

101. Silva F, Campanari S, Matteo S, Valentino F, Majone M, Villano M. Impact of nitrogen feeding regulation on polyhydroxyalkanoates production by mixed microbial cultures. *N Biotechnol.* 2017;37:90–8.

102. Johnson K, Van Geest J, Kleerebezem R, Van Loosdrecht MCM. Short- and long-term temperature effects on aerobic polyhydroxybutyrate producing mixed cultures. *Water Res.* 2010;44(6):1689–700.

103. Jiang Y, Marang L, Kleerebezem R, Muyzer G, Van Loosdrecht MCM. Effect of temperature and cycle length on microbial competition in PHB-producing sequencing batch reactor. *ISME J.* 2011;5(5):896–907.

104. Grazia G, Quadri L, Majone M, Morgan-Sagastume F. Journal of Environmental Chemical Engineering Influence of temperature on mixed microbial culture polyhydroxyalkanoate production while treating a starch industry wastewater. *J Environ Chem Eng.* 2017;5(5):5067–75.

105. Inoue D, Suzuki Y, Sawada K, Sei K. Polyhydroxyalkanoate accumulation ability and associated microbial community in activated sludge-derived acetate-fed microbial cultures enriched under different temperature and pH conditions. *J Biosci Bioeng.* 2018;125(3):339–45.

106. Montiel-Jarillo G, Carrera J, Suárez-Ojeda ME. Science of the total environment enrichment of a mixed microbial culture for polyhydroxyalkanoates production: Effect of pH and N and P concentrations. *Sci Total Environ.* 2017;583:300–7.

107. Villano M, Beccari M, Dionisi D, Lampis S, Miccheli A, Vallini G, Majone M. Effect of pH on the production of bacterial polyhydroxyalkanoates by mixed cultures enriched under periodic feeding. *Process Biochem.* 2010;45(5):714–23.

108. Dias JML, Lemos PC, Serafim LS, Oliveira C, Eiroa M, Albuquerque MGE, Ramos AM, Oliveira R, Reis MAM. Recent advances in polyhydroxyalkanoate production by mixed aerobic cultures: From the substrate to the final product. *Macromol Biosci.* 2006;6(11):885–906.

109. Chen Z, Huang L, Wen Q, Guo Z. Efficient polyhydroxyalkanoate (PHA) accumulation by a new continuous feeding mode in three-stage mixed microbial culture (MMC) PHA production process. *J Biotechnol.* 2015;209:68–75.

110. Chinwetkitvanich S, Randall CW, Panswad T. Effects of phosphorus limitation and temperature on PHA production in activated sludge. *Water Sci Technol.* 2004;50(8):135–43.

111. Pittmann T, Steinmetz H. Potential for polyhydroxyalkanoate production on German or European municipal waste water treatment plants. *Bioresour Technol.* 2016;214:9–15.

112. Valentino F, Karabegovic L, Majone M, Morgan-Sagastume F, Werker A. Polyhydroxyalkanoate (PHA) storage within a mixed-culture biomass with simultaneous growth as a function of accumulation substrate nitrogen and phosphorus levels. *Water Res.* 2015;77:49–63.

113. Moralejo-Gárate H, Kleerebezem R, Mosquera-Corral A, Van Loosdrecht MCM. Impact of oxygen limitation on glycerol-based biopolymer production by bacterial enrichments. *Water Res.* 2013;47(3):1209–17.

114. Pratt S, Werker A, Lant P. Microaerophilic conditions support elevated mixed culture polyhydroxyalkanoate (PHA) yields, but result in decreased PHA production rates. *Water Sci Technol.* 2012;65(2):243–6.

115. Satoh H, Iwamoto Y, Mino T, Matsuo T. Activated sludge as a possible source of biodegradable plastic. *Water Sci Technol.* 1998;38(2):103–9.

116. Huang L, Chen Z, Wen Q, Lee D. Enhanced polyhydroxyalkanoate production by mixed microbial culture with extended cultivation strategy. *Bioresour Technol.* 2018;241:802–11.

117. Fradinho JC, Reis MAM, Oehmen A. Beyond feast and famine: Selecting a PHA accumulating photosynthetic mixed culture in a permanent feast regime. *Water Res.* 2016;105:421–8.

118. Fradinho JC, Oehmen A, Reis M. Photosynthetic mixed culture polyhydroxyalkanoate (PHA) production from individual and mixed volatile fatty acids (VFAs): Substrate preferences and co-substrate uptake. *J Biotechnol.* 2014;185:19–27.

119. Fradinho JC, Oehmen A, Reis MAM. Effect of dark/light periods on the polyhydroxyalkanoate production of a photosynthetic mixed culture. *Bioresour Technol.* 2013;148:474–9.

120. Cui Y, Zhang H, Lu P, Peng Y. Effects of carbon sources on the enrichment of halophilic mixed microbial culture in an aerobic dynamic feeding process. *Sci Rep.* 2016;6: 30766.

121. Palmeiro-Sánchez T, Oliveira CSS, Gouveia AR, Noronha JP, Ramos AM, Mosquera-Corral A, Reis MAM. NaCl presence and purification affect the properties of mixed culture PHAs. *Eur Polym J.* 2016;85:256–65.

122. Kamm B, Gruber PR, Kamm M. *Biorefineries - Industrial Processes and Products: Status quo and Future Directions*, WILEY-VCH Verlag GmbH & Co. KGaA, Weinheim, 2006, p. 949.

123. Tsang YF, Kumar V, Samadar P, Yang Y, Lee J, Song H, Kim K, Kwon EE, Jae Y. Production of bioplastic through food waste valorization. *Environ Int.* 2019;127(January):625–44.

124. Nielsen C, Rahman A, Rehman AU, Walsh MK, Miller CD. Food waste conversion to microbial polyhydroxyalkanoates. *Microb Biotechnol.* 2017;10(6):1338–52.

125. Moita R, Ortigueira J, Freches A, Pelica J, Goncalves M, Mendes B, Lemos PC. Bio-oil upgrading strategies to improve PHA production from selected aerobic mixed cultures. *N Biotechnol.* 2013;31(4):297–307.

126. Duque AF, Oliveira CSS, Carmo ITD, Gouveia AR, Pardelha F, Ramos AM, Reis MAM. Response of a three-stage process for PHA production by mixed microbial cultures to feedstock shift: Impact on polymer composition. *N Biotechnol.* 2014;31(4):276–88.

127. Bengtsson S, Pisco AR, Reis MAM, Lemos PC. Production of polyhydroxyalkanoates from fermented sugar cane molasses by a mixed culture enriched in glycogen accumulating organisms. *J Biotechnol.* 2010;145(3):253–63.

128. Amulya K, Jukuri S, Mohan SV. Sustainable multistage process for enhanced productivity of bioplastics from waste remediation through aerobic dynamic feeding strategy: Process integration for up-scaling. *Bioresour Technol.* 2015;188:231–9.

129. Chae SR, Shin HS. Effect of condensate of food waste (CFW) on nutrient removal and behaviours of intercellular materials in a vertical submerged membrane bioreactor (VSMBR). *Bioresour Technol.* 2007;98:373–9.
130. Wen Q, Ji Y, Hao Y, Huang L, Chen Z, Sposob M. Effect of sodium chloride on polyhydroxyalkanoate production from food waste fermentation leachate under different organic loading rate. *Bioresour Technol.* 2018;267:133–40.
131. Ben M, Kennes C, Veiga MC. Optimization of polyhydroxyalkanoate storage using mixed cultures and brewery wastewater. *J Chem Technol Biotechnol.* 2016;91(11):2817–26.
132. Khardenavis AA, Kumar MS, Mudliar SN, Chakrabarti T. Biotechnological conversion of agro-industrial wastewaters into biodegradable plastic, poly b-hydroxybutyrate. *Bioresour Technol.* 2007;98:3579–84.
133. Waller JL, Green PG, Loge FJ. Mixed-culture polyhydroxyalkanoate production from olive oil mill pomace. *Bioresour Technol.* 2012;120:285–9.
134. Campanari S, Augelletti F, Rossetti S, Sciubba F, Villano M. Enhancing a multi-stage process for olive oil mill wastewater valorization towards polyhydroxyalkanoates and biogas production. *Chem Eng J.* 2017;317:280–9.
135. Beccari M, Bertin L, Dionisi D, Fava F, Lampis S, Majone M, Valentino F, Vallini G, Villano M. Exploiting olive oil mill effluents as a renewable resource for production of biodegradable polymers through a combined anaerobic-aerobic process. *J Chem Technol Biotechnol.* 2009;84(6):901–8.
136. Liu H, Hall PV, Darby JL, Coats ER, Green PG, Thompson DE, Loge FJ. Production of polyhydroxyalkanoate during treatment of tomato cannery wastewater. *Water Environ Res.* 2008;80(4):367–72.
137. Anterrieu S, Quadri L, Geurkink B, Dinkla I, Bengtsson S, Arcos-Hernandez M, Alexandersson T, Morgan-Sagastume F, Karlsson A, Hjort M, Karabegovic L, Magnusson P, Johansson P, Christensson M, Werker A. Integration of biopolymer production with process water treatment at a sugar factory. *N Biotechnol.* 2013;31(4):308–23.
138. Mengmeng C, Hong C, Qingliang Z, Ngai S, Jie R. Optimal production of polyhydroxyalkanoates (PHA) in activated sludge fed by volatile fatty acids (VFAs) generated from alkaline excess sludge fermentation. *Bioresour Technol.* 2009;100(3):1399–405.
139. Morgan-Sagastume F, Hjort M, Cirne D, Gérardin F, Lacroix S, Gaval G, Karabegovic L, Alexandersson T, Johansson P, Karlsson A, Bengtsson S, Arcos-Hernández MV, Magnusson P, Werker A. Integrated production of polyhydroxyalkanoates (PHAs) with municipal wastewater and sludge treatment at pilot scale. *Bioresour Technol.* 2015;181:78–89.
140. Wijeyekoon S, Carere CR, West M, Nath S, Gapes D. Mixed culture polyhydroxyalkanoate (PHA) synthesis from nutrient rich wet oxidation liquors. *Water Res.* 2018;140:1–11.
141. Chen Y, Li M, Meng F, Yang W, Chen L, Huo M. Optimal poly (3-hydroxybutyrate/3-hydroxyvalerate) biosynthesis by fermentation liquid from primary and waste activated sludge. *Environ Technol.* 2014;35(14):1791–801.
142. Hao J, Wang H, Wang X. Selecting optimal feast-to-famine ratio for a new polyhydroxyalkanoate (PHA) production system fed by valerate-dominant sludge hydrolysate. *Biotechnol Prod Process Eng.* 2018;102:3133–43.
143. Basset N, Katsou E, Frison N, Malamis S, Dosta J, Fatone F. Integrating the selection of PHA storing biomass and nitrogen removal via nitrite in the main wastewater treatment line. *Bioresour Technol.* 2016;200:820–9.
144. Valentino F, Gottardo M, Micolucci F, Pavan P, Bolzonella D, Rossetti S, Majone M. Organic fraction of municipal solid waste recovery by conversion into added-value polyhydroxyalkanoates and biogas. *ACS Sustain Chem Eng.* 2018;6:16375–85.

145. Pozo G, Villamar AC, Martınez M, Vidal G. Polyhydroxyalkanoates (PHA) biosynthesis from kraft mill wastewaters: Biomass origin and C:N relationship influence. *Water Sci Technol.* 2011;63(3):449–56.

146. Bengtsson S, Werker A, Welander T. Production of polyhydroxyalkanoates by glycogen accumulating organisms treating a paper mill wastewater. *Water Sci Technol.* 2008;58(2):323–30.

147. Bengtsson S, Werker A, Christensson M, Welander T. Production of polyhydroxyalkanoates by activated sludge treating a paper mill wastewater. *Bioresour Technol.* 2008;99(3):509–16.

148. Dai J, Gliniewicz K, Settles ML, Coats ER, Mcdonald AG. Influence of organic loading rate and solid retention time on polyhydroxybutyrate production from hybrid poplar hydrolysates using mixed microbial cultures. *Bioresour Technol.* 2015;175:23–33.

149. Ben M, Mato T, Lopez A, Vila M, Kennes C, Veiga MC. Bioplastic production using wood mill effluents as feedstock. *Water Sci Technol.* 2011;63:1196–202.

150. Mato T, Ben M, Kennes C, Veiga MC. Valuable product production from wood mill effluents. *Water Sci Technol.* 2010;62(10):2294–300.

151. Farghaly A, Enitan AM, Kumari S, Bux F, Tawfik A. Polyhydroxyalkanoates production from fermented paperboard mill wastewater using acetate-enriched bacteria. *Clean Technol Environ Policy.* 2017;19:935–947.

152. Fauzi AHM, Chua ASM, Yoon WL, Nittami T, Yeoh HK. Enrichment of PHA-accumulators for sustainable PHA production from crude glycerol. *Process Saf Environ Prot.* 2019;122:200–8.

153. Coats ER, Watson BS, Brinkman CK. Polyhydroxyalkanoate synthesis by mixed microbial consortia cultured on fermented dairy manure: Effect of aeration on process rates/yields and the associated microbial ecology. *Water Res.* 2016;106:26–40.

154. Lemos PC, Dai Y, Yuan Z, Keller J, Santos H, Reis MAM. Elucidation of metabolic pathways in glycogen-accumulating organisms with *in vivo* nuclear magnetic resonance. *Environ Microbiol.* 2007;9(11):2694–706.

155. Tan G-Y, Chen C-L, Li L, Ge L, Wang L, Razaad I, Li Y, Zhao L, Mo Y, Wang J-Y. Start a research on biopolymer polyhydroxyalkanoate (PHA): A review. *Polymers (Basel).* 2014;6(3):706–54.

156. Sabirova J. Polyhydroxyalkanoates produced by hydrocarbon-degrading bacteria. In: Timis KN, Ed. *Handbook of Hydrocarbon and Lipid Microbiology*, Springer, Berlin Heidelberg, 2010. p. 2981–94.

157. Ishii-hyakutake M, Mizuno S, Tsuge T. Biosynthesis and characteristics of aromatic polyhydroxyalkanoates. *Polymers (Basel).* 2018;10:1–24.

158. European Commission. Towards a circular economy: A zero waste programme for Europe, 2014.

159. Vea EB, Romeo D, Thomsen M. Biowaste valorisation in a future circular bioeconomy. *Procedia CIRP.* 2018;69(May):591–6.

160. Romaní A, Michelin M, Domingues L, Teixeira JA. Valorization of wastes from agrofood and pulp and paper industries within the biorefinery concept: Southwestern Europe scenario. In: Bhaskar T, Pandey A, Venkata Mohan S, Lee D-J, Khanal S, Eds. *Waste Biorefinery*, Elsevier, Amsterdam, 2018. p. 487–504.

161. Calvo MV, Colombo B, Corno L, Eisele G, Cosentino C, Papa G, Scaglia B, Pilu R, Simmons B, Adani F. Bioconversion of giant cane for integrated production of biohydrogen, carboxylic acids, and polyhydroxyalkanoates (PHAs) in a multistage biorefinery approach. *ACS Sustain Chem Eng.* 2018;6(11):15361–73.

162. Moretto G, Russo I, Bolzonella D, Pavan P, Majone M, Valentino F. An urban biorefinery for food waste and biological sludge conversion into polyhydroxyalkanoates and biogas. *Water Res.* 2020;170:115371.

163. Koller M, Niebelschütz H, Braunegg G. Strategies for recovery and purification of poly[(R)-3-hydroxyalkanoates] (PHA) biopolyesters from surrounding biomass. *Eng Life Sci.* 2013;13(6):549–62.
164. Samorì C, Abbondanzi F, Galletti P, Giorgini L, Mazzocchetti L, Torri C, Tagliavini E. Extraction of polyhydroxyalkanoates from mixed microbial cultures: Impact on polymer quality and recovery. *Bioresour Technol.* 2015;189:195–202.
165. Valentino F, Morgan-Sagastume F, Campanari S, Villano M, Werker A, Majone M. Carbon recovery from wastewater through bioconversion into biodegradable polymers. *N Biotechnol.* 2017;37:9–23.
166. Laycock B, Arcos-Hernandez MV, Langford A, Buchanan J, Halley PJ, Werker A, Lant PA, Pratt S. Thermal Properties and Crystallization Behavior of Fractionated Blocky and Random Polyhydroxyalkanoate Copolymers from Mixed Microbial Cultures. *J Appl Polym Sci.* 2014;131:40836.
167. Hilliou L, Machado D, Oliveira CSS, Gouveia AR, Reis MAM, Campanari S, Villano M, Majone M. Impact of fermentation residues on the thermal, structural, and rheological properties of polyhydroxy(butyrate-*co*-valerate) produced from cheese whey and olive oil mill wastewater. *J Appl Polym Sci.* 2016;133(2):42818.

8 PHA Production by Mixed Microbial Cultures and Organic Waste of Urban Origin
Pilot Scale Evidence

Francesco Valentino, Marianna Villano,
Laura Lorini and Mauro Majone

CONTENTS

8.1 INTRODUCTION

The ever-increasing global population living in urban areas raises serious challenges not only for the provision of energy and materials, but also for waste management [1] and this is a particularly relevant problem since cities will always be a place of waste production. It has been estimated that waste generation in the urban areas will be approximately 1.42 kg/person/day by 2025 [2]. Several types of waste are generated in urban areas, including sewage sludge (SS) and municipal solid waste (MSW), as well as food-industry wastewater. SS is the excess sludge deriving from municipal wastewater treatment plants (WWTP) that generate two types of sludge, known as primary and secondary (or activated) sludge. The latter mainly contains microbial cells and suspended solids while the primary sludge consists of solid materials and floating grease. Therefore, SS collected from WWTP is a heterogeneous mixture of different undigested organic materials, microorganisms (also pathogens), oils, fats, inorganic materials, and moisture that needs to be disposed of [3]. Sludge treatment represents a significant fraction (approximately 50%) of the total cost required for WWTP operation. Besides that, sludge management is responsible for 40% of the total greenhouse gas emission deriving from the plants [4]. The main approaches used for SS treatment are incineration, landfill and agricultural usage but, due to the establishment of new regulations to prevent environmental hazards, these techniques are being limited in favor of the development of alternative and sustainable disposal methods [5].

As for MSW, this represents waste produced by households as well as commercial and other waste with similar composition to household waste. Actually, the MSW composition differs considerably from one municipality to another and from country to country and this variation strongly depends on socio-economic factors along with the degree of industrialization and the local climate. In general, the main component of MSW is the organic fraction (commonly more than 45% on weight basis), followed by paper, plastics, glass, metal and others. Its enormous heterogeneity makes MSW a complex waste whose management presents important challenges to the environment. Traditional management and treatment techniques consist of disposal in landfills and incineration, but both of them cause several environmental issues. In the case of landfill, the main issues are related to odor production, leachate formation, and generation of methane and carbon dioxide which increase greenhouse gas emissions and, in addition, in numerous countries MSW is disposed of in landfills without any sorting [6]. The high water content of MSW is, instead, responsible for the high energy demand of incineration [7, 8]. This is an age-old technology based on a thermal process of waste combustion for volume reduction and release of heat energy, but it is not advantageous from an energetic and economic point of view as well as in terms of environmental sustainability since it entails air pollution. The organic fraction of municipal solid waste (OFMSW) deserves particular attention. It mainly consists of food and kitchen waste and leftovers from households, restaurants, caterers and markets [2], but also incorporates green waste. The first components are generally referred to as food waste (FW) and are a serious problem in terms of waste management. Indeed, according to the UN Food and Agriculture Organization (FAO), approximately 1.3 billion tonnes of food is wasted each year, accounting for one-third of all food produced globally for human consumption [9]. Green waste primarily consists of leaves, wood cuttings from pruning, and grass collected from gardens and parks. Even though a relevant lignocellulosic fraction characterizes green waste, its composition has a wide seasonal variation (e.g. less grass and leaves are present in winter) and differs for each geographical location [10].

Management of MSW is, therefore, a demanding undertaking and issues related to public health and environmental protection promote the elaboration of stringent disposal regulations. The first rational action is to reduce waste formation combined with waste valorization through recycling and reuse or recovery of useful materials. As an example, the most commonly used technologies for the treatment and valorization of OFMSW are the composting and anaerobic digestion (AD) [11]. Composting is a natural biochemical process for the decomposition of organic matter to a stabilized and nutrient-rich end product. It involves millions of indigenous microorganisms that work under aerobic conditions, and sufficient oxygen is required for the establishment of a successful composting process [12]. The major advantages of this process are the volume reduction of mass waste and the utilization of the final product (compost) as soil fertilizer, which allows a reduction in the use of synthetic fertilizers for agricultural purposes. In general, small amounts of organic waste can be easily composted, but large amounts require mechanical aeration [13]. On the contrary, AD is an anaerobic bioprocess whereby microorganisms break down the organic matter contained in the waste for the generation of a biogas, under controlled temperature conditions and pH. Biogas mainly contains methane and carbon dioxide and it is a low-cost source of energy which can be used for electricity, heat and biofuel production. In addition to biogas, a nutrient rich digestate is also produced during AD, which can be used as soil fertilizer [8]. Anaerobic digestion is also an efficient and widely established technology for the treatment of sewage sludge. Besides that, a valorization approach that is attracting considerable and increasing attention is the conversion of municipal waste into a spectrum of biobased products ranging from biofuels to commodity chemicals and biopolymers (such as polyhydroxyalkanoates, PHA) [14]. The consideration of waste as a renewable source of valuable products is strictly associated with the concept of resource efficiency and circular bioeconomy. This is a sustainable developing model in which all materials used in the design of a biobased product are recovered, recycled and upgraded in the desired compound and the end of the life of the product is considered to promote maximum reuse of all its components as raw materials. The model of circular bioeconomy is opposed to the current fossil-based linear economy centered on four main steps: feedstock extraction–production–consumption–waste disposal [15, 16]. The transition from linear economy to circular bioeconomy is already underway, and it is becoming one of the most important themes of interest in both the scientific and political areas. A strategy for a rapid development of circular bioeconomy is the implementation of well-designed biorefineries, which allow closing loops of raw biomass materials. In this context, the urban biorefinery concept is particularly attractive to ensure an efficient and integrated conversion of different wastes of urban origin into multiple products [17]. This could allow to overcome drawbacks deriving from the fact that, even though originating from the same urban area, SS and MSW are typically handled separately.

The urban integrated biorefinery approach fits well with the generation of products of higher value than energy and compost, such as PHA.

8.2 MMC-PHA PRODUCTION IN THE URBAN BIOREFINERY MODEL

Polyhydroxyalkanoates (PHA) are biologically synthesized polyesters which serve as internal reserves of carbon and energy [18, 19]. Besides that, they are also biobased and biodegradable and are considered three-time biopolymers. Indeed, PHA can be obtained from a large range of renewable feedstock (including food waste, molasses, agro-industrial wastewater and activated sludge), especially when dealing with microbial mixed cultures (MMC), and are completely biodegradable in water and carbon dioxide (under aerobic conditions) or methane (during anaerobic digestion). Another important characteristic of PHA is the fact that they are not a single polymer but a whole family of copolymers, which feature a wide array of tunable physical and mechanical properties depending on the length and composition of the side chains. This makes PHA a valid and sustainable alternative to the most commonly used fossil-based plastics, such as poly(propylene) and poly(ethylene). For these reasons, PHA are among the main drivers of the growth of bioplastics, whose market is rapidly increasing [20]. The production of MMC-PHA involves multi-stage processes which typically include the acidogenic fermentation (AF) of the renewable feedstock, the microbial selection (commonly operated in sequencing batch reactors) and the biopolymer accumulation (in batch reactors) stages. A good performance of each stage is required to ensure the optimal performance of the overall process. In particular, AF is a key step to transform a waste into substrates that are suitable for PHA production, and the composition of the mixture of acids deriving from the fermentation process depends on the feedstock nature and primarily affects the final polymer composition and, in turn, its properties. As a consequence, PHA can be used for a broad portfolio of market applications able to face different customers' requests. These include packaging, heat sensitive adhesives, disposable utensils, agricultural films and bulk commodity plastics, as well as medical applications such as drug delivery or environmental applications consisting of the use of PHA as slow carbon release compounds in the field of advanced in situ bioremediation [21]. Based on these considerations, PHA are suitable products for the development of the urban biorefinery because they can easily adapt to the large heterogeneity of the urban waste feedstock and, also, their production technology can be simply integrated into existing biological plants for waste and wastewater treatment (such as anaerobic digestion). This also considering that the PHA production process does not generate excess sludge that needs to be handled, as the polymer can make up to about 70% of the weight of microbial cells. Another significant advantage offered by this biorefinery is a substantial reduction in PHA production cost due to the use of MMC and urban wastes as renewable no-cost substrate. Since the cost of the feedstock represents approximately half of the overall production cost [22], this is expected to enhance a rapid increase in the PHA market in the near future.

With reference to the above mentioned circular bioeconomy concept, the PHA-producing biorefinery makes it possible to close the loop through the transformation of urban biowaste streams into useful biobased plastics and the recovery of energy and nutrients from all residual side streams of the production technology that can be directed to anaerobic digestion, producing biogas and digestate, or towards compost generation for agronomic purposes [23, 24]. Also, this is a fully sustainable model from an environmental point of view because the PHA produced, replacing fossil-based plastics, contributes to reducing greenhouse gas emissions.

A deep analysis of the performance of MMC-PHA production from waste of urban origin (particularly sewage sludge, the organic fraction of municipal solid waste and food-industry wastewater) is reported in the following paragraph, with main reference to pilot-scale studies.

8.3 PILOT SCALE STUDIES FOR URBAN WASTE CONVERSION INTO PHA

It has been discussed how polyhydroxyalkanoates (PHA) can be produced as a novel adjunct to biological wastewater treatment in mixed-microbial activated-sludge systems. In a more common approach, process technologies have been developed and investigated for the production of: (a) volatile fatty acids (VFAs) by acidogenic fermentation, (b) selected/enriched biomass with PHA-storage capacity from biological wastewater treatment, and (c) PHA from the further biological treatment of different waste and residual organic streams.

Feast and famine environments have been created under alternating aerobic conditions [25, 26], even though anoxic [27–30] and anaerobic [31] conditions have been successfully explored making this technology technically feasible for integrating the ideas of biomass selection and production to a wide range of existing infrastructures for wastewater treatment, including even municipal wastewater treatment [18].

Experience over 30 years from laboratory-scale and more recent pilot-scale studies has demonstrated the feasibility of treating different waste/wastewater with the additional benefit of producing PHA. Overall PHA yields of 10–20% (g PHA/g chemical oxygen demand removed) and biomass PHA contents of 40–70% (g PHA/g volatile suspended solids) have been consistently achieved. However, despite such attention for many years, PHA is still not making a significant impact in the global market. Generally speaking, commercial uptake has been restricted by uncertainty of supply, performance and cost. Steps are underway to establish a life cycle analysis and business models of viable value-added chains. In particular, the latter is necessary for commercial demonstration of mixed-culture PHA production as an integral part of residue management. Environmental performance assessments of mixed-culture PHA process technologies are being critically considered. Establishing a stakeholder network in which residual carbon and microbial biomass sourcing is combined with essential services of pollution control, while facilitating PHA-product-market combinations is identified as the next pivotal step for achieving the first viable examples of PHA production within a framework of residual management in a bio-based economy.

Some studies [18, 32] reviewed the challenges of scaling up PHA production from waste streams and conducted a deep analysis of the potentiality of the MMC technology, while quantifying the exponentially growing interest. Currently, as it was reported in these past reviews, there is no reference to any progress at industrial scale, even though pilot-scale examples have been considerably grown and the PHA production process has been differently designed and adapted to a various waste feedstock.

8.3.1 MUNICIPAL WASTEWATER SLUDGE

In the case of municipal wastewater treatment, the considerable amounts of primary and surplus waste activated sludge (WAS) represent a burden for treatment and disposal. However, this burden can also become an opportunity to be used as renewable resources for their further valorization and conversion into added value products. In this frame, the utilization of WWTP waste sludge as a VFA-rich feedstock for PHA synthesis has been recently applied at pilot scale, as a strategy for a new concept of WWTP design. The strategy of integrating MMC-PHA production with municipal wastewater and sludge management entail benefits of producing a functionalized biomass (with PHA-storage ability rather than waste biomass to be disposed of). The VFA sourcing from waste sludge (after fermentation step) for PHA accumulation may maximize the organic carbon utilization for PHA production.

The idea was developed at pilot scale, evaluating the potential of producing a biomass with significant PHA accumulation capacity from municipal wastewater treatment within the full context of extant daily and seasonal influent variations. The VFA-rich source was produced from the fermentation of WWTP waste sludge and integrated with the previously described MMC-PHA production from low-strength municipal wastewater [25, 33] as depicted in Figure 8.1. The use of primary sludge depended on the availability of the primary treatment of each particular WWTP facility; the secondary sludge (WAS) was considered also an option based on the flexibility of the proposed technology. The following results suggested a potential for using fermented sludge as a feedstock for PHA production with biomass produced from municipal wastewater treatment.

Pilot-scale testing comprised three units (0.5–1.0 m³) located at the Brussels North WWTP (Aquiris, Belgium): (a) sludge acidogenic fermenter, (b) aerobic sequencing batch reactor (SBR) for biomass production process and (c) aerobic polymer accumulation process. The whole system was designed based on the production of 1 kg PHA/d. The pilot units operated with actual sludge and municipal wastewater under regular process variations typical of a full-scale WWTP.

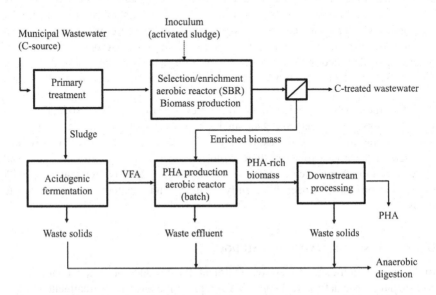

FIGURE 8.1 Schematic process flow diagram of municipal wastewater and sludge treatment in conjunction with PHA production.

The fermentation unit includes an anaerobic batch reactor and a centrifugation unit (a cylindro-conical bowl-scroll centrifuge connected to a conditioning/delivery unit of cationic polyacrylamide as coagulant/flocculant) for solid/liquid separation. Before feeding into the aerobic SBR, a wastewater delivery line with a drum filter (130–600 μm) for solid/fiber removal was adopted. The SBR treating the filtered wastewater operated under a fully aerobic feast-famine regime (organic loading rate (OLR), 3.0 g; chemical oxygen demand (COD)/L/d; short hydraulic retention time (HRT), of 3.0 h; sludge retention time (SRT), 1–2 d; 12 cycles per day). Similar operating conditions were adopted in a more controlled laboratory-scale study conducted in parallel [26]. The filtered wastewater feeding the SBR was mainly characterized by 200–700 mg COD/L, 30–270 mg COD_{SOL}/L, 18–65 mg COD_{VFA}/L, 19–76 mg N/L and 2–9 mg P/L. After accumulation, the PHA-rich biomass was conveyed to the thickening units (settler and diffused air floatation), dewatering centrifuge and an oven drying unit.

The sludge acidogenic fermentation (pH 5.5–6.5) was accomplished for both mesophilic (35 and 42°C) and thermophilic (55°C) conditions, even though fermentation at 42°C was preferred due to the higher VFA yield (0.27–0.33 g COD_{VFA}/g VS) and VFA conversions of solubilized COD (0.85–0.90 g COD_{VFA}/g COD_{SOL}). Both mesophilic and thermophilic fermentation products were mainly dominated by acetic (28–38%, COD basis), butyric (15–26%) and propionic (13–23%) acid. This VFA spectrum was primarily regulated by the sludge characteristics, such as carbohydrates, fat and protein compositions, rather than by the fermenter's operating conditions.

Performances were also robust in the feast-famine SBR, treating the readily biodegradable COD (RBCOD) of the municipal wastewater, despite the daily and seasonal variations in wastewater characteristics, temperature and the poor content of VFA (conventionally used for the MMC-PHA production technology in both aerobic steps). Feast-to-aerobic-cycle length ratios below 0.20 suggested a good selective pressure for PHA-storing organisms. Conditions for a stronger selection was achieved by discharging the more slowly biodegradable soluble COD with the effluent immediately after the feast period and establishing a more stringent famine as described in [33]. However, the biomass-producing unit process requires further COD and nutrient removal (the anammox-based N removal post-treatment has been suggested to be included). The accumulation capacity of the selected consortium was quantified up to 0.39 g PHA/g VSS (volatile suspended solids). Even though the nutrient levels were considered a challenge for the maximization of the PHA biomass content, the production of biomass with 0.50 g PHA/g VSS was considered to be realistically achievable within the typically available carbon flows at municipal waste management facilities. Even nutrients in excess were associated with increased productivities due to concurrent PHA storage and active biomass growth [34]. Due to the presence of 3-C unit VFAs, fermented sludge liquors yield PHA copolymers of 3HB and 3HV at 26–34 wt.-% 3HV, with weight average molar masses likely around 500 kDa, and high thermal stability (T_d = 291°C).

In order to enhance the accumulation potential of the biomass selected with unfermented municipal wastewater, a series of aerobic feast-famine acclimation cycles applied prior to PHA accumulation was evaluated as a strategy in the particular context of sludge/municipal wastewater valorization [35]. The biomass enriched during the treatment of municipal wastewater was exposed to aerobic feast-famine acclimation cycles with fermented waste-sludge liquor. Biomass acclimation led to more than doubling of the specific VFA uptake and PHA storage rates, and specific PHA productivities during the accumulation stage. The biomass PHA content increased from 0.39 to 0.46 g PHA/g VSS. A similar bacterial community structure during acclimation indicated that a physiological rather than a genotypic adaptation occurred in the biomass.

The necessity of a post-treatment for nutrient removal was taken into account in a parallel study [36], where a novel process treating sludge reject water that integrates the side-stream biological nitrogen removal (via nitrite) with the selection of MMC-PHA storing biomass was examined in a small pilot-scale reactor (30 L). The approach was based on the MMC selection in an SBR by the alternation of aerobic feast periods for ammonia conversion to nitrite followed by anoxic famine for denitritation driven by internally stored PHA as carbon source. The integration of PHA production with conventional nitrification-denitrification process was previously investigated for the treatment of sugar beet factory waters [37], by operating an anoxic feast and aerobic famine phase. However, in the sludge treatment line of municipal WWTP, the nutrient loads are typically higher, and side-stream treatment technology able to integrate the conversion of fermented sludge to PHA with nutrient removal via nitrite is much more attractive. The technology allowed the reduction of the nutrient loads to the main line of the WWTPs, lowering operational costs, with the recovery of PHA as a valuable resource, giving further economic advantages to the WWTP. The schematic overview of the process is depicted in Figure 8.2.

The biomass was able to accumulate up to 0.19–0.21 g PHA/g VSS, by using both sewage sludge or primary sludge fermentation liquids (COD/N/P = 100:7.8–9.7:0.1–2.1), with observed yields around 0.40 COD_{PHA}/COD_{VFA}. The presence of nutrients lowered the COD driven for PHA production, as confirmed in parallel accumulations performed with synthetic VFA mixture (0.44 g PHA/g VSS; COD/N/P = 100:0:0). Limiting further the presence of nutrients during the accumulation has been proposed as a necessary

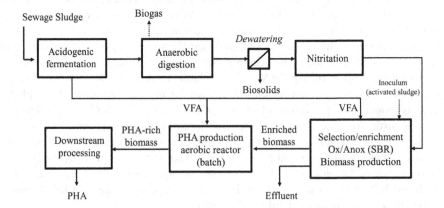

FIGURE 8.2 Process scheme for the combined selection of MMC-PHA storing biomass and nitrogen removal from sludge reject water by applying the aerobic/anoxic feast/famine regime and nitrogen removal via nitrite.

step for process optimization purposes. In terms of PHA composition and properties, the 3HB represents the major part of monomers (56–60 wt.-%), while percentages of 3HV were close to 40%. A smaller fraction of 2HH (less than 5%) was also quantified. The PHA produced was composed of long molecular chains; the high molecular weight (650–740 kDA) and the narrow chain length distribution (Đ 1.22–1.35) supported the idea of thermoplastic applications. Also, the low crystallinity in combination with a low T_g (from –1.1 to –0.5°C) indicate biopolymers with amorphous characteristics.

8.3.2 THE ORGANIC FRACTION OF MUNICIPAL SOLID WASTE (OFMSW) AND FOOD INDUSTRY WASTEWATER

Although many types of wastewater can be used for the production of PHA by the MMC technology, high concentrations of fermentable COD, relatively low nutrient levels (especially nitrogen) and solid concentrations, as well as low toxicity, promotes the technical process feasibility. In this frame, food industry effluents may be considered the most suitable substrates for waste-based PHA production. Leachate from the composting industry and the organic fraction of municipal solid waste (OFMSW) are also interesting streams for their valorization trough PHA synthesis, but additional challenges should be considered for their relatively high nitrogen content and the presence of solids.

Despite these positive inputs, pilot and demonstration plant information is not widely spread, considering that the interest in particular for food waste conversion into PHA is relatively recent.

The first pilot-scale example for the generation of PHA from milk and ice cream processing wastewater was designed in a continuous mode reactor system. The pilot-scale study was conducted at a dairy industry in Nagpur (Maharashtra state in India), operating at industrial site conditions with raw wastewater [38]. The integrated platform consisted of (1) a three-step acidogenic fermentation reactor, (2) a conventional activated sludge production reactor and (3) a PHA synthesis reactor, for the accumulation within the biomass (up to 0.43 g PHA/g VSS). The final biomass was harvested while treating the raw dairy wastewater to meet the disposal limits and simultaneously reducing the generation of sludge to be disposed of. The fermentation reactor operated optimally at pH 6.0, HRT 3.3 d, food to microorganism ratio (F/M) of 0.18. The VFA production approximately counted for 33.8% of the loaded COD. The activated sludge production reactor operated under stable conditions at pH 8.2 ± 0.2, F/M ratio of 0.87, very low average HRT of 0.7 d, SRT of 3.0 d and VSS concentration of 1.8 g/l (average). The HRT was the most important factor for the PHA synthesis from the two described reactors. The optimal HRT was established at 1.8 d. The SRT in this reactor was considered similar to HRT for reducing the retention of PHA-rich sludge in the system. This was necessary to avoid the consumption of accumulated PHA for biomass growth.

In a subsequent work, the enrichment of a PHA-producing MMC on industrial wastewater in a pilot installation set-up was discussed [39]. Experiments were conducted with wastewater from a candy bar factory (Mars, Veghel, the Netherlands). The pilot platform was conventionally designed as a three-step process comprising anaerobic fermentation (in two steps), enrichment and accumulation reactors. The Mars factory wastewater was pre-treated in a flotation-based fat separation unit before entering into the pilot system. Two fermentation reactors operated in series: an upflow sludge blanket (USB) type reactor (V = 60 L; pH 4.5, HRT 4 h, SRT around 4 days), and a second reactor (V = 1.5 m³, pH 4.5, HRT 4 d, T 40°C) to maximize the conversion of the fermentable COD to VFA. To keep the reactor effluent nitrogen depleted (favorable for the accumulation reactor operation only) the target COD/N mass ratio was set at 300:1, by adding nitrogen source in the form of urea. The aerobic SBR enrichment reactor (V = 0.2 m³) was operated with a cycle length of 12 h and HRT of 1.0 d. The reactor T was maintained at 30°C. The aerobic fed-batch accumulation reactor (V = 0.2 m³) was operated at the same T of SBR; the substrate was dosed differently from one batch to another depending on the carbon source availability and the amount of PHA that could be potentially accumulated. The selected microbial community was able to accumulate 0.70 g PHA/g VSS and it was dominated by the species *Plasticicumulans acidivorans*, the same organism as enriched in laboratory investigation under similar regimes [40]. The yield over the whole process (including anaerobic pre-treatment, enrichment and accumulation steps) was estimated equal to 0.30 g COD_{PHA}/gCOD. A significant part of the influent COD (0.11 g active $biomass_{COD}$/gCOD) was used for biomass production in the enrichment step. As suggested by the authors, overall yield and related process performances may be further optimized by directing the fermentation to produce more VFA (at the expense of ethanol), since the VFA influent fraction was only around 64% of the soluble COD. In terms of economic viability, the use of chemicals for pH control has been preliminarily discussed for the anaerobic fermentation and enrichment reactor, as well as the dosage of allylthiourea (ATU) to prevent the conversion of ammonium to nitrate. As previously discussed [25], the authors underlined the necessity of having clean effluent water in accordance with local legislation. To accomplish this obligation, the process required a reduction of residual biodegradable COD, by means of an aerated post-treatment unit operation.

More recently, an integrated multistage pilot-scale process in which the OFMSW was used as a valued source for PHA and biogas production has been described [23]. The OFMSW came from the source-sorted collection inside the Treviso municipality (northeast Italy). The OFMSW was mechanically pre-treated (for plastic and inert material removal) and the squeezed fraction was used for experimental purposes. The efficient separate collection ensured a high volatile organic content (90% of the total solids), meaning that the fraction of putrescible and fermentable material was high enough to support the first fermentation process. Technical and economical feasibilities of the multi-step approach have been demonstrated, providing a possible upgrade to traditional biowaste management (currently based on AD). A pH-controlled OFMSW fermentation stage produced a liquid VFA-rich stream having a COD_{VFA}/COD_{SOL} ratio of 0.90; the pH fermentation value (above 5.0) was controlled by the buffering digestate recirculation from the AD stage. The volume of recirculated digestate was automatically dosed based on the on-time measured pH, preventing unavoidable pH decreases due to VFA production and ensuring a stable fermentation activity (average VFA around 16.0 g COD_{VFA}/L). After a solid/liquid separation step (a coaxial centrifuge equipped with a filter bag 5.0 μm porosity), the liquid part was used in the following aerobic stages for biomass and PHA production. The solid fraction, instead, was valorized into biogas through AD (specific biogas production, SGP = 0.71 m³/kg VS; biogas composition 65 vol.-% CH_4 and 35 vol.-% CO_2), obtaining energy and minimizing secondary flux waste generation.

The anaerobic line consisted of a 0.2 m^3 continuous stirred tank reactor (CSTR) thermophilic fermenter (HRT 3.3 d, OLR 20.5 kg VS/m^{3}d, T 55°C) and a 0.76 m^3 CSTR thermophilic digester (HRT 12.7 d, OLR 3.9 kg VS/(m^3·d), T 55°C). The aerobic PHA line consisted of a 0.14 m^3 SBR (HRT = SRT 1.0 d, cycle length 6 h, OLR 2.5–3.0 g COD$_{SOL}$/(L d)) and 0.1 m^3 fed-batch reactor (OLR 10–14 g COD$_{SOL}$/(L d)).

Reliable biomass enrichment was demonstrated by a stable feast-famine regime, with a feast/cycle length ratio often below 0.1 h/h. The selected consortium accumulated PHA up to 55 wt.-%. Compared to the traditional AD process, this multi-step approach for OFMSW valorization and management demonstrated for the first time the technical and economic possibility to recover both electrical energy and added-value PHA. An overall yield of 3.7% kg PHA/kg VS (untreated organic solids) was also estimated.

8.3.3 THE INTEGRATION OF MUNICIPAL WASTEWATER TREATMENT WITH OFMSW AND FOOD INDUSTRY WASTE VALORIZATION

Waste production, processing and disposal are increasing challenges for urban areas. In this context, local biorefineries can use waste from surrounding industries and municipalities in a symbiotic manner. Today, there are very few examples of facilities that can convert biological sludge and the OFMSW (or other food industry waste/wastewater) into anything other than compost and energy. The pilot-scale examples are still few and of recent development.

A pilot-scale demonstration process (0.5–0.8 m^3) was designed for the biological treatment of municipal wastewater for carbon and nitrogen removal while producing PHA at Leeuwarden WWTP in Friesland, the Netherlands [41]. The process comprised steps for pre-denitrification, nitrification and post-denitrification and included integrated fixed film activated sludge (IFAS) with biofilm carrier media to support nitrification. The wastewater treatment performance achieved, in line with European standards, was quantified for total chemical oxygen demand (83% removal) and total nitrogen (80% removal) while producing a biomass that was able to accumulate up to 0.49 g PHA/g VSS with fermented residues coming from the local greenhouse tomato production industry, as feedstock. Robust performance in wastewater treatment and enrichment of PHA-producing biomass was demonstrated under realistic conditions including influent variability during 225 days of operation. The IFAS system was found to be advantageous since maintaining nitrification on the biofilm allowed for a relatively low SRT (2 days) for the suspended biomass in the bulk phase. Lower SRT has advantages in higher active fraction in the biomass which leads to higher PHA productivity and content. The process proposed by the authors is depicted in Figure 8.3. The possibility to use wastewater for biomass enrichment means that infrastructure and operations that are anyway necessary for wastewater treatment is utilized to produce a PHA-storing biomass. Hence, there is no need for a parallel treatment line with dedicated reactors for the production of such biomass. With this scheme, the feedstock with higher VFA levels can be fully conveyed to PHA synthesis giving larger overall potential PHA production (which is something already demonstrated in previous works, with different process schemes and goals [26, 33]). The outcomes of this demonstration showed that PHA production can be readily integrated with carbon and nitrogen removal from municipal wastewater.

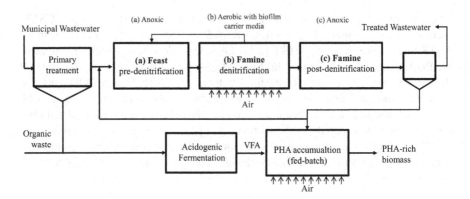

FIGURE 8.3 Process for PHA production from municipal wastewater treatment with biological carbon and nitrogen removal in continuous mode of operation.

Within the concept of sludge/food waste management and municipal wastewater treatment, the work described in [42] was centered on processing surplus activated sludge biomass from the Bath full-scale municipal wastewater treatment plant in the Netherlands to produce PHA (PHARIO project: "Stepping stone to a sustainable value chain for PHA bioplastic using municipal activated sludge". http://phario.eu/). Full-scale surplus activated sludge was fed to a pilot facility to produce PHA-rich biomass using fermented volatile fatty acid (VFA) rich liquors from candy factory wastewater or primary sludge sources. The PHA production involved in the most-applied three-step process: a batch pilot scale fermenter (1.2 m³) operated on site at Rioolwaterzuiveringsinstallatie (RWZI) Bath; a full-scale WWTP process for active biomass production (RWZI Bath) and a fed-batch pilot-scale (0.5 m³) reactor for PHA accumulation. The candy factory wastewater was incubated at 37°C for up to seven days with a controlled pH (5.5–6.0). Suspended solids from the fermented stream were removed by means of pre-settling with added flocculent polymer followed by a 10 μm filtration step. Primary sludge was fermented for six days at 37°C. The sludge matrix was self-buffering and the pH was maintained between 4.8–5.5. The fermenter batch was first discharged to a holding tank and then pumped to a centrifuge decanter screw at 10 rpm. Cationic polymer was added in-line for coagulation/flocculation to improve the solids/liquid separation. Prior to each accumulation, the biomass was subjected to an acclimation phase [35]. This acclimation comprised a sequence of three feast-famine cycles. The feast in each cycle was generated by stimulating the biomass close to maximal respiration with a substrate pulse at 0.2 g COD/L. A famine phase was set to be four times longer than the feast period. After the last acclimation famine period, the accumulation process was automatically started and maintained 16–20 h in duration. Fifty-two batches out of 59 were successful. On average, RWZI Bath activated sludge produced PHA-rich biomass at 0.41 g PHA/g VSS (meaning that 1 kg of Bath sludge can be expected to reliably accumulate 0.7 kg PHA). The authors underlined the ways to further improve the PHA accumulation potential for a municipal WWT activated sludge, with feasible expectation of up to 0.60 g PHA/g VSS. The extracted polymer had an average weight molecular mass more variable than expected, but nominally in the order of

500 kDa. The observed variability was a consequence of a secondary downstream effect, and mostly related to the chosen temperature of the drying process. Within PHARIO activities, it has been found that full-scale municipal WWTP can serve as process units that supply functionalized PHA-producing biomass (activated sludge) even without modifications to the WWTP. The life cycle assessment (LCA) estimated that PHARIO PHA will have a 70% lower environmental footprint compared to current PHA plastics due to the use of the third-generation feedstock.

The possibility of integrating MMC-PHA production into infrastructures typically adopted for biowaste residuals and wastewater treatment has been also proposed by [43], in particular in the context of urban environment and urban organic waste recovery. Although the OFMSW and sewage sludge (SS) originate from the same urban area and contain similar organic matter, they are collected separately and handled with different technologies. For this reason, the authors proposed a combined OFMSW-SS treatment (at pilot scale), by using the PHA-MMC technology. The pilot-scale platform was previously described [23] and it was operated with the same feedstock mixture usually conveyed to the parallel full-scale AD process. The source-sorted collection of OFMSW was made throughout the whole Treviso municipality and transferred to the full-scale WWTP after its squeezing and homogenization. The squeezed OFMSW was then mixed with SS produced in the WWTP OFMSW and SS at a volumetric fraction of 30–45% and 55–70% for OFMSW and SS, respectively. The fermentation process was successfully conducted under mesophilic (37-42°C) and thermophilic (55°C) condition with the same SRT (six days, equal to HRT). Despite the lowest average VFA concentration (19 g COD/L), the lowest applied temperature gave the highest COD_{VFA}/COD_{SOL} of 0.75, a crucial parameter for an efficient selection/enrichment of a PHA-accumulating consortium. This condition was chosen as the most suitable for the production of a VFA-rich stream amenable for the aerobic PHA line. The SBR and the fed-batch accumulation reactors were used as previously described [23]. The overall process PHA yield was estimated at 6.5 wt-% with respect to the volatile solids (VS) of the initial untreated waste stream. The biogas production from the solid-rich overflow is still possible and it is under evaluation. Preliminary analysis demonstrated that the introduction of a designed PHA line required 60% of the influent organic solids for the fermentation step and the production of the liquid VFA-rich stream. The residual VS amount (overflows) was represented by the sum of the 40% unfermented solids and the solid-rich pre-fermented fraction ("cake"), discharged from the solid/liquid separation unit. In ad-hoc designed full-scale platform (approximately 70,000 Person Equivalent, PE) where the overflow AD treatment is included, biogas can be also produced. A revenue of 134,591 €/year has been estimated from the sale of generated electricity (13.03 MWh/day). In theory, taking into account an average PHA value of 4.0 €/kg [44], an additional revenue of 180,000 €/year has to be considered, potentially making the biorefinery platform a more profitable solution if compared to the traditional AD process in the Treviso full-scale plant, where a maximum (but not always achievable) income of 245,000 €/year from the electrical energy produced was estimated [45].

Table 8.1 summarizes the main results from all the pilot-scale experiences with MMC-PHA production technologies. Previous laboratory-scale studies which

TABLE 8.1

Summary of Main Results Obtained in Fed-Batch Accumulations and Comparison with Previous Studies Focused on Food Waste Valorization

VFA-rich stream	Scale	PHA content [g PHA/g VSS]	Polymer composition [% 3HB/3HV, g/g]	$Y_{P/S}^{SBR}$ [COD/COD]	q^{Pbatch} [mg P/(gX$_a$/h)] COD basin	PHA productivity [g PHA/(L·h)]	Overall PHA yield [g PHA/kg VS]	Ref.
Milk and ice cream wastewater	Pilot	0.39–0.43		0.25				[38]
Municipal wastewater sludge	Pilot	0.27–0.34	70–75/25–30	0.18				[33]
Mars candy factory wastewater	Pilot	0.70 ± 0.05	84/16	0.37		0.5	0.3**	[39]
Municipal wastewater sludge	Pilot	0.27–0.38	66–74/26–34			0.1–0.14		[25]
Municipal wastewater sludge	Pilot	0.19–0.21	57/41 + 2% (HH)	0.3–0.4			0.11**	[36]
Municipal wastewater sludge	Pilot	0.19–0.46	75/25	0.34	23–91			[35]
Municipal sludge and tomato industry wastewater	Pilot	0.45–0.49	51–58/42–49	0.30–0.39	35–28			[41]
Pre-treated OFMSW	Pilot	0.39–0.52	7–13/93–87	0.36	255–436	0.28–0.49	37	[23]
Municipal sludge and candy factory wastewater	Pilot	0.41	58–100/0–42	0.4–0.42				[42]

TABLE 8.1 (CONTINUED)

Summary of Main Results Obtained in Fed-Batch Accumulations and Comparison with Previous Studies Focused on Food Waste Valorization

VFA-rich stream	Scale	PHA content [g PHA/g VSS]	Polymer composition [% 3HB/3HV, g/g]	$Y_{P/S}^{SBR}$ [COD/COD]	q^{Pbatch} [mg P/(gX$_a$/h)] COD basin	PHA productivity [g PHA/(L·h)]	Overall PHA yield [g PHA/kg VS]	Ref.
OFMSW and municipal sludge	Pilot	0.43–0.46	87–90/10–13	0.22–0.36	279–301	0.29–0.36	67	[44]
Leachate	Laboratory	0.78^{b,c}						[46]
Percolate*	Laboratory and pilot	0.41–0.48	54–57/46–43	0.44–0.52		0.18–0.29	63–114	[47]
OFMSW*	Laboratory	0.11		0.08	20			[48]
OFMSW-PS^d mixture*		0.09		0.003	11			
Food waste (FW)*	Laboratory	0.24	60/40	0.09–0.17	11–42			[49]

b biomass enriched with synthetic VFA mixture (90 vol-%) and leachate (10 vol-%); c biomass enriched with synthetic VFA mixture (75 vol-%) and leachate (25 vol-%); d primary sludge; *fermented feedstock; **g COD$_{PHA}$/g COD$_{(0)}$

already presented the idea of MMC technology integration with waste/wastewater treatment facilities or upgrade of existing Anaerobic Digestion (AD) plants are also reported.

8.4 CONCLUSIONS AND OUTLOOK

The inclusion of MMC-PCHA production in the urban biorefinery model is an optimal option for the valorization of municipal waste. The sustainable management of the huge amount of waste deriving from urban areas, along with the generation of a renewable product suitable for different market applications, are the principal factors that make this approach attractive and in rapid development. This also considering that the biopolymer produced is not a recycled material, but a virgin product newly synthesized by selected mixed microorganisms. The microbial origin coupled to the biobased nature and biodegradability property allow consideration of PHA as a three-time biopolymer, widening the interest in its production. There are two main benefits related to PHA production from urban waste, consisting in the fact that processes for PHA production can easily adapt to the large heterogeneity and variable composition that characterize such waste and can also be simply integrated into existing biological plants for waste and wastewater treatment, with great advantage from a practical point of view. These potentialities have driven the establishment of pilot-scale plants, essential to verify the feasibility of the process. Data collected from available pilot-scale studies are encouraging both in terms of urban waste conversion into PHA and polymer content in the microbial cells, which attains values of 50% (on a weight basis). The latter is a crucial parameter for the economical evaluation of the process since it significantly impacts the polymer extraction cost. Along this line, the search for efficient and sustainable extraction procedures to be applied for PHA recovery from mixed cultures is one of the principal objectives to be developed in future pilot-scale researches. Also, a deep analysis of the effect of impurities deriving from waste feedstock on the physical and mechanical properties of the final polymer as well as the study of the variability of its composition are required to define which are the appropriate market destinations of the produced PHA.

Overall, the information offered by pilot-scale studies is essential in order to establish, in the near future, a massive production of MMC-PHA from urban waste at industrial level.

ACKNOWLEDGMENT

This work was financially supported by the RES URBIS (RESources from URban BIo-waSte) Project (GA No 730349) in the frame of the European Horizon 2020 (Call CIRC-05-2016) program.

REFERENCES

1. Satchatippavarn S, Martinez-Hernandez E, Pah Hang MY, *et al.* Urban biorefinery for waste processing. *Chem Eng Res Des* 2016; 107: 81–90.

2. Maina S, Kachrimanidou V, Koutinas A. A roadmap towards a circular and sustainable bioeconomy through waste valorization. *Curr Opin Green Sustain Chem* 2017; 8: 18–23.
3. Bora AP, Gupta DP, Durbha KS. Sewage sludge to bio-fuel: A review on the sustainable approach of transforming sewage waste to alternative fuel. *Fuel* 2020; 259: 116262.
4. Gherghel A, Teodosiu C, De Gisi S. A review on wastewater sludge valorisation and its challenges in the context of circular economy. *J Clean Prod* 2019; 228: 244–263.
5. Cieslik BM, Namiesnik J, Konieczka P. Review of sewage sludge management: standards, regulations and analytical methods. *J Clean Prod* 2015; 90: 1–15.
6. Vaverková MD. Landfill impacts on the environment—review. *Geosciences* 2019; 9: 431.
7. Nakatsuka N, Kishita Y, Kurafuchi T, *et al.* Integrating wastewater treatment and incineration plants for energy- efficient urban biomass utilization: A life cycle analysis. *J Clean Prod* 2020; 243: 118448.
8. Nayak A, Bhushan B. An overview of the recent trends on the waste valorization techniques for food wastes. *J Environ Manage* 2019; 233: 352–370.
9. Morone P, Koutinas A, Gathergood N, *et al.* Food waste: Challenges and opportunities for enhancing the emerging bio-economy. *J Clean Prod* 2019; 221: 10–16.
10. Inghels D, Dullaert W, Bloemhof J. A model for improving sustainable green waste recovery. *Resour Conserv Recy* 2010; 110: 61–73.
11. Yingqun Ma Y Liu Y. Turning food waste to energy and resources towards a great environmental and economic sustainability: An innovative integrated biological approach. *Biotechnol Adv* 2019; 37: 107414.
12. Dhamodharan K, Varma VS, Veluchamy C, *et al.* Emission of volatile organic compounds from composting: A review on assessment, treatment and perspectives. *Sci Total Environ* 2019; 695: 1133725.
13. Abdel-Shafy HI, Mansour MSM. Solid waste issue: Sources, composition, disposal, recycling, and valorization. *Egypt J Petrol* 2018; 27: 1275–1290.
14. Liguori R, Amore A, Faraco V. Waste valorization by biotechnological conversion into added value products. *Appl Microbiol Biotechnol* 2013; 97: 6129–6147.
15. Karan H, Funk C, Grabert M, *et al.* Green bioplastics as part of a circular bioeconomy. *Trends Plant Sci* 2019; 24: 237–249.
16. Jacquet N, Haubruge E, Richel A. Production of biofuels and biomolecules in the framework of circular economy: A regional case study. *Waste Manage Res* 2015; 33: 1121–1126.
17. Strazzera G, Battista F, Herrero Garcia N, *et al.* Volatile fatty acids production from food wastes for biorefinery platforms: a review. *J Environ Manage* 2018; 226: 278–288.
18. Valentino F, Morgan-Sagastume F, Campanari S, *et al.* Carbon recovery from wastewater through bioconversion into biodegradable polymers. *New Biotechnol* 2017; 37: 9–23.
19. Reis MAM, Albuquerque M, Villano M, *et al.* Mixed culture processes for polyhydroxyalkanoate production from agro-industrial surplus/wastes as feedstocks. In: Moo-Young M, Ed., *Comprehensive Biotechnology*, 2nd ed., vol. 6. Amsterdam: Elsevier, 2011. p. 669–683.
20. Thakur S, Chaudhary J, Sharma B, *et al.* Sustainability of bioplastics: Opportunities and challenges. *Curr Opin Green Sustain Chem* 2018; 13: 68–75.
21. Pierro L, Matturro B, Rossetti S, *et al.* Polyhydroxyalkanoate as a slow-release carbon source for in situ bioremediation of contaminated aquifers: From laboratory investigation to pilot-scale testing in the field. *New Biotechnol* 2017; 37: 60–68.
22. Tsang YF, Vanish Kumar V, Samadar P, *et al.* Production of bioplastic through food waste valorization. *Environ Int* 2019; 127: 625–664.

23. Valentino F, Gottardo M, Micolucci F, *et al.* Organic fraction of municipal solid waste recovery by conversion into added-value polyhydroxyalkanoates and biogas. *ACS Sustainable Chem Eng* 2018; 6: 16375–16385.

24. Valentino F, Moretto G, Gottardo M, *et al.* Novel routes for urban bio-waste management: A combined acidic fermentation and anaerobic digestion process for platform chemicals and biogas production. *J Cleaner Prod* 2019; 220: 68–375.

25. Morgan-Sagastume F, Hjort M, Cirne D, *et al.* Integrated production of polyhydroxy-alkanoates (PHAs) with municipal wastewater and sludge treatment at pilot scale. *Bioresource Technol* 2015; 181: 78–89.

26. Valentino F, Morgan-Sagastume F, Fraraccio S, *et al.* Sludge minimization in municipal wastewater treatment by polyhydroxyalkanoate (PHA) production. *Environ Sci Pollut Res* 2015; 22: 7281–7294.

27. Li Z, Wang S, Zhang W, *et al.* Nitrogen removal from medium- age landfill leachate via post-denitrification driven by PHAs and glycogen in a single sequencing batch reactor. *Biores Technol* 2014; 169: 773–777.

28. Chen H, Yang Q, Li X, *et al.* Post-anoxic denitrification via nitrite driven by PHB in feast-famine sequencing batch reactor. *Chemosphere* 2013; 92: 1349–1355.

29. Dionisi D, Renzi V, Majone M, *et al.* Storage of substrate mixtures by activated sludges under dynamic conditions in anoxic or aerobic environments. *Wat Res* 2004; 38: 2196–2206.

30. Dionisi D, Majone M, Tandoi V, *et al.* Sequencing Batch Reactor: influence of periodic operation on performance of activated sludges in biological wastewater treatment. *Ind Eng Chem Res* 2001; 40: 5110–5119.

31. Bengtsson S. The utilization of glycogen accumulating organisms for mixed culture production of polyhydroxyalkanoates. *Biotechnol Bioeng* 2009; 104: 698–708.

32. Rodriguez-Perez S, Serrano A, Pantion AA, *et al.* Challenges of scaling up PHA production from waste stream. A review. *J Environ Manag* 2018; 205: 2015–2030.

33. Morgan-Sagastume F, Valentino F, Hjort M, *et al.* 2014. Polyhydroxyalkanoate (PHA) production from sludge and municipal wastewater treatment. *Water Sci Technol* 2014; 69: 177–184.

34. Valentino F, Karabegovic L, Majone M, *et al.* Polyhydroxyalkanoate (PHA) storage within a mixed-culture biomass with simultaneous growth as a function of accumulation substrates nitrogen and phosphorus levels. *Wat Res* 2015; 77: 49–63.

35. MorganSagastume F, Valentino F, Hjort M, *et al.* Acclimation process for enhancing polyhydroxyalkanoate accumulation in activated-sludge biomass. *Waste Biomass Valorisation* 2019; 10: 1065–1082.

36. Frison N, Katsou E, Malamis S, *et al.* Development of a novel process integrating the treatment of sludge reject water and the production of polyhydroxyalkanoates (PHAs). *Environ Sci Technol* 2015; 49: 10877–10885.

37. Anterrieu S, Quadri L, Geurkink B, *et al.* Integration of biopolymer production with process water treatment at a sugar factory. *New Biotechnol* 2014; 31: 308–323.

38. Chakravarty P, Mhaisalkar V, Chakrabarti T. Study on poly-hydroxyalkanoate (PHA) production in pilot scale continuous mode wastewater treatment system. *Bioresour Technol* 2010; 101: 2896–2899.

39. Tamis J, Luzkova K, Jiang Y, *et al.* Enrichment of Plasticicumulans acidivorans at pilot-scale for PHA production on industrial wastewater. *J Biotechnol* 2014; 192: 161–169.

40. Johnson K, Jiang Y, Kleerebezem R, *et al.* Enrichment of a mixed bacterial culture with a high polyhydroxyalkanoates storage capacity. *Biomacromol* 2009; 10: 670–676.

41. Bengtsson S, Karlsson A, Alexandersson T, *et al.* A process for polyhydroxyalkanoate (PHA) production from municipal wastewater treatment with biological carbon and nitrogen removal demonstrated at pilot-scale. *New Biotechnol* 2017; 35: 42–53.

42. Werker A, Bengtsson S, Korving L, *et al.* Consistent production of high quality PHA using activated sludge harvested from full scale municipal wastewater treatment – PHARIO. *Water Sci Technol* 2018; 78: 2256–2269.
43. Valentino F, Moretto G, Lorini L, *et al.* Pilot-scale polyhydroxyalkanoate production from combined treatment of organic fraction of municipal solid waste and sewage sludge. *Ind Eng Chem Res* 2019; 58: 12149–12158.
44. Valentino F, Moretto G, Gottardo M, *et al.* Novel routes for urban bio-waste management: A combined acidic fermentation and anaerobic digestion process for platform chemicals and biogas production. *J Clean Prod* 2019; 220: 368–375.
45. Moretto G, Ardolino F, Piasentin A, *et al.* Integrated anaerobic codigestion system for the organic fraction of municipal solid waste and sewage sludge treatment: an Italian case study. *J Chem TechnolBiot* 2019. doi:10.1002/jctb.5993.
46. Korkakaki E, Mulders M, Veeken A, *et al.* PHA production from the organic fraction of municipal solid waste (OFMSW): Overcoming the inhibitory matrix. *Wat Res* 2016; 96: 74–83.
47. Colombo B, Favini F, Scaglia B, *et al.* Enhanced polyhydroxyalkanoate (PHA) production from the organic fraction of municipal solid waste by using mixed microbial culture. *Biotechnol Biofuels* 2017; 10: 201.
48. Basset N, Katsou E, Frison N, *et al.* Integrating the selection of PHA storing biomass and nitrogen removal via nitrite in the main wastewater treatment line. *Bioresource Technol* 2016; 200: 820–829.
49. Amulya K, Jukuri S, Venkata Mohan S. Sustainable multistage process for enhanced productivity of bioplastics from waste remediation through aerobic dynamic feeding strategy: Process integration for up-scaling. *Bioresource Technol* 2015; 188: 231–239.

9 Production Quality Control of Mixed Culture Poly(3-Hydroxbutyrate-co-3-Hydroxyvalerate) Blends Using Full-Scale Municipal Activated Sludge and Non-Chlorinated Solvent Extraction

Alan Werker, Simon Bengtsson, Peter Johansson, Per Magnusson, Emma Gustafsson, Markus Hjort, Simon Anterrieu, Lamija Karabegovic, Tomas Alexandersson, Anton Karlsson, Fernando Morgan-Sagastume, Luc Sijstermans, Martin Tietema, Etteke Wypkema, Yede van der Kooij, Alexandra Deeke, Cora Uijterlinde and Leon Korving

CONTENTS

9.1 INTRODUCTION

In a recent pilot-scale project, PHARIO (in Dutch, PHA uit RIOolwater) [1, 2], a principal objective was to assess technical feasibility, economic viability and environmental performance for industrial-scale production of commercial quality polyhydroxyalkanoates (PHA). The PHARIO scenario entailed direct accumulation of PHA in surplus municipal activated sludge fed with volatile fatty acid rich feedstocks derived from regionally available, and acidogenic fermented, wastewater or other residual organic matter [3–6]. In the downstream, PHA was recovered by achieving a high thermal stability of the polymers in the biomass [7], followed by extraction with non-chlorinated solvents [8]. Studies of technical feasibility, economic viability and environmental performance all suggest that such mixed microbial culture (MMC) PHA production is motivated, at least in principle [3, 6, 9–13]. However, production can only be motivated in practice if a well-defined product can be predictably, repeatedly and reliably made. The extensive experimental data collected from the PHARIO investigation enabled deepened analyses and considerations of upscaled polymer product quality control that were undertaken for the purpose of this study. The purpose of this work was to critically assess potential for upscaled MMC polyhydroxyalkanoate (PHA) commercial production quality control.

The research literature specifically addressing anticipated scale-up challenges for MMC PHA production, recovery and product quality control is lacking [14–17]. Product quality goes hand in hand with perceptions of product value. Higher quality raw materials and products are deemed to have greater economic value. Waste activated sludge from municipal wastewater treatment is perceived as low-quality. It is a waste byproduct to be managed, reduced and disposed of [18]. Notwithstanding that waste activated sludge is, in the present day, abundant, there is an anticipated scepticism that using such a raw material in a bioprocess can contribute to the procurement of commercial grade platform chemicals. While the MMC research literature provides numerous examples that MMC PHA may produce relevant quantities and different qualities of PHA [14, 16, 17, 19–25], the underlying meaning of that quality is not generally discussed because the typically small quantities of material produced in this research have not been produced for any specific context of application. The quality and properties of the polymers reported are also not generally extrapolated to an industrial process that will be required to control quality with a maintained source of supply of raw materials.

We sought to couple the trends of measured polymer properties and production variability observed during PHARIO to the corresponding changes of process and feedstock, amidst the underlying and uncontrolled elements of microbiological variability that may be expected with a municipal activated sludge [26, 27]. The PHARIO pilot PHA production efforts ensued over four seasons of WWTP operations from which grab samples of full-scale surplus activated sludge were routinely sourced for PHA accumulations. From the experience of using a full-scale activated sludge over an extended campaign of pilot-scale production operations, we wished to generate an unvarnished perspective of the ability to engineer mixed culture PHA product quality in a bioprocess to produce a PHA-rich biomass, and to recover, in downstream chemical processing with non-chlorinated solvent extraction, a commercial grade polymer.

For this work, copolymer blends of poly(3-hydroxybutyrate-*co*-3-hydroxyvalerate), PHBV, were accumulated nominally twice per week and batch-wise over ten months at kilogram pilot-scale. The mixed microbial culture biomass was the full-scale municipal surplus activated sludge from the Bath wastewater treatment plant in the Netherlands. Fresh activated sludge batches were taken from the full-scale waste secondary sludge management line. These batches were fed aerobically with volatile fatty acid (VFA) rich substrates, based on just-in-time feed-on-demand methods [28], to the maximum extant PHBV biomass accumulation potential. The VFA-rich feedstocks were either well-defined mixtures of acetic and propionic acids, pilot-scale acidogenic fermented industrial wastewater or municipal primary sludge. The resultant batches of PHA-rich biomass were acidified, dewatered, oven dried, crushed to granulate and recovered by non-chlorinated solvent extraction at laboratory and pilot-scales using either 2-butanol or acetone. Fifty-nine production batches formed the basis from which to explore polymer recovery and potential for up-scaled production quality control. In total, 19 kilograms of polymer comprising varying grades of PHBV were produced. Systematic considerations of process and polymer properties were made and are presented herein towards generating insight for upscaled MMC PHA production processes with control of polymer type, molecular weight, and thermal and mechanical properties.

9.2 MATERIALS AND METHODS

9.2.1 BIOMASS AND FEEDSTOCK SOURCES

In the normal routine of production operations, two PHA accumulation batches were performed each week over ten months of pilot operations (July 2015 to March 2016). For each batch, a grab sample of fresh gravity belt thickened (50 to 70 gTS/kg) surplus full-scale municipal activated sludge was obtained. The activated sludge came from the Bath WWTP, The Netherlands (Waterboard Brabantse Delta), a large plant (470,000 PE) with ten parallel treatment trains, that receives wastewater via relatively long sewer pressure lines. The wastewater is pretreated by screening and primary clarifiers. The bioprocess configuration is with pre-denitrification and nitrification (Modified Ludzack-Ettinger, MLE, process). Solids retention time was typically around 17 days and on average 14.2 tons of dry solids (tDS) waste activated sludge was produced per day. Aeration basin temperatures were maximum in August (23°C), minimum in February (10°C), and 17 ± 4°C on average. Removal of P was mainly by chemical precipitation (0.42 mol-Me/mol-P_{in}). Influent wastewater quantity and quality over the period was 110.7 ± 62.1 ML/d, 518 ± 162 mg-sCOD/L (soluble chemical oxygen demand), 217 ± 73 mg-BOD_5/L (5-day biochemical oxygen demand), 222 ± 55 mg-TSS/L (total suspended solids), 46 ± 13 mg-TN/L (total nitrogen), and 8 ± 2 mg-TP/L (total phosphorus). The influent WWTP water quality may be considered to be typical for Europe [29].

The wastewater contains contributions from the industries in the area, such as the central sludge incineration facility Slibverwerking Noord-Brabant (SNB) that discharges an effluent containing a condensate that is typically rich in volatile fatty acids. The SNB effluent is first treated on site by steam strippers and Demon® technology operating in parallel. Organic loading contribution from SNB had been estimated to be not more than 2–3% of the total COD loading to the WWTP. There is anaerobic digestion on site of WWTP Bath and the reject streams from sludge thickening and dewatering are also returned to the main headworks.

The activated sludge grab sample was assessed directly for solids content and about one kilogram of thickened activated sludge (as dry volatile solids) was delivered to the pilot accumulation process along with dilution water. The accumulation pilot process comprised a working aeration volume of about 400 L coupled to a clarifier volume of 120 L [3]. The aerated volume was mechanically stirred, and air was supplied via a blower with coarse bubble aeration. Active pumping to and from the clarifier (20 L/min) maintained a short solids residence time in the clarifier. Depletion of dissolved oxygen for the biomass fraction recirculated through the clarifier resulted in periodic oxygen limiting kinetics for fractions of the biomass in the process. Oxygen limitation is known to prolong the accumulation processes [30]. Therefore, time for the fractions of biomass in the clarifier, at any given time, was considered as an accumulation dead time. Otherwise dissolved oxygen levels were targeted to be at least 1 mg O_2/L in the main reactor vessel.

PHA was accumulated in the biomass at a set operating temperature of 25°C by semi-continuous supply of a VFA-rich feedstock. VFA feedstocks used in the project were either a fermented carbohydrate-rich process effluent delivered from a local candy factory, centrate from fermented primary municipal sludge or defined mixtures of acetic and propionic acids.

Candy factory wastewater was delivered batch-wise once per week to Bath WWTP in 1 m³ transport containers. A pilot-scale fermenter was used to maximize the fermentation product content in a well-stirred volume of 1200 L. The candy factory process water was supplemented with dilution water to about 18 g/L soluble COD and incubated at 37°C for up to 7 days with pH between 5.5 and 6.0. A 3.5-day fermentation time was sufficient, but seven days were given conservatively and for simplicity with respect to the working logistics. The fermenter pH was maintained with a feedback controller and pulse-wise additions of concentrated NaOH (45 wt.-%). The suspended solids at the end of each batch (nominally about 1.2 gTSS/L) were settled and retained within the process at an SRT of between seven and ten days. Suspended solids in the fermentation effluent were removed by settling with added flocculant polymer (Flopam FO 4800 SH, SNF Floerger), followed by a Hydrotech pilot drum filter model 801 (HDF 801, with 10 μm screen). Suspended solids levels in and out of the HDF were on average 0.6 and 0.2 g/L, respectively. The fermented water contained on average a soluble organic content of 16 gCOD/L. Decrease in COD across the fermentation process was primarily due to water entrainment from the pilot HDF backwash. The feedstock nutrient balance COD:N:P was adjusted to 100:0.5:0.1 (by weight) by chemical additions of NH_4Cl and KH_2PO_4 as necessary. The feedstock total suspended solids content with respect to soluble organic matter was 0.011 ± 0.010 gTSS/gCOD.

Primary sludge was also delivered fresh to Bath WWTP from the De Dommel Tillburg facility batchwise in 1 m³ transport containers. Solids content was variable, but on average 45 gTSS/L. The delivered primary sludge was fermented in the same 1200 L fermenter as above with continuous mechanical stirring for six days at 37°C, with pH monitoring but with no pH control. The primary sludge matrix was self-buffering and the pH remained between 4.8 and 5.5. No solids were retained for the primary sludge fermentation between batches. The fermented sludge was first discharged to a holding tank and then under automation control pumped to a centrifuge decanter with rotating bowl (3000 rpm) and a relative velocity between bowl and screw of 10 rpm. Cationic polymer (Flopam FO4800SH, SNF/FLOERGER) was added in-line for coagulation/flocculation. The polymer addition was controlled by PLC to 50 g-polymer/kg-TSS. Centrate was collected in a holding tank and iron chloride was added from concentrated solution (44 wt.-%) to precipitate excess soluble phosphorus based on a molar dosing ratio of 1.42:1 $Fe:PO_4^{3-}$. The centrate was further processed through a pilot Hydrotech drum filter (HDF 801, with 18 μm screen) to produce feedstock batches for the pilot-scale accumulation. This feedstock soluble organic content was about 7 gCOD/L with a COD:N:P balance of 100:5:0.1 (by weight). The suspended solids content was 0.018 ± 0.022 gTSS/gCOD.

So-called "synthetic" feedstocks were generated by diluting and blending proportions of concentrated acetic and propionic acids in IBC containers to about 10 gCOD/L. A targeted COD:N:P (by weight) of 100:1:0.05 nutrient balance was obtained by chemical additions of NH_4Cl and KH_2PO_4. The pH-value was adjusted to 5 by addition of NaOH (33 or 45 wt.-%). The synthetic feedstock contained negligible if any suspended solids.

9.2.2 PHA-RICH BIOMASS PRODUCTION

Prior to each accumulation the process biomass was subject to a substrate acclimation phase [31, 32]. Acclimation comprised a sequence of three feast and famine cycles. The feast in each of these cycles was generated by stimulating the biomass to a feast respiration with a pulse of substrate of 200 mgCOD/L. The feast duration was measured based on changes in dissolved oxygen concentration. A period of famine was provided to last four times as long as the time of feast. The acclimation phase was designed as a controlled assessment of the biomass respiration response from batch to batch over the ten months of operations, and to provide all the biomass batches with the same perturbation history to a new feedstock before the onset of the accumulation.

After the acclimation, the accumulation process was started automatically. The 59 accumulations were conducted on average at 25.5 ± 1.1°C. They were sustained using semi-continuous feedstock supply based on just-in-time feed-on-demand based on biomass respiration control [28]. The control objectives were to sustain a prolonged period of near maximal feast with relatively small pulse wise inputs of substrate. The volume of each pulse targeted a maximum soluble COD concentration of 200 mgCOD/L. The accumulation process was typically maintained from 16 to 20 hours. Process flow rates and volumes, as well as dissolved oxygen and pH levels were monitored and logged.

In separate experiments, oxygen uptake rates were evaluated by dosing selected masses of 50 to 100 mgCOD/L of substrate (acetate, propionate and ethanol), individually and in combination, to known volumes and concentrations of activated sludge suspended solids in a well-mixed and aerated vessel. Oxygen transfer coefficients were estimated from mass balance considerations following a timed sequence of with, without and with aeration [33].

The accumulation process was terminated at a preselected time, with exceptions of abnormal events detected by the process automatic control. Due to the logistics of access to the site, accumulation termination occurred in the middle of the night. After accumulation termination, the aeration was turned off and the biomass was collected in the main process volume and allowed to settle by gravity. With return of operating personnel to the site in the morning, and now under manual control, 100 to 150 L of thickened mixed liquor containing PHA-rich biomass were pumped over to a 200 L holding tank and the mixed liquor pH was adjusted to 2 by titration with concentrated H_2SO_4 [7]. Following acidification, the solids were further thickened by floatation and the thickened solids were then dewatered by means of a filter bag centrifuge (at 980 × g with filter bags defined with 7 L/dm²/ min at 200 Pa) after adding dewatering chemicals (Flopam EM 840 TBD). The main batch was dewatered on average to about 19% dry solids. The sludge cake was transferred to drying trays and dried at 70°C for 24 hours. The dried biomass was granulated and sifted to produce a semi-product of defined particle size (0.71 to 2 mm in diameter). Over the course of the process operations including fermentation, accumulation and downstream processing, selected samples were obtained for basic water quality and solids analyses. From these analyses the process mass balances were evaluated and potential influences for the process operations on the recovered polymer product quality were examined. Mixed liquor grab samples taken during operations were dewatered to about 10% DS in 50 mL aliquots by centrifugation at laboratory-scale (4200 × g for 5 minutes) and then dried at 70°C (unless otherwise stated).

The PHA production thereby comprised 59 batches annotated as Ann with A01 to A59, accumulated with three different types of feedstocks, candy factory wastewater (C feedstock), primary sludge fermentate (P feedstock) and selected VFA mixtures (S feedstock). PHA production was in two main phases after the initial commissioning of the facility. The first phase focused on using C-feedstocks over six months. The second phase utilized P-feedstocks over four months. S-feedstocks were interspersed over the entire production campaign in order to monitor for any changes in the biomass performance over the ten months.

9.2.3 PHA RECOVERY BY SOLVENT EXTRACTION

A benchmark for quality of the polymer "in-the-biomass" was assessed by recovering the polymer following methods of extraction that were developed to cause negligible, if any, changes to the polymer in the process [34]. The benchmark extraction assessment of the polymer quality in the biomass was with acetone extraction at 125°C for one hour. Biomass, with a grain size of less than 0.71 mm was weighed (10 g/L polymer loading) into a test tube (KIMAX 12 mL) and 10 mL of acetone was added. During extraction, test tubes were vortex mixed every five minutes for the first 15 minutes and then every 15 minutes. The tubes were cooled for four minutes including one minute of centrifugation (3500 × g). The solvent was carefully decanted into a beaker where the polymer was precipitated by adding at least five times the solvent volume with deionized water while stirring with a magnetic stirrer. The supernatant was filtered under vacuum and the polymer solids were dried overnight at 70°C.

Laboratory (1 L) and pilot (10 L) PHA extractions were undertaken on selected batches to supply a range of material types for investigations related to the quality and opportunities for commercial developments for these polymers. The 1 and 10 L reactors were thermostated autoclaves (Buchiglass) wherein the biomass was retained in the process by means of a custom metal sieve basket. The biomass solids in the 1 L reactor were agitated by mechanical mixing. An external recirculation loop driven by a positive displacement pump was used for mixing the 10 L system. The pump was operated at constant revolutions per minute, but flow rates changed proportionally due to changes in solvent viscosity. Solvent viscosity changed due to temperature and the dissolved polymer concentration and its average molecular mass. Temperatures and mass flow rates of mixing or recirculation were monitored and logged. The pilot-scale recoveries were undertaken as previously disclosed [8] and with 2-butanol. All pilot-scale extractions and loadings were performed with a standard protocol for reproducibility and comparison of an influence of polymer type. A solvent loading (maximum polymer concentration for 100% extracted polymer) of 50 g/L polymer was applied. A 52-minute heating program from 50 to 140°C was used and the average solvent temperature during the whole extraction cycle was approximately 103°C. Polymer-rich solvent was discharged under pressure and separated from the retained spent biomass granulate. The cooled solvent formed a gel and the excess solvent in this matrix was pressed out by either a custom manual 1 L laboratory-scale basket press or a similar hydraulic (10 L scale) mechanical press that retained the polymer cake. Recovered polymers were rinsed to remove residual extraction solvent as was necessary to compensate for the non-industrial process nature of the laboratory and piloting pressing equipment. Recovered polymers were air dried at 70°C.

The extraction conditions targeted on average about 87% of extractable polymer based on extraction kinetic evaluations [8, 34]. Extraction kinetics were evaluated in a similar way to the acetone benchmark extractions but now using 150 mg of

PHA-rich biomass with the same grain size of 0.71 to 2 mm with 10 mL of 2-butanol over 45 minutes under isothermal conditions at selected temperatures and vortex mixing every five minutes for the first 15 minutes and then every 15 minutes. Standard test-tubes with Teflon screw caps were used, and tare, wet and dry weights along the process were taken for mass balance calculations. The polymer-rich solvent was separated from the biomass suspended solids and the extracted polymer purity was evaluated by thermal gravimetric analysis (TGA) and Fourier transform infrared spectroscopy (FTIR) [34].

9.2.4   COPOLYMER BLEND FRACTIONATION

Selected recovered polymers were fractionated at laboratory scale. 200 mg PHA were dissolved in 20 ml chloroform at 100°C for 3–5 minutes. Any non-dissolved solids were removed by centrifugation (3500 × g for five minutes) followed by decantation of the supernatant to a clean test tube. Aliquots of 1 mL n-hexane were added to the polymer-chloroform solution to the point where an initiation of precipitate formation was observed. The mixture was then left to stand for 24 hours to allow for precipitate formation. Precipitated solids were removed by centrifugation (3500 × g for 5 minutes) followed by decantation of the supernatant to a clean test tube. The procedure of adding n-hexane in 1 mL increments was continued until the next precipitate formation, and precipitation and separation were repeated as above. This fractionation procedure was undertaken until no further precipitation was observed. The recovered polymer fractions were dried 60°C for 72 hours and analyzed (size exclusion chromatography (SEC), differential scanning calorimetry (DSC), FTIR and TGA) as previously described [34].

9.2.5   COPOLYMER BLENDING

Manufactured copolymer blends of PHBV were generated at test tube scale by combining selected recovered polymers in different proportions. All blends were made by adding dry weighed polymers to a total mass of 300 mg. The polymer mass could be dissolved to a well-mixed solvent-polymer solution in 8 mL 2-butanol at 140°C in seven minutes. This solution was then decanted into a Petri dish, the solvent was evaporated, and the polymer was air dried at 50°C over 24 hours. The blended PHBVs were then analysed by DSC [34].

9.2.6   MECHANICAL TESTING

Mechanical testing of selected batches of the recovered copolymers was undertaken in order to assess for the consistency of the material behaviour as a function of polymer type in the range from 20 to 40% weight 3HV in PHBV. The recovered polymers were melt processed into test elements for evaluating the material mechanical properties. Test elements were aged for two weeks at room temperature prior to mechanical testing.

Tensile test dog bones were made by hot press. A grab sample of 0.9 g of recovered PHBV powder was loaded into a dog bone mould with a thickness of 2 mm, width of

5 mm and a gauge length of 15 mm. The loaded mould was placed in the hot press at 180°C. Pressure was subsequently increased to 30 bars. Once pressure had been applied the mould was kept at 180°C for five minutes. After five minutes the mould was cooled by water to 70°C before releasing the pressure and removing the mould. Six test bars were made for each sample. Tensile testing was carried out according to ASTM D638 standard method using an Instron 430I tensile testing machine. A small preload (<5 N) was applied to the specimen at a crosshead speed of 0.1 mm/min in order to eliminate any bending. Tensile force was then applied with a cross head speed of 10 mm/min and the load-extension curve of the specimen was recorded.

Flexural test bars were similarly made by hot press. A grab sample of 0.5 g of recovered PHBV powder was loaded into a test bar mould with a width of 6 mm, a depth of 1 mm and a length of 40 mm. The same hot press cycle used for the dog bones was applied. Flexural testing was carried out according to ASTM D790 standard method using an Instron 430I tensile testing machine. The specimen was deflected until rupture occurred in the outer surface or until a maximum strain of 5.0% was reached, whichever came first.

9.2.7 ANALYSES

Polymer analyses included DSC, TGA, SEC, melt rheology, attenuated total reflection (ATR) FTIR (ATR-FTIR) and gas chromatography with flame ionization detector/mass spectrometer (GC-FID/MS) all as previously described [34, 35]. Analysis of PHA-rich biomass was also assessed by GC-FID/MS, ATR-FTIR, and TGA [34]. Water quality and solids analyses (COD, nitrogen, phosphorus, total suspended solids, volatile suspended solids volatile fatty acids and alcohols) were performed as previously described [3]. Trace metals and selected polyaromatic hydrocarbons (PAH) contaminant analyses of PHA-rich biomass and recovered polymers were performed externally by an accredited laboratory (Eurofins Environment Testing Sweden AB). Metals and PAH were assessed according to ISO 11466/EN13346 mod ICP-AES and ISO 18287:2008 mod, respectively. Statistical analyses and multilinear regression of results were performed with GraphPad Prism Version 8.

Polymer decomposition and decomposition rates were quantified with respect to the estimated polymer weight average molecular mass (M_w). The molecular mass degradation over a period of time was defined by Equation 9.1:

$$N_s\left(t\right) = \frac{M_{w(t=0)}}{M_{w(t)}} - 1 \tag{9.1}$$

When N_s is equal to 1, the polymer average molecular mass has been reduced to half of the original value. Monte Carlo modelling (data not shown) of random chain scission degradation for lognormal polymer distributions suggests that the random scission rate is directly proportional to the rate change in N_s. Therefore, the decomposition rate $r_s(t)$ for the polymers was quantified by the rate change of N_s. A constant decomposition rate under relatively constant environmental conditions was typically observed and estimated when trends with more than two M_w measurements as a function of time were available. An average decomposition rate is reported when only two M_w measurements were made on selected samples as a function of time according to Equation 9.2:

$$r_s(t) = \frac{dN_s}{dt} \approx \frac{\Delta N_s}{\Delta t} \qquad (9.2)$$

The weight average molecular mass half-life ($t_{1/2}$) is defined as the time to reach one half of the initial weight average molecular mass ($N_s(t_{1/2}) = 1$).

9.3 RESULTS AND DISCUSSION

9.3.1 OVERALL POLYMER PRODUCTION PERFORMANCE

Fifty-nine batch-wise PHA accumulations were performed twice weekly as part of the pilot plant operations starting from 18 June 2015 and these were spread out relatively evenly until 24 March 2016. There were seven batches where polymers from resultant PHA-rich biomass were not further processed for this assessment due to mechanical or unforeseen upsets in the piloting operations and control. Out of the 52 remaining accumulations, 20 were made with defined mixtures of acetic and propionic acids, (S-feedstocks), 27 were made with the supply of 26 batches of fermented candy wastewater filtrate (C-feedstocks), and five were made with the supply of 12 batches of fermented primary sludge centrate (P-feedstocks).

Fresh full-scale gravity-thickened waste activated sludge was the active MMC source used directly [5] for pilot-scale accumulations without further enrichment. The yield of PHBV on substrate was 40 ± 10% (gCOD/gCOD). Active biomass growth did occur to varying degrees due to the presence of nutrients (nitrogen and phosphorus) in the feedstock, but this growth appeared to be concurrent to polymer storage [36]. In our experience, partial nutrient limitation stimulates improved PHA accumulation potential, and so even the S-feedstocks included added nitrogen and phosphorus. The average PHA content of the biomass expressed as volatile suspended solids (VSS) reached on average 40 ± 5% (gPHA/gVSS). Polymer content of the biomass determined by TGA [34] was used to develop a partial least squares (PLS) model predicting the PHA content by FTIR for more rapid assessment with this biomass [37].

The accumulations were performed with constant process volume, meaning that the influent volume inputs displaced an equal volume that was the effluent flow from the bioprocess process clarifier. According to mass balances for all the accumulations, the consumed COD was on average 76 ± 10% of the COD supplied to the accumulation process. Operating strategies to minimize "substrate leakage" in the effluent exist but these were not implemented as part of the investigation. The pH-value typically increased asymptotically to a final value of 7.3 to 8.9 (mean ± standard deviation of 8.5 ± 0.4) due to accumulation of excess cations with the VFA consumption.

Thickened activated sludge biomass was used as the starting material to simplify the material flow from the wastewater treatment plant. The biomass was diluted to the starting biomass concentration. Typically, in the Netherlands thickened sludge (nominally 4% DS) is routinely transported for further processing (drying and incineration). Sludge transport in this case included thickened waste activated sludge, mixtures of primary and secondary sludge and digester sludge. So, sludge transport is part of the Dutch national logistics of sludge management. Therefore, the use of thickened solids served to demonstrate that dewatered waste-activated sludge solids were suitable as the biomass source within this existing Dutch context of solid waste transport and management. In separate experiments, thickened and fresh surplus activated sludge from Bath WWTP was shipped to Sweden. In off-site laboratory-scale PHA accumulation bioassays, the PHA accumulation potential was directly comparable to the experience for the same biomass processed on site at pilot scale (unpublished work). Therefore, where VFA-rich substrate supplies come from separate sites, one may consider the collective regional logistics and economy for the supply chain of transporting biomass and/or substrate to an optimally located site of PHA-accumulation. The transport of water should be minimized as it adds both economic and environmental costs [1, 9].

It was estimated that 95% of the PHA accumulation potential was asymptotically reached after 14 hours at 25 °C. Accumulation times applied were in the range from 11 to 24 hours (mean ± standard deviation of 17 ± 3 hours). There is no requirement that the temperature should be 25 °C and equally successful results have been demonstrated for similar mixed culture accumulations over a wide temperature range from 15 to 30 °C independent of the conditions of temperature during enrichment [38].

A trade-off exists as a function of temperature with respect to kinetics, yields, and oxygen transfer efficiency. Higher temperatures increase the accumulation kinetics but decrease the yield on substrate. Oxygen transfer efficiency is also reduced at higher temperatures and this may limit the process capacity. Lower temperatures mean slower accumulation, but with better yield in substrate utilization and better oxygen transfer efficiencies. Therefore, summer and winter production operations are anticipated for balanced productivities with higher initial biomass loading during winter with longer production cycle times, and reduced biomass loading during summer with shorter production cycle times.

Optimisation and performance improvements of the full-scale enrichment biomass production or the pilot-scale PHA production were not undertaken during this investigation because the main goal was to produce polymer in a consistent manner as routinely and as often as possible with the biomass that was directly available. Notwithstanding, it was considered that a consistently greater enrichment for the PHA storing phenotype, and a higher degree of PHA accumulation potential (PAP) could be readily achieved for the Bath surplus activated sludge, and for municipal activated sludge in general over a wide range of process configurations. Parallel investigations at the time [1], in a survey of 15 Dutch WWTPs, suggested that minor process changes to elicit a periodicity, but not necessarily with strict regularity, of a more stringent environment of "feast" in the full-scale process are conducive to increased PAP. That feast stimulation relies on the presence of readily biodegradable organic matter (RBCOD). However, this RBCOD does not need to be dominated by volatile fatty acids. One of the 15 WWTPs, exhibiting such interpreted periodic feast conditions had a PAP of 52% (gPHA/gVSS). PAP from a municipal biomass in excess of 50% gPHA/gVSS was therefore considered to be realistic to expect even for municipal wastewater with relatively low VFA content [3, 6]. The expected level of PHA storage capacity by direct accumulation is lower than what has been achieved with more functionalized MMCs [12, 39], but it is in agreement and similar in levels with many of the previous studies on diverse mixed cultures and municipal organic wastewater [11, 17]. Thus, it is interesting to note that an enrichment biomass may be produced on a low grade dilute organic wastewater stream like municipal wastewater, and the PAP may still be similar in level to biomass specifically enriched on fermentation waters with significantly higher VFA content. In this way, we have shown that in principle, VFAs may be directed from fermented streams for polymer production without the need to sacrifice VFAs for a specific PHA-accumulating biomass production step.

Acclimation [31, 32] is interpreted to influence the biomass physiological state towards an improved PAP response without generating any further enrichment [40]. It was noteworthy to observe a consistent performance of a full-scale municipal activated sludge biomass in polymer production quantity over four seasons of production even though acclimation of the biomass before accumulation was not optimized. This consistency suggested that the enrichment conditions in the full-scale WWTP due to pre-denitrification anoxic feast were robust. On average 997 ± 160 g of input activated sludge volatile solids to the pilot process produced 710 ± 133 g of PHA contained in a PHA-rich biomass. The 52 batches of PHA-rich biomass formed the basis to evaluate the upstream and downstream MMC PHA quality management and control.

The downstream processing of the biomass involved dewatering to about 1% DS, mixed liquor acidification with sulphuric acid to pH 2, dissolved air flotation to nominally 4% DS, addition of dewatering agents, centrifugation to 18 ± 3% DS, and drying at 70°C. The dried solids were then granulated and sieved to a defined particle size distribution (0.71 to 2 mm) for extraction.

Benchmark (10 mL acetone extraction) polymer recovery was made for all the respective main batches of pilot processed and dried biomass (so-called "Main Batch" polymer quality). This benchmark provided an estimate of the extant quality of the polymer in the biomass after accumulation and downstream processing to a dried biomass, but before extraction. Reference polymer-in-biomass quality assessments were also made from biomass samples taken just after acidification but before solids thickening by floatation (so-called "Reference" polymer quality). The Reference polymer assessments were made to assess for the possibility that the downstream handling was in some way detrimental to the polymer quality. In addition, 23 × 10 L pilot extractions were undertaken from selected batches in order to supply a range of material types for mechanical testing as well as further investigations related to opportunities for commercial developments for these polymers (so-called "Pilot Extraction" polymer quality). The 23 × 10 L scale extractions were representative of the range of types of PHBV blends produced over a range of average 3HV contents. Detailed results from mechanical testing of specimens made from the pure polymers (20 to 45 wt.-% 3HV) are reported in this work. Results of compounding bioplastic formulations in comparison to the lower 3HV content copolymer blends have been presented elsewhere [1].

9.3.2 PRODUCTION OF DIFFERENT COPOLYMER TYPES

At Bath WWTP, about 5200 tDS surplus biomass is managed per year, making it a potential raw material supplier to generate 2800 tPHA per year given a parallel annual regionally available supply of about 6500 tCOD as VFA-rich substrate. The COD composition of the feedstocks used in PHARIO was characterized with respect to VFAs (acetic, butyric, propionic, valeric, hexanoic and heptanoic acids), as well as ethanol and other dissolved COD. Other COD was the organic content remaining with respect to the total soluble COD. Figure 9.1 shows the variations of organic compositions that were used for PHA accumulations by design (S-feedstocks), due to inherent variability (P-feedstocks), and due to a mix of uncontrolled factors (C-feedstocks).

The C-feedstock fermentation results were the most variable. This variation was due to conditions that were beyond our control. The wastewater management at source permitted for significant but preventable degrees of fermentation generating ethanol within the storage tank at the factory. The C-fermentation outcomes were also linked to known challenges, in general, of fermenting a carbohydrate-rich stream. The dump-feed batch fermentation pilot process for this work was not selected, a priori, for this kind of wastewater where the possible variability in fermentation performance may be otherwise constrained in scale up [41, 42]. The expressed C-feedstock variation in the work became an important benefit towards observing the polymer type and molecular mass as functions of the feedstock composition. The continuous stirred tank reactor (CSTR) fermentation method for the P-feedstock was a more suitable or well-matched bioprocess method, and the effluent fermentate quality was relatively consistent. The P-feedstock compositional variation was used to reference an inherent influence of feedstock quality variation on polymer quality at full scale. S-feedstock compositional variations were applied as an experimental control.

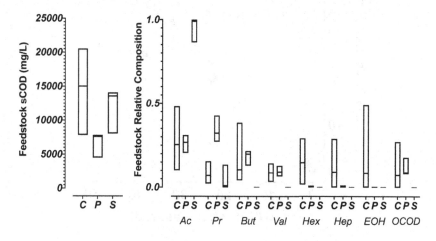

FIGURE 9.1 Range and median of organic strength (left) and compositional content (right) for C-, P- and S-feedstocks used for pilot-scale PHA production. Ac (acetic acid), Pr (propionic acid), But (butyric acid), Val (valeric acid), Hex (hexanoic acid), Hep (heptanoic acid), EOH (ethanol), OCOD (other chemical oxygen demand).

The metabolism for MMC conversion of short-chain-length VFAs into PHBV copolymers has been previously described [43]. Longer chain hexanoic and heptanoic fatty acids were considered to be metabolized via ß-oxidation and to contribute to 3HB and 3HV monomers [44, 45]. Ethanol may become oxidized to acetic acid and then to 3HB [12, 45, 46]. Other COD (if assumed as carbohydrate) has been previously observed to be metabolized to 3HB via the Embden-Meyerhof Pathway (EMP) [47, 48]. Feed-on-demand methods provided substrate in successive frequent doses based on respiration, so the individual feedstock components were expected to be consumed concurrently and contribute to a distribution of polymer chains that were random copolymer blends of poly(3-hydroxybutyrate-*co*-3-hydroxyvalerate) or PHBV [49]. The blends of individual polymer chains are expected to be characterized by distributions of both 3HV content and molecular mass.

The resultant PHBV type, expressed by the average 3HV content, was correlated to the feedstock organic composition. The average 3HV content (molar basis) could be predicted based on expected conversions of substrates to 3HB and 3HV [47] only if conversion coefficients were adjusted in weighting to a best fit by multilinear least squares regression analysis. An empirical model also directly defined a function relating the average relative 3HV content (COD basis), to the relative 3HV contributing feedstock components, and this approach provided a direct prediction by multilinear regression (Figure 9.2, $R^2 = 0.99$ and RMSE of 2 wt.-%). The COD model also suggested a statistically significant contribution to 3HV from the other COD fraction that was not anticipated within the assumptions of the initially applied metabolic

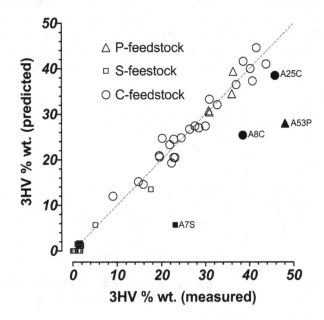

FIGURE 9.2 Measured versus predicted weight average 3HV content of PHBV copolymer blends as a function of the substrate composition (three distinct types of feedstocks). Four outliers (filled symbols) were excluded from the model due to noted abnormal accumulation termination conditions.

model from a previous work [47]. A second order factor for propionic acid content was statistically significant and further, but slightly, improved the model fit (from $R^2 = 0.98$ and RMSE of 2.5 wt.-%) expressed without this second order term. The second order term is indicative of a shift in metabolic response with increasing propionic acid content [43]. Model parameters with estimated standard errors were as follows according to Equation 9.3:

$$f_{3HV} = \alpha_3 \cdot f_3 + \beta_3 \cdot f_3^2 + \alpha_5 \cdot f_5 + \alpha_7 \cdot f_7 + \alpha_O \cdot f_O \qquad (9.3)$$

where f_{3HV} is the 3HV fraction (COD basis) of the PHBV, f_i (I = 3, 5, 7) is the feedstock soluble COD for propionic, valeric, and heptanoic acids, and f_o is the other unknown COD soluble fraction of the feedstock. The best fit coefficients with standard errors were: α_3 (1.413 ± 0.142), β_3 (−2.144 ± 0.402), α_5 (1.055 ± 0.132), α_7 (0.892 ± 0.073) and α_O (0.295 ± 0.048).

The fitting function was limited to the substrate components that were identified to be the significant contributing factors to 3HV content (propionic, valeric and heptanoic acids, plus other COD). Model results were improved by omitting four outliers that came from accumulations A7(S), A8(C), A25(C) and A53(P). These accumulations were all with abnormal terminations caused by errors in operations or loss in substrate supply. Model predictions were also improved when applied to S-, C- or P-feedstocks, independently. Thus, factors of the bioprocess and the presence of feedstock components influence the conversion of substrate into copolymer blends. The manner of feed supply, accumulation time and pH are known to potentially influence the biomass response, and the type and distribution of polymer produced [49–53]. However, additional parameters to the model, such as normalized variation of the average pH (from 7 to 8) and/or accumulation time (11 to 24 hours), were not found to be statistically significant factors influencing the predicted polymer type. Consistent accumulation conditions and bioprocess methods gave a deterministic outcome of the accumulated average copolymer blend composition. Accumulation conditions are nevertheless important to control as the literature and the four excluded outliers suggest. The bioprocess conditions in combination with the feedstock will modulate the type of polymer produced. A different activated sludge biomass fed in exactly the same way may not produce exactly the same type of copolymer blend [54, 55], but it is anticipated that the polymer type will be similarly influenced by the feedstock quality. This work supports the expectation that diverse mixed cultures producing PHA are robust [56, 57]. The model is empirical, and specific to the range of feedstocks compositions that the biomass was disposed to. It cannot be assumed to express a generically valid relationship.

It was most noteworthy that for an MMC PHA production process, a predictable outcome of average 3HV content could be based just on the substrate organic composition for three distinct feedstocks. The full-scale activated sludge MMC was a "black-box" over ten months but it was nevertheless a resilient and predictable active ingredient for polymer production. Consistently applied bioprocess conditions produced a predictable PHBV type that was a simple function of the individual substrate contributions. If the copolymer composition can be predicted by the feedstock composition for a given MMC, it suggests that even full-scale surplus municipal activated sludge biomass, that is not specifically enriched nor optimised for the PHA storing phenotype, may still be a suitable raw material to supply for commercial production of PHBV.

Some degree of variability of the organic composition for fermented feedstocks may be unavoidable in practice. The twelve fermentations of distinct batches of primary sludge (P) resulted in mean percent composition with standard deviation for VFAs (acetic, propionic, butyric, valeric, hexanoic and heptanoic acids), ethanol and other COD of 26.3 ± 4.0, 35.1 ± 6.3, 15.8 ± 3.2, 7.6 ± 1.6, 0.1 ± 0.3, 0.1 ± 0.2, 0.3 ± 0.4 and $14.7 \pm 6.3\%$ (COD basis), respectively. Variability of the P-feedstock components used for the polymer type prediction (Equation 9.3) was not strongly correlated. Propionic acid correlated negatively with other COD content (Pearson correlation coefficient of -0.745). Thus, a greater degree of acidogenic fermentation contributed to a relative increase in propionic acid content. Increased levels of valeric acid were similarly marginally positively correlated to increased hexanoic and heptanoic acid content, but these longer chain fatty acids were only minor components for the P-feedstock (Figure 9.1). From stochastic normal distribution modelling of the observed individual VFA variability (1000 points), it was predicted that a fermented primary sludge would produce a PHBV copolymer blend with average 3HV content of $32 \pm 3\%$wt. The experimentally observed 3HV content window from four P-feedstock accumulations was 33 ± 3 wt.-%. (neglecting A53(P) which was abnormally terminated due to an upset in the feed supply). Therefore, if the bioprocess conditions for the accumulation are consistently applied, the polymer type for a given feedstock source may be expected to be within a relatively narrow window. Feedstock batches or sources can be adjusted by blending to produce a targeted copolymer type based on a model that can be evolved with increasing certainty as the database for the MMC input-output polymer type transfer function grows. The polymer type window can furthermore be narrowed and tuned to targeted properties by selective master batch blending during polymer recovery as will be further described below.

9.3.3 COPOLYMER DISTRIBUTION AND THERMAL PROPERTIES

The copolymer blend type is being characterized by the average weight fraction of 3HV in the PHBV. PHBV is isodimorphic [58] due to an ability of 3HV monomer units to be accommodated in the crystal lattice of poly-3HB, and 3HB monomer units to be accommodated in a crystal lattice of poly-3HV. However, 3HV units tend to distort the poly-3HB crystal lattice, where 3HB units are accommodated in a poly-3HV lattice [24]. Melt temperatures are decreased by the disturbance to the crystal packing, and a minimum melting temperature is observed in the region of 50 wt.-% 3HV in PHBV. Thus, the polymers exhibit a eutectic point for crystallinity and melting temperatures in the neighbourhood of 50 wt.-% 3HV. The PHBV copolyesters made in this investigation were with a pre-eutectic average 3HV wt.-% content. Previous research with post-eutectic (average wt.-% of 3HV >50%) random copolymers of PHBV suggested that the type of polymer produced may not always be indicative of an ability to modulate polymer properties by changing co-monomer content [59]. Therefore, considerable effort was placed within this investigation to characterize, in contrast, pre-eutectic PHBV systematic changes and variability in properties (crystallinity, molecular weight, melt characteristics and thermal stability) that may be more indicative of quality control with respect to critical aspects for processing and application [60]. It was a goal to confirm the extent to which the produced PHBV polymer type could be classified by the average 3HV content, and if the polymer physico-chemical properties were predictable from that classification.

The average 3HV content is often quantified by biomass digestion, analyte extraction and gas chromatography [35], but we found that the 3HV content for the respective batches of PHBVs was also predictable from the estimated DSC weight average melt temperature [34]. This relationship between polymer type and melt properties requires that the thermal and melt history of the samples being assessed are identical. Glass transition temperatures were dependent on the average 3HV content, and these measurements were based on a standardized thermal history with rapid quenching at −50°C/min after a second melt ramp. The melting characteristics were derived from a history of melt (185°C) and quench cycle (up and down at 10°C/min), and then sample aging for two weeks at room temperature. Aging the sample for two weeks at 25°C was to conservatively allow time for secondary crystallization and the polymer to reach an equilibrium microstructure [61]. The glass transition temperature measurements were based on a microstructure with attempted maximized amorphous content, while the melting temperatures were based on a maximized crystalline microstructure.

The influence of aging on crystallinity (inferred by the melt enthalpy peak, ΔH), and quenching in generating an amorphous microstructure (inferred by glass transition ΔCp), are shown in Figures 9.3 and 9.4. The glass transition ΔC_p is related to

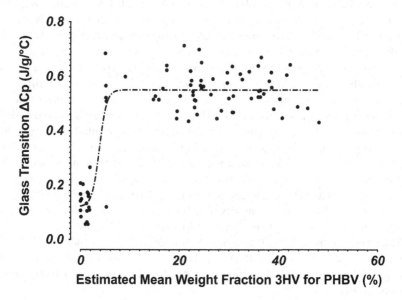

FIGURE 9.3   Change in heat capacity (ΔCp) over the glass transition exhibited as a function of copolymer blend average 3HV percent weight content, and with a −50°C/min quench from the polymer melted at 185°C [1].

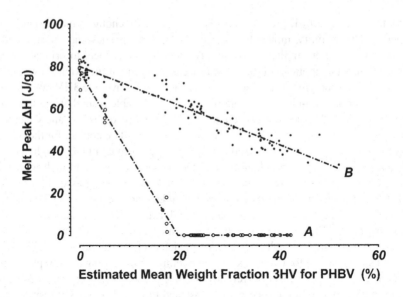

FIGURE 9.4 Melt peak enthalpy (ΔH) exhibited with or without two weeks aging at room temperature as a function of copolymer blend average 3HV percent weight content, and with a −10°C/min quench from the polymer melted at 185°C: A. Melt and quench at −10°C/min to −70°C, warm at 10°C/min to melt with no aging. B. Melt and quench −10°C/min to −70°C, warm at 10°C/min to 25°C and age for two weeks at 25°C, warm from 25°C at 10°C/min to melt [1].

the amorphous fraction of the solid. A quench rate of −50°C/min will provide very little time for the melt to crystallize upon cooling. Thus, a rapid quench is intended to maximize the amorphous fraction. The data suggested that, above a mean 3HV weight content of about 5%, the quenched polymer morphology is anticipated to be 100% amorphous ($\Delta Cp \approx 0.55$ J/g/°C). Below a mean 3HV content of about 5%, ΔC_p is considerably lower ($\Delta Cp \approx 0.12$ J/g/°C) suggesting that the material still formed a semi-crystalline material even with rapid quenching. Based on the observed presence or absence of a melt peak, with a melt ramp after, at a quenching rate of −10°C/min from the melt, a mean weight content above 20% 3HV resulted in an amorphous polymer (based on the subsequent melt ΔH). Crystallization rates have been shown to attenuate more rapidly for blend compositions greater than about 11 wt.-% 3HV [62]. A tendency for the lower 3HV materials to more rapidly crystallize is why PHB has application as a nucleating agent for the higher 3HV copolymer blends [63]. The polymers behaved in a consistent way as a function of average 3HV content given comparisons after a standardized thermal history.

Therefore, we used the polymer melt characteristics with standardized thermal history as an indication of the effective polymer type. These methods relied on the extent of 3HV-dependent crystallinity that developed given a consistently applied thermal history [64].

Discrimination of polymer type based on differences in crystallinity can also be sensed by FTIR analyses. FTIR analysis of the same polymer samples with PLS modelling also predicted the polymer average 3HV content. The PLS model loadings were weighted to normalized spectral changes between 790 and 1070 cm^{-1}, and between 2832 and 3082 cm^{-1}. These regions are expected to reflect relative differences of crystalline versus amorphous fractions of the polymer [65]. The kinetics of change for a solvent cast film of a 14% 3HV PHBV were evaluated in a previous study [66]. In this previous work, spectral changes in the region from 1690 to 1700 cm^{-1} (C = O bands) were interpreted to follow the crystallization kinetics. Different regions of FTIR spectra may indicate changes in crystallization of a given polymer, or changes in crystallinity between different copolymer blends. The FTIR signal response to changes in polymer crystallinity can also be sensed for polymer granules *in situ* in the biomass [67].

For this investigation, a second PLS model for polymer content in the biomass was generated based on PLS loadings from spectral regions 905 to 1348 cm^{-1}, 1490 to 1810 cm^{-1} and 2800 to 3051 cm^{-1}, and based on the experience of previous work for mixed cultures [37]. If the non-polymer chemical content (i.e. proteins, carbohydrates, lipids and genetic material) of activated sludge is relatively constant, then a given model may have generic application from its calibration. However, our experience is that such FTIR models of PHA content may need to be calibrated for specific biomasses. FTIR methods are always indicative, but quantification (absolute or relative) has been found to be reliable only within the narrowed context of either a specific MMC biomass source, or even within a specific accumulation [34].

Preservation of the polymer for downstream processing and extraction have been anticipated to be improved by driving the polymer to a more crystalline form [68–70]. As discussed further, optimal solvent recovery conditions are sensitive to the amount as well as the average type of polymer in the biomass. Robust and rapid methods for assessing polymer type and condition in the biomass are thus desirable for setting production operating conditions and the recovered product quality control. Based on this work and the literature, FTIR analyses should indicate PHA content, as well as PHA type and condition in the biomass, so long as the evaluation is based on thermal history or context of the samples managed in a systematic way. Further development in the application of FTIR for MMC process monitoring and control in scale up is recommended.

The aged polymer melt peak ΔH was predictable from the logarithm of the weight average melt temperature ($r^2 = 0.96$ and RMSE of 3.0 J/g), and without any statistically significant influence of molecular mass in the range from 250 to 1200 kDa (Equation 9.4),

$$\Delta H = A + B \cdot \log\left(T_m\right) \tag{9.4}$$

where T_m is the weight average blend melt temperature (°C), and with constants and standard errors ($A = 200 \pm 8$, $B = 126 \pm 4$). The melting of these polymers followed a distinctive broadening as a function of 3HV content and the melting trends could generally be modelled by an asymmetric sigmoid as a function of mean 3HV content (Figure 9.5, Equation 9.5):

$$f_m = \frac{1}{\left(1 + n \cdot \exp\left(-k \cdot \left(T - T_{50}\right)\right)\right)^{1/n}} \tag{9.5}$$

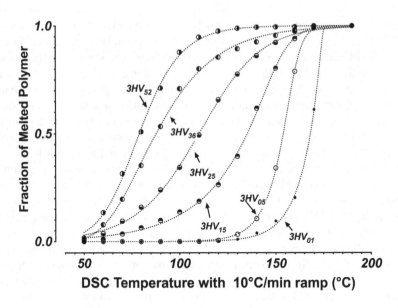

FIGURE 9.5 Example typical trends of the copolymer blend fractional melt as a function of DSC temperature for samples that were melted (185°C), quenched (–10°C/min) and aged (two weeks at room temperature) [1].

where f_m is the fraction of polymer melted as a function of DSC temperature (T with heat rate 10°C/min) in degrees Celsius, and k, n and T_{50} are fitting parameters expressing the polymer melting trend. The fitting parameters k, T_{50} and n are functions of the copolymer blend 3HV content.

The polymer typically melted over a range of temperatures wherein the median temperature decreased and the range increased as the copolymer type went from lower to higher average 3HV contents (Figure 9.6) up to near eutectic levels ($\approx$ 50 wt.-% 3HV, [71]). This suggests that with increasing 3HV content to the eutectic level, the co-polymer blends comprised a similarly wide distribution of copolymer types once the average 3HV content was above about 10 wt.-% 3HV. The melt and crystallization characteristics of these full-scale activated sludge PHBV blends were directly correlated to the weight average 3HV content and the weight average 3HV content was directly predictable from the feedstock composition.

Feedstock was supplied by pulse-wise additions at relatively low concentrations of substrate (maximum 200 mg COD/L) based on established feed-on-demand methods to maintain near maximal polymer accumulation kinetics [28]. Thus, it was assumed but not explicitly confirmed that batches of random copolymer blends were produced and these batches comprised distributions with respect to both 3HV content and molecular mass [24, 49, 51, 59].

The blend nature of the PHBV produced was evaluated in further detail. The 3HV and molecular weight distributions in the copolymer may or may not be distinct between different organisms in MMCs. Different bacteria may have distinct substrate preferences and metabolism for PHA storage [54–56, 72–74]. The individual polymer granules are nevertheless extracted from the biomass as a whole and

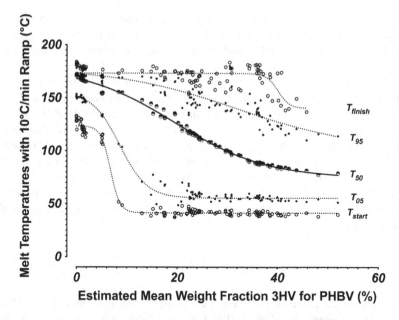

FIGURE 9.6 Summarized DSC results showing the estimated start and finish, as well as 5, 50 and 95% fractional melt temperatures for melt, quench and aged samples, as a function of estimated average weight 3HV content in batches of pilot accumulated PHBV copolymer blends [1].

the mixture of polymers from the diversity of microorganisms forms a homogenous blend mixture dissolved in the solvent. Thus, it was natural to wonder if the potential heterogeneity of polymer accumulated by different organisms would result in challenges to produce a consistent polymer product. Are the components of the blended mixture of individual storage granules compatible with one and another?

When more than one type of polymer is blended, the T_g of the mixture will depend on the relative amount of both components [75]. If the mixture is homogenous in nature, then the blend is expected to exhibit just one glass temperature (T_g) value. If a blend exhibits more than one T_g value, then the mixture is inhomogeneous, and components of the microstructure are phase separated. Inhomogeneous polymer blends may not possess good mechanical properties due to a potential for weakness and crack propagation at the phase boundaries. A single glass transition temperature (based on the thermal history for a maximized amorphous fraction [34]) was observed for all the respective 52 batches of recovered polymers suggesting that the batches of MMC polymer blends were miscible. The estimated T_g midpoint was correlated to a function of the onset T_g temperature, the weight average melt temperature, and the inverse of the weight average molecular mass ($r^2 = 0.97$ with RMSE of 0.5°C) as anticipated with slight modification to the Fox-Flory equation (Equation 9.6a) [76]:

$$T_{g(mp)} = A \cdot T_{g(s)} + B \cdot T_m + C \cdot T_m^2 + \frac{D}{M_w} \tag{9.6a}$$

Where $T_{g(mp)}$ is the glass transition mid-point temperature (°C), $T_{g(s)}$ is the glass transition start temperature (°C), T_m is the weight average melt (maximized crystallinity) temperature (°C), and M_w is the weight average molecular mass. Regression constants (A, B, C and D) were 0.402 ± 0.046, -0.027 ± 0.006, $3.3e-4 \pm 3.8e-5$ and -712 ± 135, respectively. All parameters in the regression analysis were statistically significant (P <0.0001). Inclusion of the onset T_g temperature improved the regression fit (from $r^2 = 0.93$ with RMSE of 0.9°C). The added information of $T_{g(s)}$ could indicate an influence of the polymer distribution on the glass transition. The added parameter could also compensate for measurement error from an uncertainty of determining the glass transition temperature mid-point. Correcting the estimated $T_{g(mp)}$ by the estimated $T_{g(s)}$ provided a direct relationship ($R^2 = 0.94$ with RMSE of 0.5°C) between estimated glass transition temperature with the polymer melt temperature and weight average molecular mass (Equation 9.6b):

$$T_{g(mp)} - A \cdot T_{g(s)} = B \cdot T_m + C \cdot T_m^2 + \frac{D}{M_w} \qquad (9.6b)$$

Since the weight average melt temperature was, furthermore, a function of the blend average 3HV content, the MMC PHBV T_g was in effect a function of average 3HV content with secondary contribution from the blend weight average molecular mass. Glass transition temperatures were found to be sensitive to molecular mass towards generating a T_g downshift of 1 to 7°C within the range of M_w that are commonly reported for these polymers (in the order of 1000 down to 100 kDa). Therefore, the weight average T_m estimated the weight average 3HV content (with no detected influence of M_w in the 250 to 1200 kDa range), while T_g was found to be sensitive to both 3HV and weight average molecular mass over a narrower temperature range. The average 3HV content was directly predictable from the feedstock composition, and, as will be discussed further, weight average molecular mass was predicted to be influenced by controllable factors of the accumulation methods and downstream biomass management.

The fact that the blend of extracted PHBV was miscible (one observed glass transition temperature), suggested that individual copolymers produced by different microorganisms in the activated sludge cannot be markedly different, or that even a diversity of PHA storage metabolism in MMCs can still effectively generate a distribution of sufficiently compatible copolymer blends that bridge to one and another.

Chloroform-hexanol blend fractionations were performed in four experiments in order to obtain a deepened impression of the blend nature of selected batches of recovered copolymer blends. In each experiment, three fractions could be readily separated, and each of the separated fractions were miscible unto themselves as these fractions were also characterized by a single detectable glass transition temperature (see Table 9.1). Individual fractions could be defined by respective average copolymer content, molecular mass, thermal properties by DSC, and an amount of hexanol to precipitate the respective fractions.

The tendency for fractions to precipitate as a function of the hexanol volume fraction in the hexanol-chloroform solution was found to be a linear function of the

TABLE 9.1

Hexane-Chloroform Fractionation of Selected Batches of Activated Sludge Produced PHBV Copolymer Blends

Fraction	Hexane [vol.-%]	3HV [wt.-%]	T_g (°C)	M_w [kDa]
Parent		19	1.27	555
1	59.1	18	2.68	724
2	62.5	19	1.44	551
3	76.9	25	−0.86	107
Parent		21	0.45	373
1	61.5	17	2.70	495
2	64.9	20	1.35	489
3	70.1	29	−0.64	251
Parent		24	−0.53	206
1	59.1	9	2.72	256
2	64.0	21	0.87	297
3	72.3	33	−0.50	179
Parent		42	−2.35	291
1	62.5	37	−0.71	383
2	69.0	41	−1.87	363
3	73.5	n.a.	n.a.	125

n.a. = not available due to limited amount of sample for analysis

respective precipitated fraction molecular weight, and T_g difference (r^2 = 0.92 and RMSE of 1.7 vol.-% hexanol). This T_g difference was the difference between the T_g of the precipitated copolymer blend fraction and the T_g of the parent copolymer blend (Equation 9.7):

$$f_{hex} = A + B \cdot M_w + C \cdot \left(T_{g_i} - T_{g_p} \right) \tag{9.7}$$

Where f_{hex} is the vol.-% hexanol in the solvent, M_w is the separated fraction weight average molecular mass (kDa), T_{g_i} is the estimated ith separated fraction (i = 1 to 3) glass transition temperature (°C), and T_{g_p} is the estimated parent copolymer blend glass transition temperature (°C). Estimated statistically significant coefficients based on multilinear regression of the data from the four pooled experiments (Table 9.1) were 72.71 ± 1.35 (A), −0.14 ± 0.004 (B), and −2.63 ± 0.43 (C), respectively.

A difference in measured T_g between respective precipitated fractions and the parent copolymer blend will be due to the difference in average 3HV content and weight average molecular mass. A model based on 3HV (estimated from T_m) instead of T_g difference (Equation 9.8) gave a similar outcome (R^2 = 0.89 and RMSE of 1.9 vol-% hexanol) to Equation 9.7:

$$f_{hex} = A + B \cdot M_w + C \cdot \left(f_{3HV_i} - f_{3HV_p} \right) \tag{9.8}$$

Where f_{3HVi} and f_{3HVp} are the average weight 3HV content for the fraction and parent, respectively. Estimated statistically significant coefficients based on multilinear regression of the data from the four pooled experiments in this formulation were 72.560 ± 1.556 (A), −0.018 ± 0.004 (B), and 0.541 ± 0.104 (C). Therefore, the tendency for fractionation depended on the distribution of molecular mass in the polymer chains as well as the average copolymer content. The precipitated fractions were found not to differ by more than approximately 16 wt.% average 3HV content from their respective parent copolymer blend.

Since three principal fractions were always precipitated, one could speculate that the accumulation was accomplished by three predominant phenotypes of PHA storing bacteria in the culture producing, on average, distinctly different but nevertheless miscible blends of PHBV as an ensemble of copolymers. However, compositional fractionation and a chemical composition distribution are similarly observed even for pure culture short-chain-length-PHA production [71] and the number of fractions observed may depend on the degree of refinement taken with hexanol-chloroform precipitation [49]. Therefore, the nature of the polymer chemical distribution observed with a full-scale activated sludge is similar to copolymer blends found to be produced by pure cultures.

Given that the copolymer blends could be fractionated, and that the fractions were miscible copolymer blends, it was of interest to confirm if miscible, as well as immiscible blends, could be artificially produced. These experiments were undertaken as positive and negative controls to ideas of master batch blending accomplished during extraction, and to give confirmation of the T_g measurement sensitivity to resolve more than one T_g value, if it existed.

Blends of selected binary and ternary mixtures of different copolymer blends were made in an analogous way to the pilot study solvent extraction methods. Weighed proportions of well-characterized copolymer blends were combined by dissolving the polymer mixtures to a homogenous solution in 2-butanol followed by solvent evaporation and drying. Mixtures were combined in proportions to reach selected targeted average 3HV copolymer blends. The Fox equation estimates the T_g of a homogeneous polymer blend as a function of the respective component weight fractions (w_a and w_b) to form blend $(_{ab})$ T_g values as follows according to Equation 9.9:

$$\frac{1}{T_{g,ab}} = \frac{w_a}{T_{g,a}} + \frac{w_b}{T_{g,b}} \qquad w_a + w_b = 1 \qquad (9.9)$$

Where T_g is the glass transition temperature in Kelvin, w_i is the component weight fraction.

In a first set of experiments, eight binary 50:50 blends of PHBV were made. PHBV with $3HV_{01}$ (i.e. 1 wt.-% 3HV) was blended in a series of eight cases with PHBV containing up to 40 wt.-% ($3HV_{08}$, $3HV_{09}$, $3HV_{17}$, $3HV_{24}$, $3HV_{25}$, $3HV_{31}$, $3HV_{39}$ and $3HV_{40}$). A shift in T_g values followed the Fox equation up to critical conditions of an immiscible blend (two observed T_g values). The critical 3HV blend content ($3HV_{cr}$) was estimated by the midpoint between the last observed miscible (single T_g), and the first observed immiscible (2 T_g) blends. An estimated critical 3HV difference ($\Delta 3HV_{cr}$) of 12.5 wt.-% was the difference between the parent 3HV content and the

3HV content at the critical point ($3HV_{cr}$) for onset of immiscible blending. Similarly, six binary 50:50 blends of PHBV with 14 wt.-% of 3HV ($3HV_{14}$) were made with PHBV containing 17 to 40 wt.-% ($3HV_{17}$, $3HV_{25}$, $3HV_{30}$, $3HV_{31}$, $3HV_{37}$ and $3HV_{40}$). In this second case, a $\Delta 3HV_{cr}$ of 10.0 wt.-% was estimated.

A set of 11 binary blends were also constructed with 0:100 10:90, 30:70, 40:60, 45:55, 50:50, 55:45, 60:40, 70:30, 90:10 and 100:0 wt.-% proportions from four distinct accumulation batches. Eleven copolymer blends were made with $3HV_{08}$ and $3HV_{41}$ as well as with $3HV_{14}$ and $3HV_{37}$. Once again, the blend T_g trends followed the Fox equation up to a point where the blends were no longer miscible given the appearance of two detectable T_g values (Figure 9.7). The estimated $\Delta 3HV_{cr}$ with respect to the dominant proportion was 11.3 ± 2.5 wt.-%.

From all of these binary blending experiments, the estimated average critical $\Delta 3HV_{cr}$ between the blend 3HV composition and the parent 3HV content was 11.3 ± 2.1 wt.-% 3HV (Figure 9.7). This outcome agrees very well with previous literature based on copolymer blends of PHBV produced by pure cultures [77, 78]. Thus, these MMC PHBVs appear to exhibit thermal and blend properties that are consistent with already well-established experience from the pure cultures producing the same type of polymers [71].

In a final set of blending experiments, ternary mixtures were produced. A ternary 1:1:1 mixture of ($3HV_{01}$, $3HV_{14}$ and $3HV_{31}$) formed a miscible copolymer blend (T_g

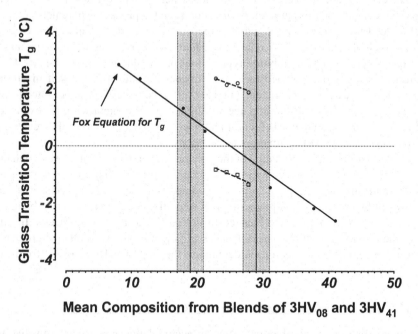

FIGURE 9.7   Glass transition temperature(s) of 2-butanol compounded mixtures of 3HV08 and 3HV41 with estimated (shaded) regions of 3HVcr indicated with respect to the dominant component. Blends were with a single glass transition temperature (•) up to the point where the mixed polymers became immiscible to exhibit two glass transition temperatures (open symbols).

of 2.1°C), while a similar mixture of ($3HV_{01}$, $3HV_{17}$ and $3HV_{39}$) was biphasic (T_g of −2.5 and 4.4°C). $3HV_{14}$ with a maximum $\Delta 3HV$ with respect to $3HV_{31}$ of 8.5 wt.-% could bridge $3HV_{01}$ to $3HV_{31}$ to form a miscible copolymer blend. In contrast, $3HV_{17}$ with a maximum $\Delta 3HV$ of 11 wt.-% with respect to $3HV_{39}$ could not bridge $3HV_{01}$ to $3HV_{39}$. The results of the ternary blending agreed with expectations based on the estimated average critical difference found from the binary mixture experiments.

The results from fractionation and blending clearly demonstrated that accumulation batch to batch shifts in average 3HV content due to variability in substrate composition or accumulation conditions may be absorbed during the PHA recovery steps. Batches of PHA-rich biomass containing known mass and type of PHBVs can be purposefully "master batched" in selected proportions before solvent extraction of PHBV to produce a well-defined average 3HV content and distribution. A predicted homogenous copolymer blend type can then be extracted from the pre-mixed biomass batches. PHBV copolymer blends can be combined and other compounding agents may even also be introduced during such solvent recovery methods [8]. Miscible blends of materials can be generated during extraction based on the principle that a critical 3HV difference should not be exceeded, or it should be bridged by other components. Failure to achieve a miscible blend can be readily identified because two T_g temperatures become measurable, and conditions of master batching strategy may be evaluated and optimised a priori at test tube scale.

The polymer melting and thermal characteristics were predictable from the average 3HV content and this parameter can be controlled as part of the polymer accumulation (feedstock) or during the polymer recovery (pre-extraction master batching). Since manipulation of the 3HV content was found to be independent of the biomass, the quality control of the polymer type should be feasible to achieve for MMC at industrial scale. Different feedstocks or feedstock variations may result in differences in average copolymer content, but these differences may be compensated for in the downstream. Notwithstanding, the variation of 3HV content with, for example, fermented primary sludge was found to be relatively low (± 3 wt.-% 3HV content), given consistent applied accumulation conditions.

Compounding and combination of the polymers to targeted average 3HV contents as part of the PHA recovery process offers significant potential for cost savings as well as a means to preserve other quality parameters of the polymer such as molecular weight by avoiding a compounding melt step after extraction. Different biomass batches may be combined and blended to suit particular application needs. In this way, the quality control for master batches of polymer type with other ingredients can be strategically engineered on demand at the step of PHA recovery from the biomass.

9.3.4 FACTORS INFLUENCING WEIGHT AVERAGE MOLECULAR MASS

Polymer properties in the biomass after an accumulation process are sensitive to the conditions during downstream processing. Recent investigations show nicely the influence of biomass handling on interpreted crystallinity and a coalescence of the polymer granules in the biomass [67]. Therefore, influence of the downstream processing on polymer quality is as important to the methods applied to influence the polymers during the bioprocess. Biomass drying and non-chlorinated solvent extraction methods require disposing the biomass to elevated temperatures. The polymer will become severely degraded if the thermal stability of the PHA in the biomass (as measured by TGA [7, 34]) is low.

Acidification of the biomass before final dewatering was found to be a reliable means to reproducibly increase the thermal stability of the polymer in the biomass. Acidification resulted in minor solubilization of the biomass in terms of COD (31 ± 21 mgCOD/gTSS), nitrogen (1.5 ± 0.6 gN/gTSS), and phosphorus (4.8 ± 2.2 gP/gTSS). Buffering capacity that created acid demand for pH adjustment came from the dissolved solids (82 ± 16 meq/L) as well as the suspended solids (6.2 ± 1.5 meq/gTSS). Acid demand for pH adjustment can be reduced with the degree of dewatering applied before the acidification. Acidification is interpreted to reduce the polymer associated cationic content of the PHA-biomass wherein especially divalent cations have been shown to catalyse the polymer thermal degradation [79–83]. Even recovered polymers exhibiting low thermal stability can be washed with acids, including organic acids, resulting in significant increase of the PHA decomposition temperature [84].

Reference grab sample (including post acidification, lab-scale centrifugation and drying), and Main Batch grab sample (including post acidification, thickening by dissolved air floatation, dewatering chemicals addition, pilot centrifugation and drying) polymer-in-biomass as well as recovered benchmark polymer qualities were compared. The degree of dewatering was higher for the Main Batch versus the Reference solids (18 ± 3 versus 10 ± 2% DS) due to the addition of dewatering chemicals for the Main Batch and this resulted in a slight but statistically significant decrease (α = 0.05) in the dry solids ash content (10 ± 2 versus 13 ± 2%). However, the polymer-in-biomass Main Batch thermal decomposition temperature (T_d) was slightly but significantly (α = 0.05) lower than the Reference T_d (284.5 ± 1.5 versus 286.3 ± 1.1°C). Polymer-in-biomass T_d was, in turn, lower than the Pilot Extracted polymer T_d (289.6 ± 1.0°C). The polymer thermal stability is known to be influenced by the biomass [85] but a lower dried biomass ash content due to improved dewatering did not necessarily improve thermal stability. This suggests that the nature and proximity of cation association with the polymer are more important than the presence of cations per se. The dewatering chemicals may also have had some minor influence on the polymer-in-biomass thermal stability. No statistically significant difference (α = 0.05, paired t-test) could be found between the Main Batch versus Reference biomass PHA content (gPHA/gVS) nor the respectively assessed Main Batch versus Reference benchmark weight average molecular mass of the polymer-in-biomass.

The Main Batch polymer-in-biomass, before drying, was not absolutely stable. Parallel aliquots of the wet dewatered pilot PHA-rich biomass were incubated isothermally at 4 and 20°C in air for 29 days. The estimated molecular mass decomposition rate (r_s) was approximately constant over time and dependent on the storage temperature (0.130 ± 0.008 days^{-1} at 20°C and 0.011 ± 0.006 days^{-1} at 4°C). Therefore, the molecular weight half-life was about 8 and 87 days for 20 and 4°C storage, respectively. Polymer content also decreased concurrently following first order kinetics with a half-life of 11 and 210 days at 20 and 4°C, respectively. Concurrent loss of molecular weight and total mass is indicative of bulk degradation and consumption of the polymer in the biomass [86]. A relatively constant molecular mass with loss of content would suggest surface degradation of the polymer granules. Therefore, these results suggest that bulk polymer degradation can ensue after accumulation in wet storage, even after acidification to pH 2.

In contrast, the polymer within oven dried PHA-rich biomass was found to be stable for extended periods of time. Similarly, produced and dried MMC PHA-rich biomass main batches, from a previous piloting investigation, were re-evaluated following the same methods, and after three years of dark storage at room temperature. For eight re-evaluated batches of stored biomass, there was no statistically significant ($\alpha = 0.05$) difference in the PHA content. The average estimated molecular mass decomposition rate was slow in all cases, but measurably sensitive to PHBV type. The average decomposition rates were 0.001, 0.126 and 0.210 years^{-1} for 3HV average contents of 0, 17 and 43 wt.-%. The apparent worst case scenario for the closer to eutectic 3HV content still suggested a molecular weight half-life of 4.8 years. It was not surprising that the less crystalline (higher 3HV content) polymers-in-biomass had shorter shelf-life [87]. Biodegradation of PHBVs is influenced by polymer type which in turn influences controlling factors of enzymatic attack through the development of crystallinity, microstructure and surface morphology [88]. Drying the biomass at 70°C may not have completely inactivated depolymerase activity. Intra- and extracellular depolymerases are reported to often appear to exhibit optimal activity in the thermophilic range and alkaline pH [89]. The dried biomass is hygroscopic, and moisture content from Main Batch TGA assessments was 2.9 ± 0.8% weight. The amorphous fraction of the polymer in the biomass would be selectively more susceptible to any residual, albeit low, levels of possible enzymatic activity [90, 91]. Stored solids moisture content and temperature will therefore influence the shelf life of inventoried PHA-rich biomass.

The dewatered wet PHA-rich biomass should be dried as soon as possible. These results of dried storage, in combination with evidence of wet storage decomposition activity after acidification, further suggested that drying at 70°C was insufficient to completely eliminate residual enzyme activity, and higher drying temperatures should be used to decrease the risk of undue polymer decomposition in storage. Dried batches of PHA-rich biomass can be inventoried for reasonable lengths of time, and different grades can be drawn from, mixed and combined selectively to pre-defined master batches of defined extracted PHBV types. The more stable this semi-product is in storage, the better. The degree of stability depends on the type of copolymer and the conditions of storage. Storage and drying methods were consistently applied for this study. Even if the conservatively applied 70°C drying is interpreted in hindsight not to have been optimal, the outcomes of quality were still consistent and predictable.

The interpretation that acidification was insufficient to completely eliminate all depolymerase activity was further supported by an additional set of drying experiments. Since dewatered acidified polymer-in-biomass was sensitive to temperature, an initial increase in temperature during drying may promote some degree of biologically mediated polymer decomposition. This decomposition rate will increase until temperatures exceed the enzyme activity optimum and will ensue until the enzymes become thermally inactivated if temperatures become high enough. Thus, biomass drying conditions should be applied to raise the bulk dewatered solids temperatures as quickly as possible. One method to heat more quickly is to apply a faster initial heating gradient by using a higher initial drying temperature. We found that higher initial drying temperatures (120 or 170°C) for 15 minutes before continued drying at between 70 and 100°C resulted with up to double the benchmark weight average molecular mass when compared to the same biomass batch dried isothermally at 70°C.

The molecular mass is therefore sensitive to the post accumulation, dewatering, pH and thermal history. The wet mass distribution and morphology, alongside time, temperatures and heat transfer rates are all important factors. Since all of these factors are controllable, they can be engineered and optimized explicitly towards maintaining the M_w. For this investigation, the standardized 70°C drying protocol, plus the directed side experiments, nevertheless enabled to identify these factors contributing to the observed product molecular weight variability.

The Main Batch weight average molecular weights were variable and ranged from 250 to 1200 kDa. In hindsight, isothermal drying at 70°C, as described above, was considered to have reduced the as-accumulated M_w by about half. As also mentioned above, there was no statistically significant difference between the Main Batch versus Reference biomass benchmark weight average molecular mass of the polymer-in-biomass. The Main Batch (about 2 kg solids at 18% DS) was dried as a spread-out cake, and the Reference biomass sample (about 1 g solids at 10% DS) was a centrifuge pellet in a 50 mL Falcon sample test tube. The paired differences between Reference and Main Batch samples were variable (mean difference 5 kDa with a standard deviation of 129 kDa), supporting the interpreted influence of uncontrolled factors of drying on molecular weight loss. However, this degree of variability could not adequately explain an outcome for the estimated polymer-in-biomass M_w range of 1000 kDa for the pilot production Main Batches.

The weight average molecular masses, on average, were significantly different ($\alpha = 0.05$) between the respective feedstocks (from low to high) at 513 ± 166, 713 ± 115 and 931 ± 137 kDa, for the C-, S- and P-feedstocks. Polymerization involves steps of chain initiation, chain propagation and chain termination [92, 93]. Conditions promoting a higher probability for chain termination during PHA accumulation will produce polymers of lower average molecular mass. In previous work [28], it was found that frequent sufficiently large, albeit brief, events of reduced respiration during MMC accumulation led to polymers of lower average molecular mass. It was interpreted that polymerization limiting kinetics should be maintained during PHA accumulation in order to sustain minimal chain termination probability in the bioprocess, and, in so doing, optimize the product towards a maximum possible average molecular mass. Periodic interruptions such as events of respiration inhibition, or substrate starvation, may result in more frequent events of chain termination. More frequent events of chain termination will lead, on average, to reduced molecular mass development. Conditions of the accumulation bioprocess may have a significant influence on the polymer molecular mass.

For this investigation, the experimental data of feedstock concentration and composition, alongside accumulation specific parameters or outcomes (accumulation time, polymer content and type, non-PHA biomass growth, average pH) were considered for possible correlation to the logarithm of polymer weight average molecular

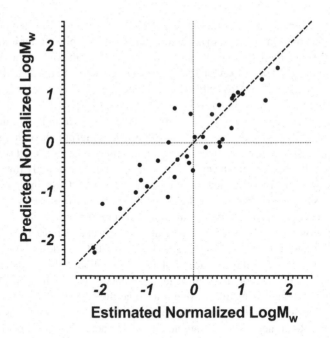

FIGURE 9.8 Assessment of measured versus predicted variability of accumulation batch to batch PHBV weight average molecular mass based on parameters of feedstock strength, feedstock composition, accumulation time, and active biomass growth. All parameters were offset and normalized by their respective mean and standard deviation.

mass. All parameters were offset and normalized by respective estimated mean and standard deviations. Variation of the normalized logarithm of M_w was found to be significantly and negatively correlated (Figure 9.8) to the variation of normalized parameters of feedstock strength, feedstock composition (just ethanol and heptanoic acid), accumulation time, and non-PHA biomass growth ($r^2 = 0.86$ and RMSE 0.37 normalized $\log M_w$) according to Equation 9.10:

$$\log\left(M_w\right) = A + B \cdot t_a + C \cdot f_{OH} + D \cdot f_{C7} + E \cdot S + F \cdot X_a \qquad (9.10)$$

Where all variables were normalized quantities, t_A is the accumulation time, f_{OH} is the COD fraction of ethanol in the feedstock, f_{C7} is the COD fraction of heptanoic acid in the feedstock, S is the feedstock COD concentration, and X_a is the estimated mass change in non-PHA VSS during the accumulation. The normalization factors (mean and standard deviations) were 2.77 ± 0.17 LogkDa (LogMw), 17.5 ± 3.0 h (t_A), 0.06 ± 0.10 (f_{OH}), 0.05 ± 0.08 (f_{C7}), 13.2 ± 3.6 gCOD/L (S), and 104 ± 226 mg (X_a). The estimated mean and standard deviation of the fit parameters to Equation 9.10 were 0.241 ± 0.087 (A), −0.209 ± 0.092 (B), −0.348 ± 0.081 (C), −0.424 ± 0.101 (D), −0.558 ± 0.111 (E), −0.198 ± 0.081 (F).

Higher values of the respective parameters within their ranges of variability resulted in a reduction of the product average molecular mass. Accumulation average pH (in the range from 7 to 9) and polymer type (in the range from 0 to 41 3HV wt.-%) were not found to exert any detectable influence the product polymer molecular mass. Pure culture research suggests that alkaline conditions are favourable for polymer average molecular mass development [94, 95]. For MMCs, the optimal pH for accumulation depends on the environment in which the MMC is selected. In our experience, an accumulation response in this WWTP activated sludge mixed culture (selected at pH 7–7.5) becomes significantly inhibited below pH 6. The results of this investigation are commensurate with expectations of previous work where changes in feedstock did not influence the outcomes of the polymer average molecular mass [96]. Full-scale activated sludge as well as specifically developed MMC enrichment cultures exhibit robust performance for polymer quality over a wide range of conditions and feedstocks.

As the PHA content approaches the maximum potential for the biomass, conversion yields decrease [38, 53], and active biomass growth can ensue given availability of nutrients [36]. The presence of nitrogen has been implicated in previous work looking at the influence of culture conditions on the polymer molecular mass [95, 96]. In this work, active biomass increase was not to the detriment of the accumulated PHA content, so selective growth of the PHA-storing phenotypes was interpreted. Microbial growth can ensue while the PHA content of the dividing bacteria remains relatively constant [97]. The results would suggest that if accumulation time is drawn out beyond the time necessary to reach the biomass PHA storage potential, some sacrifice to the stored polymer molecular weight may result. Accumulation times were run conservatively for 18 ± 3 hours, but kinetic assessments suggested maximal PHA content by approximately 15 hours at 25°C. Molecular weight decrease for prolonged accumulation time may be a result of an upshift in intracellular depolymerase activity for biomass saturated with stored polymer. Molecular mass has been shown to decrease more rapidly than polymer mass in a biomass shift from feast to famine growth on polymer [96]. An onset of turnover for the stored polymer [98] may influence existing and continued molecular weight development under conditions of combined growth and storage. The influence on the PHA molecular mass already stored would be an increase if decomposition was selective to the lower molecular mass fraction in the distribution [99], but this was not observed. Alternatively, given continued polymer storage with concurrent growth and storage, an increase in the chain termination probability may be a consequence of divisions in metabolic carbon flux. Mitigation of the interpreted negative impact of extended accumulation time, and active growth, on the average molecular mass requires that the accumulation process be terminated consistently as the biomass maximum PHA content is reached.

The consumption of ethanol was significantly slower than for acetate (Figure 9.9) and this outcome has also been found previously for MMCs enriched on municipal [100] and industrial [12] wastewater. Ethanol is typically converted to PHB [12, 101] but the additional steps of conversion for both ethanol and heptanoic acid may draw from the flux of carbon directed to polymerization, and in this way an increased probability for chain termination may result. Separate respiration experiments were illustrative of how acetate consumption could be followed by more slowly converted substrates, such as ethanol (Figure 9.9). Acetic acid is interpreted to be a common principal driver for stimulating a PHA accumulation response in MMCs. Addition of

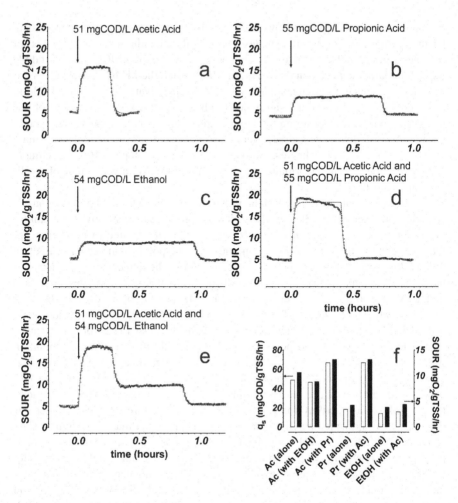

FIGURE 9.9 Respiration response of the WWTP activated sludge biomass from conditions of endogenous respiration and after the addition and consumption of pulses (a) acetic acid (Ac), (b) propionic acid (Pr), (c) ethanol (EtOH), (d) acetic and propionic acids and (e) acetic acid and ethanol. Estimated average rates of specific substrate uptake (white bars) and oxygen uptake (black bars) from the trends (a-e) are also summarized (f).

acetic acid had a positive influence on respiration and the conversion rate of propionic acid (Figure 9.9). The conversion rate of ethanol was not stimulated by the addition of acetic acid and thus, conversion of ethanol was not coupled to acetic acid in the same way as acetic and propionic acids. A respiration feed-on-demand trigger waiting on residual ethanol depletion may therefore lead to unintended frequent events of disruption in acetate supply. This can lead to reduced average molecular mass due to increased chain termination as previously observed [28]. Pure culture PHA production similarly reports a negative influence of alcohols on molecular weight during accumulation [102, 103]. A potential for the negative effect of increased ethanol or heptanoic acid in the mixed liquor for molecular mass development in mixed cultures requires further investigation. In the end, maximal polymer molecular mass seems to demand that alcohols and longer chain fatty acids be of negligible concentration in the feedstock.

Acidogenic fermentation conditions, such as pH control, can be applied to mitigate ethanol production in the first place [104]. We found as part of side experiments (data not shown) that ethanol levels may be reduced by conversion to acetic acid during the fermentation through the introduction of a suitable electron sink [105] to the fermentation bioprocess.

A statistically significant influence of feedstock strength on molecular mass was surprising because the pulse wise substrate volumetric additions were adjusted to achieve a peak COD concentration of 200 mg/L in all cases, and independent of feedstock strength. An influence of feedstock strength on M_w may have been coincidental because C-, S-, and P-feedstocks (15.6 ± 2.6, 11.8 ± 2.3 and 7.1 ± 1.5 gCOD/L) were coincident on average to feedstock specific polymer outcomes of average molecular mass (513 ± 166, 713 ± 115 and 931 ± 137 kDa). However, the feedstock strength remained marginally but still statistically significant in the evaluation of just the normalized C-feedstock data in isolation ($R^2 = 0.84$ and RMSE 0.40 normalized LogM_w with n = 25), and higher feedstock concentration was not correlated to ethanol content in the feedstock. The influence of feedstock strength is postulated to relate to the manner of feeding because many, even brief, periodic events will periodically increase the polymer chain termination probability to lower the average molecular mass [28]. For the piloting methods that were applied for this investigation, higher average polymer molecular mass was obtained when ethanol and longer chain fatty acid levels in the feedstock were relatively low, when accumulation times were not unduly prolonged beyond the reach of maximum extant PHA-content, and when feedstock sCOD concentrations were under 10 gCOD/L.

The polydispersity index (Đ) from the benchmark evaluation of the Main Batch polymer quality (10 mL at 10 g/L in acetone) was 1.71 ± 0.09 (Figure 9.10). Similarly, for pilot extraction polymer quality (10 L at 50 g/L with 2-butanol) Đ was 1.72 ± 0.17. Đ for step growth linear polymerization, as is the interpreted case for PHA polymerization in biomass [92, 93], is expected to approach 2 [106]. Pure culture and mixed culture investigations found Đ values in the range from 1.3 to 2.2 [94, 96, 107], but values in the neighbourhood of higher than 3 have also been reported [24]. Some of the observed and published variation in Đ was anticipated to be due to measurement errors. Practical challenges and uncertainties exist in the determination of the number average molecular mass (M_n) by SEC. M_n estimation is more sensitive than M_w to errors in background determination and baseline drift. The fact that M_w from SEC was found to be very well correlated to melt rheology dynamic viscosity measurements suggests that the biomass produced a polymer with a consistently similar Đ [108]. Broadening of the molecular mass distribution is, at the same time, to be expected with extensive thermal decomposition by random chain scission [109]. In this work, the trend of average estimated Đ was not apparently influenced by the weight average molecular mass over the range of the batches produced and recovered (250 to 1200 kDa). The average Đ was not influenced by accumulation conditions that resulted in lower average molecular mass. The Đ values on average were also not altered due to molecular mass loss with the pilot-scale extraction methods.

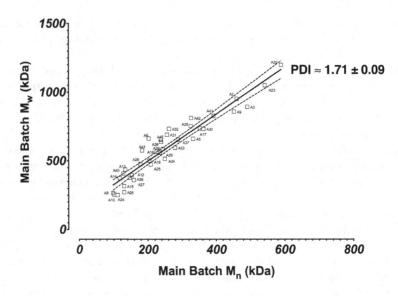

FIGURE 9.10 Trend of weight versus number average molecular masses by size exclusion chromatography for benchmark evaluations of respective batches of activated sludge PHBVs. The trend slope provides an indication of average polydispersity index value (Ð) [1].

The kinetics of 10 L pilot-scale extractions could be monitored based on the relative changes of viscosity-dependent mixing recirculation flow rates given constant pump frequency, and correction for the influence of temperature. Solvent viscosity decreases with temperature but increases with the concentration of dissolved polymer [110, 111]. For the same dissolved polymer concentration, viscosity also increases with the polymer molecular mass. For a given solvent, extraction temperatures decrease with an increase of average 3HV content towards the eutectic due to decrease in crystallinity. Figure 9.11 shows typical trends of extraction progress and temperature as a function of time in two cases (A45 with 1 %wt. 3HV and A24 with 36 wt.-% 3HV). Both extractions were made with a solvent loading of 50 gPHA/L. The solvent loading was defined as the theoretical maximum concentration of PHA in the solvent given 100% extraction efficiency. How much biomass to add to the extraction system to give this loading was determined from the extractable polymer content estimated by TGA [34]. The increase in specific mixing flow rate relates to the progress of extraction with increase of polymer concentration in the solvent. This value gives an indication of changes in the specific viscosity. The difference in the asymptotic extraction trends given the same temperature program in time illustrates the influence of different copolymer type on the extraction kinetics. Less crystalline materials are melted into the solvent at lower temperatures. The trends for A24 extraction occur over a broader temperature range, and this reflects the relatively broader distribution of copolymer composition at the A24 higher level of 3HV content. Specific flow rates are asymptotic to a plateau, and the residual upward trend by the end of extraction thus shows that not all polymer was recovered. The pilot 10 L 2-butanol extractions resulted, on average, in very a minor PHBV difference.

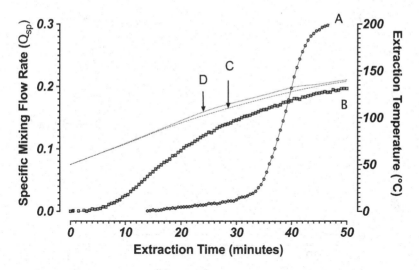

FIGURE 9.11 Typical trends showing progress of pilot-scale PHBV extraction as a function of time and solvent (2-butanol) temperature due to changes in solvent viscosity (dependent on flow rate) as the polymer becomes dissolved in solution. Specific flow rates are corrected for influence of temperature on the extraction solvent viscosity: A. Qsp for A45 was with 1% average wt. 3HV and M_w of 551 kDa, B. Qsp for A34 was with 36% average wt. 3HV in PHBV and M_w of 251 kDa, C. extraction temperature with time for A45, and D. extraction temperature with time for A24.

The 10 L extracted polymers were on average with 0.8 wt.-% higher 3HV content with respect to the Main Batch benchmark (10 mL) acetone extractions. Therefore, a minor amount of low 3HV content polymer (PHB) was not recovered for the conditions (time and temperature) applied at 10 L. The solution viscosity (specific flow rate) differences in magnitude (given the same conditions of solvent loading and approximate extent of extraction) reflect the differences in the polymer average molecular mass for A24 and A45. The pilot extracted weight average molecular masses for A24 and A45 were 251 and 551 kDa, respectively. Extraction extent, polymer type and polymer molecular mass can all be observed and optimised, case-by-case, through direct or indirect viscosity measurement during solvent extraction [8].

Extraction conditions targeted about 87% recovery (for PHB) based on kinetics estimated from a mass balance with isothermal extractions performed at laboratory (test tube) scale [8, 34]. Figure 9.12 illustrates typical results for the influence of 3HV content and extent of polymer recovery for the same granulate size distribution, but with different biomass samples containing a range of PHBV copolyesters, extracted at different isothermal temperatures over 45 minutes. From these trends, the first order kinetics of extraction as a function of time and temperature were estimated. An evaluation of isothermal polymer recovery kinetics shows how higher temperatures are required to recover copolymer blends of the more crystalline polymers with lower average 3HV content in 2-butanol. A higher average extraction temperature, or longer extraction time, may be expected to reduce losses of lower 3HV content co-polymers in the blend. However, the polymer decomposition rate resulting in loss

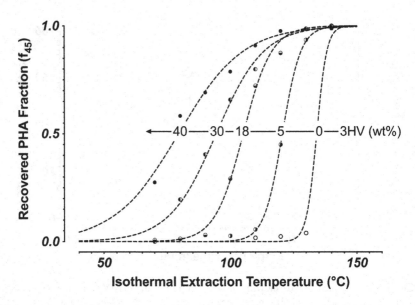

FIGURE 9.12   Fraction of extracted PHBV after 45 minutes as a function of isothermal extraction temperature and copolymer blend average 3HV percent weight content [1].

in polymer molecular mass during extraction has also been found to be dependent on time, temperature, concentration, solvent type, as well as factors related to the biomass that can influence polymer thermal and chemical stability [7, 8, 34, 112]. Pushing the extraction process in time to reach 100% polymer recovery will come with some sacrifice to the molecular mass. However, as discussed further below, this degree of sacrifice is predictable, if thermal stability is consistent from batch to batch. A balance of demands between productivity and product specification can therefore be made depending on the economy of the polymer and its intent in application.

The extraction parameter settings that were consistently applied for an acid stabilized biomass (T_d = 284.5 ± 1.5°C), resulted in an estimated 21 ± 2% loss of M_w. This loss was independent of the starting weight average molecular mass, the type of polymer, and the variation of the amount of polymer in the biomass. The consistently applied conditions of extraction (maximum 140°C, with average process temperature of 103°C, over nominally 52 minutes), for a consistently stabilized polymer-in-biomass, translated to a relatively constant average decomposition rate of 0.257 ± 0.039 h^{-1}. The half-life of the polymer molecular mass under these conditions of extraction is estimated to be 3.9 hours. In contrast, four replicate experiments disposing different Main Batches of PHA-rich biomass to isothermal incubation in air at 170°C over three hours resulted in an estimated decomposition rate of 0.011 ± 0.009 min^{-1}, meaning a M_w half-life of about 1.5 hours. Thus, control of molecular weight at industrial production can be achieved based on the predictability of how the polymer is affected during recovery.

Melt viscosity is a critical parameter for melt processing of polymers. The melt viscosity is positively correlated to the polymer molecular mass and negatively correlated to temperature. So, viscosity levels can be modulated by temperature control once the polymers are melted and at temperatures or over times before any crystallization occurs. At the same time, the maintenance of processability relates to the degree of stability of molecular weight during processing as a function of temperature, additives and processing conditions. A suitable molecular weight range imparting the right level of viscosity and melt strength characteristics to the polymer in the melt can depend on the type of processing. Polymers will degrade in molecular mass at rates dependent on temperatures (and shear rates) applied during melt processing, and the thermal stability of the polymer. Rates of decomposition may furthermore be influenced by the presence or absence of non-polymer chemical impurities that can influence the polymer chemical stability. Industrial process conditions, including the introduction of additives, need to be optimized to the polymer, its viscosity and how these might change due to decomposition in the melt. Ideally, melt decomposition rates should be low with respect to the requisite processing times and temperatures, and these rates should be a reproducible property of the polymer. Consistent decomposition rates permit for establishing routine protocols in processing that mitigate and/or can be applied to purposefully manipulate the end point average molecular mass after processing. A material that behaves consistently is deemed to have a greater breadth of opportunities for applications and will demand greater value.

Polymer melt stability was quantified by the loss of melt dynamic viscosity at 180°C over five minutes. Melt dynamic viscosity was calibrated to the polymer weight average molecular mass, and the rate of M_w loss was quantified as the linear decomposition rate at 180°C in the melt [34, 113]. Reference and Main Batch benchmark recovered samples resulted in polymers exhibiting a consistent decomposition rate at 180°C in the melt of 0.108 ± 0.003 min^{-1}. The pilot extracted polymers gave very similar but slightly lower melt decomposition rates of 0.090 ± 0.002 min^{-1}. The melt stability was independent of the starting M_w, and the results indicated for recovered polymers of similar processing behaviour except for the differences in initial molecular mass already discussed. Decomposition rates (r_s) decreased, for the polymer still in a melted state, by 66 and 96% as the temperature was decreased to 170 and 160°C, respectively. Dynamic viscosity in the melt increased by 49 and 60% in the temperature decrease to 170 and 160°C. This consistent observed performance of solvent recovered MMC PHA melt stability was independent of the type of copolymer blend.

Water and solvent-based methods have been investigated principally at laboratory scale for MMC PHA recovery and purification [17, 19, 22, 114]. In addition, selective biological methods are being advocated [115–117]. Organic solvent and aqueous based methods for PHA recovery both have merits and challenges. Both are likely to be more than suitable technically in some cases to reach similar specific property and economic targets for commercial PHA production. One or the other may be better suited to the task of recovery depending on the larger context of specifications, alongside economic and environmental performance. Hybrid recovery solutions are also possible. All methods require systematic evaluations of polymer quality at pilot and demonstration scale and over an extended period of routine MMC polymer production because no commercial-scale experience exists today for MMC PHA production and recovery. In this work, a so-called PHA-poor non-chlorinated solvent was used for the polymer recovery. These solvents require elevated temperatures often above the solvent boiling point to dissolve PHA. The more crystalline the PHA the greater the severity of extraction conditions with respect to solvent, temperatures, and extraction time. Thus, thermal and chemical stability of the polymer in the biomass is essential [7]. Literature suggests that there are many non-chlorinated solvents which can extract PHA from dried PHA-rich biomass at elevated temperatures [118]. Examples of PHA-poor solvents include common alcohols [8], acetone [119], dimethyl carbonate [120], cyclohexanone [121] and propylene carbonate [122]. These PHA-poor solvents, when sufficiently heated, will diffuse into the PHA matrix, break up crystallinity and dissolve the resulting amorphous polymer mass into solution [123]. The best type of PHA-poor solvent depends on the polymer type, the polymer condition in the biomass, and the methods applied in post extraction processing. Decomposition rates of the polymer during extraction are also solvent-dependent. Experimental results of extraction with freeze-dried biomass cannot be generally extrapolated to expectations of performance with oven dried biomass. Solvents that work well with PHB are not necessarily also optimal with a copolymer blend of PHBV with 40% average weight 3HV. Extraction results and methods for a 200 kDa M_w copolymer are not necessarily the same as for a 2000 kDa M_w copolymer. The type of biomass, polymer, and condition of the polymer in the biomass can all influence the extent and quality of the extracted polymer. Given a suitable solvent, biomass particulate matter can be separated from the viscous polymer-solvent solution, and the extracted polymer can then be recovered from the solvent upon cooling [8]. For this investigation, acetone and 2-butanol were used as the extraction solvents. The applied conditions of recovery (temperature, concentration and time) gave predictable outcomes and could be tuned to type of polymer and indicated for opportunities for master batching of polymer-type. Decomposition rates of the polymer during extraction were manageable and predictable, and the recovered polymer was with a consistent thermal stability and polydispersity (Đ). These outcomes were obtained with disclosed methods and processes that mimicked those anticipated for commercial-scale production.

9.3.5 Fate of Inorganic and Organic Impurities

Pilot extracted polymers were recovered to greater than 98% purity. A preliminary evaluation of selected chemical impurities in the biomass and recovered polymer was made towards an indication for fate of contaminants that can be present in municipal activated sludge [124–129]. Polymers and plastics are regulated in Europe in different directives and regulations that depend on the context of the materials, the products and/or the intent of use. For example, metals are considered under Registration, Evaluation, Authorisation and Restriction of Chemicals (REACH, EC no 1907/2006) [130], for packaging and packaging waste (Directive 94/62/EC) [131] and on plastic materials and articles intended to come into contact with food (EU Regulation 10/2011) [132]. Priority substances such as polyaromatic hydrocarbons (PAHs) are considered under REACH EU no 1272/2013 [133] as an amendment to EC no 1907/2006.

Accumulation A50 (S-feedstock) was studied with respect to the fate of contaminants coming into the process with the municipal activated sludge. A50 was made from 1041 gVSS starting activated sludge (73% gVSS/gTSS) fed with a 99:1 (COD basis) acetic to propionic acid in the feedstock. An estimated 1650 gVSS of dried PHA-rich biomass was produced (2 wt.-% 3HV in PHBV) with 40% gPHA/gVSS and 90 % gVSS/gTSS after downstream processing and drying. Thus, PHA was accumulated with negligible change in initial non-PHA (activated sludge biomass) as volatile suspended solids (VSS).

A general evaluation for selected PAHs and metal ions was made for grab samples of:

- the input activated sludge (before PHA accumulation, 7.6% dry solids),
- the same batch of activated sludge after accumulation with downstream processing but before drying (19.2% dry solids), and
- the oven dried recovered polymer after 1 L-scale 2-butanol solvent extraction, pressing to remove excess solvent, and rinsing with 2-butanol to remove residual extraction solvent.

Table 9.2 summarizes the measured cations and PAHs. Total sums were made by assuming substances below the method detection limits were present at half their detection limit. Not all regulated substances were assessed, but the outcomes were indicative of chemical fate through the piloting process.

The input activated sludge fed to the pilot processes contained measurable quantities of metal ions and PAHs as might be expected for a municipal wastewater treatment process handling wastewater from both municipal and industrial sources. The measured cations accounted for about 35% of the initial input activated sludge ash content. For the post-accumulation PHA-rich biomass, the total ash content represented 21% of the measured cations. Accounting for the PHA content of the biomass, it is estimated that about 65% the input cations entering with the activated sludge were removed from the post accumulation biomass. Acidification of the biomass prior to final dewatering and drying is directed to reduce the extent of cations associated with the polymer towards improving polymer-in-biomass thermal stability before drying. Acidification and dewatering should reduce the overall metal ion content of the PHA-rich-biomass suspended solids [134, 135]. Cation carryover to an organic solvent was anticipated to be low. The recovered polymer contained about 0.04% grams of measured cations per gram polymer and only 1% of the available cations in the PHA-rich biomass were carried over to the polymer.

Directive 94/62/EC requires that the sum of the amounts of lead, cadmium, mercury and hexavalent chromium be less than 100 mg/kg in packaging and packaging wastes. Lead and cadmium were measured, and both were under the requirement by at least an order of magnitude. Similarly, REACH (EC no 1907/2006) requires that cadmium in polyesters be less than 100 mg/kg. Food contact demands (EU Regulation 10/2011) are naturally more stringent and require that arsenic, chromium, lead, copper and zinc be less than 1, 1, 1, 5 and 100 mg/kg. The arsenic detection limit was higher than 1 mg/kg, and lead levels were estimated to be at the borderline of this regulation. Therefore, in this example, specific regulated metal content for this municipal activated sludge could limit the recovered polymer application involving

TABLE 9.2

PAH and Cation Analysis of A50 Input Activated Sludge, Accumulated PHA-Rich Biomass, and 2-Butanol Recovered PHBV. An Acetic and Propionic Acid Feedstock Was Used (99:1 COD Basis), Producing a PHBV with average 2 wt.-% 3HV

Entity	Input Activated Sludge	PHA-rich Biomass	Recovered PHBV	Units
PHA Content	<0.03	0.40	>0.98	gPHA/gVS
Ash Content	0.27	0.10	n.a.	g/gTS
Volatile Solids	1041	1650	642	g
Benzo(a)anthracene	0.13	<0.03	<0.081	mg/kg
Chrysene	0.13	<0.03	<0.081	mg/kg
Benzo(b, k)fluoranthene	0.27	0.078	0.096	mg/kg
Benzo(a)pyrene	0.13	0.052	<0.081	mg/kg
Indeno(1,2,3-cd)pyrene	0.13	<0.03	<0.081	mg/kg
Dibenzo(a,h)anthracene	<0.03	<0.03	<0.081	mg/kg
Naphthalene	0.066	<0.03	<0.081	mg/kg
Acenaphthylene	0.066	<0.03	<0.081	mg/kg
Acenaphthene	<0.03	<0.03	<0.081	mg/kg
Fluorene	0.13	<0.03	<0.081	mg/kg
Phenanthrene	0.13	0.052	0.13	mg/kg
Anthracene	0.066	<0.03	<0.081	mg/kg
Fluoranthene	0.27	0.078	<0.081	mg/kg
Pyrene	0.2	0.052	<0.081	mg/kg
Benzo(g,h,i)perylene	0.066	<0.03	<0.081	mg/kg
Total PAHs	1.00	<0.3	0.45	mg/kg
Aluminium (Al)	33,000	3900	57	mg/kg
Arsenic (As)	15	3.9	<4.6	mg/kg
Barium (Ba)	260	130	<19	mg/kg
Lead (Pb)	77	36	5.4	mg/kg
Boron (B)	31	5.6	<11	mg/kg
Iron (Fe)	29,000	15,000	210	mg/kg
Cadmium (Cd)	1.6	0.28	<0.19	mg/kg
Calcium (Ca)	25,000	1200	<91	mg/kg
Cobolt (Co)	8.5	3.7	<0.46	mg/kg
Copper (Cu)	450	300	5.4	mg/kg
Magnesium (Mg)	5000	570	<91	mg/kg
Manganese (Mn)	370	60	<3.7	mg/kg
Molybdenum (Mo)	35	23	<1.9	mg/kg
Nickel (Ni)	26	9.2	3.2	mg/kg
Selenium (Se)	4.6	2.4	<0.91	mg/kg
Silver (Ag)	4	2.5	<0.46	mg/kg
Tin (Sn)	22	3.2	2.8	mg/kg
Zinc (Zn)	1000	180	37	mg/kg
Total Cations	94,305	21,430	433	mg/kg

food contact if those regulated metals were not removed to a slightly greater extent before or after the extraction process. Metal levels associated with the polymer may be reduced by an aqueous acidic wash of the recovered polymer [84]. Application of organic volatile acids can be advantageous in this regard as they may be evaporated and recovered towards collecting the metals, leaving no residue in the polymer, and enabling reuse of process chemicals.

Non-polar organic chemicals are more likely to be co-extracted in an organic solvent together with the polymer compared to cations. Some non-polymer microbial constituents that become co-extracted can be of functional benefit. Microbial lipids that are co-extracted in the solvent have been shown to be useful as compounding agents to improve the mechanical properties of the polymer [136]. The residual extraction solvent that is evaporated from recovered polymer will leave the non-volatile, soluble impurities with the polymer. Measured PAH levels were low in the source biomass (less than 2 ppm in total). Levels were reduced considerably in the biomass after the accumulation process and most substances in the PHA-rich biomass were below the detection limit. With respect to the non-polymer volatile solids, the PAH absolute levels were reduced by 60%. Based on just the two detectable PAHs in the polymer (benzo[b k]fluoranthene, and phenanthrene), 61% of the maximum possible carry-over from the PHA-rich biomass was estimated. Therefore, the degree of organic contamination of the PHA-rich biomass is important to monitor and control to avoid undue carry-over of unwanted organic impurities by solvent extraction. These impurities may be reduced for recovery by a preliminary solvent rinse before extraction [8]. In this case, the PAH levels were already sufficiently low. REACH (EU No 1272/2013) [133] amendment to EC no 1907/2006 [130] requires that plastic articles in general should not contain more than 1 mg/kg of eight listed PAHs (benzo[a]anthracene, chrysene, benzo[b]fluoranthene, benzo[k]fluoranthene, benzo[j]fluoranthene, benzo[e]pyrene, benzo[a]pyrene and dibenzo[a,h]anthracene). Plastic articles likely to come into habitual contact with children should not contain more than 0.5 mg/kg of these listed PAHs. All the listed PAHs that were measured more than met the requirements for application of the polymers recovered from municipal activated sludge in child contact applications. This preliminary evaluation supports a potential for MMC PHA produced from municipal activated sludge to readily meet and more than exceed regulatory requirements for even the more stringently controlled applications. However, that potential and any need for additional but feasible steps of processing, that will add to the costs of recovery, do rely on the levels of unwanted impurities present in the PHA-rich-biomass to start with. These considerations related to both polymers and additives used in plastic are no different from the management of toxicity of virgin or recycled polymers, additives and plastics derived from fossil crude oil [137, 138]. In principle, this evaluation also shows that the fate of contaminants may be readily evaluated and so any risks and remedies can be determined based on evaluation of the production of raw materials (activated sludge and accumulation feedstock), the quality of the accumulated PHA-rich biomass, and all these factors juxtaposed the processing costs and specific requirements and commercial intent for the polymers. This is no different than any other well-accepted production process taking in raw materials for conversion into fine chemicals.

9.3.6 COPOLYMER BLEND DISTRIBUTION AND MECHANICAL PROPERTIES

Mechanical test elements were made from eight distinct batches of pilot produced and recovered PHBV copolyesters in the average 3HV range from 20 to 45 wt.-% (Table 9.3). With 3HV contents from approximately 15 to 40% (Figure 9.6), polymer blends, with the same thermal history (melt and quench and aged at room temperature for two weeks to maximize crystallinity), consistently melted in the range starting from approximately 40 and ending at 180°C. At the same time, the weight average melting temperatures, and the crystallinity decreased as a function of the average 3HV content (Figure 9.4). Therefore, in the average 3HV range from 20 to 45% weight, the copolymer blends were interpreted to contain a similar range of 3HV content, but in different proportions. The objective of mechanical testing was to evaluate how shifts in the average copolymer content modulated the mechanical properties given an otherwise similar overall range in distribution of copolymer type. For the selected batches, weight average molecular masses were distributed between 200 and 600 kDa (average 400 kDa, median 426 kDa), so a potential influence from molecular mass on the mechanical properties was examined as well.

Factors that can influence polymer mechanical properties include the extent and distribution of crystallinity, molecular mass, method of processing, composition of the polymer, and temperature [60]. All mechanical test samples were processed and aged in the same manner, and average copolymer content in the blend was found to directly correlate to extent of crystallinity. Aging was applied to enable the materials to fully develop their crystallinity [61]. The samples tested were in a progression of 3HV content towards the PHBV eutectic point. The mechanical properties were anticipated to be modelled by the general function according to Equation 9.11 [60, 139, 140]:

$$\log(P) = A + B \cdot f_{3HV} + \frac{C}{M_w} \tag{9.11}$$

Where P is a given mechanical property, f_{3HV} is the average weight percent of 3HV in PHBV, and M_w is the weight average molecular mass (kDa). Limiting characteristic

TABLE 9.3

Accumulation Recovered PHBV Batches Used for Pure Polymer Mechanical Testing

Accumulation Batch	Average 3HV [wt.-%]	M_w [kDa]
A09	21	694
A16	22	426
A10	24	232
A15	24	276
A18	30	513
A17	31	440
A13	35	221
A14	42	332

values of mechanical properties were anticipated as the average copolymer blend content increases from pure PHB to a eutectic average copolymer blend composition in the neighbourhood of 50 wt.-% 3HV in PHBV.

An equation with limiting conditions (Equation 9.12) was estimated assuming that the property values would not change significantly within the region of the eutectic. Mechanical property values at 42 wt.-% average 3HV content were used to estimate an assumed limiting property value in approach to the eutectic. Thus, it was assumed that from this limiting value up to the eutectic, the further increase in average 3HV content at least up to the eutectic point would have marginal if any further influence on properties:

$$\log(P) = A + B \cdot f_{3HV} + \frac{C}{M_w} \qquad 20 \leq f_{3HV} \leq f_{\lim}$$

(9.12)

$$\log(P) = A + B \cdot f_{\lim} + \frac{C}{M_w} \qquad f_{\lim} \leq f_{3HV} \leq f_e$$

Where f_{3HV} is the polymer average 3HV content, f_{lim} is the limiting average 3HV content for influence on properties, and f_e is the eutectic 3HV content (percent weight). Trends of tensile modulus, tensile strength, elongation at break, flexural modulus and flexural stress at 3.5% strain are shown in Figure 9.13 and estimated parameter values are given in Table 9.4. Based on this assumption, it was predicted that the mechanical properties become insensitive to increasing average 3HV contents towards the eutectic at an f_{lim} value of 38 ± 4 average weight percent 3HV. An influence of molecular mass was only found to be statistically significant for the case of tensile strength with a "C" coefficient of −19 kDa·log(MPa). This suggests a threshold region for an influence of weight average molecular mass between 100 and 300 kDa. The outcome is in close agreement to results of molecular weight influence on tensile strength for pure culture PHBV [141]. Tensile strength is predicted to increase by only 2% from 1000 to 2000 kDa but decrease by 16 and 23% from 1000 to 200 and 100 kDa. Figure 9.13 shows the trend for the average molecular mass, of 400 kDa, while model fit values based on the respective specimen M_w are also indicated.

Replicate specimens exhibited a degree of variation as might be expected for mechanical testing. Increased absolute variability with increased elongation to break was understood to be due to the progressively increased probability of an influence of a test sample defect as the material became stretched plastically. It was a challenge to manufacture, at small scale and with limited materials, identical test specimens without any mould defects. Notwithstanding, the mechanical property trends exhibited a clear and principal influence of copolymer blend average 3HV content in the range from 20 to 40 wt.-% 3HV. Thus, pre-eutectic copolymer PHBV mechanical properties can be modulated by manipulation of the copolymer distribution components either as part of the accumulation process (feedstock control) or during the extraction process (PHA-rich biomass master batching).

In contrast, previous work examining properties of post-eutectic PHBV random copolymer blends did not suggest a systematic shift in properties as a function of 3HV content with average 3HV contents from 55 to 75 wt.-% [59]. In this previous study,

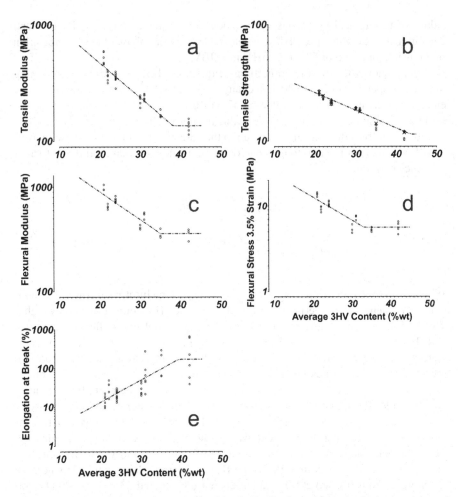

FIGURE 9.13 Influence of copolymer blend distribution on mechanical properties of medium to high pre-eutectic PHBVs. Property values at 3HV of 42% were used to represent the limiting eutectic property value with the data fit by least squares regression to Equation 4. Individual sample values are shown in the scattergram. Parameter values are reported in Table 9.4. An influence of M_w was significant for tensile strength (b). The trends line in (b) shows the trend for a 400 kDa M_w and the Xs indicate the model fit based on the sample to sample values of Mw.

properties were not significantly influenced by a change in 3HV content (elasticity modulus of 907 ± 42 MPa, 6 ± 1%). Further evaluation of these materials did reveal distinctions of crystallization behaviour between pre- and post-eutectic PHBVs [142]. Therefore, the pre- and post-eutectic average 3HV weight PHBV blends are understood to behave as two distinctly different types of polymers in response to changes of average copolymer composition. In a recent study, evaluations were made of an MMC PHA with 27% average weight 3HV in PHBV produced at pilot scale with an acetic and propionic acid blended feedstock [143]. Tensile modulus (940 MPa),

TABLE 9.4

With Reference to Equation 9.4, Estimated Model Parameter Values with Standard Errors for the Trends Shown in Figure 9.13. Property (P) Values at f_{lim} are Estimated from Sample A14 Average Values (9.3). Flexural Stress is at 3.5% Strain. The Parameter f_{lim} Was Estimated Based on the Best-Fit Values of A, B, and C from Equation 9.4. Parameter Values A or C Were Excluded if They Were Found to be Statistically Not Significantly Different than Zero (n.s.)

Property (P)	Tensile Modulus (MPa)	Tensile Strength (MPa)	Elongation at Break (%)	Flexural Modulus (MPa)	Flexural Stress (MPa)
A	3.272 ± 0.0484	1.767 ± 0.0192	n.s.	3.493 ± 0.072	1.641 ± 0.076
B	-0.030 ± 0.002	-0.015 ± 0.0007	0.057 ± 0.001	-0.027 ± 0.003	-0.027 ± 0.003
C	n.s.	-18.99 ± 3.161	n.s.	n.s.	n.s.
f_{lim} % 3HV	38	43	39	35	33
P at f_{lim}	136	11	172	356	6

tensile strength (19 MPa) and elongation at break (13%) were in range of agreement to this work considering possible differences due to extrusion process methods of the test sample preparation in this case. Pure culture PHBV with 22 wt.-% average 3HV aged test specimens [144] resulted in markedly lower tensile strength (15.6 MPa) and higher elongation at break (182%). In this pure culture work, weight average molecular mass was significantly higher (1600 kDa) and the melt range suggested a potentially narrower copolymer blend distribution (80 to 140°C) compared to the one tested in this investigation (40 to 180°C). A higher molecular mass should increase tensile strength, but the interpreted narrower compositional distribution again suggests an influence from the copolymer content on mechanical properties with pre-eutectic PHBV copolymer blends. Thus, master batch blending (or fractionation) to influence the copolymer compositional distribution is interpreted to be important to quality control for regulation of mechanical properties at industrial-scale production for pre-eutectic PHBVs. Production control for consistency of average molecular mass and thermal stability of molecular mass during processing is similarly considered to be critical for reproducible processability and mechanical properties.

The trends of mechanical properties could not be extrapolated to the limit of PHB. Mechanical testing on PHBV with 1% weight 3HV extruded specimens gave tensile modulus and strength of 3100 and 31 MPa [143]. This work suggests an extrapolated lower value for tensile modulus and a higher value for tensile strength. Below 20% average 3HV content for the as-produced polymers, the crystallinity increases similarly, but the compositional distribution is interpreted to narrow significantly based on the melt peak temperature range (Figure 9.6). It seems likely that properties for commercial materials will need to be tuned by batch blending towards modulating the breadth of the distribution for these lower 3HV PHBVs. The expressed opportunities for property modulation and quality control through master batch blending of MMC PHBVs are recommended to be developed further.

9.4 CONCLUSIONS AND OUTLOOK

9.4.1 FURTHER OUTLOOK

Recovery of PHA from biomass is more than simply comparing yield and purity [19, 22], notwithstanding the important coupling of these two parameters to the economics of recovery and production efficiency. Economics of recovery cannot be properly understood without anchoring such discussion in the economics of the polymers within a targeted application. Quality is not tied to a specific polymer purity (or impurity), average molecular mass, melting point, crystallinity or even PHA type. These properties are inextricably linked to a spectrum of needs for a number of possibly meaningful PHA application areas [15, 145–153, 154–159]. Despite the breadth of innovation for applications with PHA in the literature, it seems likely that successful applications for MMC PHA will not evolve in the market until there is security for an established material supply. In this work it was found that a PHBV with nominally 32 wt.-% average 3HV can be reliably produced using fermentate from primary sludge as feedstock. Both primary and municipal activated sludges are significantly available raw materials, so these results are relevant for commercial-scale production of a specific polymer type. A 32% 3HV PHBV is not currently commercially available. It has anticipated desirable material properties for applications, but it crystallizes slowly making it more challenging to process than PHB. Processing of semi-crystalline polymers is often with the goal of manipulating the crystalline development due to needs in processing but also for the polymer application. This development can involve chemical additives but also the processing conditions and application of forces during polymer crystallization can be critical for establishing unique and attractive mechanical properties. Such technique development and optimization take significantly more PHBV than can be produced at pilot scale. Innovation to target commercial exploitation for the unique nature and properties of these moderate sub-eutectic PHBVs and to solve particular processing challenges will take tens, if not hundreds or more, kilograms of supply. This level of supply would take at least a demonstration-scale facility because it is not just the supply of the polymer, but it is also the development of the application and economy with respect to the polymer quality and its management that must be reproducibly maintained within the context of commercial production with a defined and commercially available supply of raw materials.

Particular polymer properties can be critical to compounding, processing and/or the product application. Without specifying and tuning the properties to a specific set of needs for compounding, processing or application, one can only meaningfully consider the ability to regulate and control properties in general. Therefore, examples of any given MMC PHA quality from reported pilot or laboratory-scale work may not be indicative of the industrial-scale commercial potential if these data are not coupled to demonstration of ability for production quality control in properties. The ability to control the polymer quality in production for commercial applications at industrial scale will most certainly lead to the evolution of downstream market demands. Therefore, MMC PHA have real commercial value if the material properties can be understood to be readily controllable to within well-defined product quality windows. In this investigation it was found that the MMC PHA material properties including copolymer type, blend composition, thermal properties, molecular mass, thermal stability, impurities and mechanical properties have potential to be controlled both within the accumulation bioprocess and in the downstream chemical recovery process.

Much optimistic as well as pessimistic speculation surrounds the idea to scale up MMC PHA production methods after two decades of advances through scientific and engineering efforts up to pilot-scale. The optimistic outlook is that polymer product quality demands from MMC PHA production and recovery will (somehow) be robustly met. PHA value chains are thereby expected to be able to viably support a reliable amount of polymer supply to market, and to meet downstream commercial and regulatory standards for, what are, despite much creative drive and ongoing innovations and positive examples in principle, still currently undefined, commercial products and services in practice. Since no commercial-scale mixed culture PHA production exists today, commercial products and services are difficult to nurture without a stable source of polymer supply from which links to opportunities and market can be evolved and, sometimes most unexpectedly, discovered [160].

Speculative scepticism for MMC PHA production advocates a doubt for success with MMC PHA at commercial scale. Product quality control difficulties are anticipated by sceptics given the often unspecified, dynamic and diverse nature of PHA-storing bacteria possibly present within enriched MMCs [161]. How can one be sure of the polymers that will be produced without full knowledge and control of the "microbial producers present" in contrast to PHA production by pure culture methods? In addition, input organic residual (waste) feedstock sources are expected to have unpredictable supply and composition. Acidogenic fermentation product control advances [41, 162–167], but the effluent VFA compositions may not be possible to keep absolutely well-controlled in all cases. A window of variability will always exist. The sceptic's negative criticism implies that MMC polymers will be of little value because of an underlying assumption that there will be a deficit in quality control with MMC PHA production. However, and to the contrary, this investigation does not support the sceptic's hypothesis. Within a "worst case scenario" of an MMC that was not in any way purposefully optimized and enriched for PHA production, principal properties followed consistent and predictable trends, and this lends support to the real-world potential for industrial quality control management with scaled up MMC PHA production.

The context of MMC PHA production is envisioned to be a collection of several regional PHA-rich biomass production facilities supplying a centralized polymer recovery facility. A commercially relevant quantity of PHA supply to the market is expected to be derived from the cumulative recovery of polymer from a multiple of regional PHA-rich biomass production sources. Therefore, the polymer recovery activities may be expected to serve several MMC biomass production sources, using substrates derived from several types of municipal and industrial organic residuals and waste streams, and generating a wide variety of biomass matrices containing a mix of possible polymer types and molecular weights. Recovery facilities are likely to be more centralized and with a larger capacity than any individual PHA-rich biomass production facility due to economies of scale. Thus, the recovery methods may need to be able to accommodate a range of polymer types and content for a diversity of biomass sources. The methods applied in this investigation do meet these needs. Conditions of extraction are tuneable based on polymer-type and biomass granulate dependent kinetics. These kinetics can be estimated from simple standardized test tube scale evaluations [34] that were reliably applied to pilot-scale recovery methods [8].

Specific economically well-trimmed solutions for full-scale polymer recovery methods and conversion processes to market cannot be tackled a priori without knowing the details, range and context of the value chain(s) involving the full-scale feedstocks, the enrichment biomass sources, the resultant polymers and the application intent. Notwithstanding, the goal in this work was to derive general principles, strategies and outlook, related to the biomass, the polymers and the process chain. Capital cost for polymer recovery infrastructure is expected to be the most significant bottleneck for launching industrial-scale MMC PHA production, and, therefore, it is currently the most important factor to de-risk towards engaging regional stakeholder support and investment. Since a great opportunity exists with a significant mass of potential raw material supply for MMC PHA production coming from the municipal sector (organic fraction of municipal solids wastes as well as primary and secondary wastewater treatment sludges), some form of long-term public/private cooperation will become essential for success in scaled-up MMC PHA commercial production.

9.4.2 CONCLUSIONS

From this investigation it can be concluded that:

- A full-scale municipal activated sludge is a robust raw material for MMC production of poly(3-hydroxybutyrate-*co*-3-hydroxyvalerate) blends at pilot scale.
- Under defined conditions of PHA accumulation, the copolymer type, expressed as the weight average 3HV content, is a direct function of the feedstock organic composition.
- The thermal properties such as melting and glass transition temperatures are functions of the copolymer type and the as-accumulated copolymer blends recovered by solvent extraction are miscible.
- Miscible blend compositions, to form well-defined and specific polymer types, can be created by selective combination and extraction of biomass batches containing distinctly different copolymer types.
- The shelf life of dried PHA-rich biomass is long, meaning that batches and grades can be inventoried for generating selected master batches as part of the recovery process.
- Compounding ingredients can be introduced before or during the extraction process in order to form homogenous formulations and avoid a melt compounding step.
- Optimal conditions for extraction of a specific polymer type, and the extent of molecular weight loss during extraction are predictable and can be engineered to specific targets.

- Molecular mass of the accumulated polymers is decreased during accumulation by the presence of ethanol and longer chain length fatty acids in the feedstock.
- Molecular mass of the accumulated polymers is understood to be optimal when the accumulation is not unduly prolonged beyond the time of PHA saturation in the biomass.
- Molecular mass is sensitive to handling after accumulation such that activities, times and temperatures before, during and after drying should be well-controlled for reproducible outcomes.
- Polymer blends with consistent thermal melt and crystallization properties, and high thermal stability can be produced.
- The levels of contamination of the raw materials will influence the recovery process and methods necessary to reach specific targets of impurities for the recovered polymers.
- The polymer mechanical properties in the mid to high pre-eutectic range are modulated in a predicable way by changes to the copolymer distribution.

ACKNOWLEDGEMENTS

PHARIO was financially supported by subsidy from the Topsector Energy program of the Dutch Ministry of Economic Affairs (TKI Biobased Economy) and contributions by the PHARIO project partners: Veolia Water Technologies, the Dutch water authorities Brabantse Delta, DeDommel, Fryslân and Scheldestromen, STOWA, KNN and Slibverwerking Noord-Brabant. PHARIO was a team effort and the author list, comprising the PHA production technology developers, and project management team, unfortunately does not include all contributors that made the project deliverables and the overall project a great success. The authors most gratefully acknowledge: Kruger AS: Hans Erik Madsen, Carlos Llobet Pérez; Hydrotech: Carles Pellicer-Nacher; Waterschap Brabantse Delta: Lennert De Graaf, Martijn Gebraad, Gijs Doornbusch, Louise Johann-Deusing, Sean van der Meulen, Kees van Hoof, Ruben de Wild, Danny Tak, Leon Maas, Jan van Eekelen, Tomas van Eekeren, Eric Groenewald, Jack Eversdijk, Levien van Dixhoom; Waterschap De Dommel: Doy Schellekens, Peter van Horne, Victor Claessen, Aad Oomens; KNN Advies and KNN Bioplastics: Yme Flapper, Peter Dijkstra, Cindy Visser, Onno de Vegt; Waterschap Scheldestromen: Jo Nieuwlands, Avans Hogeshool: Koen van Beurden, Jack van Schijndel; Pezy Group: Jan Hoekstra, Joop Onnekink, Abel Hartlief, Thijs Feenstra; Biomer: Urs Hänggi; Bioplastech: Kevin O'Connor, Ramesh Padama; Wageningen University and Research Centre: Gerald Schennink, Richard op den Kamp, Hans Mooibroek, Frans Kappen; Veolia: Jacob Bruus, Eric Train, Sebastien Bessenet, Yves Ponthieux, Corrine Jamot, Carina Roselius, Gitte Andersen, Stig Stork. The section on Fate of Cation and Organic Impurities was inspired and influenced from information and insights on European regulatory frameworks studied as part of the EU Horizon 2020 RESURBIS project (Grant 730349, resurbis.eu). Alan Werker is grateful to RESURBIS coordinator, Prof. Mauro Majone, for the opportunity of involvement to follow RESURBIS developments as an external advisor.

REFERENCES

1. Bengtsson S, Werker A, Visser C, *et al. PHARIO Stepping Stone To A Sustanable Value Chain For PHA Bioplastics Using Municipal Activated Sludge.* STOWA, Amersfoort, The Netherlands, 2017.
2. Werker A, Hjort M, Bengtsson S, *et al.* Consistent production of high quality PHA using activated sludge harvested from full scale municipal wastewater treatment – PHARIO. *Water Sci Technol* 2018;78(11):2256–2269.
3. Morgan-Sagastume F, Hjort M, Cirne D, *et al.* Integrated production of polyhydroxy-alkanoates (PHAs) with municipal wastewater and sludge treatment at pilot scale. *Bioresour Technol* 2015;181:78–89.
4. Arcos-Hernández M, Montaño-Herrera L, Murugan Janarthanan O, *et al.* Value-added bioplastics from services of wastewater treatment. *Water Pract Technol* 2015;10(3):546.
5. Arcos-Hernandez M V., Pratt S, Laycock B, *et al.* Waste activated sludge as biomass for production of commercial-grade polyhydroxyalkanoate (PHA). *Waste Biomass Valorization* 2013;4(1):117–127.
6. Bengtsson S, Karlsson A, Alexandersson T, *et al.* A process for polyhydroxyalkanoate (PHA) production from municipal wastewater treatment with biological carbon and nitrogen removal demonstrated at pilot scale. *N Biotechnol* 2017;35:42–53.
7. Werker AG, Johansson P, Magnusson P, *et al.* Method for recovery of stabilized polyhydroxyalkanoates from biomass that has been used to treat organic waste. 2013;(US20130203954A1).
8. Werker A, Johansson P, Magnusson P. Process for the extraction of polyhydroxyalkano-ates from biomass. 2015;(US20150368393A1).
9. Morgan-Sagastume F, Heimersson S, Laera G, *et al.* Techno-environmental assessment of integrating polyhydroxyalkanoate (PHA) production with services of munici-pal wastewater treatment. *J Clean Prod* 2016;137:1368–1381.
10. Dietrich K, Dumont MJ, Del Rio LF, *et al.* Producing PHAs in the bioeconomy — Towards a sustainable bioplastic. *Sustain Prod Consum* 2017;9:58–70.
11. Valentino F, Moretto G, Lorini L, *et al.* Pilot-scale polyhydroxyalkanoate production from combined treatment of organic fraction of municipal solid waste and sewage sludge. *Ind Eng Chem Res* 2019. doi:10.1021/acs.iecr.9b01831.
12. Tamis J, Lužkov K, Jiang Y, *et al.* Enrichment of Plasticicumulans acidivorans at pilot scale for PHA production on industrial wastewater. *J Biotechnol* 2014;192(Part A):161–169.
13. Troschl C, Meixner K, Drosg B. Cyanobacterial PHA production—review of recent advances and a summary of three years' working experience running a pilot plant. *Bioengineering* 2017;4(2):26.
14. Rodriguez-Perez S, Serrano A, Pantión AA, *et al.* Challenges of scaling-up PHA pro-duction from waste streams. A review. *J Environ Manage* 2018;205:215–230.
15. Visakh PM. Polyhydroxyalkanoates (PHAs), their blends, composites and nanocom-posites: State of the art, new challenges and opportunities. In: Roy I, Visakh PM, edi-tors. *Polyhydroxyalkanoate (PHA) based Blends, Composites and Nanocomposites.* RSC Green Chemistry, United Kingdom, 2015: 1–17.
16. Kourmentza C, Plácido J, Venetsaneas N, *et al.* Recent advances and challenges towards sustainable polyhydroxyalkanoate (PHA) production. *Bioengineering* 2017;4(2):55.
17. Valentino F, Morgan-Sagastume F, Campanari S, *et al.* Carbon recovery from wastewa-ter through bioconversion into biodegradable polymers. *N Biotechnol* 2017;37:9–23.
18. Zhang L, Xu C, Champagne P, *et al.* Overview of current biological and thermo-chem-ical treatment technologies for sustainable sludge management. *Waste Manag Res* 2014;32(7):586–600.

19. Kosseva MR, Rusbandi E. Trends in the biomanufacture of polyhydroxyalkano-ates with focus on downstream processing. *Int J Biol Macromol* 2018. doi:10.1016/j.ijbiomac.2017.09.054.

20. Raza ZA, Abid S, Banat IM. Polyhydroxyalkanoates: Characteristics, production, recent developments and applications. *Int Biodeterior Biodegrad* 2018;126(October 2017):45–56.

21. Nikodinovic-Runic J, Guzik M, Kenny ST, *et al*. Carbon-rich wastes as feedstocks for biodegradable polymer (polyhydroxyalkanoate) production using bacteria. *Adv Appl Microbiol* 2013;84:139–200.

22. Jacquel N, Lo C-WW, Wei Y-HH, *et al*. Isolation and purification of bacterial poly(3-hydroxyalkanoates). *Biochem Eng J* 2008;39(1):15–27.

23. Możejko-Ciesielska J, Kiewisz R. Bacterial polyhydroxyalkanoates: Still fabulous? *Microbiol Res* 2016;192(2016):271–282.

24. Laycock B, Halley P, Pratt S, *et al*. The chemomechanical properties of microbial poly-hydroxyalkanoates. *Prog Polym Sci* 2014;39(2):397–442.

25. Raza ZA, Abid S, Banat IM. Polyhydroxyalkanoates: Characteristics, production, recent developments and applications. *Int Biodeterior Biodegrad* 2018;126:45–56.

26. Werker AG. An evaluation of full-scale activated sludge dynamics using microbial fatty acid analysis. *Water Sci Technol* 2006;54(1):11–19.

27. Muszyński A, Tabernacka A, Miłobedzka A. Long-term dynamics of the microbial community in a full-scale wastewater treatment plant. *Int Biodeterior Biodegrad* 2015;100:44–51.

28. Werker A, Bengtsson S, Karlsson A. Method for accumulation of polyhydroxyalkano-ates in biomass with on-line monitoring for feed rate control and process termination. 2010;(US8748138B2).

29. Pons MN, Spanjers H, Baetens D, *et al*. Wastewater characteristics in Europe - A sur-vey. *Eur Water Manag Online* 2004;70(4):1–10.

30. Pratt S, Werker A, Morgan-Sagastume F, *et al*. Microaerophilic conditions support elevated mixed culture polyhydroxyalkanoate (PHA) yields, but result in decreased PHA production rates. *Water Sci Technol* 2012;65(2):243–246.

31. Werker A, Bengtsson S, Hjort M, *et al*. Process for enhancing polyhydroxyalkanoate accumulation in activated sludge biomass. 2017;(US 2017 / 0233271 A1).

32. Morgan-Sagastume F, Valentino F, Hjort M, *et al*. Acclimation process for enhanc-ing polyhydroxyalkanoate accumulation in activated-sludge biomass. *Waste Biomass Valorization* 2019;10(4):1065–1082.

33. Garcia-Ochoa F, Gomez E, Santos VE, *et al*. Oxygen uptake rate in microbial pro-cesses: An overview. *Biochem Eng J* 2010;49(3):289–307.

34. Chan CM, Johansson P, Magnusson P, *et al*. Mixed culture polyhydroxyalkanoate-rich biomass assessment and quality control using thermogravimetric measurement meth-ods. *Polym Degrad Stab* 2017;144:110–120.

35. Werker A, Lind P, Bengtsson S, *et al*. Chlorinated-solvent-free gas chromato-graphic analysis of biomass containing polyhydroxyalkanoates. *Water Res* 2008;42(10–11):2517–2526.

36. Valentino F, Karabegovic L, Majone M, *et al*. Polyhydroxyalkanoate (PHA) storage within a mixed-culture biomass with simultaneous growth as a function of accumula-tion substrate nitrogen and phosphorus levels. *Water Res* 2015;77:49–63.

37. Arcos-Hernandez M V., Gurieff N, Pratt S, *et al*. Rapid quantification of intracellu-lar PHA using infrared spectroscopy: An application in mixed cultures. *J Biotechnol* 2010;150(3):372–379.

38. De Grazia G, Quadri L, Majone M, *et al*. Influence of temperature on mixed microbial culture polyhydroxyalkanoate production while treating a starch industry wastewater. *J Environ Chem Eng* 2017;5(5):5067–5075.

39. Johnson K, Kleerebezem R, van Loosdrecht MCM. Influence of the C/N ratio on the performance of polyhydroxybutyrate (PHB) producing sequencing batch reactors at short SRTs. *Water Res* 2010;44(7):2141–2152.

40. Valentino F, Brusca AA, Beccari M, *et al.* Start up of biological sequencing batch reactor (SBR) and short-term biomass acclimation for polyhydroxyalkanoates production. *J Chem Technol Biotechnol* 2013;88(2):261–270.

41. Tamis J, Joosse BM, van Loosdrecht MCM, *et al.* High-rate volatile fatty acid (VFA) production by a granular sludge process at low pH. *Biotechnol Bioeng* 2015;112(11):2248.

42. Zoetemeyer RJ. *Acidogenesis of Soluble Carbohydrate-Containing Wastewaters.* University of Amsterdam, The Netherlands, 1982.

43. Lemos PC, Serafim LS, Reis MAMM. Synthesis of polyhydroxyalkanoates from different short-chain fatty acids by mixed cultures submitted to aerobic dynamic feeding. *J Biotechnol* 2006;122(2):226–238.

44. Chung Wook C, Yoon Seok K, Young Baek K, *et al.* Articles : Isolation of a *Pseudomonas* sp. strain exhibiting unusual behavior of poly (3 - hydroxyalkanoates) biosynthesis and characterization of synthesized polyesters. *J Microbiol Biotechnol* 1999;9(6):847.

45. Gottschalk G. *Bacterial Metabolism.* 2nd edition. Springer-Verlag: New York, 1985.

46. Alderete JE, Karl DW, Park CH. Production of poly(hydroxybutyrate) homopolymer and copolymer from ethanol and propanol in a fed-batch culture. *Biotechnol Prog* 1993;9(5):520–525.

47. Gurieff N. *Production of Biodegradable Polyhydroxyalkanoate Polymers Using Advanced Biological Wastewater Treatment Process Technology.* University of Queensland, Australia, 2007.

48. Jian Y, Yingtao S. Metabolic carbon fluxes and biosynthesis of polyhydroxyalkanoates in *Ralstonia eutropha* on short chain fatty acids. *Biotechnol Prog* 2004;20(4):1015–1024.

49. Laycock B, Arcos-Hernandez M V., Langford A, *et al.* Thermal properties and crystallization behavior of fractionated blocky and random polyhydroxyalkanoate copolymers from mixed microbial cultures. *J Appl Polym Sci* 2014;131(19). doi:10.1002/app.40836.

50. Villano M, Beccari M, Dionisi D, *et al.* Effect of pH on the production of bacterial polyhydroxyalkanoates by mixed cultures enriched under periodic feeding. *Process Biochem* 2010;45(5):714–723.

51. Laycock B, Arcos-Hernandez M V., Langford A, *et al.* Crystallisation and fractionation of selected polyhydroxyalkanoates produced from mixed cultures. *N Biotechnol* 2014;31(4):345–356.

52. Janarthanan OM, Laycock B, Montano-Herrera L, *et al.* Fluxes in PHA-storing microbial communities during enrichment and biopolymer accumulation processes. *N Biotechnol* 2016;33(1):61–72.

53. Montano-Herrera L, Laycock B, Werker A, *et al.* The evolution of polymer composition during PHA accumulation: The significance of reducing equivalents. *Bioengineering* 2017;4(1):20.

54. Wang X, Carvalho G, Reis MAM, *et al.* Metabolic modeling of the substrate competition among multiple VFAs for PHA production by mixed microbial cultures. *J Biotechnol* 2018;280(October 2017):62–69.

55. Albuquerque MGE, Carvalho G, Kragelund C, *et al.* Link between microbial composition and carbon substrate-uptake preferences in a PHA-storing community. *ISME J* 2013;7(1):1–12.

56. Carvalho G, Pedras I, Karst SM, *et al.* Functional redundancy ensures performance robustness in 3-stage PHA-producing mixed cultures under variable feed operation. *N Biotechnol* 2018;40(October 2016):207–217.

57. Queirós D, Fonseca A, Rossetti S, *et al.* Highly complex substrates lead to dynamic bacterial community for polyhydroxyalkanoates production. *J Ind Microbiol Biotechnol* 2017;44(8):1215–1224.

58. Bluhm TL, Hamer GK, Marchessault RH, *et al*. Isodimorphism in Bacterial Poly(β-hyd roxybutyrate-co-β-hydroxyvalerate). *Macromolecules* 1986;19(11):2871–2876.
59. Arcos-Hernández MV, Laycock B, Donose BC, *et al*. Physicochemical and mechanical properties of mixed culture polyhydroxyalkanoate (PHBV). *Eur Polym J* 2013;49(4):904–913.
60. Seymour RB, Carraher CE. *Structure-Property Relationships in Polymers*. 1st edition. Plenum Press, New York, 1984.
61. Bloembergen S, Holden DA, Hamer GK, *et al*. Studies of composition and crystallinity of bacterial poly(β-hydroxybutyrate-*co*-β-hydroxyvalerate). *Macromolecules* 1986;19(11):2865–2871.
62. Yoshie N, Saito M, Inoue Y. Effect of chemical compositional distribution on solid-state structures and properties of poly(3-hydroxybutyrate-co-3-hydroxyvalerate). *Polymer (Guildf)* 2004;45(6):1903–1911.
63. Cal AJ, Grubbs B, Torres LF, *et al*. Nucleation and plasticization with recycled low-molecular-weight poly-3-hydroxybutyrate toughens virgin poly-3-hydroxybutyrate. *J Appl Polym Sci* 2018;47432:47432.
64. Menczel JD, Prime RB. *Thermal Analysis of Polymers: Fundamentals and Applications*. John Wiley, New Jersey, 2008 doi:10.1002/9780470423837.
65. Hu Y, Zhang J, Sato H, *et al*. Multiple melting behavior of poly(3-hydroxybutyrate-co -3-hydroxyhexanoate) investigated by differential scanning calorimetry and infrared spectroscopy. *Polymer (Guildf)* 2007;48(16):4777–4785.
66. Kansiz M, Domínguez-Vidal A, McNaughton D, *et al*. Fourier-transform infrared (FTIR) spectroscopy for monitoring and determining the degree of crystallisation of polyhydroxyalkanoates (PHAs). *Anal Bioanal Chem* 2007;388(5–6):1207–1213.
67. Sedlacek P, Slaninova E, Enev V, *et al*. What keeps polyhydroxyalkanoates in bacterial cells amorphous? A derivation from stress exposure experiments. *Appl Microbiol Biotechnol* 2019;104(4):1905–1917.
68. Yu J, Plackett D, Chen LXLL. Kinetics and mechanism of the monomeric products from abiotic hydrolysis of poly[(*R*)-3-hydroxybutyrate] under acidic and alkaline conditions. *Polym Degrad Stab* 2005;89(2):289–299.
69. Jian Y. Recovery and purification of polyhydroxyalkanoates. 2008;(US 2008/0220505 A1).
70. Porter M, Yu J. Monitoring the in situ crystallization of native biopolyester granules in *Ralstonia eutropha* via infrared spectroscopy. *J Microbiol Methods* 2011;87(1):49–55.
71. Yoshie N, Yoshie I. Structure, composition and solution properties of PHAs. In: Steinbuchel A, editor. *Biopolymers - Polyesters II Properties and Chemical Structure*. Wiley-VCH, Germany, 2002: 133–156.
72. Pardelha F, Albuquerque MGE, Reis MAM, *et al*. Flux balance analysis of mixed microbial cultures: Application to the production of polyhydroxyalkanoates from complex mixtures of volatile fatty acids. *J Biotechnol* 1970;162(2–3):336–345.
73. Carvalho G, Oehmen A, Albuquerque MGE, *et al*. The relationship between mixed microbial culture composition and PHA production performance from fermented molasses. *N Biotechnol* 2014;31(4):257–263.
74. Dias JML, Oehmen A, Serafim LS, *et al*. Metabolic modelling of polyhydroxyalkanoate copolymers production by mixed microbial cultures. *BMC Syst Biol* 2008;2:1–21.
75. Fried JR. *Polymer Science and Technology*. 3rd edition. Prentice Hall: New York, 2014.
76. Shojaeiarani J, Bajwa DS, Rehovsky C, *et al*. Deterioration in the physico-mechanical and thermal properties of biopolymers due to reprocessing. *Polymers (Basel)* 2019;11(1):1–17.
77. Organ SJ, Barham PJ. Phase separation in a blend of poly(hydroxybutyrate) with poly(hydroxybutyrate-*co*-hydroxyvalerate). *Polymer (Guildf)* 1993;34(3):459–467.

78. Organ SJ. Phase separation in blends of poly(hydroxybutyrate) with poly(hydro xybutyrate-co-hydroxyvalerate): Variation with blend components. *Polymer (Guildf)* 1994;35(1):86–92.
79. Kim KJ, Doi Y, Abe H. Effects of residual metal compounds and chain-end structure on thermal degradation of poly(3-hydroxybutyric acid). *Polym Degrad Stabil* 2006;91. doi:10.1016/j.polymdegradstab.2005.06.004.
80. Aoyagi Y, Yamashita K, Doi Y. Thermal degradation of poly[(R)-3-hydroxybutyrate], poly[ε-caprolactone], and poly[(S)-lactide]. *Polym Degrad Stab* 2002. doi:10.1016/ S0141-3910(01)00265-8.
81. Kawalec M, Adamus G, Kurcok P, et al. Carboxylate-induced degradation of poly(3-hydroxybutyrate)s. *Biomacromolecules* 2007;8(4):1053–1058.
82. Kim KJ, Doi Y, Abe H. Effect of metal compounds on thermal degradation behavior of aliphatic poly(hydroxyalkanoic acid)s. *Polym Degrad Stab* 2008;93(4):776–785.
83. Abe H. Thermal degradation of environmentally degradable poly(hydroxyalkanoic acid)s. *Macromol Biosci* 2006;6(7):469–486.
84. Lundmark S, Persson P, MPampos K, et al. Förfarande för rening av ett poly(hydroxialkanoat). 2011;(SE535673C2).
85. Kopinke FD, Remmler M, Mackenzie K, et al. Thermal decomposition of biodegradable polyesters - I. Poly(beta-hydroxybutyric acid). *Polym Degrad Stab* 1996;52:25–38.
86. Laycock B, Nikolić M, Colwell JM, et al. Lifetime prediction of biodegradable polymers. *Prog Polym Sci* 2017;71:144–189.
87. Mergaert J, Anderson C, Wouters A, et al. Biodegradation of polyhydroxyalkanoates. *FEMS Microbiol Lett* 1992;103(2–4):317–321.
88. Arcos-Hernandez M V., Laycock B, Pratt S, et al. Biodegradation in a soil environment of activated sludge derived polyhydroxyalkanoate (PHBV). *Polym Degrad Stab* 2012;97(11):2301–2312.
89. Jendrossek D, Handrick R. Microbial degradation of polyhydroxyalkanoates. *Annu Rev Microbiol* 2002;56(1):403–432.
90. Weng YX, Wang XL, Wang YZ. Biodegradation behavior of PHAs with different chemical structures under controlled composting conditions. *Polym Test* 2011;30(4):372–380.
91. Morse MC, Liao Q, Criddle CS, et al. Anaerobic biodegradation of the microbial copolymer poly(3-hydroxybutyrate-co-3-hydroxyhexanoate): Effects of comonomer content, processing history, and semi-crystalline morphology. *Polymer (Guildf)* 2011;52(2):547–556.
92. Penloglou G, Roussos A, Chatzidoukas C, et al. A combined metabolic/polymerization kinetic model on the microbial production of poly(3-hydroxybutyrate). *N Biotechnol* 2010;27(4):358–367.
93. Kawaguchi Y, Doi Y. Kinetics and mechanism of synthesis and degradation of poly(3-hydroxybutyrate) in *Alcaligenes eutrophus*. *Macromolecules* 1992;25(9):2324–2329.
94. Yeom SH, Yoo YJ. Effect of Ph on the molecular-weight of poly-3-hydroxybutyric acid produced by alcaligenes Sp. *Biotechnol Lett* 1995;17(4):389–394.
95. Yoon JS, Park SK, Kim YB, et al. Culture conditions affecting the molecular weight distribution of poly(3-hydroxybutyrate-co-3-hydroxyvalerate) synthesized by *Alcaligenes* sp. SH-69. *J Microbiol* 1996;34(3):279–283.
96. Serafim LS, Lemos PC, Torres C, et al. The influence of process parameters on the characteristics of polyhydroxyalkanoates produced by mixed cultures. *Macromol Biosci* 2008;8(4):355–366.
97. Pfeiffer D, Wahl A, Jendrossek D. Identification of a multifunctional protein, PhaM, that determines number, surface to volume ratio, subcellular localization and distribution to daughter cells of poly(3-hydroxybutyrate), PHB, granules in *Ralstonia eutropha* H16. *Mol Microbiol* 2011;82(4):936–951.

98. Taidi B, Mansfield DA, Anderson AJ. Turnover of Poly(3-hydroxybutyrate) (PHB) and its influence on the molecular-mass of the polymer accumulated by *Alcaligenes eutrophus* during batch culture. *Fems Microbiol Lett* 1995;129(2–3):201–205.

99. Shimizu H, Tamura S, Shioya S, *et al*. Kinetic-study of poly-D(-)-3-hydroxybutyric acid (Phb) production and its molecular-weight distribution control in a fed-batch culture of alcaligenes-eutrophus. *J Ferment Bioeng* 1993;76(6):465–469.

100. Beccari M, Dionisi D, Giuliani A, *et al*. Effect of different carbon sources on aerobic storage by activated sludge. *Water Sci Technol* 2002;45(6):157–168.

101. Dionisi D, Renzi V, Majone M, *et al*. Storage of substrate mixtures by activated sludges under dynamic conditions in anoxic or aerobic environments. *Water Res* 2004. doi:10.1016/j.watres.2004.01.018.

102. Hyakutake M, Tomizawa S, Mizuno K, *et al*. Alcoholytic cleavage of polyhydroxyalkanoate chains by Class IV synthases induced by endogenous and exogenous ethanol. *Appl Environ Microbiol* 2014;80(4):1421–1429.

103. Thomson NM, Hiroe A, Tsuge T, *et al*. Efficient molecular weight control of bacterially synthesized polyesters by alcohol supplementation. *J Chem Technol Biotechnol* 2014;89(7):1110–1114.

104. Strazzera G, Battista F, Garcia NH, *et al*. Volatile fatty acids production from food wastes for biorefinery platforms: A review. *J Environ Manage* 2018;226(8):278–288.

105. Lovley DR, Phillips EJP, Lonergan DJ, *et al*. Fe (III) and S0 reduction by Pelobacter carbinolicus. *Appl Environ Microbiol* 1995;61(6):2132–2138.

106. Flory PJ. *Principles of Polymer Chemistry*. Cornell University Press, New York, 1953.

107. Bengtsson S, Pisco AR, Johansson P, *et al*. Molecular weight and thermal properties of polyhydroxyalkanoates produced from fermented sugar molasses by open mixed cultures. *J Biotechnol* 2010. doi:10.1016/j.jbiotec.2010.03.022.

108. Yamaguchi M, Arakawa K. Effect of thermal degradation on rheological properties for poly(3-hydroxybutyrate). *Eur Polym J* 2006;42(7):1479–1486.

109. Xiang H, Wen X, Miu X, *et al*. Thermal depolymerization mechanisms of poly(3-hydroxybutyrate-*co*-3-hydroxyvalerate). *Prog Nat Sci Mater Int* 2016;26(1):58–64.

110. Collyer AA, Clegg DW (eds.). *Rheological Measurement*. 2nd edition. Springer, London, 1998.

111. Pamies R, Hernández Cifre JG, del Carmen López Martínez M, *et al*. Determination of intrinsic viscosities of macromolecules and nanoparticles. Comparison of single-point and dilution procedures. *Colloid Polym Sci* 2008;286(11):1223–1231.

112. Montano-Herrera L, Pratt S, Arcos-Hernandez M V., *et al*. In-line monitoring of thermal degradation of PHA during melt-processing by Near-Infrared spectroscopy. *N Biotechnol* 2014;31(4):357–363.

113. Malengreaux C. *Rheology and Thermal Stability of Polyhydroxyalkanoates*. Lund University: Lund, 2008.

114. Kunasundari B, Sudesh K. Isolation and recovery of microbial polyhydroxyalkanoates. *Express Polym Lett* 2011;5(7):620–634.

115. Kunasundari B, Murugaiyah V, Kaur G, *et al*. Revisiting the single cell protein application of *Cupriavidus necator* H16 and recovering bioplastic granules simultaneously. *PLoS One* 2013;8(10):1–15.

116. Ong SY, Zainab-L I, Pyary S, *et al*. A novel biological recovery approach for PHA employing selective digestion of bacterial biomass in animals. *Appl Microbiol Biotechnol* 2018;102(5):2117–2127.

117. Kunasundari B, Arza CR, Maurer FHJJ, *et al*. Biological recovery and properties of poly(3-hydroxybutyrate) from *Cupriavidus necator* H16. *Sep Purif Technol* 2017;172:1–6.

118. Kurdikar DL, Strauser FE, Solodar AJ, *et al*. High temperature PHA extraction using PHA-poor solvents. 2000;(US6087471). http://www.google.com/patents/US6087471.

119. Koller M, Bona R, Chiellini E, *et al.* Extraction of short-chain-length poly-[(*R*)-hydroxyalkanoates] (*scl*-PHA) by the 'anti-solvent' acetone under elevated temperature and pressure. *Biotechnol Lett* 2013;35(7):1023–1028.
120. Samorì C, Abbondanzi F, Galletti P, *et al.* Extraction of polyhydroxyalkanoates from mixed microbial cultures: Impact on polymer quality and recovery. *Bioresour Technol* 2015;189:195–202.
121. Rosengart A, Cesário MT, de Almeida MCMD, *et al.* Efficient P(3HB) extraction from *Burkholderia sacchari* cells using non-chlorinated solvents. *Biochem Eng J* 2015;103:39–46.
122. Fiorese ML, Freitas F, Pais J, *et al.* Recovery of polyhydroxybutyrate (PHB) from *Cupriavidus necator* biomass by solvent extraction with 1,2-propylene carbonate. *Eng Life Sci* 2009;9. doi:10.1002/elsc.200900034.
123. Miller-Chou BA, Koenig JL. A review of polymer dissolution. *Prog Polym Sci* 2003;28(8):1223–1270.
124. Sun S, Jia L, Li B, *et al.* The occurrence and fate of PAHs over multiple years in a wastewater treatment plant of Harbin, Northeast China. *Sci Total Environ* 2018;624:491–498.
125. Fatone F, Di Fabio S, Bolzonella D, *et al.* Fate of aromatic hydrocarbons in Italian municipal wastewater systems: An overview of wastewater treatment using conventional activated-sludge processes (CASP) and membrane bioreactors (MBRs). *Water Res* 2011;45(1):93–104.
126. Turovskiy IS, Mathai PK. *Wastewater Sludge Processing.* John Wiley & Sons Limited, New Jersey, 2006.
127. Carberry JB, Englande AJ. *Sludge Characteristics and Behavior.* NATO ASI Series No. 66, Martinus Nijhoff Publishers, Boston, 1983.
128. Wiechmann B, Dienemann C, Kabbe C, *et al.* Sewage sludge management. *J Hydrol* 1997;200:198–221.
129. Mininni G, Mauro E, Piccioli B, *et al.* Production and characteristics of sewage sludge in Italy. *Water Sci Technol* 2019;79(4):619–626.
130. European Commission. Regulation (EC) no 1907/2006 of the European parliament and of the council. *Off J Eur Union* 2006;L396/1.
131. European Commission. European Parliament and council directive 94/62/EC on packaging and packaging waste. *Off J Eur Communities* 1994;L365/10.
132. European Commission. Commission regulation (EU) No 10/2011 of 14 January 2011 on plastic materials and articles intended to come into contact with food. *Off J Eur Union* 2011;L12/1.
133. European Commission. Commission Regulation (EU) 1272/2013 amending annex XVII to REACH as regards polycyclic aromatic hydrocarbons. *Off J Eur Union* 2013;L328/69. http://eur-lex.europa.eu/LexUriServ/LexUriServ.do?uri=OJ:L:2013:328:0069:0071:EN:PDF.
134. Hammaini A, González F, Ballester A, *et al.* Biosorption of heavy metals by activated sludge and their desorption characteristics. *J Environ Manage* 2007;84(4):419–426.
135. Zhou Y, Zhang Z, Zhang J, *et al.* New insight into adsorption characteristics and mechanisms of the biosorbent from waste activated sludge for heavy metals. *J Environ Sci (China)* 2016;45:248–256.
136. Werker A, Hernandez M, Laycock B, *et al.* Method of producing polyhydroxyalkanoate compounded plastics having improved mechanical properties. 2015;(US20150291768A1).
137. Hahladakis JN, Velis CA, Weber R, *et al.* An overview of chemical additives present in plastics: Migration, release, fate and environmental impact during their use, disposal and recycling. *J Hazard Mater* 2018;344:179–199.
138. Groh KJ, Backhaus T, Carney-Almroth B, *et al.* Overview of known plastic packaging-associated chemicals and their hazards. *Sci Total Environ* 2019;651:3253–3268.

139. Ogawa T. Effects of molecular weight on mechanical properties of polypropylene. *J Appl Polym Sci* 1992;44(10):1869–1871.
140. Flory PJ. Tensile strength in relation to molecular weight of high polymers. *J Am Chem Soc* 1945;67(11):2048–2050.
141. Luo S, Grubb DT, Netravali AN. The effect of molecular weight on the lamellar structure, thermal and mechanical properties of poly(hydroxybutyrate-*co*-hydroxyvalerates). *Polymer (Guildf)* 2002;43(15):4159–4166.
142. Langford A, Chan CM, Pratt S, *et al*. The morphology of crystallisation of PHBV/PHBV copolymer blends. *Eur Polym J* 2019;112:104–119.
143. Chan MC, Vandi LJ, Pratt S, *et al*. Understanding the effect of copolymer content on the processability and mechanical properties of polyhydroxyalkanoate (PHA) / wood composites. *Compos Part A Appl Sci Manuf* 2019;124(May):105437.
144. Savenkova L, Gercberga Z, Bibers I, *et al*. Effect of 3-hydroxy valerate content on some physical and mechanical properties of polyhydroxyalkanoates produced by Azotobacter chroococcum. *Process Biochem* 2000;36(5):445–450.
145. Boyandin AN, Kazantseva EA, Varygina DE, *et al*. Constructing slow-release formulations of ammonium nitrate fertilizer based on degradable poly(3-hydroxybutyrate). *J Agric Food Chem* 2017;65(32):6745–6752.
146. Majeed Z, Ramli NK, Mansor N, *et al*. A comprehensive review on biodegradable polymers and their blends used in controlled-release fertilizer processes. *Rev Chem Eng* 2015;31(1):69–95.
147. Bourbonnais R, Marchessault RH. Application of polyhydroxyalkanoate granules for sizing of paper. *Biomacromolecules* 2010;11(4):989–993.
148. Kalia VC. *Biotechnological Applications of Polyhydroxyalkanoates*. Springer, Singapore, 2000.
149. Wu CS, Liao HT, Cai YX. Characterisation, biodegradability and application of palm fibre-reinforced polyhydroxyalkanoate composites. *Polym Degrad Stab* 2017;140:55–63.
150. Bugnicourt E, Cinelli P, Lazzeri A, *et al*. Polyhydroxyalkanoate (PHA): Review of synthesis, characteristics, processing and potential applications in packaging. *Express Polym Lett* 2014;8(11):791–808.
151. Ivanov V, Stabnikov V, Ahmed Z, *et al*. Production and applications of crude polyhydroxyalkanoate-containing bioplastic from the organic fraction of municipal solid waste. *Int J Environ Sci Technol* 2014;12(2):725–738.
152. Ramachandran H, Kannusamy S, Huong KH, *et al*. Blends of polyhydroxyalkanoates (PHAs). In: Roy I, MVP, editors. *Polyhydroxyalkanoate (PHA) based Blends, Composites and Nanocomposites*. RSC Green Chemistry, United Kingdom, 2015: 66–97.
153. Gumel AM, Annuar MSM. Nanocomposites of polyhydroxyalkanoates (PHAs). In: Roy I, Visakh PM, editors. *Polyhydroxyalkanoate (PHA) based Blends, Composites and Nanocomposites*. RSC Green Chemistry, United Kingdom, 2015: 98–118.
154. Khandal D, Pollet E, Avérous L. Polyhydroxyalkanoate-based multiphase materials. In: Roy I, Visakh PM, editors. *Polyhydroxyalkanoate (PHA) based Blends, Composites and Nanocomposites*. RSC Green Chemistry, United Kingdom, 2015: 119–140.
155. Yee LH, Foster LJR. Polyhydroxyalkanoates as packaging materials: Current applications and future prospects. In: Roy I, Visakh PM, editors. *Polyhydroxyalkanoate (PHA) based Blends, Composites and Nanocomposites*. RSC Green Chemistry, United Kingdom, 2015: 183–207.
156. Luis P, Barreto M, Cardoso R, *et al*. Packaging applications of polyhydroxyalkanoates (PHAs). In: Roy I, Visakh PM, editors. *Polyhydroxyalkanoate (PHA) Based Blends, Composites and Nanocomposites*. RSC Green Chemistry, United Kingdom, 2015: 208–226.

157. Vandi LJ, Chan CM, Werker A, *et al*. Wood-PHA composites : Mapping opportunities. *Polymers (Basel)* 2018;10(751):1–15.
158. Gogotov IN, Gerasin VA, Knyazev Y V, *et al*. Composite biodegradable materials based on polyhydroxyalkanoate. *Appl Biochem Microbiol* 2010;46(6):659–665.
159. Chan CM, Vandi LJ, Pratt S, *et al*. Composites of wood and biodegradable thermoplastics: A review. *Polym Rev* 2018;58(3):444–494.
160. Christensen CM. *The Innovator's Dilemma: When New Technologies Cause Great Firms to Fail*. Harvard Business School Press: Boston, MA, 1997.
161. Morgan-Sagastume F. Characterisation of open, mixed microbial cultures for polyhydroxyalkanoate (PHA) production. *Rev Environ Sci Biotechnol* 2016;15(4):593–625.
162. Silva FC, Serafim LS, Nadais H, *et al*. Acidogenic fermentation towards valorisation of organic waste streams into volatile fatty acids. *Chem Biochem Eng* Q 2013;27(4):467–476.
163. Garcia-Aguirre J, Esteban-Gutiérrez M, Irizar I, *et al*. Continuous acidogenic fermentation: Narrowing the gap between laboratory testing and industrial application. *Bioresour Technol* 2019;282(6):407–416.
164. Feng K, Li H, Zheng C. Shifting product spectrum by pH adjustment during long-term continuous anaerobic fermentation of food waste. *Bioresour Technol* 2018;270(12):180–188.
165. Dahiya S, Mohan SV. Selective control of volatile fatty acids production from food waste by regulating biosystem buffering: A comprehensive study. *Chem Eng J* 2019;357(2):787–801.
166. Zhou M, Yan B, Wong JWC, *et al*. Enhanced volatile fatty acids production from anaerobic fermentation of food waste: A mini-review focusing on acidogenic metabolic pathways. *Bioresour Technol* 2018;248:68–78.
167. Esteban-Gutiérrez M, Garcia-Aguirre J, Irizar I, *et al*. From sewage sludge and agri-food waste to VFA: Individual acid production potential and up-scaling. *Waste Manag* 2018;77(2018):203–212.

Part V

Industrial Aspects

10 Economics and Industrial Aspects of PHA Production

José Geraldo da Cruz Pradella

CONTENTS

10.1 INTRODUCTION

Polyhydroxyalkanoates (PHA), as a family of diverse biodegradable, biocompatible biopolyesters [1], have gone through many years of efforts towards commercialization. However, despite the efforts of academia and industry, PHA production and commercialization remain at low scale. In this chapter, commercialized PHA will be reviewed in terms of the history of its economic achievements, pointing out remaining bottlenecks and the perspectives of its competitiveness with synthetic polymers.

10.2 A BRIEF HISTORY OF PHA

10.2.1 P(3HB) AND P(3HB-*CO*-3HV)

10.2.1.1 WR Grace

The patents USA 3,036,959 and USA 3,044,942 assigned to W.R. Grace & Co in 1962, were probably the first ones that helped design the industrial production of poly-3-hydroxybutyrate (P(3HB)). In this process P(3HB) production was organo-trophic accomplished by growing *Bacillus megaterium* in unbalanced semi-synthetic culture media comprising of 20 g/L glucose, 0.5 g/L $(NH_4)_2$ SO_4 and other salts, while *Rhodosporodium rubrum* was used for auxotrophic CO_2/H_2 growth and P(3HB) biosynthesis. After cell biomass collection, the process of recovering poly-3-hydroxybutyric acid from a bacterial cell mass containing this polyester comprised of drying the bacterial cell mass, dispersing it in methylene chloride/ethanol solution to extract the poly-3-hydroxybutyric acid, separating the methylene chloride/ethanol/polyester solution from the cell residue, and recovering the polyester product from solution [2, 3].

10.2.1.2   ICI

The well-known attempt of Imperial Chemical Industries Biological, England started in the 1970s due to the oil crisis and sudden price rise. The copolymer P(3HB-*co*-3HV) was produced using a *Cupriavidus necator* (formerly *Alcaligenes eurtophus*) UV-mutant unable to grow in propionic acid in a fed-batch high-cell-density process. It comprises a first growth phase in a well-balanced synthetic culture media growing using glucose as a carbon source at a certain biomass concentration (20 to 30 g dry weight/L). At this point phosphorus culture media deprivation deviated cell metabolism to the polyester biosynthesis and P(3HB-*co*-3HV) with controlled flux of glucose and propionic acid to the fermenter. The bioprocess was completed when PHA reached 70 to 80% of the total dry weight. A solvent extraction methodology, similar to that previously proposed [3] was developed but soon abandoned as it was evaluated to be very complicated and expensive mainly due to the low product yield from the carbon source (glucose and propionic acid), and solvent recovery step. A recovery route based on solubilization of the non-PHA part of the cell biomass was therefore set up. Cells were harvested by heat shock and flocculated. The rendered biomass was submitted to enzyme and detergent washes to solubilize cell components leaving behind water insoluble purified biopolymer. A limited range of copolymers were made and blended to give the required 3HV content. The powder was melted, extruded, and made into chips, which were fabricated through conventional polymer processing equipment. In the middle of the 1980s ICI commercialized a copolymer by the trade name Biopol® worldwide [4]. By that time, they had the intention to scale-up its production from 300 t/year to 5000 t/year, thereby decreasing the price of Biopol® from US$ 16/kg to about US$ 7–8/kg (https://www.icis.com/explore/resources/news/1991/09/23/25670/ici-reduces-cost-ups-capacity-for-biopol/). At that point, ICI was the biggest worldwide initiative to produce Biopol® P(3HB) and 3HV copolymer. However, in the 1990s glucose raw material was about US$ 2,000/t and the price of solvent (or lytic enzymes) to perform polymer extraction and purification was also very high. In 1990, the agricultural and pharmaceutical businesses of ICI, including Biopol®, were spun-off as Zeneca Ltd. In 1996, Monsanto, USA, bought all patents concerning biopolymer from Zeneca. Since this acquisition, the emphasis at Monsanto was on producing Biopol® and related copolymers in plants, aiming to improve their cost of production and properties for different end-use applications [5]. In 2001, Monsanto stopped their biopolymer research program, and the commercial Biopol® business was sold to Metabolix, USA. In 2009 a Metabolix and Archer Daniels Midland (ADM) joint venture (as Telles) opened the largest PHA plant (trade name Mirel) in Iowa, USA, with a production capacity of 50,000 t/year. A wet corn milling plant belonging to ADM and adjacent to the microbial fermentation facility separates corn into grains, starches, and sugars. The sugar stream was used as carbon feedstock for the PHA-producing bacteria [6]. However, in 2012, ADM withdrew from Telles and the joint venture with Metabolix was discontinued. Biopolymer Mirel production was downsized and nowadays is integrated into the technological platform of Y10 Bioscience, USA (https://www.yield10bio.com/).

10.2.1.3 PCD

Other companies motivated by the 1970s oil crisis, such as Petrochemie Danubia (PCD) and Chemie Linz in Austria also started development production programs using a growth associated P(3HB) production strain of *Alcaligenes latus*. However, in the first half of 1986, crude oil prices fell from US$ 100 to about US$ 12 per barrel and continued at a low price for the next decade making PHA uncompetitive with conventional petrochemical plastics. This fact was probably important in discouraging those companies from continuing to explore both P(3HB) and its copolymer with 3-hydroxyvaleric acid. In 1994, the strain and patent of PCD were acquired by Dr Urs Hänggi to become the company Biomer, in Germany. Plasticizers, nucleants, and additives were mixed with the biopolymer to create a different formulation with improved properties but maintained biodegradability [7].

10.2.1.4 PHB Industrial

In the 1990s the sugar price in the international market was commercialized at US$ 200 to US$ 300 per metric tons. The possibility of using sucrose as carbon feedstock led Copersucar Technology Center (CTC), Institute of Technological Research of São Paulo State (IPT) and São Paulo State University to engage in P(3HB) and P(3HB)-*co*-3HV research [8]. Moreover, isoamyl alcohol, a byproduct of ethanol fermentation, was disclosed to be used as the solvent phase for PHA extraction from microbial biomass [9]. The technology was transferred to the Brazilian company PHB Industrial S.A. by the end of 1995. Nowadays they are able to run an industrial plant using raw sucrose from sugarcane to produce either P(3HB) and P(3HB)-*co*-3HV by *Cupriavidus necator* DSM 545 and a mutant strain of *Burkholderia sacchari*. The set-up combines a train of seed fermenters (0.1 m^3 and 1.0 m^3) followed by the production fermenter of 13 m^3. The set-up was integrated with Usina da Pedra, Serrana, SP, Brazil, a traditional bioethanol sugar mill. Former pilot plant experiments ran at IPT led to the production of 70 g/L of P(3HB) and at a productivity of 1.7 g/(L·h) and 3.1 kg sucrose per kg P(3HB), using *Cupriavidus necator* DSM 545 in a fed-batch two-phase fermentation. Economic analysis indicated that P(3HB) production cost for 10,000 t/year using this technology was estimated to be about US$ 2.25–2.75 per kg, depending on the sugar cost [10]. Today price and production capacity of the various biopolymer grades commercialized under the trade name Biocycle are not disclosed. Concerning P(3HB)-*co*-3HV production, it should be mentioned that monomer incorporation to a PHA chain demands the use of a C3 or C5 carbon source such as propionic or valeric acid. Competitive P(3HB)-*co*-3HV cost may be achieved with the use of the selected mutagenic strain *B. sacchari* IPT 189 [11].

10.2.1.5 TianAn Biopolymer

In 2000, TianAn Biopolymer in NingBo, China, started a 2,000 t/year P(3HB-*co*-3HV) industrial scale production from dextrose from corn or cassava as the carbon feedstock and *C. Cupriavidus* strain. The process is based on a proprietary non-solvent extraction and purification technology and the biopolymer under the trade name ENMAT with various grades are commercialized at about US$ 4.0 per kg [7, 12].

Bio-on, an Italian company, started its activities in 2007. The main products are PHA powder to be used in the cosmetics industry and environmental waste treatment, such as oil spills. Since 2017 it has operated an industrial plant of 1000–2000 t/year in the Emilio Romana region, mainly using sugar beet residue as carbon feed stock http://www.bio-on.it/production.php#p2.

10.2.2 P(3HB-*co*-4HB)

10.2.2.1 Tianjin Green Bioscience

In 2009, Tianjin Green Bioscience, in a joint venture with DSM, gradually scaled up their fermentation from pilot to a fermenter of 150 m³. The company started with strain development followed by equipment design and included the development of downstream processing conditions to facilitate the application of their product. *Ralstonia eutropha* selected strain is used to produce P(3HB-*co*-4HB). With the addition of 1,4-butanediol in different amounts, 4-hydroxybutyrate can be varied in the copolymer SoGreen™, thus modifying its physical-chemical properties for various applications. Tianjin Green Bioscience facilities have a PHA production capacity of 10,000 t/year of (http://www.tjgreenbio.com).

10.2.3 P(3HB-*co*-*mcl*-3HA)

10.2.3.1 Procter & Gamble

The general molecular structure of the P(3HB-*co*-3HA*mcl*) Nodax™ class PHA copolymers is a random copolymer of predominantly (*R*)-3-hydroxybutyrate (3HB) and other (*R*)-3-hydroxyalkanoate (3HA) comonomer units, i.e., a P(3HB-*co*-3HA) copolymer. Examples of such 3HA units with medium-chain-length (*mcl*) side groups include (*R*)-3-hydroxyhexanoate (3HHx), (*R*)-3-hydroxyoctanoate (3HO), (*R*)-3-hydroxydecanoate (3HD), and (*R*)-3-hydroxyoctadecanoate (3HOd). Such PHA copolymers typically consist of at least 50 mol-% 3HB and at least 2 mol-% secondary *mcl*-3HA units [13].

Procter & Gamble (Cincinnati, OH, USA) initiated the early development of this class of PHA copolymers around the late 1980s. During several decades of intensive research effort, Procter & Gamble accumulated a broad base of intellectual property associated with this class of material [14–16]. In 2007, Meridian took over the Nodax™ technology from Procter & Gamble for the full commercialization of this class of bioplastics. In 2014 Danimer Scientific, a company specializing in PLA formulation, merged with Meridian as MHD, and in 2016 the name was changed to Danimer Scientific. Danimer Scientific nowadays produces Nodax™ from canola oil as carbon feedstock, and presumably *Aeromonas* sp. as microorganism. The company claims that it uses a proprietary extraction process in which the biomass is removed to isolate the final purified PHA product. The resulting PHA is then dried, producing the clean white powder ready to be pelletized. The absence of toxic chemicals is one of the qualities that make PHA particularly appealing as a biopolymer for food contact and medical purposes. By the end of 2018, Danimer was planning to acquire a former biotech processing plant in Kentucky, USA, and scheduled to start a large-scale commercial production of the Nodax™ copolymers in 2019 (https://danimerscientific.com/pha-the-future-of-biopolymers/).

10.2.3.2 Kaneka

Another branch of research on P(3HB)-*co*-*mcl*-3HA was taken over by Kaneka (Kanegafuchi Kagaku Kogyo) in Japan. Shiotani and co-workers [17] isolated strains of *Aeromonas hydrophila* and *Aeromonas cavie* for the first time and were able to produce PHA containing a 3-hydroxybutyrate (3HB) unit and a 3-hydroxyhexanoate (3HHx) unit, a three-component copolymer containing at least a 3-hydroxybutyrate (3HB) unit and a 3-hydroxyhexanoate (3HHx) unit, and a four-component copolymer containing at least a 3-hydroxybutyrate (3HB) unit and a 3-hydroxyhexanoate (3HHx) unit. In 2001 Procter & Gamble, which previously owned the patents for P(3HB-*co*-*mcl*-PHA), licensed Kaneka to develop business on the P(3HB-*co*-HHx) copolymer. Today Kaneka claims to produce and commercialize 1000 t/year of its PHBM™ material expanding capacity to 5000 t/year by the end of 2019 (http://www.kaneka.co.jp/en/business/material/nbd_001.html).

Overall, by the beginning of 2019 a literature and web survey of PHA producers demonstrated that a few companies that started their activities by the 1990s still remain in business, while fresh competitors have appeared in the arena since 2000. However, PHA total installed capacity worldwide still remains very low with total maximum installed capacity, at the best, around 30,000 t/year, Table 10.1. In comparison with other biodegradable and non-biodegradable bioplastics, PHA accounted for less than 2% of the installed global capacity in 2018 (Figure 10.1).

10.3 PHYSICAL PROPERTIES

One of the reasons that has been claimed for low PHA utilization is its poor physical-mechanical properties for packing and other plastics applications. Recently a very interesting survey of the packing, mechanical, and physical properties of commercially available PHA was conducted, together with other biodegradable and non-biodegradable biopolymers [18], summarized in Table 10.2. The evaluated biopolymers were polylactic acid (PLA), several polyhydroxyalkates: poly-3-hydroxybutyrate 3HB (PHB), P(3HB-*co*-4HB) (PHBH) and P(3HB-*co*-3HV) (PHBV) - thermoplastic starch (TPS), polybutylene adipate terephthalate (PBAT), polybutylene succinate (PBS), poly(ε-caprolactone) (PCL) and biobased poly(ethylene) (BioPE), and, as a future biopolymer candidate, the thermoplastic polyurethane elastomer (TPU). The pure materials were processed after drying at 60°C to reduce humidity in a flat film

TABLE 10.1

Commercial PHA Plant and Company Activity by the End of 2019

Company	Location	Process	Capacity	Products	Trade name	Source
Biomer	Germany	*Alcaligenes latus*, C-source sucrose	50 t/year *	P(3HB)	Biomer TM	http://www.biomer.de/IndexE.html
Polyferm Canada	Canada	*Pseudomonas* sp., C source sugar/plant oil	not known	*mcl*-PHA	VersaMerTM	https://www.polyfermcanada.com/index.html
PHB Industrial S.A.	Brazil	*Cupriavidus necator*; *Burkholderia sacchari*, C source sucrose, solvent extraction	100 t/year	P(3HB) P(3HB-*co*-3HV)	Biocycle®	http://www.biocycle.com.br/site.htm
Y10/Metabolix	USA	Lab scale, cells plant	not known	PHA	n.d.	https://www.yield10bio.com/
Tepha Inc.	USA		not known	P(4HB)	TephaFlexTM	https://www.tepha.com/
Danimer Scientific	USA	Canola oil as carbon source	not known	P(3HB-*co*-3HHx)	NodaxTM	https://danimerscientific.com/pha-the-future-of-biopolymers/
Kaneka	Japan	Vegetable oil as carbon source	1000 t/year to 5000 t/year by 2019	P(3HB-*co*-3HHx)	PHBHTM	http://www.kaneka.co.jp/en/business/material/nbd_001.html
Tianan Biologic Materials Co.	China	*Cupriavidus necator* strain, C source glucose, non-solvent extraction	2000 t/year	P(3HB) P(3HB-*co*-3HV)	ENMATTM	http://www.tianan-enmat.com/index.html#

(Continued)

TABLE 10.1 (CONTINUED)
Commercial PHA Plant and Company Activity by the End of 2019

Company	Location	Process	Capacity	Products	Trade name	Source
Tianjin GreenBio Materials Co.	China	*Cupriavidus necator* C sources glucose + 1-4 butanodiol	10,000 t/year	P(3HB-*co*-3HV) P(3HB-*co*-4HB)	SoGreenTM	http://www.tjgreenbio.com
Bluepha Co., Ltd	China	GMO *Halomonas* sp.; glucose as C source; non-solvent extraction	not known	P(3HB-*co*-4HB) and others	BluePHA	http://en.bluepha.com/
Shandond Ecomman Technol. Co., Ltd.	China	sugar and glucose as C source	not known	P(3HB-*co*-4HB)	AmbioTM	https://ecomannbruce.en.ecplaza.net/
Cardia Bioplastics	Australia	Not specified	not known	PHA modified starch composites		http://www.cardiabioplastics.com/
Newlight Technologies	USA	CH₄ + air, Newlight's 9Xbiocatalyst	not known	PHA	AirCarbonTM	https://www.newlight.com/
NaturePlast	France	not known	not known	PHA	not known	http://natureplast.eu/
Bio-On	Italy	Sugarbeet and others, non-solvent extraction method	up to 2000 t/year	P(3HB) P(3HB-*co*-3HV)	MinervTM	http://www.bio-on.it/production.php#p2

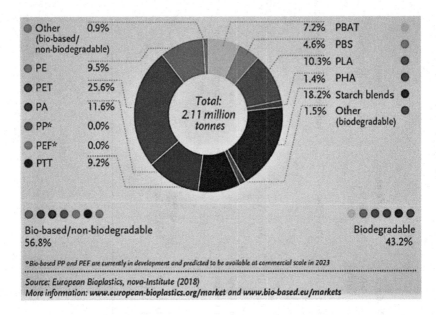

FIGURE 10.1 Global production capacities of bioplastics 2018 (adapted from https://www .european-bioplastics.org/market/).

extrusion plant type E30 M (30 mm screw, 30 D barrel length, 300 mm nozzle width) manufactured in house. The temperature profile was set at 60–150–165–165–160–160°C in order to minimize heat-induced degradation and the rotational speed was set at 30 rpm. The cast films had a thickness of ~70–80 µm.

The analysis performed on the crystallinity showed it had the highest values for TPS, PHBV with variable 3HV content, PBAT, PHB, and PCL, around 68, 67, 52, 50, and 45% respectively. PLA and PHBH films had low crystallinity, about 23 and 15–16%, respectively. The mechanical analyses highlighted the high tensile strength (between 20 and 100 MPa) and Young modulus (>600 MPa) but an extremely low elongation at break (<20%) of PLA and PHA films. In contrast, these values were quite different from biopolymers PBAT, TPS, PBS, and BioPE that presented tensile strength around 40 MPa and an elongation at break of 600 to 1200%.

On the other hand, most of the PHA materials have barrier properties against water vapor much lower than plastics, for the conventional plastics poly(ethylene) low and high-density (PE-LD; PE-HD), poly(propylene) (PP) and BioPE and acceptable values for O2 permeability are shown in Figure 10.2.

Recent improvements in PHA's mechanical properties with use of fillers, nucleant agents, plasticizers, and better understanding of processing parameters have brought this kind of biopolymer closer to the packing and disposables market [19, 20].

10.4   COST AND ECONOMICS

The "Bioplastic Market Data 2018" [21] states that the latest market data compiled by European Bioplastics in cooperation with the research institute nova-Institute,

TABLE 10.2

Summary of Origin and Assessed Physical Mechanical Characteristics of Current Commercially Available Biopolymers, Adapted from [18]

Type of biopolymer	Company	Grade	Crystallinity [%]	Average molecular weight M_w [g/mol]	Tensile strength σ [MPa] (*)	Elongation at break [%] (°)
PLA	NatureWorks LLC	202D	0.23	177,604	70–75	5
PHB	Biomer	P209	0.50	211,368	12–17	4
PHBV3	Tianan Biologic Material Co., Ltd.	Enmat Y1000P	0.67	233,270	35–39	3
PHBV7	PHB Industrial S.A.	Biocycle	0.64	233,370	23–30	3
PHBV11	Nature Plast	PHI 002	0.65	203,018	17–23	3
PHBHB13	Shandond Econman Technol. Co., Ltd.	EM 40000	0.16	166,126	41–49	5
PHBHB18	Shandond Econman Technol. Co., Ltd.	EM 20010	0.15	158,295	38–40	4
TPS	Novamont SPA	Mater-Bi PT P.C.CS	0.68	40,830	25–31	900–1100
PBAT	BASF SE	Ecoflex F Blend C1200	0.52	62,991	38–42	1000–1200
PBS	Mitsubishi Chemical Corp.	GS Pla DF 92WN	0.17	105,342	15	600–900
PCL	Perstorp Holding AB	Capa 6500	0.45	64,264	25–32	900–110
TPU	BASF Polyurethanes	Elastollan 880A 13 N	0.02	73,384	42	600–950
BioPE	Braskem S.A.	SLL218 LLDPE	0.31	–	31	1200–1300

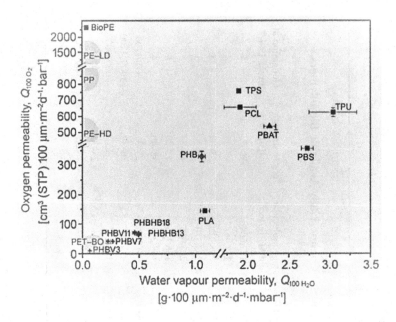

FIGURE 10.2 Oxygen and water vapor permeability for biopolymers and conventional plastics, adapted from [18].

global bioplastics production capacity is set to increase from around 2.11 Mt in 2018 to approximately 2.62 Mt in 2023, represented mainly by packing and disposable goods. With the amelioration of PHA's physical and mechanical processing properties, at this point, the main roadblock for its usage intensification in the huge market of packing and disposable utensils chains remain in its production and commercialization costs. Table 10.2 displays prospective economic studies and real production price of PHA from different raw materials and technologies. Over time, clearly it is demonstrated a selling price decreasing with time from US$ 15–17/kg of Biopol® in the 1990s, down to US$ 3– 5/kg nowadays for commercialized PHA (Biocycle, SoGreen, PHBH, Nodax, Minerv, Ambio, ENMAT, among others).

On the other hand, preliminary economic evaluations of the "Next Generation Industrial Biotechnology" [22] or the "second generation feedstock" [23] bioprocess claimed to attain an even more reduced production cost around US$ 1.5–2.0/kg PHA (Table 10.3). Those new bioprocesses are microbial biopolyester production endowed with: a) use of organic wastes as feedstock, b) non-sterilized open bioreactors set-up, c) biosynthesis provided by extremophiles and/or mixed culture microorganisms, d) environmental benign low-cost downstream PHA recovery processes.

Table 10.3 Evolution of industrial PHA technology and economic analysis

The "learning curve" obtained from the data in Table 10.3 is presented in Figure 10.3. It shows that the commercial price of US$ 2.0/kg of P(3HB) or less seems to be possible to achieve at relatively small scale and low investment cost. However, only continuous industrial and academic efforts during the next decade to come will help to accomplish it. It is believed that the pursuit in this direction must be multiplied to

TABLE 10.3

Evolution of Industrial PHA Technology and Economic Analysis

Organism	Feedstock	Cost [US$/ tonne]	Feed or broth conditioning	PHA	PHA productivity [kg/(m³·h]	DSP characteristics	CAPEX [US$]	Price US$ kg⁻¹	Capacity [t/y]	Year	Ref.
Alcaligenes eutrophus	glucose	500	feed sterilization	P(3HB-*co*-HV)	0.7–0.8	Enzyme + tensoactive washing	not known	16.0	300	1991	[24]
Alcaligenes eutrophus	glucose	500	feed sterilization	P(3HB-*co*-HV)	0.7–0.8	Enzyme + tensoactive washing	not known	7.0	5000	1991	[24]
r *E coli*	glucose	500	feed sterilization	P(3HB)	2.2	SDS lysis+hypochlorite sol. washing	26,917,000	6.14	2850	1997	[25]
Cupriavidus necator DSM545	sucrose from sugarcane	300	Upgrade to glucose+ fructose syrup + feed sterilization	P(3HB)	1.4	Isoamyl alcohol extraction	9,000,000	12.5– 15	100	1998	(*)
Alacaligenes latus DSM 1123	sucrose	300	feed sterilization	P(3HB)	4.94	SDS lysis+hypochlorite sol. washing	395,871,000	2.6	100,000	1998	[26]
r *Rasltonia eutropha*	soyben oil	300	feed sterilization	P(3HB-*co*-3HH)	2.0–3.5	SDS lysis+hypochlorite sol. washing	n.d	3.5– 4.5	5,000	2003	[27]

(Continued)

TABLE 10.3 (CONTINUED)
Evolution of Industrial PHA Technology and Economic Analysis

Organism	Feedstock	Cost [US$/tonne]	Feed or broth conditioning	PHA	PHA productivity [kg/(m³·h)]	DSP characteristics	CAPEX [US$]	Price US$ kg⁻¹	Capacity [t/y]	Year	Ref.
Cupriavidus necator DSM545	sucrose from sugarcane	300	upgrade to glucose+ fructose syrup + feed sterilization	P(3HB)	1.4	Isoamyl alcohol extraction	38,000,000	2.75	10,000	2006	[10]
Cupriavidus necator	glucose	300	feed sterilization	P(3HB-co-HV)	n.d.	Non-solvent extraction	n.d.	4.40	2000	2008	[28]
Cupriavidus necator	waste biodiesel glycerol	55.4	upgrade to pure glycerol + feed sterilization	P(3HB)	1.0	Chemical / enzymatic digestion + H_2O_2 washing	n.d.	1.94	422	2011	[29]
Rhodospirillum rubrum	syngas from gasified switchgrass	55	gasification followed CO+H2 fermentation	P(3HB)	n.s	SDS lysis + hypochlorite sol. washing	55,457,679	1.62	4000	2009	[30]
Haloferax mediterranei DSM 1411	rice-ethanol stillage	null	desalinization of fermented broth	P(3HB-co-HV)	0.14	SDS lysis + solvent washing	6,015,695	2.05	1890	2015	[31]

(Continued)

TABLE 10.3 (CONTINUED)
Evolution of Industrial PHA Technology and Economic Analysis

Organism	Feedstock	Cost [US$/tonne]	Feed or broth conditioning	PHA	PHA productivity [kg/(m³·h)]	DSP characteristics	CAPEX [US$]	Price US$ kg⁻¹	Capacity [t/y]	Year	Ref.
Selcted mixed bacteria culture	industrial waste water	null	none	P(3HB)	n.s.	Alkali + surfactant	n.d.	1.55	1500	2015	[32]
Cupriavidus neccator	slaughtering wastes	null	Upgrade to fatty acids esters + feed sterilization	P(3HB)	0.9	Mechanical lysis + separation	73,783,879	1.56	10.000	2017	[33]

(*) http://www.biocycle.com.br/imprensa_ing_01.htm, n.d.: not disclosed

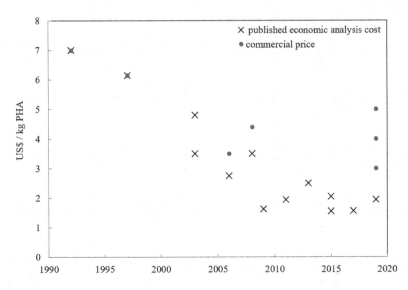

FIGURE 10.3 Learning curve of PHA production: (.) price from suppliers; (X) cost production from literature (data from Table 10.3).

challenge the commercial arena in a future scenario of petrochemical resins priced in the range of US$ 0.8–1.0/kg.

The "Bioplastic Market Data 2018" report [21] states that the latest market data compiled by European Bioplastics in cooperation with the research institute nova-Institute, global bioplastics production capacity is set to increase from around 2.11 Mt in 2018 to approximately 2.62 Mt in 2023, represented mainly by packing and disposable goods. Nowadays, PHA shares only 2% of this market, and today's selling price is the main obstacle for this class of bioproducts. With the increasing demand for bioplastics we believe that focus on new technologies for PHA industrial production must be undertaken. Low-cost raw residues as feedstock, specialized strains free from the necessity of bioprocess sterilization, and free-solvent downstream process seem to be an interesting bioroute to decrease PHA cost to facilitate its penetration into the commodity plastic market.

REFERENCES

1. Holmes, P.A. "Application of PHB: A microbially produced biodegradable thermoplastic," *Phys. Technol.*, vol. 32, no. 16, pp. 32–36, 1985.
2. Baptist, J.N. "Process for preparing poly-β-hydroxy butyric acid," *US patent 3,036,95*, USPTO, USA, 1962.
3. Baptist, J.N. "Process for preparing poly-β-hydroxy butyric acid," *US patent 3,044,942*, USPTO, USA, 1962.
4. Byrom, D. "Industrial production of copolymer from alcaligenes eutrophus." In: (Ed.) Dawes, E.A. *Novel Biodegradable Microbial Polymers*, NATO ASI Series (Series E: Applied Sciences), vol. 186, pp. 113–117, 1990, Springer, Dordrecht.
5. Asrar, J., Gruys, K. J. "Biodegradable polymer (Biopol®)." In: (Ed.) Steinbuchel, A. *Biopolymers Online*, pp. 53–68, 2005, Wiley Online.

6. DiGregorio, B.E."Biobased performance bioplastic: Mirel," *Chem. Biol.*, vol. 16, no. 1, pp. 1–2, 2009.

7. Rivero, C.P., Hu, Y., Kwan, T.H. *et al.* "Bioplastics from solid waste," In (Eds.) Wong, J.W.-C., Tyagi, R.D. and Pandey, A. *Current Developments in Biotechnology and Bioengineering*, Elsevier, pp. 1–26, 2017.

8. Bueno Netto, C.L., Craveiro, A.M., Pradella, J.G.C. *et al.* "Processo para produzir polihidroxialcanoatos a partir de açúcares extraídos de cana de açúcar," *BR patent PI 9103116-A*, INPI, Brazil, 1993.

9. Pradella, J.G.C., Rissi, L., Matsubara, R.S.M. "Processo de separação de polihidroxial-canoatos de biomassa de microorganismos ou de plantas," *BR patent PI0705509-9 A2*, INPI, Brazil, 2007.

10. Vaz Rossell, C.E., Mantelatto, P.E., Agnelli, J.A.M., Nascimento, J. "Sugar-based biorefinery – technology for integrated production of poly(3-hydroxybutyrate), sugar, and ethanol," In: (Eds.) Kamm, B., Gruber, P.R., Kamm, M. *Biorefineries – Industrial Processes and Products: Status Quo and Future Directions*, vol. 1, pp. 209–226, 2008, John Wiley & Sons, Inc.

11. Rocha, R.C.S., Da Silva, L.F., Taciro, M.K., Pradella, J.G.C. "Production of poly(3-hyd roxybutyrate-co-3-hydroxyvalerate) P(3HB–co–3HV) with a broad range of 3HV content at high yields by *Burkholderia sacchari* IPT 189," *World J. Microbiol. Biotechnol.*, vol. 24, no. 3, pp. 427–431, 2008.

12. G. Q. Chen, "A microbial polyhydroxyalkanoates (PHA) based bio- and materials industry," *Chem. Soc. Rev.*, vol. 38, n° 8, pp. 2434–2446, 2009.

13. Noda, I., Lindsey, S.B., Caraway, D. "Nodax™ class PHA copolymers: Their properties and applications." In: (Ed.) Chen, G.Q. *Plastics from Bacteria. Microbiology Monographs*, vol. 14, pp. 237–254, 2010, Springer, Berlin, Heidelberg.

14. Noda, I. "Biodegradable copolymers and plastic articles comprising biodegradable copolymers," *US Patent 5,498,692 A*, 1996.

15. Noda, I. "Films and absorbent articles comprising a biodegradable polyhydroxyalkano-ate comprising 3-hydroxybutyrate and 3-hydroxyhexanoate comonomer units," *US patent 5,990,271*, 1999.

16. Noda, I., Schechtman, L.A. "Solvent extraction of polyhydroxyalkanoates from biomass," *US Patent 5,942,597 A*, 1999.

17. Shiotani, K.G. "Copolymer and method for producing thereof," *US Patent 5,292,860*, 1994.

18. Jost, V. "Packaging related properties of commercially available biopolymers – An overview of the status quo," *Express Polym. Lett.*, vol. 12, no. 5, pp. 429–435, 2018.

19. Bugnicourt, E., Cinelli, P., Lazzeri, A., Alvarez, A.V. "Polyhydroxyalkanoate (PHA): Review of synthesis, characteristics, processing and potential applications in packaging," *Express Polym. Lett.*, vol. 8, no. 11, pp. 791–808, 2014.

20. Poltronieri, P., Kumar, P. "Polyhydroxyalkanoates (PHAs) in industrial applications," In: (Eds.) Martínez, L., Kharissova, O., Kharisov, B. *Handbook of Ecomaterials*, Cham: Springer International Publishing, 2018, pp. 1–30.

21. European Bioplastics and Institute for Bioplastics and Biocomposites, "Global produc-tion capacities of bioplastics 2017," p. 1, 2013.

22. Yu, L., Wu, F., Chen, G. "Next-generation industrial biotechnology-transforming the current industrial biotechnology into competitive processes," *Biotechnol. J.*, vol. 14, no. 9, p. 1800437, 2019.

23. Koller, M., Braunegg, G. "Advanced approaches to produce polyhydroxyalkanoate (PHA) biopolyesters in a sustainable and economic fashion," *EuroBiotech J.*, vol. 2, no. 2, pp. 89–103, 2018.

24. ICIS. "ICI reduce cost, up capacity for Biopol®," (https://www.icis.com, January 2020).

25. Choi, J., Lee, S.Y. "Factors affecting the economics of polyhydroxyalkanoate produc-tion by bacterial fermentation," *Appl. Microbiol. Biotechnol.*, vol. 51, no. 1, pp. 13–21, 1999.

26. Lee, S.Y., Choi, J. "Effect of fermentation performance on the economics of poly(3-hydroxybutyrate) production by *Alcaligenes latus*," *Polym. Degrad. Stab.*, vol. 59, no. 1–3, pp. 387–393, 1998.

27. Akiyama, M., Tsuge, T., Doi, Y. "Environmental life cycle comparison of polyhydroxy-alkanoates produced from renewable carbon resources by bacterial fermentation," *Polym. Degrad. Stab.*, vol. 80, no. 1, pp. 183–194, 2003.

28. Shen, L., Haufe, J., Patel, M.K. "Product overview and market projection of emerging bio-based plastics PRO-BIP 2009 Utrecht, The Netherlands," n°. June, 2009.

29. Naranjo, J.M., Posada, J.A., Higuita, J.C., Cardona, C.A. "Valorization of glycerol through the production of biopolymers: The PHB case using Bacillus megaterium," *Bioresour. Technol.*, vol. 133, pp. 38–44, 2013.

30. Choi, D., Chipman, D.C., Bents, S.C., Brown, R.C. "A techno-economic analysis of polyhydroxyalkanoate and hydrogen production from syngas fermentation of gasified biomass," *Appl. Biochem. Biotechnol.*, vol. 160, no. 4, pp. 1032–1046, 2010.

31. Bhattacharyya, A., Jana, K., Haldar, S. *et al.*, "Integration of poly-3-(hydroxybutyrate-c o-hydroxyvalerate) production by *Haloferax mediterranei* through utilization of still-age from rice-based ethanol manufacture in India and its techno-economic analysis," *World J. Microbiol. Biotechnol.*, vol. 31, no. 5, pp. 717–727, 2015.

32. Fernández-Dacosta, C., Posada, J.A., Kleerebezem, R. *et al.* "Microbial community-based polyhydroxyalkanoates (PHAs) production from wastewater: Techno-economic analysis and ex-ante environmental assessment," *Bioresour. Technol.*, vol. 185, pp. 368–377, 2015.

33. Shahzad, K., Narodoslawsky, M., Sagir, M. *et al.* "Techno-economic feasibility of waste biorefinery: Using slaughtering waste streams as starting material for biopolyes-ter production," *Waste Manag.*, vol. 67, pp. 73–85, 2017.

11 Next Generation Industrial Biotechnology (NGIB) for PHA Production

Fuqing Wu and Guo-Qiang Chen

CONTENTS

11.1 INTRODUCTION

Microbial synthesis of polyhydroxyalkanoates (PHA) has been developed for applications in the packaging, medicine, pharmacy, agriculture, and food industries [1, 2]. However, its high production cost limits the large-scale applications of PHA. The bacterial chassis of current industrial biotechnology is the most important factor leading to a high cost [3]. The high cost of using a common chassis is mainly due to the necessity for sterilization during fermentation, consumption of large amounts of fresh water and energy, discontinuous processes, and the high cost of substrates [4]. Therefore, the development of user-friendly chassis is an important way to reduce these burdens. 'Next Generation Industrial Biotechnology (NGIB)' was first proposed aiming to overcome the disadvantages of the current industrial biotechnology [5]. NGIB is based on extremophiles, especially halophiles grown without contamination under unsterile conditions in seawater (Figure 11.1) [3].

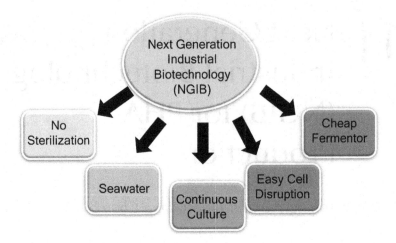

FIGURE 11.1 Water and energy saving *H. bluephagenesis* based NGIB has been successfully developed operated in an open and continuous way with convenient downstream processing means [3].

11.2   CHASSIS FOR NGIB

Common chassis of industrial biotechnology, including *Escherichia coli*, *Pseudomonas putida*, *Lactobacillus* spp. or *Lactococcus* spp., *Bacillus subtilis*, *Corynebacterium glutamicum*, *Streptomyces coelicolor*, and *Saccharomyces cerevisiae* must be grown under sterilization to prevent possible microbial contaminations. Chassis of the next generation of industrial biotechnology (NGIB) could be extremophile bacteria (no archaea) and bacteria able to consume special substrates such as methanol, carbon dioxide (CO_2), or syngas. Extremophiles are microorganisms that live under relatively extreme conditions such as high or low temperature, pH, osmotic pressure, atmospheric pressure, radiation, presences of heavy metal and/or organic solvents et al. that are not suitable for the survival of other microorganisms [6]. The harsh growth conditions for extremophiles effectively prevent the growth of non-extremophile organisms. Some extremophiles can grow fast in the presence of inexpensive or toxic substrates or in the absence of sufficient water, they are candidates for NGIB provided they can be grown fast enough for industrial processing.

Halophiles are extremophiles that require salt (NaCl) for growth, they are divided into six subgroups including non-halophile (<0.2 M), slight halophile (0.2–0.5 M), moderate halophile (0.5–2.5 M), borderline extreme halophile (2.5–4.0 M), extreme halophile (4.0–5.9 M) and halo-tolerant (non-halophile tolerating 2.5 M salt) [7, 8]. Many halophiles are grown under high pH providing an additional barrier for the growth of common bacteria. These halophiles as PHA producers have the following advantages: (1) the cultures do not require sterilization, reducing energy consumption and process complexity; (2) seawater can be used to replace freshwater for growth; (3) suitable for continuous cultures that significantly improve production efficiency; (4) cells can be disrupted by short-term hypoosmotic treatment to reduce product purification costs; (5) plastic, ceramic, and even cements can be used to

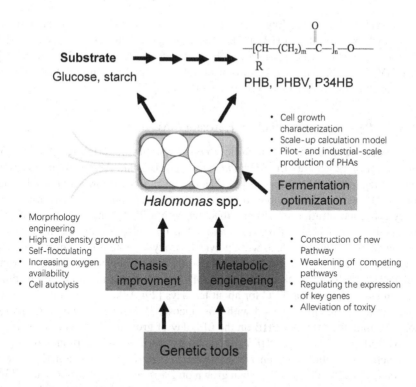

FIGURE 11.2 Advantages of next generation industrial biotechnology (NGIB).

construct fermentor vessels. *Halomonas bluephagenesis* strain TD01 has been successfully engineered to produce chemicals with a higher efficiency (Figure 11.2) [3].

11.3 PRODUCTION OF PHA BY HALOPHILES

11.3.1 PRODUCTION OF PHA BY *HALOMONAS* SPP.

Halophilic bacteria are suitable microorganisms for NGIB as the high salt in the medium inhibits the growth of non-halophilic microorganisms. The family *Halomonadaceae* is composed of 123 different halophilic or halotolerant bacterial species including 90 *Halomonas* species [9]. PHA-producing halophilic bacteria were isolated from hypersaline conditions including deep-sea [10], salt lakes [11, 12], saline soil [13], and hypersaline ponds [14]. It was reported that 32 *Halomonas* species are known to synthesize PHA [15]. *Halomonas boliviensis* was isolated in the soil sample around the lake Laguna Colorada and it can tolerate a wide range of NaCl concentrations (0–25% w/v), temperature (0–45°C) and pH (6–11) in the environment [13]. This bacterium is able to produce 50–80 wt.-% PHB contents from various carbon sources including volatile fatty acids (VFAs), mono- and disaccharides as well as starch hydrolysate [16–18]. The highest PHB yield of 41 wt.% was attained after 24 hours of cultivation in air-lift reactors (ALRs) [17]. *Halomonas bluephagenesis* TD01 was isolated from Aydingkol Lake in Xinjiang, China [12]. *H. bluephagenesis* TD01 maintained contamination-free under open and continuous conditions for 14 days [12]. Engineered *H. bluephagenesis* TD01 was used to produce poly(3-hydroxybutyrate) (PHB) [12], copolymer poly(3-hydroxybutyrate-*co*-3-hydroxyvalerate) (PHBV) consisting of 3-hydroxybutyrate (3HB) and 3-hydroxyvalerate (3HV) [19, 20], and poly(3-hydroxybutyrate-*co*-4-hydroxybutyrate) (P34HB) of 3-hydroxybutyrate and 4-hydroxybutyrate (4HB) [21, 22].

11.3.2 PRODUCTION OF PHA BY HALOPHILIC ARCHAEA

The production of PHA by halophilic *archaea* was first reported in 1972 [23]. To date a total of 12 genera of the *Halobacteriaceae* (Domain Archaea) families were detected regarding PHA synthesis [24–27]. Huang et al. used low-cost extruded rice bran (ERB) as a carbon source to produce PHA by *Haloferax mediterranei* [28]. A dry cell concentration of 140 g/L together with a PHA concentration of 78 g/L was obtained from a fed-batch fermentation [28]. The halophilic archaeon *Haloferax mediterranei* accumulates 55.6 wt.% poly(3-hydroxybutyrate-*co*-3-hydroxyvalerate) using rice bran and corn starch [15]. *Haloarcula marismortui* was able to accumulate 21% PHB in its cell dry mass (CDM) when cultured in a minimal medium with excessive glucose at 37°C for about ten days [29]. When grown on 20% NaCl synthetic medium supplemented with 2% glucose, *Halogeometricum borinquense* TN9 accumulates 14 wt.-% PHB on the fifth day of growth [30]. A *Natrinema* species 1KYS1 produced 53% PHBV when starch was used as a carbon source [31]. Extremely halophilic archaeon *Halogeometricum borinquense* strain E3 showed maximum 74% PHBV of CDM when grown on 2% glucose as a carbon source [32]. Extremely halophilic archaeon *Natrinema ajinwuensis* RM-G10 accumulated 61% PHBV consisting of 13.93 mol-% 3HV at 72 hours in repeated batch cultures [33].

11.4 GENETIC TOOLS FOR HALOPHILE ENGINEERING

In order to improve PHA production, it is necessary to develop genetic tools for regulating gene expression [3]. A series of tools have been developed for *Halomonas bluephagenesis* TD01(Table 11.1). A conjugation procedure for plasmid transformation and gene knockout has been developed based on a markerless gene replacement method with the help of a suicide vector pRE1126I-SceI [34]. A plasmid pSEVA341 with high-copy number was constructed for the inducible expression of multiple pathway genes [19]. To allow stable integration and expression of foreign genes, a chromosomal expression system was developed based on the strongly expressed *porin* gene [35]. Through randomizing the −35 and −10 elements of *porin* promoter, a constitutive promoter library with 310-fold variations of transcriptional activity was obtained, and an inducible promoter with a maximal fold induction of 230 with negligible leaky expression was constructed by integrating a lacO1 operator into the sequence 7 bp downstream of the −10 element within the *porin* promoter [36]. To reduce the processing cost, a novel cell autolysis system was developed based on the lambda phage *SRRz* genes and a synthetic ribosome binding site (RBS) allowing cell lysis after the addition of solvents or to spontaneously lyse during the stresses of downstream processing [37]. A controllable repression system based on clustered regularly interspaced short palindromic repeats interference (CRISPRi) was

TABLE 11.1

Engineering Approaches Developed for *Halomonas* spp.

	Tool Name	Function	Description
1	Chromosomal expression system	Stable expression of target genes	A suicide plasmid-mediated two-step homologous recombination system.
2	Inducible expression system	On/off switch of gene expression in *Halomonas* TD strains	The 21-bp nucleotide of lacO was inserted into the promoter P_{porin} by site-directed mutagenesis method
3	Constitutive expression system	To control the expression of key metabolic genes	By randomizing the sequence between the −35 and −10 elements, a constitutive promoter library was obtained with 310-fold variation in transcriptional activity
4	CRISPRi	A controllable gene repression system	An effective CRISPRi platform for genome editing in *Halomonas* sp. TD01
5	Novel T7-like expression systems	Transcriptional control in non-model bacteria	Expression of the cell-elongation cassette (minCD genes) and poly(3-hydroxybutyrate) (PHB) biosynthetic pathway, resulting in a 100-fold increase in cell lengths and high levels of PHB
6	CRISPRi/Cas9 system	A rapid gene editing in *Halomonas*	Deletion of *prpC* gene to produce a microbial copolymer P(3HB-*co*-3HV) with the 3HV fraction in the copolymers increased approximately 16-fold
7	P_{porin} based Promoter Library	a wide range of relative transcriptional strengths from 40 to 140,000	Promotion of *orfZ* gene encoding 4HB-CoA transferase the best one produced over 100 g/L CDM containing 80% poly(3HB-*co*-11 mol% 4HB)

established to repress expression of cell fission ring forming *ftsZ* gene [20]. To control the expression level of a target gene, three T7-like inducible expression systems, namely, MmP1, VP4, and K1F, were screened [38]. An efficient and scarless genome editing method via CRISPR/Cas9 system was invented with the highest efficiency of 100% in *Halomonas* spp. [39]. Furthermore, a low-oxygen inducible promoter was constructed to induce expression of bacterial hemoglobin encoded by gene *vgb* or PHB synthesis [40]. All the approaches secure the development of *Halomonas* spp. as chassis for NGIB targeting PHA synthesis and beyond.

11.5 ENGINEERING *HALOMONAS* SPP. FOR PHA PRODUCTION

11.5.1 Production of PHB

A high production cost is one of the limitations for PHB competitiveness for industrial and commercial applications. Several engineering approaches have been adapted to improve PHB production by *Halomonas* spp. Overexpression of the cell division inhibitor MinCD during the stationary phase in *H. bluephagenesis* TD08 elongated its shape to 1.4 times longer than the wild type, resulting in enhanced PHB accumulation from 69 wt.-% to 82 wt.-% with gravity induced cell precipitations for easy separation [19]. The overexpression of *minCD* genes and PHB pathway under a newly developed T7-like promoter, a 100-fold increase in cell lengths and high level of PHB production with up to 92% of cell dry weight were achieved [38]. PHB accumulation by *Halomonas* spp. TD01 increased by approximately 8%, when the *gltA* gene was repressed in various intensities via CRISPRi [20]. Disruption of *mreB* or *ftsZ* by a temperature-responsible plasmid expression system, 80% PHB yield increase was obtained in *Halomonas campaniensis* LS21 [41]. Increased NADH/NAD+ ratio under oxygen limitation by knocking out *etf* operon encoding an electron transport pathway resulted in 90% PHB accumulation in the cell dry weight of *H. bluephagenesis* compared to 84% in the wild type [42]. The cell growth was increased by 100% when active bacterial hemoglobin VHb was exported into the periplasm of *H. bluephagenesis* TD01 and *H. campaniensis* LS21 [40]. *H. bluephagenesis* TDHCD-R3 was a strain able to grow to high cell density and has obvious enhancement to toxic metabolites [43]. When overexpressing an optimized PHA synthesis *phaCAB* operon, TDHCD-R3 accumulated more than 90 g/L CDM containing 79% PHA when grown in a 7 L fermentor.

11.5.2 PRODUCTION OF **PHBV**

PHBV has been well investigated for its application to bone substitutes, vascular grafts, absorbable surgical sutures, medical packaging, and drug carriers because of its excellent absorption capacity, biological origin, low cytotoxicity, piezoelectricity, and thermoplasticity [44]. Many genes of *Halomonas* spp. were genetically engineered to enhance PHBV production or regulate the ratio of 3-hydroxyvalerate (3HV). Knock-down of *prpC* encoding 2-methylcitrate synthase via a newly developed CRISPRi system, the percentage of 3HV in PHBV copolymers were controllable ranging from less than 1 to 13% [20]. Deletion on *prpC* gene increased the conversion efficiency of propionic acid to 3HV monomer in random PHBV copolymers from around 10% to almost 100% [19]. Overexpression of the threonine synthesis pathway and threonine dehydrogenase allows recombinant *Halomonas* TD08 to produce PHBV consisting of 4–6 mol-% 3HV [19]. In a recent study, deletions on *prpC* and *sdhE* encoding 2-methylcitrate synthase and succinate dehydrogenase assembly factor 2, respectively, combined with overexpressing of *scpA, scpB* and *ppc*, encoding methylmalonyl-CoA mutase, methylmalonyl-CoA decarboxylase, and phosphoenolpyruvate carboxylase, respectively, the engineered *H. bluephagenesis* TY194 ($\Delta sdhE$, G7::P_{porin}-*ppc*) was found to synthesize 6.3 g/L CDM containing 65% PHBV consisting of 25 mol-% 3HV in PHBV when grown in glucose and gluconate in shake flasks [45].

11.5.3 PRODUCTION OF **P34HB**

P34HB can be used to repair damage to cardiovascular tissue, cartilage, and nerve tissue, or act as scaffold materials for tissue engineering. Overexpression of 4HB-CoA transferase encoded by *orfZ* from *Clostridium kluyveri* in *Halomonas bluephagenesis* TD01, resulted in over 70 g/L CDM containing 63% P(3HB-*co*-12 mol-% 4HB) after 48 hours under non-sterile conditions in 1 L and 7 L fermentors [21]. When overexpression of *orfZ* gene driven under a newly developed promoter based on the P_{porin} core region, the engineered strain produced over 100 g/L cell dry weight containing 80% poly(3HB-*co*-11 mol % 4HB) with a productivity of 1.59 g/(L·h) after 50 hours growth under non-sterile fed-batch conditions in a 7 L fermentor [46]. When deleted with the succinate semialdehyde dehydrogenase (gabD), the strain was grown to 26.3 g/L CDM containing 60.5% P(3HB-*co*-17.04 mol-% 4HB) in a 7 L bioreactor with glucose as the single carbon source for 60 hours under non-sterile conditions [22]. In a pilot scale-up study conducted in a 5000 L vessel, 100 g/L CDM containing 60.4% P(3HB-*co*-13.5 mol-% 4HB) was achieved after 36 hours of growth guided by a mathematical model and rational calculations[47].

11.6 MORPHOLOGY ENGINEERING FOR EASY SEPARATION

Separation of microbial cells from the cultures is the first stage for downstream processing. The common methods for cell separation include centrifugation, filtration, sedimentation, and gravity precipitation. Among them, centrifugation is commonly employed. However, when applied to an industrial scale production, it has many limitations, including complicated processes, expensive centrifugal machine, and massive energy consumption. A lot of efforts have been made to improve the separation efficiency with a lower energy cost. For microbial separation, a large cell size will truly help to decrease the separation difficulty. Several genes could efficiently improve cell sizes and accelerate the precipitation of microbial cells.

SulA is an SOS-inducible protein that interacts with *ftsZ* to inhibit cell division [48]. Over-expression of *sulA* leads to the formation of filamentary *E. coli* cells [49]. The shape changing *E. coli* accumulated more than 100% more on PHB contents and 10% increase on P34HB accumulation. Most importantly, the cells were able to precipitate after 20 minutes stillstand. Gene *mreB* plays a key role in directing the insertion of peptidoglycan precursors [50], overexpression of *mreB* resulted in larger spherical cells with 34 g/L CDM containing 86% PHB after 90 g/L glucose was consumed in 38 hours [51]. *E. coli* JM109SG (ΔmreB) overexpressing *mreB* gene under inducible expression of *sulA* could precipitate by gravity after 30 minutes without agitation [51]. When regulated the expression of *mreB* in *Halomonas campaniensis* LS21 with a temperature-responsible plasmid expression system, cells can grow normally at 30°C and gradually expand to larger spherical shapes at 37°C [41]. *H. campaniensis* LS21 cells with disrupted *ftsZ* on the genome were grown to a length of over 500 µm [41]. When combining the deletion of *minCD* genes that block the formation of cell fission rings (Z-rings) with a *phaP1* gene regulating PHA granule synthesis in *Halomonas bluephagenesis* TD01, super large PHB granules with an axial length over 10 µm were produced [52]. The deletion of *etf* gene encoding electron transfer flavoproteins in *Halomonas campaniensis* strain LS21 helps the cells to self-flocculate without changing the cell shape [53]. Most cells self-flocculated within one minute after stopping the aeration and agitation[53].

11.7 CONCLUSIONS AND OUTLOOK

PHA is considered to be an environmentally friendly bioplastic because of its biodegradability and biocompatibility as well as thermoprocessibility. It is widely used in biomedical, tissue engineering, green packaging, agricultural mulch films, and animal husbandry. The high production cost is an important factor restricting the large-scale application of PHA. Next generation industrial biotechnology (NGIB) based on extremophiles will significantly reduce production costs including reducing fermentation complexity, production equipment expenses, sources of substrates, energy, and water consumption. Several genetic tools have been developed for *Halomonas* spp. used for metabolic engineering, chassis improvement, and fermentation optimization (Figure 11.2). PHA production has been successfully conducted from a 7 L fermenter in the lab to 5000 L in industry. NGIB is an ideal way for low price and large-scale production of PHA.

REFERENCES

1. Chen GQ, Zhang J. Microbial polyhydroxyalkanoates as medical implant biomaterials. *Artif Cells Nanomed Biotechnol* 2018; 46(1):1–18.
2. Singh AK, Srivastava JK, Chandel AK, *et al.* Biomedical applications of microbially engineered polyhydroxyalkanoates: An insight into recent advances, bottlenecks, and solutions. *Appl Microbiol Biotechnol* 2019; 103(5):2007–2032.
3. Chen GQ, Jiang XR. Next generation industrial biotechnology based on extremophilic bacteria. *Curr Opin Biotechnol* 2018; 50:94–100.
4. Zhang X, Lin Y, Chen GQ. Halophiles as chassis for bioproduction. *Adv Biosyst* 2018; 2(11):1–12.
5. Wang Y, Yin J, Chen GQ. Polyhydroxyalkanoates, challenges and opportunities. *Curr Opin Biotechnol* 2014; 30:59–65.
6. Pikuta EV, Hoover RB, Tang J. Microbial extremophiles at the limits of life. *Crit Rev Microbiol* 2007; 33(3):183–209.
7. Edbeib MF, Wahab RA, Huyop F. Halophiles: Biology, adaptation, and their role in decontamination of hypersaline environments. *World J Microbiol Biotechnol* 2016; 32(8):135.
8. Ollivier B, Caumette P, Garcia JL, Mah RA. Anaerobic bacteria from hypersaline environments. *Microbiol Rev* 1994; 58(1):27–38.
9. Arahal DR, Oren A, Ventosa A. International committee on systematics of prokaryotes subcommittee on the taxonomy of Halobacteria and Subcommittee on the taxonomy of *Halomonadaceae*. Minutes of the joint open meeting, 11 July 2017, Valencia, Spain. *Int J Syst Evol Microbiol* 2017; 67(10):4279–4283.
10. Simon-Colin C, Raguenes G, Cozien J, Guezennec JG. *Halomonas profundus* sp. nov., a new PHA-producing bacterium isolated from a deep-sea hydrothermal vent shrimp. *J Appl Microbiol* 2008; 104(5):1425–1432.
11. Joshi AA, Kanekar PP, Kelkar AS, *et al.* Moderately halophilic, alkalitolerant *Halomonas campisalis* MCM B-365 from Lonar Lake, India. *J Basic Microbiol* 2007; 47(3):213–221.
12. Tan D, Xue YS, Aibaidula G, Chen GQ. Unsterile and continuous production of polyhydroxybutyrate by *Halomonas* TD01. *Bioresour Technol* 2011; 102(17):8130–8136.
13. Quillaguaman J, Hatti-Kaul R, Mattiasson B, *et al.* *Halomonas boliviensis* sp. nov., an alkalitolerant, moderate halophile isolated from soil around a Bolivian hypersaline lake. *Int J Syst Evol Microbiol* 2004; 54(3):721–725.

14. Cervantes-Uc JM, Catzin J, Vargas I, *et al.* Biosynthesis and characterization of polyhydroxyalkanoates produced by an extreme halophilic bacterium, *Halomonas nitroreducens*, isolated from hypersaline ponds. *J Appl Microbiol* 2014; 117(4): 1056–1065.

15. Quillaguaman J, Guzman H, Van-Thuoc D, Hatti-Kaul R. Synthesis and production of polyhydroxyalkanoates by halophiles: Current potential and future prospects. *Appl Microbiol Biotechnol* 2010; 85(6):1687–1696.

16. Van-Thuoc D, Quillaguaman J, Mamo G, Mattiasson B. Utilization of agricultural residues for poly(3-hydroxybutyrate) production by *Halomonas boliviensis* LC1. *J Appl Microbiol* 2008; 104(2):420–428.

17. Rivera-Terceros P, Tito-Claros E, Torrico S, *et al.* Production of poly(3-hydroxybutyrate) by *Halomonas boliviensis* in an air-lift reactor. *J Biol Res (Thessalon)* 2015; 22(1):8.

18. Garcia-Torreiro M, Lu-Chau TA, Lema JM. Effect of nitrogen and/or oxygen concentration on poly(3-hydroxybutyrate) accumulation by *Halomonas boliviensis*. *Bioprocess Biosyst Eng* 2016; 39(9):1365–1374.

19. Tan D, Wu Q, Chen JC, Chen GQ. Engineering *Halomonas* TD01 for the low-cost production of polyhydroxyalkanoates. *Metab Eng* 2014; 26:34–47.

20. Tao W, Lv L, Chen GQ. Engineering *Halomonas* species TD01 for enhanced polyhydroxyalkanoates synthesis via CRISPRi. *Microb Cell Fact* 2017; 16(48):1–10.

21. Chen X, Yin J, Ye J, *et al.* Engineering *Halomonas bluephagenesis* TD01 for non-sterile production of poly(3-hydroxybutyrate-*co*-4-hydroxybutyrate). *Bioresour Technol* 2017; 244(1):534–541.

22. Ye JW, Hu DK, Che XM, *et al.* Engineering of *Halomonas bluephagenesis* for low cost production of poly(3-hydroxybutyrate-*co*-4-hydroxybutyrate) from glucose. *Metab Eng* 2018; 47:143–152.

23. Kirk RG, Ginzburg M. Ultrastructure of two species of halobacterium. *J Ultrastruct Res* 1972; 41(1):80–94.

24. Salgaonkar BB, Braganca JM. Utilization of sugarcane bagasse by *halogeometricum borinquense* strain E3 for biosynthesis of poly(3-hydroxybutyrate-*co*-3-hydroxyvalerate). *Bioengineering (Basel)* 2017; 4(2):1–18.

25. Han J, Hou J, Liu H, *et al.* Wide distribution among halophilic archaea of a novel polyhydroxyalkanoate synthase subtype with homology to bacterial type III synthases. *Appl Environ Microbiol* 2010; 76(23):7811–7819.

26. Legat A, Gruber C, Zangger K, *et al.* Identification of polyhydroxyalkanoates in *Halococcus* and other Haloarchaeal species. *Appl Microbiol Biotechnol* 2010; 87(3):1119–1127.

27. Koller M. Polyhydroxyalkanoate biosynthesis at the edge of water activitiy-haloarchaea as biopolyester factories. *Bioengineering* 2019; 6(2):34.

28. Huang TY, Duan KJ, Huang SY, Chen CW. Production of polyhydroxyalkanoates from inexpensive extruded rice bran and starch by *Haloferax mediterranei*. *J Ind Microbiol Biotechnol* 2006; 33(8):701–706.

29. Han J, Lu Q, Zhou L, *et al.* Molecular characterization of the *phaECHm* genes, required for biosynthesis of poly(3-hydroxybutyrate) in the extremely halophilic archaeon *Haloarcula marismortui*. *Appl Environ Microbiol* 2007; 73(19): 6058–6065.

30. Salgaonkar BB, Mani K, Braganca JM. Accumulation of polyhydroxyalkanoates by halophilic archaea isolated from traditional solar salterns of India. *Extremophiles* 2013; 17(5):787–795.

31. Danis O, Ogan A, Tatlican P, *et al.* Preparation of poly(3-hydroxybutyrate-co-hydroxyvalerate) films from halophilic archaea and their potential use in drug delivery. *Extremophiles* 2015; 19(2):515–524.

32. Salgaonkar BB, Braganca JM. Biosynthesis of poly(3-hydroxybutyrate-*co*-3-hydroxy valerate) by *halogeometricum borinquense* strain E3. *Int J Biol Macromol* 2015; 78:339–346.

33. Mahansaria R, Dhara A, Saha A, *et al.* Production enhancement and characterization of the polyhydroxyalkanoate produced by *Natrinema ajinwuensis* (as synonym) identical with *Natrinema altunense* strain RM-G10. *Int J Biol Macromol* 2018; 107(B):1480–1490.

34. Fu XZ, Tan D, Aibaidula G, *et al.* Development of *Halomonas* TD01 as a host for open production of chemicals. *Metab Eng* 2014; 23:78–91.

35. Yin J, Fu XZ, Wu Q, *et al.* Development of an enhanced chromosomal expression system based on porin synthesis operon for halophile *Halomonas* sp. *Appl Microbiol Biotechnol* 2014; 98(21):8987–8997.

36. Li T, Li T, Ji W, *et al.* Engineering of core promoter regions enables the construction of constitutive and inducible promoters in *Halomonas* sp. *Biotechnol J* 2016; 11(2):219–227.

37. Hajnal I, Chen X, Chen GQ. A novel cell autolysis system for cost-competitive downstream processing. *Appl Microbiol Biotechnol* 2016; 100(21):9103–9110.

38. Zhao H, Zhang HM, Chen X, *et al.* Novel T7-like expression systems used for *Halomonas*. *Metab Eng* 2017; 39:128–140.

39. Qin Q, Ling C, Zhao YQ, et al. CRISPR/Cas9 editing genome of extremophile *Halomonas* spp. *Metab Eng* 2018; 47:219–229.

40. Ouyang P, Wang H, Hajnal I, *et al.* Increasing oxygen availability for improving poly(3-hydroxybutyrate) production by Halomonas. *Metab Eng* 2018; 45:20–31.

41. Jiang XR, Yao ZH, Chen GQ. Controlling cell volume for efficient PHB production by *Halomonas*. *Metab Eng* 2017; 44:30–37.

42. Ling C, Qiao GQ, Shuai BW, *et al.* Engineering NADH/NAD(+) ratio in *Halomonas bluephagenesis* for enhanced production of polyhydroxyalkanoates (PHA). *Metab Eng* 2018; 49:275–286.

43. Ren YL, Ling C, Hajnal I, *et al.* Construction of *Halomonas bluephagenesis* capable of high cell density growth for efficient PHA production. *Appl Microbiol Biotechnol* 2018; 102(10):4499–4510.

44. Rivera-Briso AL, Serrano-Aroca A. Poly(3-hydroxybutyrate-co-3-hydroxyvaler ate): Enhancement strategies for advanced applications. *Polymers (Basel)* 2018; 10(732):1–28.

45. Chen Y, Chen XY, Du HT, *et al.* Chromosome engineering of the TCA cycle in *Halomonas bluephagenesis* for production of copolymers of 3-hydroxybutyrate and 3-hydroxyvalerate (PHBV). *Metab Eng* 2019; 54:69–82.

46. Shen R, Yin J, Ye JW, *et al.* Promoter engineering for enhanced P(3HB– co–4HB) production by *Halomonas bluephagenesis*. *ACS Synth Biol* 2018; 7(8):1897–1906.

47. Ye JW, Huang WZ, Wang DS, *et al.* Pilot scale-up of poly(3-hydroxybutyrate-co-4-hy droxybutyrate) production by *Halomonas bluephagenesis* via cell growth adapted optimization process. *Biotechnol J* 2018; 13(5):1800074.

48. Dajkovic A, Mukherjee A, Lutkenhaus J. Investigation of regulation of FtsZ assembly by SulA and development of a model for FtsZ polymerization. *J Bacteriol* 2008; 190(7):2513–2526.

49. Wang Y, Wu H, Jiang X, Chen GQ. Engineering *Escherichia coli* for enhanced production of poly(3-hydroxybutyrate-*co*-4-hydroxybutyrate) in larger cellular space. *Metab Eng* 2014; 25:183–193.

50. Rueff AS, Chastanet A, Dominguez-Escobar J, *et al.* An early cytoplasmic step of peptidoglycan synthesis is associated to MreB in *Bacillus subtilis*. *Mol Microbiol* 2014; 91(2):348–362.

51. Jiang XR, Wang H, Shen R, Chen GQ. Engineering the bacterial shapes for enhanced inclusion bodies accumulation. *Metab Eng* 2015; 29:227–237.
52. Shen R, Ning ZY, Lan YX, *et al.* Manipulation of polyhydroxyalkanoate granular sizes in *Halomonas bluephagenesis. Metab Eng* 2019; 54:117–126.
53. Ling C, Qiao GQ, Shuai BW, *et al.* Engineering self-flocculating *Halomonas campaniensis* for wastewaterless open and continuous fermentation. *Biotechnol Bioeng* 2019; 116(4):805–815.

12 PHA Biosynthesis Starting from Sucrose and Materials from the Sugar Industry

Luiziana Ferreira da Silva, Edmar Ramos Oliveira-Filho, Rosane Aparecida Moniz Piccoli, Marilda Keico Taciro, and José Gregório Cabrera Gomez

CONTENTS

12.1 INTRODUCTION OF SUCROSE FOR PHA PRODUCTION

Studies in the 1980s by ICI (Imperial Chemical Industries, UK) on bacterial production of PHA identified glucose (from dextrose) and *Ralstonia eutropha* as promising substrate and bacterial strains to be combined for PHA production, based on product yield from carbon source, as well as feedstock costs and availability for commodity production in Europe. When considering ethanol, glucose, sucrose, methanol, acetic acid, and hydrogen as carbon or energy sources, despite sucrose seeming to be the most promising, its use was discarded due to the production of large amounts of polysaccharide by *Azotobacer vinelandii*, thus reducing the efficiency of carbon source conversion into PHA. The need for the additional processing of sucrose (chemical or enzymatic conversion of sucrose to glucose and fructose) to make it a proper feedstock for PHA production by *R. eutropha* discouraged its use [1]. However, considering the high production costs when utilizing glucose, a number of research groups focused on searching for alternative low-cost carbon sources, methanol was a possibility, but sucrose remained on the agenda. Therefore, research was directed toward: (i) isolation of new bacterial strains able to use sucrose for PHA production [2–4]; (ii) enlargement of the carbon sources spectrum used by *R. eutropha* was another important approach [5]; (iii) the use of *A. vinelandii* mutants unable to produce polysaccharide to improve PHA yield [6, 7]. Other authors also considered the relevance of the use of sucrose as a proper feedstock in reducing the cost of substrates in PHA production [8–10].

This world scenario perfectly matched the need for the sugar and ethanol industry in Brazil, the largest sugarcane producer in the world. Sugarcane is the crop containing the largest amount of sucrose. This well-established industry had great availability of sugar at relatively low cost, as well as a lot of renewable energy. Moreover, it was looking for alternative commodities to broaden its range of products and allow more favorable price conditions for such products, through the flexibility of the production system. A joint project was proposed by the largest Brazilian sugar and alcohol cooperative (COPERSUCAR), garnering the Institute for Technological Research of São Paulo (IPT) and the University of São Paulo (USP) to develop the production of PHA integrated with sugar and ethanol mills [11]. Thus, in the early 1990s, studies to develop PHA production in Brazil were triggered. The Brazilian Program for Supporting Scientific and Technological Development (PADCT) sponsored the team financially [12–15]. The Federal Funding Authority for Studies and Projects (FINEP) was its financial agent. In further years, other funding agencies, namely São Paulo Research Foundation (FAPESP) and National Council for Scientific and Technological Development (CNPq) also contributed to sponsoring the development. To accomplish the objective of developing a technology for PHA production integrated in a sugar and ethanol mill, a multidisciplinary team was in charge of different tasks performed simultaneously: (i) selection for and development of efficient PHA-accumulating bacterial strains; (ii) production of PHA from sugarcane carbohydrates; (iii) cultivation at high cell densities and development of other strategies to increase productivity; (iv) polymer extraction and purification; (v) test of polymer properties and biodegradability; (vi) technical and economic feasibility assessment; (vii) scale up to pilot plant [16].

As a result of this project there was a lot of learning, and new technologies were developed in collaboration with the industry, which were directly transferred to the private productive sector. Among the various byproducts of the sugar and alcohol mills, some advantageous solvents were identified for use in polymer separation and purification [17].

Since PHA are plastics with environmental appeal, another important economic and ecological issue is about the energy source for production. Actually, another important result of the project development was the realization that a large amount of energy is needed in this process and represents more than 10% of the production cost of poly(3-hydroxybutyrate) P(3HB) [11]. Slater and Gerngross (2000) [18] also reached a similar finding. They identified that if fossil energy is used to power the process to extract PHA from recombinant plants, its amount is higher than the amount needed to produce petrochemical plastics (considering both energy and material for the synthesis). Therefore, when selecting feedstock for PHA production it is imperative to plan the source of energy that will be used. The sugar and ethanol mill industry in Brazil is very well established and can supply not only cheap carbon sources for production of PHA, but also represents an environment with a great quantity of renewable energy available. Therefore, this scenario was ideal because it was endowed with extra energy, obtained by burning the bagasse, making possible the establishment of a sustainable process for the production of truly green plastics.

With the introduction of second-generation ethanol in the sugar and alcohol mills, a new opportunity was opened to produce PHA integrated in this model, using sugars present in sugarcane bagasse but not fermented by ethanologenic yeasts. Thus, the future of PHA production becomes even more promising using xylose, which is expected to be available in millions of tonnes. The great challenge for PHA production to consolidate also involves how society will deal with environmental and economic issues related to oil use.

12.2 USE OF MOLASSES FOR PHA PRODUCTION

Molasses is the residual syrup obtained in sugar-refining mills after repeated crystallization procedures to extract sugars from sugarcane or beet juice. When extraction by crystallization is no longer economically feasible, the resulting product is a low-grade molasses, still with relatively high sucrose content, not suitable for food or feed [19]. Although beet and sugarcane molasses compositions vary depending on features related to plant cultivation, maturity, and fertilization [20], and also to the previously applied sugar extraction protocol, they generally contain water, carbohydrates (mainly sucrose, glucose, and fructose), ashes, nitrogenous (mainly proteins) and non-nitrogenous compounds, wax, sterols, phosphatides, and vitamins [21].

If considering a mill producing only sugar, molasses could be considered as waste material that is extremely rich for the development of different bioprocesses. However, this is not true for sugar and ethanol mills in Brazil. The molasses derived from sugar production is the primary substrate for ethanol production. Therefore, this economic sector has great flexibility in producing either more sugar or more ethanol to meet the market demand, but the minimal ethanol volume production is determined by the molasses amount produced in the mill.

Different authors have pointed out molasses as a good feedstock for PHA production. The selection of bacterial strains able to produce large amounts of PHA under non-limiting growing conditions is part of this approach since molasses contains a number of nutrients. This approach could be interesting and could increase the flexibility in a system producing sugar, ethanol, and PHA.

12.3 BACTERIAL STRAINS FOR PHA PRODUCTION FROM SUCROSE

Alcaligenes eutrophus (*Ralstonia eutropha*), the model P(3HB)-producing bacterium [22] was the first choice of ICI for P(3HB) production [1]. However, it is unable to metabolize sucrose. Even though sucrose could be hydrolyzed to release glucose and fructose, using the invertase enzyme, this operation would contribute to an increase in costs, especially if a separated process of sucrose hydrolysis had to be conducted [1]. Acidic conversion of sucrose can be done in sugarcane mills, with a lower cost compared with enzyme hydrolysis; nonetheless, it is an extra step in the PHA production process. In addition, after the acid hydrolysis process, about 5% of sucrose remains unhydrolyzed, which means that the carbon source cannot be fully utilized. Therefore, the use of bacteria capable of effectively using sucrose as a carbon source becomes very important.

Some of the most studied bacteria able to produce PHA from sucrose are presented here.

12.3.1 *AZOTOBACTER VINELANDII*

Strains belonging to the *Azotobacter* genus, widely studied in the 1980s, were promising due to their ability to use sucrose and accumulate high amounts of PHA. However, this bacterium was disregarded in the selection process for industrial production of P(3HB) due to the high amount of polysaccharide produced [1]. This decision was taken because of the lack of information on the existence of mutants deficient in polysaccharide production [6]. Page and Knosp [7], using a strain unable to produce the capsule (polysaccharide) as the parental strain, developed mutant *Azotobacter vinelandii* UWD able to accumulate P(3HB) under balanced growth. The mutant had acquired a defective respiratory NADH oxidase and it was suggested to use P(3HB) production for recycling NADH during exponential growth. Mutant *A. vinelandii* UWD was subjected to low-scale experiments (100 mL in shaken 500 mL-Erlenmeyer flasks). The carbon sources tested were glucose, corn syrup, malt extract, sugarcane molasses, and beet molasses. Comparing the sucrose-containing raw materials, sugarcane molasses yielded up to 55% of cell dry mass (CDM) as P(3HB). From beet molasses, the culture reached 60%, while from malt extract 66% of CDM was attained. Those figures corresponded to a biomass production (CDM) of, respectively, 4.61, 4.57, and 4.24 g/L. When refined sucrose was tested the CDM was 3.15 g/L with 68 wt.-% P(3HB). Page [23] further reported that the same strain under shaken flask experiments, with vigorous aeration and 5% beet molasses, improved P(3HB) yield to 6.8 g/L. This improvement was attributed to the presence of substances like organic nitrogenous compounds that may have been acting as a

P(3HB)-yield-promoting factor [24]. One of the problems relevant to the use of beet molasses was the high nitrogen and phosphorus content that would not allow applying limitation of such nutrients to induce P(3HB) production [1]. Page proposed the use of *Azotobacter* as the strain of choice because of its ability of polymer accumulation simultaneous to growth. Another issue to be considered was the amount of molasses that would be possible to supply the medium without toxic effects of other constituents. Beet molasses up to 10% w/v was considered nontoxic to the strain, no color was transferred from the molasses to the product and other nutrients contained therein would stimulate bacterial growth and polymer accumulation under strong aeration. Above 10% beet molasses in solution resulted in an increase in medium viscosity, influencing oxygen transfer and contributing to foam production [24]. The addition of fish peptone to strain UWD cultures with sucrose, malt extract, beet, or sugarcane molasses (50 mL volume) indicated a significant increase in non-P(3HB) biomass formation associated with higher P(3HB) production, in the case of cultures using malt extract, sucrose, and sugarcane molasses. With the addition of fish peptone, P(3HB) production increased from 0.7 to 6.4 g/L in sucrose, from 3.2 to 6.5 g/L in sugarcane molasses, and from 1.6 to 5.9 g/L in malt extract, indicating the need for additional nitrogenous compounds on each case. This was not true for beet molasses since no difference in polymer production was detected when fish peptone was added [25]. Thereafter, experiments were conducted on bioreactor (2.5 L working volume) feeding with a beet molasses (BM) solution (50% w/v) at a rate appropriate to maintain a 5% w/v sugar content in the culture medium. Sucrose content on beet molasses was 50–52 wt.-% and other components were nitrogenous compounds (10–27%), nitrogen-free compounds (9–10 %), and ash (11–12%). Two sugar beet industry waste streams were also tested: concentrated separator byproducts (CSB) and extract molasses (EM), both containing sucrose. None of them, however, gave promising results. From the BM fed-batch experiments, however, polymer production increased up to 25 g/L at a volumetric productivity of 1–1.4 g/(L·h), sugar consumption rates ranged from 2.9 to 3.8 g/(L·h). Based on an overall $Y_{P/molasses}$ = 0.12 g/g, the authors calculated, at that point, that using BM as a feedstock would represent a substrate cost one-third lower the value considered from glucose [24]. Considering the other carbon sources tested, the author also indicated that it would be better to use CSB and EM from BM fractions as a minor addition to media containing sugars to increase PHB yield, but not as a primary substrate for PHB production [26].

The ability of *A. vinelandii* UWD in accumulating poly(3-hydroxybutyrate-*co*-3-hydroxyvalerate) P(3HB)-*co*-3HV in BM medium supplied with *n*-alkanoates (with backbone size ranging from C3 to C8) was tested in bioreactor experiments. Only valerate and heptanoate additions promoted copolymerization of 16 and 4.7 mol-% 3HV units in the copolymer [24]. Supply of nonanoate, in other experiments, revealed only traces of 3HV content, whereas from 2-pentenoate 12 mol-% 3HV were detected, despite a 56% growth inhibition. In the same set of bioreactor experiments, the strategic supply of valerate to a fed-batch culture of strain UWD in beet molasses yielded PHA at 19 to 22 g/L, containing 8.5 to 23 mol-% 3HV after 38 to 40 hours [27]. Interestingly, no 3HV content was detected when propionate was supplied, although it was a substrate for growth, according to the authors, possibly due to the presence of efficient propionate catabolic pathways. Another explanation is that *A. vinelandii* does not have 3-ketothiolase bktB [28]. According to Slater and coworkers [29] bktB is responsible for condensation of acetyl-CoA and propionyl-CoA generating 3-ketovaleryl-CoA, the precursor of 3HV monomer synthesis.

12.3.2 *ALCALIGENES LATUS*

Alcaligenes latus (currently *Azohydromonas lata*) was described as a new species by Palleroni and Palleroni [30]. It became interesting in the 1980s because of its ability to use sucrose and other low-cost substrates such as beet and sugarcane molasses, accumulating P(3HB) up to 80% of biomass during the growth phase, being also capable of incorporating 3HV units from propionate and valerate [3, 31, 32]. In the mid-1980s, the company Chemie Linz AG (Austria), through a biotechnological research unit, started the development of P(3HB) production from a strain of *A. latus* taking into account that no nutrient limitation would be needed in the process [4, 33, 34]. Biomer (Germany) also adopted a similar strategy. Cultivation of an improved mutant of *A. latus* DSM1124 was scaled up to a 15 m³ bioreactor. High cell density cultivations were achieved associated with polymer production, reaching more than 60 g/L.

During the 1990s, studies with *A. latus* were also developed in Canada [3]. *A. latus* was cultivated in chemostats supplied with a mixture of sucrose (20 g/L) and propionic acid (from 0–5 g/L) at a dilution rate of 0.15 h⁻¹. One- or two-stage cultures (two reactors in series) were performed. One stage led to 40% of CDM as PHA, with 3HV varying from 0 to 20 mol-% with an increasing supply of propionate (0–5 g/L). The two-stage chemostat production was supplied with propionic acid (8.5 g/L) and sucrose (20 g/L) and resulted in unconsumed sucrose (5.8 g/L) remaining; PHA corresponding to 43% of CDM with 18.5 mol% 3HV. In the second stage, the total PHA obtained was 58 wt % of CDM with 11 mol% 3HV content. In one-stage chemostat experiments, supplied with sucrose and replacing propionic acid by pentanoic acid, a similar amount of PHA was achieved with an increase in 3HV molar fraction to 38% [3].

Mathematical modeling of *A. latus* was also studied [35, 36]. A metabolic/polymerization kinetic model was proposed in 2010 and, in 2012 an evaluation of how nutritional and aeration conditions affect biomass production rate, P(3HB) accumulation, and molecular weight. Data of *A. latus* cultivated in sucrose were used to validate the model.

P(3HB) production by *A. latus* (ATCC 29714) was investigated by [37]. Sugar beet juice with the addition of nutrients other than sugar was tested. The best result was obtained with partial addition of nutrients, resulting in 10.3 g/L of CDM with 4.0 g/L of P(3HB) and maximum volumetric productivity of 0.22 g/(L·h).

Fed-batch cultivations of *Azohydromonas lata* DSM 1123, using products from the sugarcane industry (cane juice, 60 Brix syrup, raw cane sugar, and cane molasses) as carbon sources, were performed [38]. Optimization of the process resulted in 16.9 g/L of CDM with about 84% of P(3HB) content, a yield of sucrose in P(3HB) ($Y_{P/S}$) of 0.4 g/g and P(3HB) productivity of 0.23 g/(L·h), using a 60 Brix syrup as the carbon source in a fed batch culture.

Gahlawat and Srivastava [39] developed a kinetic mathematical model based on inhibition/limitation data of *A. latus* DSM 1124. The model was applied for designing nutrient feeding strategies (sucrose and nitrogen) to improve P(3HB) accumulation and resulted in 30.17 g/L of biomass containing 75% of P(3HB) (29.38 g/L), corresponding to a P(3HB) volumetric productivity of 0.6 g/(L·h).

Gomez and coworkers demonstrated that different *A. latus* strains (DSM 1122, 1123, and 1124) showed variable responses when cultured using sucrose and only DSM 1123 completely consumed sucrose under the conditions tested. When compared to different strains in the medium using glucose plus fructose, the strain DSM 1123 achieved the maximum theoretical efficiency expected to convert the carbon source into cells and PHB [12, 40]. Interestingly, strain DSM 1124 was shown to be unable to hydrolyze sucrose but able to consume the glucose while fructose remained in the medium in experiments providing glucose plus fructose. Koller et al. [41] reported that DSM 1124 was used industrially to produce P(3HB) during the late 1980s to early 1990s and Hängii indicated that a mutant of DSM 1124 with increased productivity had been used [4].

At the beginning of the cooperative project to establish P(3HB) production integrated with sugar and alcohol mills, *A. latus* DSM1123 was the first choice to develop the production process, however it was shown that this strain had a flocculation process that could not be controlled. Thus, cultivation with this species was discontinued due to this instability.

12.3.3 *RALSTONIA EUTROPHA* AND RECOMBINANTS

In 1994, Zhang and co-authors indicated, among *Enterobacteriaceae*, two strains as the most promising bacterial platforms for sucrose conversion to PHA: recombinant *E. coli* and *Klebsiella aerogenes* expressing PHA biosynthesis genes from *Ralstonia eutropha* ($phaCAB_{Re}$) [42]. P(3HB) was produced at 0.65 g/L.h with *K. aerogenes* 2688. However, due to the need for use of antibiotics to keep plasmids inside the cells, the cost of production would increase to prohibitive levels, according to the authors.

Sugarcane molasses was tested in *Ralstonia eutropha* (*A. eutrophus*) either in liquid or solid-state cultures, to produce PHA, however glucose was the main carbon source and molasses represented a supply of growth factors or vitamins (supplied at [0.3 to 2.5% (wt/wt)] and the PHA content ranged from 26 to 39% (wt/wt) [43, 44]. Since *R. eutropha*, the reference P(3HB)-producing strain, is not able to grow or produce this biopolymer from sucrose, metabolic engineering strategies have been employed to overcome this limitation. Genes from a sucrose-utilizing *Salmonella thyphimurium* were inserted into *R. eutropha* DSM 545 enabling this strain to grow in sucrose, accumulating up to 60% of CDM as P(3HB). Despite the maximum specific growth rate decreased from 0.30 to 0.24 h^{-1} compared to glucose [45–47], this construction was more effective than previous attempts of inserting *Bacillus subtilis* levanase-encoding genes to *R. eutropha*, which although they became able to use sucrose, showed a low efficiency, either due to the difficulty in excreting levanase or the difficulty of transporting sugars into the intracellular space [5]. Nordsiek and Bowien were also successful in constructing recombinant *R. eutropha* able to use sucrose. They cloned genes from *Paracoccus denitrificans* and *Rhodobacter capsulatus* species, taxonomically closer to *R. eutropha* ([48] apud [49]).

12.3.4 OTHER PHA-PRODUCING BACTERIA FROM SUCROSE

Besides *E. coli, A. latus, Azotobacter* sp., *Klebsiella* sp., and *R. eutropha*, another Gram-negative bacterium studied concerning PHA production from molasses was *Pseudomonas cepacia* strain G13, which was shown to accumulate up to 70 wt.-% in medium supplemented with 3% beet molasses [50]. During a screening for strains able to use sucrose to produce PHA, *P. cepacia* DSM50181 was able to hydrolyze sucrose, but unable to consume all the fructose produced. Likewise, when glucose and fructose were supplied together, the latter was not completely metabolized. In both conditions tested, 50 and 59% CDM was accumulated as PHA, respectively, representing approximately 0.20 g of P(3HB) produced per gram of carbohydrate consumed, corresponding to 50% of the maximum theoretical yield [12]. Among Gram-positive bacteria, *Bacillus* strains, namely *B. megaterium*, when supplied with sugarcane molasses as the main carbon source achieved a CDM of 72.7 g/L containing 42% of P(3HB) [51].

12.4 SETTING UP A BIOREFINERY TO PRODUCE PHA IN BRAZIL

Brazil is one of the countries with the largest production volume of sugarcane, one of the most important crops in the world [52]. In south-central Brazil, approximately 600 million tons of sugarcane were harvested in 2017/2018 [53]. As a large agro-industry, Brazilian sugarcane mills are already a biorefinery since they combine: production of ethanol, sugar at different grades, plus a number of secondary products (different alchools, for example) that also have high value and also bagasse to energy cogeneration [11]. A significant increase in production at sugar mills occurred during the 1970s when, as a result of the world oil crisis, Brazil implemented a program to produce ethanol as a fuel for vehicles, ProÁlcool, effectively combining sugar and ethanol productions [54]. In this process, after exhausting the efficiency of successive sugar extractions through crystallization from sugarcane juice, low-grade molasses, which no longer could be used as feed or food, are available to produce other biotechnological products due to their still relatively high sucrose content [19]. Usually, ethanol is produced from molasses, thus, in sugar-producing units associated with ethanol plants, molasses cannot be considered a waste product. Figure 12.1 summarizes the steps of sugar and ethanol production in Brazil.

The characteristics of Brazilian mills, associated with the availability of such low-cost and low-grade sugar sources created a favorable scenario for the integration of PHA production in Brazil [11, 55]. Although the use of molasses [1, 56] and also the integration of P(3HB) production in corn mills or sugar beet production had been previously suggested [26, 57], local conditions made it possible to start such studies in Brazil.

In the late 1980s, the then largest Brazilian cooperative of sugar and alcohol producers – COPERSUCAR – joined the Institute for Technological Research of São Paulo (IPT) and the University of São Paulo (USP) and proposed a joint project for the development of technology to produce PHA using sugarcane derivatives. Thus, in the early 1990s, studies to develop PHA production in Brazil were triggered. With this purpose, a multidisciplinary team received financial support from the Program for Supporting Scientific and Technological Development (PADCT) [12–15]. PADCT had been created in the previous decade, with resources from the World Bank and the

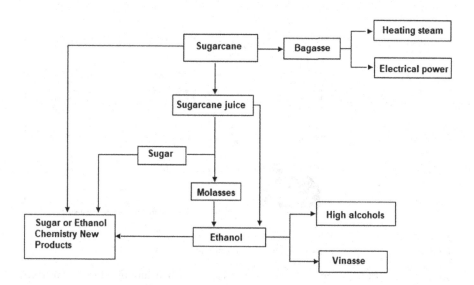

FIGURE 12.1 Steps involved in processing sugarcane to sucrose, ethanol, byproducts, and new potential products. Adapted from [55].

Brazilian Government Federal Funding Authority for Studies and Projects (FINEP) was its financial agent. PADCT was active until 2000 (http://www.finep.gov.br/a-fine p-externo/historico). In subsequent years, other funding agencies also contributed to sponsoring the development, namely São Paulo Research Foundation (FAPESP) and National Council for Scientific and Technological Development (CNPq).

A number of complementary aspects were studied according to the expertise of each team member: (i) selection for and development of efficient PHA-accumulating bacterial strains; (ii) production of PHA from sugarcane carbohydrates; (iii) cultivation at high cell densities and development of other strategies to increase productivity; (iv) polymer extraction and purification; (v) test of polymer properties and biodegradability; (vi) technical and economic feasibility assessment; (vii) scale up to pilot plant [16]. In the following years, during studies to expand the production scale, life cycle analysis studies were also developed to assess the sustainability of the entire process [41, 58].

Figure 12.2 summarizes the different topics studied during the development of PHA production from sugarcane derivatives in Brazil.

Each of these aspects is presented in this chapter. It was mandatory that the various stages of the project occurred simultaneously, thus while the search for strains was being developed, bioreactor experiments and other steps were conducted using reference strains such as *Alcaligenes eutrophus* and *Alcaligenes latus*.

12.4.1 SELECTION AND DEVELOPMENT OF EFFICIENT PHA-PRODUCING BACTERIAL STRAINS IN BRAZIL

12.4.1.1 Screening from Soil Samples and Culture Collections

The first approach adopted to obtain an efficient PHA-producing strain from sucrose was to search in culture collections worldwide for such strains. However, despite

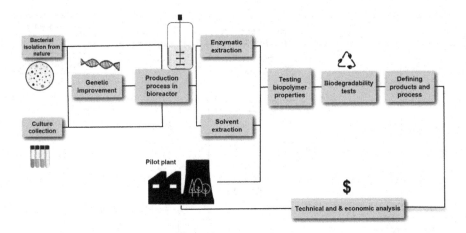

FIGURE 12.2 Steps of PHA production development in Brazil.

some of the strains being able to consume sucrose, when submitted to initial trials, either the sugar conversion to polymer was inefficient or PHA cell content was low [12, 40].

To obtain more efficient sucrose-utilizing bacteria, two additional approaches were then adopted: construction of sucrose–utilizing *Ralstonia eutropha* recombinants, and screening of new strains in nature. Simultaneously, glucose-utilizing reference strains were submitted to studies on PHA production in stirred tank reactor to accumulate knowledge in the different steps to accomplish the development of the process.

Isolation of new strains started by analyzing soil samples collected from sugarcane plantation, provided by COPERSUCAR. The first selection was performed cultivating appropriate sample dilutions in a mineral medium containing sucrose, fructose, and glucose, properly formulated to favor the growth of Gram-negative bacteria [12, 40]. Isolates were then tested for PHA accumulation using the Sudan black B staining test [59]. Some 75 PHA-accumulating isolates were quantitatively compared with reference strains (*Alcaligenes eutrophus* DSM 545, *Pseudomonas saccharophila* DSM 654, *P. cepacia* DSM 50181, *Alcaligenes latus* DSM1122, DSM 1123, DSM 1124) in a mineral medium with sucrose, glucose, and/or fructose when applicable. Table 12.1 summarizes the comparison of the strains tested.

Under the conditions tested, the sucrose-utilizing reference strains either did not consume sucrose or showed low efficiency in converting sucrose to PHA. This was the case of some of the *A. latus* strains that had been described as sucrose-utilizing [30]. Even hydrolyzing sucrose to fructose and glucose, some of those strains only used one of the resulting monosaccharides, thus reducing their efficiency to yield cells and polymer. Only *A. latus* DSM 1123 consumed sucrose completely Table 12.1). With the aim of selecting isolates capable of rapid growth and accumulating high polymer cell content, a comparison of the global cell efficiency in converting carbohydrates to cells and polymer ($Y^0_{P/C}$) was performed involving 66 bacterial strains [12] (see Figure 12.3).

TABLE 12.1

Production of P(3HB) from Carbohydrates (15 g/L) by Different Strains Cultivated in Nitrogen-Poor Mineral Salt Medium (Based on [12])

Strains	Supplied carbohydrates	Residual carbohydrates (%)			Biomass		
		S	G	F	CDM [g/L]	P(3HB) [%of CDM]	$Y^o_{P/C}$ g/g
A. eutrophus DSM 428	F	–	–	42.8	4.57	70.5	0.37
	G + F	–	90.5	0.0	3.70	67.0	031
A. eutrophus DSM 545	F	–	–	27.3	5.63	65.3	0.34
	G + F	–	1.9	0.0	6.56	76.7	0.35
A. latus DSM 1122	S	45.9	0.0	27.0	1.98	13.2	0.07
	G + F	–	0.0	102.2	2.84	33.0	0.14
A. latus DSM 1123	S	4.7	0.0	0.0	7.72	79.4	0.42
	G + F	–	0.0	0.0	7.59	76.5	0.41
A. latus DSM 1124	S	101	0.0	1.5	0.54	1.6	0.0
	G + F	_	0.0	98.8	3.19	22.5	0.10
P. cepacia DSM 50181	S	0.0	0.0	17.1	4.16	50.4	0.18
	G + F	–	0.0	11.0	5.03	59.0	0.22
P. saccharophila DSM 654	S	41.9	0.0	1.8	2.46	43.3	0.13
	G + F	–	85.2	89.5	0.48	0.0	0.0
IPT 101	G + F	_	0.0	0.0	5.59	74.6	0.29
	S	0.0	0.0	0.0	6.14	68.4	0.29
IPT044	G + F	_	1.2	1.2	5.14	60.0	0.20
	S	0.0	0.0	0.0	5.67	84.5	0.32
IPT064	G + F	_	0.0	0.0	5.83	62.2	0.23
	S	0.0	0.0	0.0	5.28	58.9	0.22
IPT048	G + F	_	0.0	0.0	5.04	65.6	0.21
	S	0.0	0.0	0.0	5.59	63.9	025

S – sucrose; F – fructose; G – glucose; CDM – cell dry mass.

Values of $Y^o_{P/C}$ are in P(3HB) produced per carbohydrate consumed. Based on [12].

Analyzing the consumption in the two conditions tested, the maximum theoretical efficiency was attained by A. latus DSM1123, A. eutrophus DSM 545, and two new isolates. However, considering the sucrose-utilizing cultures, A. latus DSM 1123 and another five isolates reached 80% and higher theoretical efficiency [12, 40].

One interesting result was that, among the isolates, 59 accumulated P(3HB) while 16 produced medium-chain-length-PHA (mcl-PHA) (data not shown in the table) [12, 40].

Isolates and reference strains were also compared concerning their ability to accumulate the copolymer of 3-hydroxybutyrate and 3-hydroxyvalerate [P(3HB-co-3HV)] from carbohydrates and propionic acid (Table 12.2).

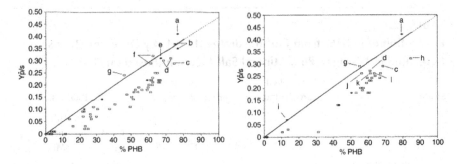

FIGURE 12.3 Comparison of reference strains (+) and soil isolates (☐) based on P3HB yield from carbohydrates ($Y^O_{P/C}$: grams of P3HB produced per gram of carbohydrates consumed). Strains were supplied with glucose plus fructose (left graph) or sucrose (right graph). The central solid lines represent the expected values when the theoretical P3HB yield ($Y^T_{P/C}$) and cell yield ($Y_{X/C}$, non-P3HB biomass) were considered. Values when supplying glucose plus fructose were $Y^T_{P/C} = 0.48$ g/g and $Y_{X/C} = 0.50$ g/g or when sucrose was the carbon source $Y^T_{P/C} = 0.50$ g/g and $Y_{X/C} = 0.52$ g/g. Points related to strains are: *A. latus* DSM 1123 (a), *A. eutrophus* DSM 545 (b), IPT-101 (c), IPT-083 (d), *A. eutrophus* DSM 428 (e), IPT-086 (f), IPT-055 (g), IPT-044 (h), *A. latus* DSM 1122 (i), IPT-076 (j), IPT-040 (k), IPT-045 (l). Adapted from [12].

TABLE 12.2
Efficiency in Converting Propionic Acid to 3HV ($Y_{3HV/Prop}$) Units from Propionic Acid by Reference Strains and Soil Isolated Bacteria in Brazil

Strains	PHA			
	% of CDM	3HB [mol-%]	3HV [mol-%]	$Y_{3HV/Prop}$ [g/g]
A. eutrophus DSM 545	71.4	96.1	3.9	0.13
A. latus DSM 1123	14.6	55.0	45.0	0.07
P. cepacia DSM 50181	38.4	97.3	2.7	0.04
IPT-040	32.3	97.1	2.9	0.05
IPT-044	51.1	97.1	2.9	0.07
IPT-045	49.4	96.2	3.8	0.08
IPT-048	44.3	96.2	3.8	0.06
IPT-055	1.5	100.0	0.0	0.00
IPT-056	30.9	98.5	1.5	0.02
IPT-076	56.8	97.1	2.9	0.10
IPT-083	56.8	96.9	3.1	0.10
IPT-086	39.0	89.9	10.1	0.09
IPT-098	17.7	94.7	5.3	0.07
IPT-101	32.3	95.4	4.6	0.05

CDM = cell dry mass from medium with glucose and propionic acid.
CDM (%): Percentage of cell dry mass accumulated as PHA; 3HB: 3-hydroxybutyrate; 3HV: 3-hydroxyvalerate; $Y_{3HV/Prop}$: 3HB yield from propionic acid.

Low efficiency in converting propionic acid to 3HV was evidenced for all strains tested, indicating the need for improvement to reduce a possible high level of propionic acid degradation by the bacteria. Byrom [6] reported such an improvement for *A. eutrophus*.

Production of P(3HB)-*co*-3HV by soil-isolated bacteria able to use sucrose was evaluated in 5 L bioreactors, considering volumetric productivity and efficiency in converting substrate into cells or product [13]. Isolates tested were IPT-044 and IPT-101. The process involved a first step of cell multiplication under balanced medium, a second step feeding sucrose at a constant rate until exhausting nitrogen and a third step feeding the bioreactor with sucrose and propionic acid at a feeding rate controlled by culture pH control demand. Online measurements (dissolved oxygen, DO) were used to define step transitions. In the accumulation step, PHA yields from carbon sources achieved were 0.18 and 0.37 g/g for isolates IPT-044 and IPT-101, respectively, the latter being similar to values reported for *A. eutrophus* (0.34 g/g) in tests performed in Brazil [13]. Volumetric productivity observed for IPT-101 (0.63 g/(L·h)) was considerably higher than the value of IPT-044 (0.23 g/(L·h)). Two disadvantages were observed: polysaccharide production by strain IPT-044, which increased medium viscosity; and phosphorous consumption even under nitrogen limitation by IPT-101, which indicated a possible polyphosphate accumulation. However, polyphosphate accumulation could be avoided by controlling phosphate supply based on bacterial conversion rate of phosphate to cells [13].

Also, in bioreactor experiments, the isolates tested showed 3HV yield from propionic acid ($Y_{3HV/Prop}$) less than 10% of the maximum theoretical yield calculated by Gomez et al. [12] as 1.35 g/g. Thus, studies directed to obtain mutants with improved $Y_{3HV/Prop}$ were developed.

Isolate IPT-044 was then identified as *Burkholderia cepacia* and IPT-101 as *Burkholderia* sp. [12]. The first one showed a broad antibiotic resistance profile, therefore was not suitable for industrial use (data not presented). Based on those previous results, *Burkholderia* sp was selected for $Y_{3HV/Prop}$ improvement. This isolate was further proved to be the new species named *Burkholderia sacchari* LMG 19450^T [60, 61], recently reclassified as *Paraburkholderia sacchari* [62].

UV mutants of *Burkholderia sacchari* unable to grow on propionic acid and still accumulating P(3HB-*co*-3HV) from carbohydrate and propionic acid were obtained. $Y_{3HV/Prop}$ increased from 0.10 g/g in the wild type up to approximately 0.35 g/g, in mutants also affected in the growth of intermediates of the α-oxidation pathway, and up to 0.80 g/g, in mutants not affected in this pathway when tested in shaken flasks [63–65] (Table 12.3).

No 3HV accumulation from unrelated carbon sources was detected, indicating a real increase in the propionic flux directed to 3HV copolymerization. In bioreactor tests, mutant IPT-189, not affected on β-oxidation, showed $Y_{3HV/Prop} = 1.20$ g/g, a yield very close to the theoretical maximum of 1.35 g/g [64]. Previous reports indicated the development of a PS-1 *A. eutrophus* mutant with the ability to incorporate increased 3HV contents to the copolymer (data on $Y_{3HV/Prop}$ in g/g were unavailable) [6] and another *A. eutrophus* mutant BK-23 reached a $Y_{3HV/Prop}$ of less than 0.6 g/g [66].

Mutant IPT-189 was further tested in bioreactor and achieved $Y_{3HV/Prop}$ values ranging from 1.25 to 1.34 g/g in cultures fed with different sucrose/propionic acid

TABLE 12.3

Efficiency of *Burkholderia sacchari* Prp Mutants in Converting Propionic Acid to 3HV Units in Medium with Sucrose and Propionic Acid

Strain	Phenotype	3HB [mol-%]	3HV [mol-%]	$Y_{3HV/Prop}$ [g/g]
B. sacchari IPT-101	Wild type	93.8	6.2	0.10
Prp mutant IPT-183	I	84.1	15.9	0.34
Prp mutant IPT-185	II	82.6	17.4	0.35
Prp mutant IPT-190	III	80.1	19.9	0.37
Prp mutant IPT-196	IV	33.2	66.8	0.78
Prp mutant IPT-189	IV	44.7	55.3	0.81

I to III – mutants affected on α-oxidation of propionate
IV – mutants non-affected on α-oxidation of propionate
3HB: 3-hydroxybutyrate; 3HV: 3-hydroxyvalerate
$Y_{3HV/Prop}$: grams of 3HV produced per gram of propionic supplied

ratios in the accumulation step. 3HV content obtained in the copolymer varied from 40 to 6.5 mol-%. A volumetric productivity of 1 g/(L·h) and 60% of biomass as PHA were observed [67]. Some results on improving *B. sacchari* concerning propionic acid use are summarized in Table 12.4.

Thus, the strains obtained were promising for industrial use with significant cost reduction. In addition, results indicated that at least two propionic acid catabolic pathways would be oxidizing propionic acid in this strain. Further studies described the presence of the 2-methylcitrate cycle (2MCC) as one of them [68]. According to the concentration of propionate available in the medium, this pathway would be more operative. This hypothesis was corroborated by the findings of Pereira and collaborators that constructed *B. sacchari* prp mutants disrupted solely on *acnM* or *acnM* and *prpC* genes from the 2MCC and demonstrated that $Y_{3HV/Prop}$ values increased with the decrease of the concentration of propionate supplied [69].

12.4.1.2 Construction of Sucrose-Utilizing *A. eutrophus* Recombinants in Brazil

In the joint project, another step developed simultaneously to the screening of natural strains was the construction of a recombinant *A. eutrophus* (*R. eutropha*) able to grow and accumulate PHA from sucrose. Since a number of bacteria consume sucrose due to a number of genes arranged as the scr-regulon, this set of genes was cloned from a strain of *Salmonella* and introduced to *A. eutrophus* DSM 545. Initial parameters evaluated under non-optimized tests showed that maximum specific growth rate decreased from 0.30 h⁻¹ in glucose to 0.24 h⁻¹ in sucrose and total biomass achieved the same levels in recombinant and wild type [45–47]. This approach was more successful than previous experiments when levanase genes from *Bacillus subtilis* were introduced in *A. eutrophus*. The strain became available to hydrolyze sucrose; however, growth was

TABLE 12.4

Improvement on *B. sacchari* Mutant Efficiency in Converting Propionic Acid to 3HV Units through the Use of Different Cultivation Strategies

Strain/mutant	Cultivation strategy	Carbon sources	3HB [mol-%]	3HV [mol-%]	$Y_{3HV/prp}$* [g/g]	Ref.
B. sacchari wt	Shaken flasks batch	Glucose/propionic acid	93.8	6.2	0.10	[63]
Prp mutant 189	Shaken flasks batch	Glucose/propionic acid	43.6	56.4	0.81	
Prp mutant 189	Bioreactor pH-stat feeding	Sucrose/propionic acid (10.3:1 w/w)	80.5	19.5	1.20	
Prp mutant 189	Sucrose: propionate various feeding ratios	Sucrose/propionic acid				[66]
		(10:1 w/w)	60.0	40.0	1.10	
		(19:1 w/w)	83.0	17.0	1.15	
		(30:1 w/w)	90.0	10.0	1.27	
		(61.5:1 w/w)	93.5	6.5	1.34	

3HB: 3-hydroxybutyrate; 3HV: 3-hydroxyvalerate; $Y_{3HV/Prp}$: 3HB yield from propionic acid.

limited either due to low efficiency on sucrose transport to the cell or because of the limited excretion of levanase to the medium [5]. An additional genetic modification improved the ability of A. *eutrophus* to incorporate 3HV units to produce P(3HB)-*co*-3HV from sucrose and propionate with 3HV contents up to 37% from a total biomass similar of the biomass obtained by the original strain [45, 46, 70].

12.4.1.3 Additional Findings and Development Resulting from the Joint Project

12.4.1.3.1 Strains Producing 3-hydroxy-4-pentenoate (3H4PE) Directly from Sucrose

Among the isolates from sugarcane plantation soil, other potential interesting strains related to PHA production were detected.

Accumulation of PHA containing the unsaturated monomer 3-hydroxy-4-pentenoate (3H4PE), directly from sucrose, was detected among strains isolated in the screening program developed in Brazil. 3H4PE was the first unsaturated monomer earlier described, being more crystalline and showing melting temperature lower than P(3HB)-*co*-3HV [71]. One isolate, *Pseudomonas* sp IPT 064 was considered able to accumulate P3BH-*co*-3H4PE, and was further identified as *Pseudomonas cepacia* [72]. This *P. cepacia* and other ten isolates showed 3H4PE accumulation. However only 5% was incorporated in the copolymer and attempts to increase this fraction by supplying 4-pentenoic acid were unsuccessful [73]. Inactivation of gene $phaC_{1Bc}$, encoding a PHA synthase, resulted in increasing the 3H4PE fraction, however the total polymer accumulated decreased expressively. The authors hypothesized that the strain possesses more than one PHA synthase, accumulating separate granules with each monomer composition, thus the product would in fact be a blend P3BH-*b*-3H4PE [74, 75].

12.4.1.3.2 Strains Accumulating Medium-Chain-Length Polyhydroxyalkanoates (mcl-PHA) Directly from Carbohydrates

The screening program developed in Brazil resulted in 75 isolates able to convert sugarcane derivatives to accumulate PHA; the majority as P(3HB) (59 isolates) and the remaining 16, probably belonging to the group of fluorescent *Pseudomonas*, produced PHA containing 3-hydroxydecanoate and 3-hydroxyoctanoate, medium-chain-length monomers [12, 40, 76]. Despite being unable to hydrolyze sucrose, *mcl*-PHA was accumulated from a mixture of glucose and fructose, carbohydrates obtained by acid hydrolysis of sucrose syrup in sugar and ethanol mills. This opened the possibility of also producing *mcl*-PHA integrated to the mills, with no need to supply related carbon sources. Interestingly, strains *P. putida* IPT046 and *P. aeruginosa* IPT171, when supplied with a similar sugar mixture (glucose and fructose), showed differences in monomer composition. IPT-046 produced *mcl*-PHA with a 3-hydroxyoctanoate (3HO) molar fraction approximately twice higher than the isolate IPT-171. On the contrary, IPT-171 showed molar fractions of 3-hydroxydodecanoate (3HDd) and 3-hydroxy-5-dodecenoate (3HDdΔ_5) two times higher when compared to IPT-046 [77] (Table 12.5).

Different species of *Pseudomonas* from the initial screening were further tested supplying glucose and fructose mixtures or food grade plant oils. A relationship

TABLE 12.5

Composition of *mcl*-PHA accumulated by two pseudomonads from a mixture of glucose and fructose (adapted from [77])

Strain	PHA composition [mol-%]					PHA in CDM [wt.-%]
	3HHx	3HO	3HD	3HDd	3HDdΔ5	
Pseudomonas putida IPT-046	3.04	28.46	63.59	2.06	2.85	49.4
Pseudomonas aeruginosa IPT-171	2.81	16.81	66.15	6.56	7.67	12.0

3HHx: 3-hydroxyhexanoate; 3HO: 3-hydroxyoctanoate; 3HD: 3-hydroxydecanoate; 3HDd: 3-hydroxydo-decanoate; 3HDdΔ5: 3-hydroxy-5-dodecenoate; CDM (%): Percentage of cell dry mass accumulated as PHA.

between the substrate supplied and molar fraction composition of PHA was defined and also the efficiency in converting precursors to specific monomers, thus contributing to the proposal of improvements to produce PHA controlling their monomer composition. In addition to monomers listed in Table 12.5, monomers of 3-hydroxy-6-dodecenoate (3HDdΔ6) were incorporated [77].

High-density cell cultures of *Pseudomonas putida* IPT-046 were studied, supplying an equimolar mixture of fructose and glucose to produce *mcl*-PHA. Fed-batch experiments testing different feeding strategies and nitrogen, or phosphate limitation were performed. The latter resulted in 50 g/L cell concentration with 63% *mcl*-PHA at a volumetric productivity of 0.8 g/(L·h) in 42 hours of fed-batch cultivation. Those productivity results were the highest for *mcl*-PHA production from simple sugars [78].

Pseudomonas putida IPT 046 was also studied in chemostat cultures with glucose and fructose as carbon sources by Taciro [79] evaluating nitrogen (N), phosphorus (P), and both nutrients (N and P) limitations. 70% of CDM as *mcl*-PHA were attained under P limitation with 0.16 g/g of yield of carbon into polymer ($Y_{P/C}$). When N was the limiting nutrient, only 40% of CDM as polymer with $Y_{P/C}$ of 0.10 g/g was achieved. The best $Y_{P/C}$ was 0.19 g/g, when N and P were simultaneously limited and 68% of CDM was accumulated as PHA. In all limitation cases studied, when the polymer content was maximum, the carbon balance showed that about 60% of the carbon consumed was directed to CO_2 production. A metabolic pathway model was proposed, and an elementary mode analysis performed to investigate possible effects of biochemical network modifications that could improve the system's conversion of substrate into polymer.

These studies then opened an additional possibility to widen the range of PHA produced integrated into a sugar and ethanol mill.

12.4.1.3.3   *Further Studies to Explore the Potential of PHA Productions of Burkholderia sacchari*

Burkholderia sacchari LMG19450 was supplied with 30 different carbon sources as co-substrates fed with glucose to evaluate the production of different monomers. It was demonstrated the ability of the strain to incorporate 3HV (up to 65 mol-%), 4HB

(9.1 mol-%), and 3HHx (1.6 mol-%) with 3HB as the main monomer in the PHA. Despite the 3HHx content of P(3HB)-*co*-3HHx could be controlled by supplying different glucose and hexanoic acid ratios, only 2% of the hexanoic acid was directed to 3HHx production [80]. An approach was developed to increase 3HHx content into the copolymer allowing a precise control on monomer molar fractions. *B. sacchari* PHA-negative mutants [81] hosting genes *phaPCJ* from *Aeromonas* sp were constructed and a two-step process to increase P(3HB)-*co*-3HHx production was developed. Contents up to 20 mol-% 3HHx in the copolymer were achieved. Cells accumulated 78% of the dry biomass as P(3HB)-*co*-3HHx at a volumetric productivity of 0.45 g/(L·h). Higher elongation at break (945%) than previous reports was observed in copolymers with 20 mol-% 3HHx [82].

A number of recent studies have involved *Burkholderia sacchari* and are summarized in Figure 12.4. This bacterium has been selected by different studies for PHA production from xylose, present in high amounts in sugarcane bagasse hydrolysates, thus opening new possibilities to expand substrate supplied in new biorefineries [83, 84]. Preliminary economic analysis indicated some targets to be achieved to make viable an industrial PHA producing process from xylose so as to integrate it with the first and second generation process for ethanol production in sugarcane mills [85]. Studies to unravel and to improve xylose catabolism in *B. sacchari* have been reported with promising perspectives [86–90]. Other successful studies reported include the use of wheat hydrolysates [91, 92] and softwood as substrate [93, 94], and also bioreactor production of 4HB from xylose or glucose combined to precursors [91, 95]. Nitrogen or phosphorous limiting conditions effects on growth and P(3HB) production yields were also recently investigated [96].

12.4.2 BIOREACTOR PROCESS DEVELOPMENT

Bioreactor process development involves growth and accumulation studies of the PHA-producing microorganism. Figure 12.5 summarizes the stages involved in this process: it starts from a freeze-preserved working bank of microorganisms from which colonies are isolated by plating in rich culture medium and subsequently transferred to a liquid medium. Cultivation in a liquid medium will constitute the bioreactor inoculum and can be performed in more than one stage, according to such features as the working volume where the final cultivation will be performed, the defined inoculum fraction, and the prevention of the lag phase in cultivation.

The initial bioreactor studies during the joint project utilized *A. eutrophus* and adopted the process in stirred tank reactors performed in two steps: (i) a balanced culture medium promoting the growth of the microorganism, and (ii) an unbalanced medium resulting from the exhaustion of some essential growth substrate, inducing accumulation of the polymer. High test molasses (HTM) was the carbon source. HTM is a sugarcane syrup (without sugar removal by crystallization) having about 85°Brix.

Since the final polymer concentration is directly proportional to the cell concentration at the end of the growth phase, in step (i) the main goal was to achieve high cell concentrations with high productivity of cell biomass and low levels of intracellular PHA.

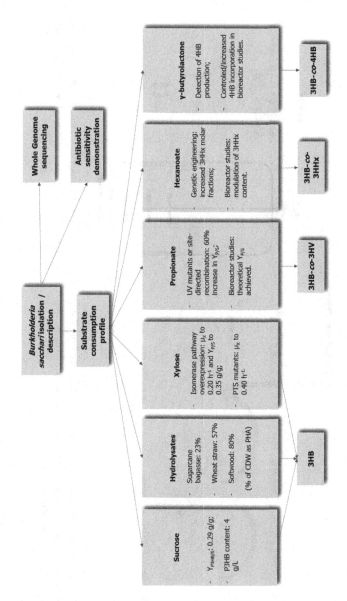

FIGURE 12.4 Different studies on PHA production using *Burkholderia sacchari* as a bacterial platform.

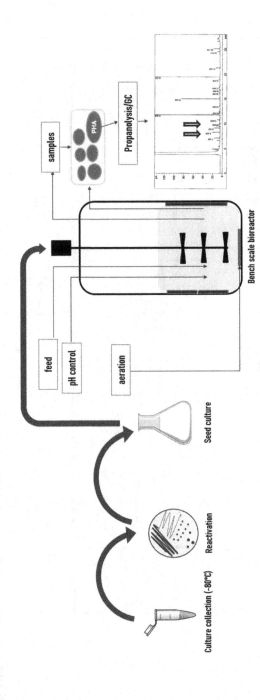

FIGURE 12.5 Bioreactor cultivation stages: starting from a working bank of microorganisms; colonies are isolated by plating, transferred to a liquid medium that constitutes the bioreactor inoculum. Periodic sampling monitors the evolution of bioreactor cultivation, where cell dry weight, substrates consumption, and product formation are evaluated.

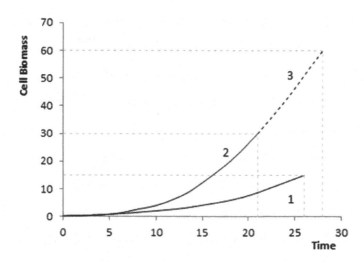

FIGURE 12.6 Illustration of the cellular growth step during the joint project. Line 1 indicates initial results when operating the bioreactor in successive batches. Line 2 refers to successive batches with a nutrient supplemented medium. Line 3 depicts a controlled feed of carbon source (limited) to avoid oxygen limitation. The final protocol combined procedures indicated by lines 2 and 3.

The evolution of step (i) is represented in Figure 12.6.

In the initial experiments, this step was operated in successive batches, obtaining cell productivity in the order of 0.6 g/(L·h) but with about 20% of PHA in biomass (line 1 in Figure 12.6). Nutrient supplementation to culture medium (line 2 in Figure 12.6) doubled cell concentration, reduced PHA content in a half and resulted in cell biomass productivity of 1.4 g/(L·h) of cell biomass. At this point, oxygen availability became the bottleneck for cell multiplication, as limitation of this nutrient restricts cell growth and induces polymer accumulation. To overcome this problem, feed of carbon source began to be controlled (limited) to avoid oxygen limitation (line 3 in Figure 12.6), resulting in a combination of procedures (lines 2 and 3 in Figure 12.6). Therefore, cell biomass productivity reached 2.1 g/(L·h) with PHA of about 5% of cell biomass, representing a 3.5× increase in biomass productivity.

In the polymer accumulation step (ii) different substrate supply strategies were tested: pulse and different constant flow rates of feeding [97].

In the process of P(3HB) production with *Ralstonia eutropha*, values of 75–80% of polymer content in cell biomass and productivity values of 1.9 g/(L·h) of polymer were reached [55].

The strategy of conducting the bioreactor cultivation to obtain the copolymer P(3HB-*co*-3HV) was similar to that employed in the production of P(3HB). The difference relied on the feeding of carbon sources during the polymer accumulation step, where propionic acid was also added.

Figure 12.7 illustrates the evolution of the bioreactor cultivations over the course of the joint project for the production of copolymer P(3HB-*co*-3HV). At the end of the second year of the project the productivity of copolymer was 0.8 g/(L·h) and

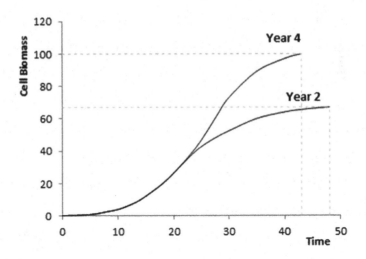

FIGURE 12.7   Illustration of bioreactor cultivation evolution for the joint project for P(3HB-
co-3HV) production.

in the fourth year 1.4 g/(L·h). A substantial improvement was observed since an
increase of 75% in P(3HB-co-3HV) productivity was reached.

12.4.2.1   Optimization of Poly(3-hydroxybutyrate) (P(3HB)) Biopolymer Production Based on a Cybernetic Mathematical Model

To improve the product productivities of P(3HB) (1.9 g/(L·h) and P(3HB-co-3HV)
(1.4 g/(L·h)), the optimization of the biopolymer production by *Ralstonia eutropha*
was performed, in a bioreactor conducted under fed-batch mode, by applying a proposed Cybernetic Structured Mathematical Model [98]. The model was described by
a system of 11 differential equations, with 12 state and eight manipulated variables,
and 61 parameters.

The mathematical model was proposed based on different phenomena observed
experimentally. Based on this model it was possible to find the optimal trajectory
of the manipulated variables that resulted in a maximization/minimization of a
given objective function subject to different constraints of the P(3HB) production
process [99].

The nonlinear programming – TOMP [100] was chosen to find the optimal substrate feed rate profile in the P(3HB) bioproduction process. From a set of initial
and constant values of the discretized control variables (u) at n intervals over time
(Equation 12.1):

$$u = \left[u_1^1, u_2^1, \ldots, u_n^1, u_n^2, \ldots, u_n^k \right] \qquad (12.1)$$

$n = 1$, number of discretization intervals
$k = 1, \ldots$ number of control intervals

$u_n{}^k$ is limited by its minimum and maximum values (Equation 12.2):

$$u_n^k \min \le u_n^k \le u_n^k \max \qquad (12.2)$$

where each control interval results in a corresponding time interval.

From the discretization, the method adopted by the algorithm TOMP followed the steps:

Maximize any objective function J subject to:

- Equations of the mathematical model
- Equation 12.1 of the discretization of u
- Equation 12.2 as a constraint of the control variable and any constraints of the state variable

Between the extremes of the time discretization intervals, the value of each control (1 k) is given by the interpolation of their respective control extreme. TOMP employed an adjustment of a curve of the type "cubic spline".

The objective function that maximizes the productivity of the P(3HB) product by manipulating only one control variable – the feed rate of sugars F1 (glucose + frutose) was employed and the P(3HB) volumetric productivity amounted to 3.11 g/(L·h). The sugar feed solution (glucose + fructose) concentration used was 600 g/L.

The optimized profile was experimentally implemented and both results: experimental and theoretical are shown in Figure 12.8 (a, b, and c).

The experimental (real) feed rate was very close to the optimal, as shown in Figure 12.8a. Even so, both sugar concentrations (glucose and fructose) do not remain at plateau as expected (4 g/L), resulting in complete consumption in 18 hours (Figure 12.8b). A direct consequence of this fact is that, although the active biomass reaches the desired concentration, the P(3HB) production starts at low rates (Figure 12.8c), since the sugar concentrations did not remain at the level required by the model. Despite that, the volumetric productivity on P(3HB) reached 2.21 g/(L·h) experimentally, after 30 hours of culture and intracellular content of P(3HB) corresponded to 62.3%. This experimental volumetric productivity is comparable to values of 2.64 g/(L·h) [101] and 2.63 g/(L·h) [102], previously reported in the literature. One possible solution to achieve the theoretical values experimentally would be to apply a closed control loop. The theoretical (3.11 g/(L·h)) and experimental P(3HB) volumetric productivity results (2.21 g/(L·h)) were 64 and 16% higher than the value of the initial protocol of the process (1.9 g/(L·h)) mentioned by Rossell et al. [55]. Likewise, the theoretical value of productivity of copolymer P(3HB-co-3HV) (10 mol-% 3HV) was 2.56 g/(L·h), 83% higher than that used by COPERSUCAR in the analysis of production cost.

12.4.2.2 Alternative Bioreactors for P(3HB) Production

In addition to the PHA production process in stirred tank bioreactors (STR), alternative reactors to reduce production costs were evaluated. P(3HB) production by *R. eutropha* with glucose in an airlift bioreactor was studied and the performance of the reactor with external circulation was tested, resulting in specific growth rates

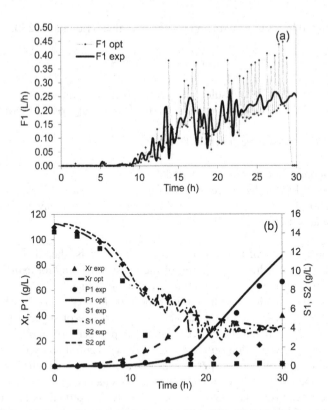

FIGURE 12.8 (a) Optimal and real substrate feed rate (F1), (b) Optimized and experimental state variables concentration profiles (S1 = glucose; S2 = fructose; Xr = residual cell biomass; P1 = P(3HB)). Adapted from [98].

values of 0.16 h⁻¹ in the exponential phase of growth and glucose yield in biomass of 0.46 g/g, values close to those obtained in STR 0.17 h⁻¹ and 0.49 g/g under similar conditions [103]. Specific growth rate values of 0.31 h⁻¹ and polymer content of about 70% of the biomass were reached at the end of the experiment in a 10 L working volume reactor with glucose and fructose as carbon sources. Excretion of organic acids (acetate and succinate) during polymer accumulation was observed for air flow rates below 30 L/min, resulting in P(3HB) contents of less than 30% of biomass [103]. Under similar conditions, oxygen transfer and consumption data, kinetic parameters, and mixing time were compared to those obtained in STR. In the airlift reactor, for aeration surface velocity of 10 m/s 50% of P(3HB) content in biomass and productivity of 0.6 g/(L·h) of polymer were reached. In STR, for the same polymer content, productivity of 0.8 g/(L·h) was obtained. Despite the higher productivity value in the mechanically stirred tank, the power consumption values in airlift showed that less energy is required for P(3HB) production, making it a good alternative to the STR.

Also using an "airlift" bioreactor, Pataquiva et al. [104] studied hydrodynamics and mass transfer in fed-batch poly(3-hydroxybutyrate) (P(3HB)) production by *B. sacchari* IPT 189. Oxygen transfer, mixing time, and hold up were considered to define the best configuration of the operated bioreactor using an experimental design. Under the studied conditions, the best results obtained were: 0.5 h^{-1} of maximum specific growth rate, total biomass of 150 g/L, a P(3HB) content of 42% of CDM and a volumetric productivity of 1.7 g/(L·h) for the polymer.

12.4.3 Downstream Process

Separation and purification studies have also been performed resulting in a PHA patent polymer recovery method using byproducts from sugar and ethanol mills. The method used alcohols higher than three carbon chain and their esters such as isoamyl alcohol, amyl acetate, isoamyl acetate, and fusel oil, all of them byproducts of the ethanol production process. The cultured broth, after centrifugation, was subjected to those solvents at high temperature, solubilizing the polymer in the organic phase. Cell debris was removed by filtration and the polymer precipitated by cooling, followed by solvent removal and drying [17].

Afterwards, Mantelato et al. [105] introduced a step of concentration of polymer/solvent suspension by microfiltration after the polymer precipitation step. This technology is used on an industrial scale.

The use of solvents has the advantage of obtaining high product purity (>99%), but involves the use of large volumes of solvents and non-solvents and the use of high temperatures that can reduce the molar mass of the polymer, changing its properties [11, 17].

The use of the enzymatic method based on bacterial commercial proteolytic enzymes (Savinase® and Esperase®) was also been studied for *R. eutropha* biomass resulting in purity up to 90% of dry biomass [106]. Zuccolo et al. [107] also conducted initial studies of using acid treatment after enzymatic lysis of cells for precipitation and biomass separation in order to increase product purity. Kapritchkoff [108] and Kapritchkoff et al. [109] expanded the studies of this approach by selecting enzymes for cell lysis and the best results were obtained with bromelain (89% product purity) and pancreatin (90% product purity). The enzymatic method described by Holmes and Lim [110] was also tested but proved expensive and resulted in low purity.

12.4.4 Assembling of a Plant for Production of P(3HB) at a Pilot Scale

In 1995, COPERSUCAR started a pilot plant to produce P(3HB), assembled in one of its associate's sugarcane mills (Usina da Pedra, in Serrana city, São Paulo State). Integrating P(3HB) to sugar and ethanol production took into account not only low-cost sugar available as substrate and large-scale fermentation expertise, but also facilities already accessible, namely, water, electricity, heating, cooling, treatment, and disposal of effluents would be used in common. Also, natural solvents yielded as byproducts of ethanol fermentation could be applied in the polymer purification steps. Another advantage was that, while sugarcane cropping and milling usually last for six months a year, polymer production could occur throughout the year, usually around 300 working days [11, 55, 111].

The process implemented at the plant aggregated the different results obtained from the aforementioned cooperative project involving the private sector, a research institute and the university [11, 17, 55, 112]. Figure 12.9 summarizes how the process was operated. In general terms, the process was based on an initial aerobic cultivation at 32–34°C, pH 6.8, in a balanced culture medium, conducted to achieve high cell density during 24–48 hours. At this point, limitation of an essential nutrient

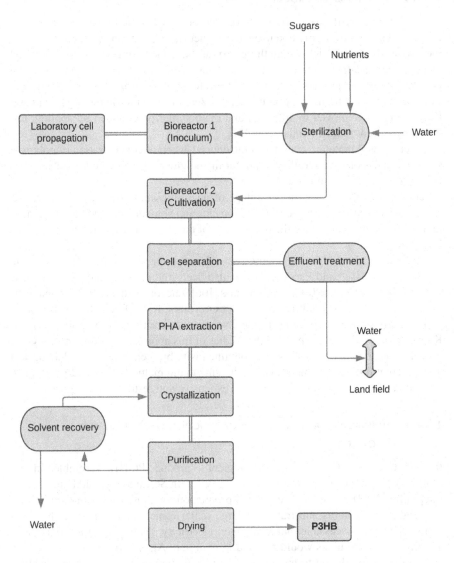

FIGURE 12.9 Steps involved in PHA production: P3HB can be produced using raw materials already available either as substrates or natural byproducts used as solvents during polymer purification plus treatment and use of effluents without harmful effects (modified from [55, 111]). The process applied combined developments made by [17, 45, 46, 112, 113].

was promoted, and the carbon source was continuously fed through a concentrated solution, so as the accumulation would take place (30–35 h) [55]. Appropriate strains were selected from those isolated or constructed during the cooperative project [12, 45, 46, 113].

The pilot plant assembled aimed at a production of five tons of P(3HB) per month [11]. Early results indicated that the process reached maximum values of cell dry mass of 150 kg/m³, with 75% of polymer, 98% cultivation process yield, P(3HB) productivity of 1.89 kg/(m³·h). Extraction was 95% efficient and P(3HB) purity (98%) with a molecular weight ranging from 250,000 Da to 1200 kDa [11, 55, 114].

From the pilot plant, data were collected to conceive scaling up and to perform technical and economic viability studies. The cost composition for a unit producing 10,000 tonnes of P(3HB) per year included 29% sugar, 20% other raw materials and chemicals, 27% equipment depreciation, 11% energy and 13% others. Steam consumption was initially considered high and attributed to the yet non-optimized pilot plant for energy employed to collect the data [11] apparently reviewed in further estimations [55]. Considering that polymer production was mostly affected by the price of sugar, studies estimated that P(3HB) would be produced at a cost less than US\$ 3.00 kg^{-1}, and that integrating polymer production to sugar mills would be the most cost-effective model, since the prices estimated were competitive. Figures made public initially estimated a total investment of US\$ 38 million required for P(3HB) production integrated to sugar mills in Brazil for a producing capacity of 10,000 tonnes per year [55].

During the year 2000, P(3HB) Industrial S/A was created along with the BIOCYCLE® brand [111].

This initiative was a very promising opportunity and industry was aware of the points to be improved, mainly energy consumption and some alternatives were proposed [11]. According to reports, sugar used for polymer production would be only 17% of the sugar produced, in the case of sucrose directly being the carbon source. This would not impact on sugar market prices nor compete for food or feed. At that point (1999–2000) the production in the central-south region of Brazil was about 17 Mt of sugar and 12 million m³ of ethanol starting from c.a. 260 Mt of sugarcane [11, 115]. Those figures more recently (2018–2019) amount 26.5 Mt of sugar and 31 million m³ of ethanol from c.a. 570 Mt of sugarcane [115].

The company reports that this approach involves a complete concept of renewability in the production of PHA including a Brazilian technology based in the use of plant-based substrate and selected organisms [111, 114].

An initial balance from pilot plant data has raised a high prospect of success for the development of an industrial production. Critical point improvements were considered relatively easily achievable goals; mainly related to the reduction of capital costs and P(3HB) production, reducing the generation of waste and energy consumption. Some points raised already in the early stages included: selection or improvements to obtain strains able to use other substrates obtained from sugarcane, increasing the percentage of intracellular polymer accumulation, efficient synthesis of co- and heteropolymers, bioreactor cultivation at more elevated temperatures – which can be critical in tropical countries; cultivation at even higher cell densities because the capacity of cultivation was then limited to 500 m^3. In addition, considering downstream processing, improvements on locally available solvent use, thermal energy and power needs reduction were points with a feasible solution [55].

Replacement of sucrose by carbohydrates released from sugarcane bagasse would be a milestone to be achieved [55, 83].

Improvement on bacterial ability to produce copolymers and also evaluation on the use of sugarcane bagasse carbohydrates were further explored.

12.5 A NEW BIOREFINERY FOR PHA PRODUCTION IN BRAZIL

One of the developments resulting from the collaborative project to produce PHA from sugarcane derivatives in Brazil was the use of sugars derived from sugarcane bagasse. Bagasse is a relatively inexpensive substrate, and for a long time has been used to generate energy in sugar and ethanol mills [11, 55]. Taking this into account, after testing bacterial isolates to the conversion of sugarcane derivatives (sucrose, glucose, fructose, and molasses) to PHA, some research groups focused on the use of sugars derived from sugarcane bagasse to produce PHA. As a major byproduct of sugar and ethanol mills in Brazil bagasse corresponds to 30% by weight of the sugarcane processed (in the 2018–2019 harvest season sugarcane processed in Brazil amounted 620,832,000 tonnes – unicadata.com.br). Despite bagasse use in energy generation there has been a surplus that could be used for non-energy applications, thus allowing the conversion to new bioproducts, including PHA. Bagasse is composed of lignin, cellulose, and hemicellulose. The latter is constituted mainly of xylose, arabinose, and glucose [116] Therefore, sugarcane bagasse is a carbohydrate-rich material and sugars therein constitute relevant carbon sources to be burnt to generate energy that could be transformed into new bioproducts. Releasing sugars from bagasse requires harsh treatments that also release a number of toxic compounds to bacteria. Detoxification procedures have been tested to transform sugarcane bagasse hydrolysates into carbon sources usable by bacteria. Bacterial strains (55 strains) were screened for the use of such detoxified sugarcane bagasse hydrolysate and high-cell-density protocols were developed, indicating the feasibility of the process of producing PHA from biomass hydrolysates [83, 84].

More recently, the flexibility and the needs of the economic scenario concerning fossil and green fuels led to the development of the production of second generation ethanol.

The idea of producing second generation ethanol utilizing sugars from sugarcane bagasse considered the use of glucose released from cellulose and also xylose released from hemicellulose. Thus, ethanologenic yeasts, hopefully able to also ferment xylose, would produce significantly more ethanol from the same planted area in Brazil. Two main bottlenecks had to be solved: to obtain ethanologenic yeasts able to efficiently ferment xylose and to release carbohydrates from sugarcane bagasse with reduced toxicity. So far, the first bottleneck has not proved easy to solve, thus cellulosic ethanol has been proposed. It has proved difficult to find ethanol-producing yeasts showing similar productivities to those when using glucose. Thus, after bagasse hydrolysis, pentoses remain as a promising carbon source to generate new bioproducts integrated to sugar and ethanol mills. Ethanol could be then produced by yeasts from glucose, released from cellulose, and a xylose-rich secondary stream would be available to produce second generation PHA [116]. Other bioproducts could be also part of this model biorefinery.

12.6   CONCLUSIONS AND OUTLOOK

The Brazilian sugar and alcohol sector has, for several decades, been composed of very large units operating as autonomous industrial complexes. Its main products are sugar, ethanol, bagasse, and energy. Several other inputs are generated, such as solvents and molasses that also constitute products in this chain and which may return as inputs at other points in this production. This model, combined with sugarcane cultivation near the mills, and integration with waste treatment and disposal, has made this model considered a green production cycle [117, 118]. Allowing this cycle to include new bioproducts would be a beneficial integration and a sustainable biorefinery model.

The PHA production model developed in Brazil is an example of a new bioproduct inserted in this biorefinery that could expand, using, for example, various bagasse derivatives (pentoses, levulinic acid) to diversify the composition of PHA or generate other bioproducts and precursors of molecules of industrial interest. This development has been recognized as a prime example of integration of academia and industry to develop a sustainable process to develop a truly green plastic [11, 41].

In addition, the scientific and technological development model in Brazil of PHA production is an excellent example of cooperation involving industry, university, and research institutes bringing promising results. A huge number of students and professionals with complementary expertise were among the human resources developed that are also a very relevant result from the cooperative project.

With the introduction of second generation ethanol into the sugar and alcohol mills, a new opportunity has been opened to produce PHA integrated in this model, using sugars present in sugarcane bagasse but not fermented by ethanologenic yeasts and other byproducts such as levulinic acid to generate 3HV, in addition to the solvents and other facilities already available. Thus, the future of PHA production becomes even more promising using xylose, which is expected to be available in millions of tonnes. The great challenge for PHA production to consolidate also involves how society will deal with the environmental and economic issues related to oil use. The price of oil has been one of the main factors motivating the investment in biorefineries Brazil, which have the industrial units (sugar and alcohol plants) that are the most suitable embryos for the development of biorefineries. Once the appropriate scenario is reached, this potential model can be implemented and PHA may then be one of the products of these industrial units.

Political and economic reasons and particular interests, however, have not yet allowed the expansion of this production model and the search for partners has been one of the strategies disclosed by the incorporated company. Producing PHA as proposed has not generated a demand capable of spurring the leap to real industrial scale production, which perhaps depends on determining its actual applications to satisfy the potentially involved sectors of society.

A similar cooperation of these different sectors must converge in a single effort, taking into consideration the environmental perspectives for the future, without the need to leave out of the scenario the intrinsic nature of the financial benefits to the industry.

ACKNOWLEDGMENTS

Initial developments were proposed by Brazilian cooperative of sugar and alcohol (COPERSUCAR) and supported by the Program for Supporting Scientific and Technological Development (PADCT) and the Federal Funding Authority for Studies and Projects (FINEP). Throughout the years, other agencies sponsored further developments: São Paulo Research Foundation (FAPESP), National Council for Scientific and Technological Development (CNPq 140321/2017-9 EROF Scholarship; CNPq 309086/2018-3 LFS productivity fellowship, CNPq 309134/2015-3 JGCG productivity fellowship), and Coordination for the Improvement of Higher Education Personnel (CAPES).

Research was developed at the University of São Paulo (USP), the Institute for Technological Research of São Paulo (IPT), and COPERSUCAR mainly in collaboration with a number of partners mentioned in the text.

REFERENCES

1. Byrom D. Polymer synthesis by microorganisms: Technology and economics. *Trends Biotechnol* 1987; 5(9): 246–250.
2. Braunegg G, Bogensberger B. Zur Kinetik Des Wachstums und der Speicherung von poly-D(–)-3-hydroxybuttersäure bei *Alcaligenes latus*. *Acta Biotechnol* 1985; 5(4): 339–345.
3. Ramsay BA, Lomaliza K, Chavarie C, Dubé B, Bataille P, Ramsay JA. Production of poly-(beta-hydroxybutyric-co-beta-hydroxyvaleric) acids. *Appl Environ Microbiol* 1990; 56(7): 2093–2098.
4. Hänggi UJ. Pilot scale production of PHB with *Alcaligenes latus*. In: Dawes EA, ed. *Novel Biodegradable Microbial Polymers*, Springer Netherlands, Dordrecht. 1990: 65–70.
5. Friehs K, Lafferty RM. Cloning and expression of the levanase gene in *Alcaligenes eutrophus* H16 enables the strain to hydrolyze sucrose. *J Biotechnol* 1989; 10(3–4): 285–291.
6. Byrom D. Industrial production of copolymer from *Alcaligenes eutrophus*. In: Dawes EA, ed. *Novel Biodegradable Microbial Polymers*, Springer Netherlands, Dordrecht. 1990: 113–117.
7. Page WJ, Knosp O. Hyperproduction of poly-beta-hydroxybutyrate during exponential growth of *Azotobacter vinelandii* UWD. *Appl Environ Microbiol* 1989; 55(6): 1334–1339.
8. Yamane T. Yield of poly-D(-)-3-hydroxybutyrate from various carbon sources: A theoretical study. *Biotechnol Bioeng* 1993; 41(1): 165–170.
9. Page WJ. Waste sources for polyhydroxyalkanoate production. Proceedings of the 1996 International Symposium on Bacterial Polyhydroxyalkanoates, National Research Council of Canada, Davos, Switzerland. 1996: 56–65.

10. Braunegg G, Genser K, Bona R, Haage G, Schellauf F, Winkler E. Production of PHAs from agricultural waste material. *Macromol Symp* 1999; 144(1): 375–383.
11. Nonato RV, Mantelatto PE, Rossell CE. Integrated production of biodegradable plastic, sugar and ethanol. *Appl Microbiol Biotechnol* 2001; 57(1–2): 1–5.
12. Gomez JGC, Rodrigues MFA, Alli RCP *et al.* Evaluation of soil gram-negative bacteria yielding polyhydroxyalkanoic acids from carbohydrates and propionic acid. *Appl Microbiol Biotechnol* 1996; 45(6): 785–791.
13. Gomez JGC, Fontolan V, Alli RC *et al.* Production of P(3HB)-co-3HV by soil isolated bacteria able to use sucrose. *Rev microbiol* 1997; 28(Suppl. 1): 43–48.
14. Silva LF da, Gomez JGC, Rocha RCS, Taciro MK, Pradella JGda C. Produção biotecnológica de poli-hidroxialcanoatos para a geração de polímeros biodegradáveis no Brasil. *Quím Nova* 2007; 30(7): 1732–1743.
15. Gomez JGC, Bueno Netto CL. Producao de poliésteres bacterianos. In: Lima Ude A, Aquarone E, Borzani W, Schmidell W, eds. *Biotecnologia Industrial - Volume 3 - Processos Fermentativos e Enzimáticos*. 1st ed., Editora Edgard Blücher Ltda, Sao Paulo. 2001: 219–248.
16. Gomez JGC, Rodrigues MFA, Alli RCP *et al.* Polímeros bacterianos para plásticos biodegradáveis. Anais do 2o Congresso Brasileiro de Polímeros. *ABPol – Associação Brasileira de Polímeros* 1993; 1: 351–356.
17. Derenzo S, Matsubara RMS, Mantelatto PE *et al.* Brazilian patent PI9302312-0 Processo de extração de biopolímeros. Instituto Nacional da Propriedade Industrial, INPI, Rio de Janeiro, 1995.
18. Gerngross TU, Slater SC. How green are green plastics? *Sci Am* 2000; 283(2): 37–41.
19. Solaiman DKY, Ashby RD, Foglia TA, Marmer WN. Conversion of agricultural feedstock and coproducts into poly(hydroxyalkanoates). *Appl Microbiol Biotechnol* 2006; 71(6): 783–789.
20. Cardozo NP, Sentelhas PC. Climatic effects on sugarcane ripening under the influence of cultivars and crop age. *Sci Agric* 2013; 70(6): 449–456.
21. Clarke MA. SYRUPS. In: *Encyclopedia of Food Sciences and Nutrition*, Luiz Trugo and Paul M Finglas eds, Academic Press, Cambridge, MA: 5711–5717. 2003.
22. Reinecke F, Steinbüchel A. *Ralstonia eutropha* strain H16 as model organism for PHA metabolism and for biotechnological production of technically interesting biopolymers. *J Mol Microbiol Biotechnol* 2009; 16(1–2): 91–108.
23. Page W. Production of poly-β-hydroxybutyrate by *Azotobacter vinelandii* strain UWD during growth on molasses and other complex carbon sources. *Appl Microbiol Biotechnol* 1989; 31: 329–333.
24. Page W. Production of polyhydroxyalkanoates by *Azotobacter vinelandii* UWD in beet molasses culture. *FEMS Microbiol Lett* 1992; 103(2–4): 149–157.
25. Page W. Production of poly-β-hydroxybutyrate by *Azotobacter vinelandii* UWD in media containing sugars and complex nitrogen sources. *Appl Microbiol Biotechnol* 1992; 38: 117–121.
26. Page WJ. Suitability of commercial beet molasses fractions as substrates for polyhydroxyalkanoate production by *Azotobacter vinelandii* UWD. *Biotechnol Lett* 1992; 14(5): 385–390.
27. Page WJ, Manchak J, Rudy B. Formation of poly(hydroxybutyrate-*co*-hydroxyvalerate) by *Azotobacter vinelandii* UWD. *Appl Environ Microbiol* 1992; 58(9): 2866–2873.
28. Segura D, Vargas E, Espín G. Beta-ketothiolase genes in *Azotobacter vinelandii*. *Gene* 2000; 260(1–2): 113–120.
29. Slater S, Houmiel KL, Tran M *et al.* Multiple beta-ketothiolases mediate poly(beta-hydroxyalkanoate) copolymer synthesis in *Ralstonia eutropha*. *J Bacteriol* 1998; 180(8): 1979–1987.

30. Palleroni NJ, Palleroni AV. *Alcaligenes latus*, a new species of hydrogen-utilizing bacteria. *Int J Syst Bacteriol* 1978; 28(3): 416–424.

31. Chen GQ, Zhang G, Park SJ, Lee SY. Industrial scale production of poly(3-hydroxyb utyrate-co-3-hydroxyhexanoate). *Appl Microbiol Biotechnol* 2001; 57(1–2): 50–55.

32. Chen GQ, König KH, Lafferty RM. Production of poly-D(-)-3-hydroxybutyrate and poly-D(-)-3-hydroxyvalerate by strains of *Alcaligenes latus*. *Antonie Van Leeuwenhoek* 1991; 60(1): 61–66.

33. Hrabak O. Industrial production of poly-β-hydroxybutyrate. *FEMS Microbiol Lett* 1992; 103: 251–255.

34. Lafferty RM, Korsatko B, Korsatko W. Microbial production of poly-β-hydroxybutyric acid. In: Rehm HJ, Reed G (eds) *Biotechnology*, vol 6B. VCH, Weinheim, 1988, 135–176.

35. Penloglou G, Roussos A, Chatzidoukas C, Kiparissides C. A combined metabolic/ polymerization kinetic model on the microbial production of poly(3-hydroxybutyrate). *N Biotechnol* 2010; 27(4): 358–367.

36. Penloglou G, Chatzidoukas C, Kiparissides C. Microbial production of polyhydroxybu tyrate with tailor-made properties: An integrated modelling approach and experimental validation. *Biotechnol Adv* 2012; 30(1): 329–337.

37. Wang B, Sharma-Shivappa RR, Olson JW, Khan SA. Production of polyhydroxybu tyrate (PHB) by *Alcaligenes latus* using sugarbeet juice. *Ind Crops Prod* 2013; 43: 802–811.

38. Wisuthiphaet N, Napathorn SC. Optimisation of the use of products from the cane sugar industry for poly(3-hydroxybutyrate) production by *Azohydromonas lata* DSM 1123 in fed-batch cultivation. *Process Biochem* 2016; 51(3): 352–361.

39. Gahlawat G, Srivastava AK. Model-based nutrient feeding strategies for the increased production of polyhydroxybutyrate (PHB) by *Alcaligenes latus*. *Appl Biochem Biotechnol* 2017; 183(2): 530–542.

40. Gomez JGC. Isolamento e caracterizacao de bacterias produtoras de polihidroxialca noatos. Master dissertation, Universidade de São Paulo, Brazil, 1994.

41. Koller M, Maršálek L, Miranda de Sousa Dias M, Braunegg G. Producing microbial polyhydroxyalkanoate (PHA) biopolyesters in a sustainable manner. *N Biotechnol* 2017; 37(A): 24–38.

42. Zhang H, Obias V, Gonyer K, Dennis D. Production of polyhydroxyalkanoates in sucrose-utilizing recombinant *Escherichia coli* and *Klebsiella* strains. *Appl Environ Microbiol* 1994; 60(4): 1198–1205.

43. Beaulieu M, Beaulieu Y, Melinard J, Pandian S, Goulet J. Influence of ammonium salts and cane molasses on growth of *Alcaligenes eutrophus* and production of polyhydroxy butyrate. *Appl Environ Microbiol* 1995; 61(1): 165–169.

44. Oliveira FC, Freire DMG, Castilho LR. Production of poly(3-hydroxybutyrate) by solid-state fermentation with *Ralstonia eutropha*. *Biotechnol Lett* 2004; 26(24): 1851–1855.

45. Vicente EJ. Brazilian patent PI9806581-5 Cepa transgênica de *Alcaligenes eutrophus* e método de obtenção de cepa transgênica de *Alcaligenes eutrophus*, Instituto Nacional da Propriedade Industrial, INPI, Rio de Janeiro, 2000.

46. Vicente EJ. Brazilian patent PI9805116 Cepa mutante de *Alcaligenes eutrophus*, cepa transgênica de mutante de *Alcaligenes eutrophus* e método de obtenção, Instituto Nacional da Propriedade Industrial, INPI, Rio de Janeiro, 1998.

47. Fava ALB. Clonagem e expressão do regulon scr em *Alcaligenes eutrophus* visando a produção de polihidroxibutirato a partir de sacarose. Mater dissertation, Universidade de São Paulo, Brazil, 1997.

48. Nordsiek G, Bowien B. Construction of sucrose-utilizing strains of *Alcaligenes entro phus*. *Forum Mikrobiologie* 1990; 13: 213.

49. Steinbüchel A. Polyhydroxyalkanoic acids. In: Byrom D, ed. *Biomaterials*, Palgrave Macmillan UK, London. 1991: 123–213.
50. Celik GY, Beyatli Y. Determination of poly-beta-hydroxybutyrate (PHB) in sugarbeet molasses by *Pseudomonas cepacia* G13 strain. *Zuckerindustrie* 2005; 130: 201–203.
51. Kulpreecha S, Boonruangthavorn A, Meksiriporn B, Thongchul N. Inexpensive fedbatch cultivation for high poly(3-hydroxybutyrate) production by a new isolate of *Bacillus megaterium*. *J Biosci Bioeng* 2009; 107(3): 240–245.
52. Figueroa-Rodríguez KA, Hernández-Rosas F, Figueroa-Sandoval B, Velasco-Velasco J, Aguilar Rivera N. What has been the focus of sugarcane research? A bibliometric overview. *Int J Environ Res Public Health* 2019; 16: 3326.
53. UNICA - União da Indústria de Cana-de-açúcar - Final report of 2017/2018 harvest season - South-Central region of Brazil. http://unicadata.com.br/listagem.php?idMn=102.
54. Goldemberg J, Coelho ST, Guardabassi P. The sustainability of ethanol production from sugarcane. *Energy Policy* 2008; 36(6): 2086–2097.
55. Rossell CEV, Mantelatto PE, Agnelli JAM, Nascimento J. Sugar-based biorefinery–technology for integrated production of poly(3-hydroxybutyrate), sugar, and ethanol. In: Kamm B, Gruber PR, Kamm M, eds. *Biorefineries-Industrial Processes and Products*, Wiley-VCH Verlag GmbH, Weinheim, Germany. 2005: 209–226.
56. Yamane T, Fukunaga M, Lee YW. Increased PHB productivity by high-cell-density fed-batch culture of *Alcaligenes latus*, a growth-associated PHB producer. *Biotechnol Bioeng* 1996; 50(2): 197–202.
57. Hänggi UJ. Requirements on bacterial polyesters as future substitute for conventional plastics for consumer goods. *FEMS Microbiol Rev* 1995; 16(2–3): 213–220.
58. Harding KG, Dennis JS, von Blottnitz H, Harrison STL. Environmental analysis of plastic production processes: Comparing petroleum-based polypropylene and polyethylene with biologically-based poly-beta-hydroxybutyric acid using life cycle analysis. *J Biotechnol* 2007; 130(1): 57–66.
59. Schlegel HG, Lafferty R, Krauss I. The isolation of mutants not accumulating poly-beta-hydroxybutyric acid. *Arch Mikrobiol* 1970; 71(3): 283–294.
60. Brämer CO, Vandamme P, da Silva LF, Gomez JG, Steinbüchel A. Polyhydroxyalkanoate-accumulating bacterium isolated from soil of a sugar-cane plantation in Brazil. *Int J Syst Evol Microbiol* 2001; 51(5): 1709–1713.
61. Alexandrino PMR, Mendonça TT, Guamán Bautista LP *et al*. Draft genome sequence of the polyhydroxyalkanoate-producing bacterium *Burkholderia sacchari* LMG 19450 isolated from Brazilian sugarcane plantation soil. *Genome Announc* 2015; 3: e00313–15.
62. Sawana A, Adeolu M, Gupta RS. Molecular signatures and phylogenomic analysis of the genus *Burkholderia*: Proposal for division of this genus into the emended genus *Burkholderia* containing pathogenic organisms and a new genus *Paraburkholderia* gen. nov. harboring environmental species. *Front Genet* 2014; 5: 429.
63. Oliveira MC, Anjos MD, Silva LF *et al*. Obtenção de mutantes bacterianos deficientes na síntese de poli-3-hidroxibutirato. *Congresso Nacional de Genética*, Águas de Lindóia, São Paulo, Brasil, 1998.
64. Silva LF, Gomez JG, Oliveira MS, Torres BB. Propionic acid metabolism and poly-3-hydroxybutyrate-co-3-hydroxyvalerate (P(3HB)-*co*-3HV) production by *Burkholderia* sp. *J Biotechnol* 2000; 76(2–3): 165–174.
65. Silva LF. Estudo do catabolismo de propionato em *Burkholderia* sp visando o aumento da eficiencia na producao de P(3HB)-*co*-3HV - Um plastico biodegradavel. Doctoral thesis, Universidade de São Paulo, Brazil, 1998.
66. Lee SY. Bacterial polyhydroxyalkanoates. *Biotechnol Bioeng* 1996; 49(1): 1–14.

67. Rocha RCS, da Silva LF, Taciro MK, Pradella JGC. Production of poly(3-hydroxyb utyrate-co-3-hydroxyvalerate) P(3HB−co−3HV) with a broad range of 3HV content at high yields by *Burkholderia sacchari* IPT 189. *World J Microbiol Biotechnol* 2008; 24(3): 427–431.
68. Brämer CO, Silva LF, Gomez JGC, Priefert H, Steinbüchel A. Identification of the 2-methylcitrate pathway involved in the catabolism of propionate in the polyhydroxyalkanoate- producing strain *Burkholderia sacchari* IPT101 and analysis of a mutant accumulating a copolyester with higher 3-hydroxyvalerate cont. *Appl Environ Microbiol* 2002; 68(1): 271–279.
69. Pereira EM, Silva-Queiroz SR, Cabrera Gomez JG, Silva LF. Disruption of the 2-methylcitric acid cycle and evaluation of poly-3-hydroxybutyrate-co-3-hydroxyvalerate biosynthesis suggest alternate catabolic pathways of propionate in *Burkholderia sacchari*. *Can J Microbiol* 2009; 55(6): 688–697.
70. Sartori DM, Vicente EJ. Obtenção de um mutante de *Alcaligenes eutrophus* melhorado geneticamente para a produção do co-polímero polihidroxibutirato-polihidroxivalerato (PHB-PHV). Master dissertation, Universidade de São Paulo, Brazil, 1998.
71. Valentin HE, Berger PA, Gruys KJ *et al.* Biosynthesis and Characterization of Poly(3-hydroxy-4-pentenoic acid). *Macromolecules* 1999; 32(22): 7389–7395.
72. Rodrigues MFA, da Silva LF, Gomez JGC, Valentin HE, Steinbüchel A. Biosynthesis of poly (3-hydroxybutyric acid-co-3-hydroxy-4-pentenoic acid) from unrelated substrates by *Burkholderia* sp. *Appl Microbiol Biotechnol* 1995; 43(5): 880–886.
73. Rodrigues MFA, Oliveira MC, Gomez JGC *et al.* Anais do; VI Seminário de Hidrólise Enzimática de Biomassas, Maringá, Brasil, 1996. Anais do: VI Seminário de Hidrólise Enzimática de Biomassas, Anais do SHEB 1996.
74. Rodrigues MF, Valentin HE, Berger PA *et al.* Polyhydroxyalkanoate accumulation in *Burkholderia* sp.: A molecular approach to elucidate the genes involved in the formation of two homopolymers consisting of short-chain-length 3-hydroxyalkanoic acids. *Appl Microbiol Biotechnol* 2000; 53(4): 453–460.
75. de Andrade Rodrigues MF, Vicente EJ, Steinbüchel A. Studies on polyhydroxyalkanoate (PHA) accumulation in a PHA synthase I-negative mutant of *Burkholderia cepacia* generated by homogenization. *FEMS Microbiol Lett* 2000; 193(1): 179–185.
76. Gomez JGC. Produção por *Pseudomonas* sp. de polihidroxialcanoatos contendo monômeros de cadeia média a partir de carboidratos: Avaliação da eficiência, modificação da composição e obtenção de mutantes. Doctoral thesis, Universidade de São Paulo, Brazil, 2000.
77. Silva-Queiroz SR, Silva LF, Pradella JGC, Pereira EM, Gomez JGC. PHA(MCL) biosynthesis systems in *Pseudomonas aeruginosa* and *Pseudomonas putida* strains show differences on monomer specificities. *J Biotechnol* 2009; 143(2): 111–118.
78. Diniz SC, Taciro MK, Gomez JGC, da Cruz Pradella JG. High-cell-density cultivation of *Pseudomonas putida* IPT 046 and medium-chain-length polyhydroxyalkanoate production from sugarcane carbohydrates. *Appl Biochem Biotechnol* 2004; 119(1): 51–70.
79. Taciro MK. Processo contínuo de produção de polihidroxialcanoatos de cadeia média (PHAmcl) sob limitaçao múltipla de nutrientes. Doctoral thesis, Universidade de São Paulo, Brazil, 2008.
80. Mendonça TT, Gomez JGC, Buffoni E *et al.* Exploring the potential of *Burkholderia sacchari* to produce polyhydroxyalkanoates. *J Appl Microbiol* 2014; 116(4): 815–829.
81. Filipov MCO, Silva LF, Pradella JGC, Gomez JGC. Obtenção de mutantes de *Burkholderia* sp IPT 101 deficientes na despolimerização de polihidroxibutirato (P(3HB)). *XX Congresso Brasileiro de Microbiologia 1999*, SBM, Salvador, Bahia, Brazil, 1999: 274.

82. Mendonça TT, Tavares RR, Cespedes LG *et al.* Combining molecular and bioprocess techniques to produce poly(3-hydroxybutyrate-co-3-hydroxyhexanoate) with controlled monomer composition by *Burkholderia sacchari*. *Int J Biol Macromol* 2017; 98: 654–663.

83. Silva LF, Taciro MK, Michelin Ramos ME, Carter JM, Pradella JG, Gomez JG. Poly-3-hydroxybutyrate (P(3HB)) production by bacteria from xylose, glucose and sugarcane bagasse hydrolysate. *J Ind Microbiol Biotechnol* 2004; 31(6): 245–254.

84. Silva LF, Gomez JGC, Taciro MK *et al.* Brazilian patent PI02073560 Processo de produção de PHB e seu copolímero PHB-*co*-HV a partir de hidrolisados de bagaço de cana-de-açúcar.

85. Raicher G. Análise Econômica da Produção de Polímeros Biodegradáveis no contexto de uma Biorefinaria a partir de cana-de-açúcar. Doctoral thesis, Universidade de São Paulo, Brazil, 2011.

86. Lopes MSG, Gomez JGC, Silva LF. Cloning and overexpression of the xylose isomerase gene from *Burkholderia sacchari* and production of polyhydroxybutyrate from xylose. *Can J Microbiol* 2009; 55(8): 1012–1015.

87. Lopes MSG, Gosset G, Rocha RCS, Gomez JGC, Ferreira da Silva L. PHB biosynthesis in catabolite repression mutant of *Burkholderia sacchari*. *Curr Microbiol* 2011; 63(4): 319–326.

88. Guamán LP, Oliveira-Filho ER, Barba-Ostria C, Gomez JGC, Taciro MK, da Silva LF. *xylA* and *xylB* overexpression as a successful strategy for improving xylose utilization and poly-3-hydroxybutyrate production in *Burkholderia sacchari*. *J Ind Microbiol Biotechnol* 2018; 45(3): 165–173.

89. Guamán LP, Barba-Ostria C, Zhang F, Oliveira-Filho ER, Gomez JGC, Silva LF. Engineering xylose metabolism for production of polyhydroxybutyrate in the non-model bacterium *Burkholderia sacchari*. *Microb Cell Fact* 2018; 17(1): 74.

90. Raposo RS, de Almeida MCMD, de Oliveira M da, da Fonseca MM, Cesário MT. A *Burkholderia sacchari* cell factory: Production of poly-3-hydroxybutyrate, xylitol and xylonic acid from xylose-rich sugar mixtures. *N Biotechnol* 2017; 34: 12–22.

91. Cesário MT, Raposo RS, de Almeida MD *et al.* Production of poly(3-hydroxybutyrate-*co*-4-hydroxybutyrate) by *Burkholderia sacchari* using wheat straw hydrolysates and gamma-butyrolactone. *Int J Biol Macromol* 2014; 71: 59–67.

92. Cesário MT, Raposo RS, de Almeida MCMD, van Keulen F, Ferreira BS, da Fonseca MM. Enhanced bioproduction of poly-3-hydroxybutyrate from wheat straw lignocellulosic hydrolysates. *N Biotechnol* 2014; 31(1): 104–113.

93. Dietrich K, Dumont M-J, Orsat V, Del Rio LF. Consumption of sugars and inhibitors of softwood hemicellulose hydrolysates as carbon sources for polyhydroxybutyrate (PHB) production with *Paraburkholderia sacchari* IPT 101. *Cellulose* 2019; 26: 7939–7952.

94. Dietrich K, Dumont M-J, Schwinghamer T, Orsat V, Del Rio LF. Model study to assess softwood hemicellulose hydrolysates as the carbon source for PHB production in *Paraburkholderia sacchari* IPT 101. *Biomacromolecules* 2017; 19(1): 188–200.

95. Raposo RS, de Almeida MCMD, da Fonseca MMR, Cesário MT. Feeding strategies for tuning poly (3-hydroxybutyrate-*co*-4-hydroxybutyrate) monomeric composition and productivity using *Burkholderia sacchari*. *Int J Biol Macromol* 2017; 105(1): 825–833.

96. Oliveira-Filho ER, Silva JGP, de Macedo MA, Taciro MK, Gomez JGC and Silva LF Investigating nutrient limitation role on improvement of growth and poly(3-hydroxybutyrate) accumulation by *Burkholderia sacchari* LMG 19450 From xylose as the sole carbon source. *Front. Bioeng. Biotechnol.* 2020, 7: 416. doi: 10.3389/fbioe.2019.00416.

97. Taciro MK, Teixeira RP, Pradella JGC, Netto CLB, Simões DA. Poly-3-hydroxybutyrate-co-3-hydroxyvalerate accumulation at different strategies of substrate feeding. *Rev Microbiol* 1997; 28: 49–53.
98. Piccoli RAM, Quiroz LHC, Fleury AT *et al.* Optimization of polyhydroxyalkanoates bioproduction, based on a cybernetic mathematical model. *Braz J Chem Eng*, 2020 (in press).
99. Piccoli RAM. Otimização do processo de produção de copolímeros de polihidroxialcanoatos por via fermentativa, baseada num modelo matemático estruturado. Doctoral thesis, Universidade de São Paulo, Brazil, 2000.
100. Kraft D. Algorithm 733; TOMP---Fortran modules for optimal control calculations. *ACM Trans Math Softw* 1994; 20(3): 262–281.
101. Lee YW, Yoo YJ. High cell density of *Alcaligenes eutrophus* and PHB production by optimization of medium compositions. *Korean J Appl Microbiol Biotechnol Lett* 1994; 14: 811–816.
102. Lee JH, Hong J, Lim HC. Experimental optimization of fed-batch culture for poly-beta-hydroxybutyric acid production. *Biotechnol Bioeng* 1997; 56(6): 697–705.
103. Tavares LZ, da Silva ES, da Cruz Pradella JG. Production of poly(3-hydroxybutyrate) in an airlift bioreactor by *Ralstonia eutropha*. *Biochem Eng J* 2004; 18(1): 21–31.
104. Pradella JGda C, Taciro MK, Mateus AYP. High-cell-density poly (3-hydroxybutyrate) production from sucrose using *Burkholderia sacchari* culture in airlift bioreactor. *Bioresour Technol* 2010; 101(21): 8355–8360.
105. Mantelatto PE, Duzzi AM, Sato T *et al.* EP 1853 713 B1 (WO2005052175) process for recovering polyhydroxyalkanoates ('PHAS') from cellular biomass, European Patent Office, München, 2005.
106. Zuccolo M, Ribeiro AMM, Ogaki Y, Alli RCP, Bueno Netto CL. Utilização de enzimas comerciais no estudo de lise celular de *Alcaligenes eutrophus* DSM 545. *IV SHEB - Seminário de Hidrólise Enzimática de Biomassa*, Maringá, 1994.
107. Zuccolo M, Carter JM, Morita D *et al.* Separação de um poliéster biodegradável por precipitação ácida. *V SHEB - Seminário de Hidrólise Enzimática de Biomassa*, Maringá, 1996.
108. Kapritchkoff FM. Recuperação e purificação de polihidroxibutirato de *Ralstonia eutropha* por via enzimática, doctoral thesis, University of Sao Paulo, Brazil, 2003.
109. Kapritchkoff FM, Viotti AP, Alli RCP *et al.* Enzymatic recovery and purification of polyhydroxybutyrate produced by *Ralstonia eutropha*. *J Biotechnol* 2006; 122(4): 453–462.
110. Holmes PA, Lim GB. US patent US4910145A Separation process, USPTO, Alexandria, VA, 1990.
111. Nonato RV. Sustainable polymer from sugar cane - BIOCYCLE®, 2012. http://www.fapesp.br/eventos/2012/07/Biopolymers/ROBERTO.pdf.
112. Bueno Netto CL, Craveiro AM, Pradella JGC *et al.* Brazilian patent PI 9103116–8 Processo paraproduzir polihidroxialcanoatos a partir de açúcares extraídos da cana-de-açúcar, Instituto Nacional da Propriedade Industrial, INPI, Rio de Janeiro, 1993.
113. Silva LF, Gomez JGC. Brazilian patent PI98065572 Processo de obtenção e mutante de *Burkholderia* sp mais eficiente na utilização de propionato para produção de copolímero biodegradável, Instituto Nacional da Propriedade Industrial, INPI, Rio de Janeiro, 1998.
114. Biocycle website. http://www.biocycle.com.br/site.htm.
115. Brazilian Sugarcane Industry Association – UNICA. https://www.unicadata.com.br/index.php?idioma=2.

116. Silva LF, Taciro MK, Raicher G *et al.* Perspectives on the production of polyhydroxyalkanoates in biorefineries associated with the production of sugar and ethanol. *Int J Biol Macromol* 2014; 71: 2–7.
117. Macedo IC. Greenhouse gas emissions and energy balances in bio-ethanol production and utilization in Brazil. *Biomass Bioenergy* 1996; 14: 77–81.
118. Seabra JEA, Macedo IC, Leal MRLV. Greenhouse gases emissions related to sugarcane ethanol. In: Luis Augusto Barbosa Cortez (ed) *Sugarcane Bioethanol — R&D for Productivity and Sustainability*. Editora Edgard Blücher, Sao Paulo, 2014: 291–300.

13 LCA, Sustainability and Techno-Economic Studies for PHA Production

*Khurram Shahzad, Iqbal Muhammad Ibrahim
Ismail, Nadeem Ali, Muhammad Imtiaz Rashid,
Ahmed Saleh Ahmed Samman, Mohammad Reda
Kabli, Michael Narodoslawsky, and Martin Koller*

CONTENTS

13.1   INTRODUCTION

The sustainability of a new process in development can be measured by comparing diverse production methodologies using Life Cycle Assessment (LCA)-based ecological footprint methodology known as Sustainable Process Index (SPI) [1, 2]. Such evaluation includes the ecological footprint comparison of unit polymer production utilizing alternate energy (heat and electricity) provision resources as well as the impact of the geographical location of the bioprocessing unit. Production of "bioplastics" from inexpensive starting materials is a prime example for the implementation of the SPI tool.

Nowadays, use of plastic materials is pivotal in our daily life. These materials are used for the provision of various services like packaging of food, low density supplies, chemical resilient commodities, specialized niche structures in electronic fabrication as well as their usage in the medical fields as implants and scaffolds, etc., have opened new horizons for research and development in academia and industry [3–8].

Being bio-based and biodegradable, the environmental potential and benefits of bioplastics like polyhydroxyalkanoates (PHA) fascinate numerous researchers, especially in recent decades. However, such bioplastics have an inherent economic drawback compared to their fossil-based counterparts, namely higher raw material costs, technologically immature production, and small-scale production facilities. Therefore, established plastics from the petro-industry still have the economic advantages on their side. This makes it clear that for successful market penetration of PHA and other biopolymers, advantages in other fields different to economics are indispensably needed in order to compensate for this; in any case, as long as fossil resources are still available at dumping prices, the ecological performance of PHA is the one and only selling point for industry and the end-consumer for switching to these biomaterials. Nevertheless, it is the combination of ecological superiority and economic competitiveness which is needed to outperform petro-plastics on the market!

A thorough study of the available literature makes clear that PHA and other biopolymers do not innately outperform petro-plastics in terms of sustainability; just being bio-based and biodegradable is not enough to claim that a material is *per se* of higher sustainability than established recalcitrant plastics. Rather, it is the holistic life cycle of a (bio)material which needs to be considered before a reliable, straightforward assessment of its sustainability can be done. An LCA for PHA production needs to consider the entire production process starting from the raw material generation, the biotechnological conversion of the raw material towards the end-product, the energy supply, the management of side streams of raw material production and the bioprocess itself, the downstream process for product recovery from microbial biomass, and the treatment of spent PHA after its use as a plastic item. In this context, special emphasis needs to be devoted to the raw material aspects; of course, it is no challenge anymore to use first generation feedstocks like sugars or edible oils for PHA biosynthesis; however, these processes are already more or less optimized, and unambiguously uneconomic. Most of all, it is definitely not sustainable to misdirect materials of nutritional significance towards bioplastic production. Hence, both economic and ethical considerations call for alternative raw materials to be used as ethically clear, inexpensive feedstocks for biotechnological PHA production; they must not display any interference with the food or feed supply chain.

In the context of such inexpensive materials, which are out of scope for food purposes and typically need to be disposed of in an often expensive way, different organic side streams of the animal processing industry, only recently materialized. Animal processing for meat production is a steadily growing industrial branch in various, especially economically emerging, global regions. In normal practice, the animal residue materials arising from the slaughtering facilities are considered as waste material with no market price. These materials usually require further investment for their appropriate disposal in compliance with the strict environmental legislation. According to the process material flow diagram shown in Figure 13.1 the slaughtering process produces three main material streams, namely meat, rendering

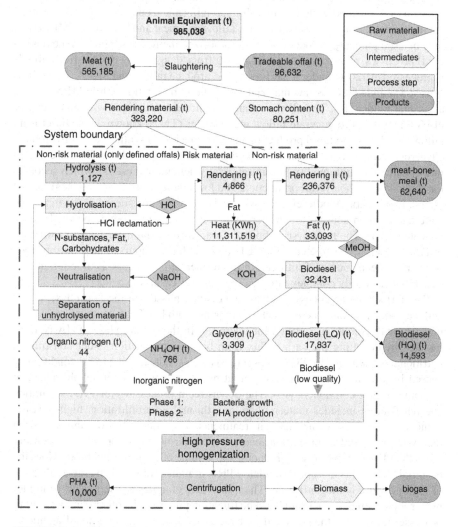

FIGURE 13.1   The detailed process design and waste fraction distribution and their relevant material flow for 10,000 t PHA production (modified from [20]).

residues (blood, bones, and excessive fat), and non-rendering residues including colostrum, manure, and digestive tract material [9]. The process design and development were carried out using real time animal slaughtering products and waste fractions composition data obtained from a Master's thesis [10]. The key sub-processes shown in the process design flowsheet include hydrolysis, rendering, biodiesel production, biotechnological fermentation, and related downstream processing for PHA recovery from biomass and PHA purification. The process design reveals that there are two rendering processes (rendering I and rendering II) considered based on the type of the material to be processed. As per the European Union regulation # 1774/2002 [11], the raw material for rendering is further divided into two main categories, namely risk material and non-risk material. The body parts of animals, infected or confirmed with transmissible spongiform encephalopathy (TSE) will be handled in rendering I. The products, i.e. tallow and meat and bone meal (MBM) can only be used for energy production in an authorized incineration plant. In the rendering II processes, all risk-free material arising from slaughtering of TSE-free animals is processed at 120°C under 3 bar pressure to produce tallow and MBM. This tallow is used as edible fat or as raw material for biodiesel production, while MBM is used as animal fodder [12]. The alkaline transesterification technique with an efficiency of 98% is utilized to generate tallow methyl ester (TME) known as biodiesel from tallow [13]. In the biodiesel production process, glycerol is synthesized as byproduct. Glycerol is produced at the rate of 0.1 kg/kg biodiesel produced [14]. Biodiesel is further classified as high quality (unsaturated fatty acid methyl esters), which accounts for about 45% of the overall production, and low quality (saturated fatty acid methyl esters). Low quality biodiesel fractions and glycerol are considered as the main raw materials (carbon source) for the fermentation process to produce polyhydroxyalkanoate (PHA). The hydrolysis subprocess transforms offal material into a mixture of amino acids when treated with HCl at 120°C for six hours. These amino acids serve as a cheap source of nitrogen in the fermentation process as an alternate for typically used costly complex nitrogen sources, e.g. casamino acids [12].

The fermentation process involves two distinct phases: in the first phase, a predefined concentration of active bacterial biomass with low PHA content is generated under growth conditions optimal for microbes. In the second phase, PHA is accumulated by deviating the carbon flux from biomass production under constrained nutritional conditions to PHA biosynthesis. This nutritional stress condition is caused by limiting one or more essential nutrients, e.g. nitrogen or phosphate or any other minor component [15]. The downstream processing for PHA separation and purification includes material handling through microfiltration, high pressure homogenization, and centrifugation treatments. The microfiltration unit filters out the excessive liquid fermentation media ("spent fermentation broth") and concentrates the PHA-rich biomass by reducing the volume of fermentation broth. Now, the bacterial cell wall is disrupted to release PHA granules from cells by the application of high-pressure homogenization. Then, the PHA granules are separated from biomass rubble (cell debris) through a centrifugation process. The other complementary processes included in the process design are anaerobic digestion for handling microbial biomass debris, manure, digestive tract, belly content to produce biogas, and a wastewater treatment unit.

The solid waste collection and transport to the handling facility is the key cost factor which varies between 50 and 80% for developing countries, while in under-developed countries, it has been reported to amount to up to 80–90% of the overall waste management cost [16, 17]. The animal residue collection cost is calculated by assuming a waste collection radius of about 75 km, which means 150 km traveling per trip for waste collection. The international unit for cargo transport is "tonne kilometer (tkm)", which is demarcated as "a unit of measure of goods transport which represents the transport of one tonne by road over one kilometer" [18]. The commercial expense for animal residue collection and transportation is simulated at 0.41 €/(t·km) [19]. This cost of material collection and delivery to the facility for further handling is integrated into the overall system boundaries established for the investigation.

The current study provides techno-economic as well as sustainability feasibility, including material collection, material and energy flow for the key subprocesses (hydrolysis, rendering, biodiesel production, fermentation, and downstream processing, biogas production, and wastewater treatment) calculated for a 10,000 t/year PHA production facility.

13.2 ECONOMIC ANALYSIS

It is essential to perform the analysis of economic viabilities of a new technology to be delivered to the industry for successful implementation already at an early stage of process development. Normally, the cost of any product or service consists of three main components, i.e. direct, fixed, and general costs. The expenses which directly impact the production cost, i.e. raw material price, labor, and utility expenses, constitute the direct costs. The costs which are independent of the facility operation or production rate has to be taken into account for calculations even when the production facility is non-operational. It includes facility depreciation, insurance, and taxes. The latter section of the costs includes administrative costs, research and development funds, and funds for maintenance of the plant's operation units [21]. A detailed economic analysis of the sub-processes based on investment costs, operating costs, and sensitivity analysis is given in the next sections.

13.2.1 INVESTMENT COST

An overview of the direct and fixed cost is shown in Table 13.1, according to the data obtained from project partners and a literature review.

13.2.1.1 Investment for Hydrolysis, Rendering, Biodiesel Production, and Fermentation

The investment for the chemical hydrolysis reaction unit includes the total estimated price for the chemical reaction vessel, an acid recovery facility, and installments related to the chopping of the offal material. High-quality stainless-steel material was considered for the fabrication of the reaction vessel due to the extremely corrosive nature of the acidic material used in the reaction process. The estimated cost for the reaction vessel includes the price of high-quality steel required for fabrication of

TABLE 13.1

Investment Cost Overview of the Sub-Processes (Modified from [20])

Processing Units	Investment [€]
Hydrolysis Facility	120,000
Rendering Unit	25,600,000
Biodiesel Production Unit	17,920,000
Fermentation Facility	7,650,700
Downstream Processing facility	1,295,000
WWT Plant	8,352,994
Biogas Generation Facility	11,949,186
Infrastructure	896,000
Total	**73,783,879**

the vessel along with the price of other accessories (50% of the steel price) to control the reaction process. The estimated investment for hydrolysis was 28,516 € considering a steel price of 760 €/t [22].

According to the data obtained from one of the partners in the venture named Argus Umweltbiotechnologie GmbH (located in Berlin, Germany), an estimated cost of 60,000 € was considered for the microfiltration unit to be used for acid recovery, related to a fermentation media processing capacity of 1 m³/h. Similarly, the investment cost for the chopping installment was obtained from a commercial chopper producer company based in Italy.

The chopping unit price varies between 30,000 € and 55,000 € depending on the material processing capacity from 3000 to 5000 kg/h. The investment cost data for a rendering plant and biodiesel production facility were acquired from the Technical Director of the company partner Argent Energy (UK) Ltd. (located in Scotland, UK) through personal communication. Similarly, the investment costs for the fermentation were simulated by measuring the plant size and calculating its development cost [12]. The downstream processing investment cost consists of installation of the membrane filtration unit, homogenization unit, and centrifugation unit. This data was also provided by the project partner Argus Umweltbiotechnologie GmbH.

13.2.1.2 Wastewater Treatment Plant (WWTP) Feasibility

In order to accomplish a feasibility study of a WWTP, it was necessary to calculate the annual water consumption and discharge. The overall amount of wastewater to be processed includes wastewater discharge coming from the rendering facility, process water discharged from the biodiesel production unit and wastewater released from downstream the processing unit for a 10,000 t/year PHA production facility. The annual estimated wastewater to be processed in the WWTP amounts to 227,863 m³/year at a rate of 690 m³/day. For the work presented in the chapter at hand, it was assumed that "the wastewater generated in PHA production facility has similar composition to the one discharged in the breweries".

TABLE 13.2

Volume of WWTP Subunits Compared to WWTP in Vienna [20]

Plants	Population Equivalent P.E.	Primary Settler V [m³]	Aeration tank V [m³]	Final Settler V [m³]	Sludge Thickener
WWTP in Vienna	4,000,000	28,415	42,000	200,000	13,500
ANIMPOL	115,082	818	1,208	5,754	388

The literature review reveals that the value of biological oxygen demand (BOD) for the wastewater discharge from the brewery is about 1.8 kg/hL of beer. This value is similar to the wastewater having 9 kg of BOD/m³ [23]. This value was used to compute the BOD pressure per day, with BOD load value of 6214 kg/day. The capacity of wastewater is usually stated in the units of population equivalent (PE). It is demarcated as "per capita" wastewater discharge of 0.2 m³/day or BOD load of 54 g/(inhabitant/day) for the sewage water discharge. The computed BOD load value was used to calculate the wastewater input and its relevant PE value for this project, which is 115,082.37 inhabitants [24, 25]. A comparative breakdown of WWTP size based on PE value was made considering a standard WWTP facility in Vienna, as shown in Table 13.2.

It is assumed that 1 kg of BOD requires 2 kg of O_2. Utilizing this value surface aeration power is calculated, which amounts to 464 kW. The methodology described by Haandel and Lubbe [26] was used to calculate the cost for each subunit of the WWTP. The overall computed investment cost, for the subunits including primary settler, aeration tank, final settler, sludge thickener, and kW installed, amounted to 3,554,466 €. It was assumed that the extra installments related to industrial WWTPs, including dissolved air flotation setup, oil separator, and extra refining units such as carbon adsorption and advanced oxidation units, cost around 35% of the core cost, which is 1,244,063 €. Other relevant costs, which include project planning, feasibility study, management, and civil works, account for 50 and 100% of the overall development cost for the mega and small projects, respectively. This project was presumed to be a small facility, and 100% of the core cost was considered as supplementary investment cost. This leads to the overall WWTP investment cost value to 8,352,994 €.

13.2.1.3 Investment for the Biogas Plant

The simulated data revealed that 4,808,525.36 m³/year of biogas will be generated at an hourly generation rate of 601 m³. The investment cost for a biogas plant with 250 Nm³/h biogas production capacity was considered as baseline investment [27]. For 601 m³/h biogas production, the investment cost was extrapolated using linear function, except for the fermenter cost, where a factor of 0.70 was utilized. The overall computed investment cost for the biogas plant was 3,212,350 €.

13.2.1.4 Infrastructure

Infrastructure cost for the industrial facility has been considered for an area of about 900 m² internal floor area. It includes the cost breakdown from the feasibility report to the turnkey construction facility [28]. The calculated infrastructure construction cost is about £700,000, which is equivalent to about 896,000 € considering £1 = 1.28 €.

13.2.1.5 Depreciation

The lifespan of the facility is considered to be 20 years. Using a linear function for all sub-processes, the investment cost annual depreciation is calculated as shown in Table 13.3.

13.2.1.6 Operational Cost

The running costs for the rendering and biodiesel units were obtained from one of the partner groups in the consortium. The operating costs for the hydrolysis, fermentation, WWTP and biogas plant, and other sub-processes were calculated according to the above described method. The simulated investment costs for the hydrolysis, fermentation process, biogas plant, and WWTP were used for calculating the operational estimates for individual sub-processes. The comprehensive breakdown of operative costs for the fermentation process is given in Table 13.4.

The overall operational cost for PHA production is the sum of this calculated operational estimate along with the cost of chemicals used in the hydrolysis and

TABLE 13.3
Biorefinery Facility Depreciation
Breakdown (Modified from [20])

Processing facility	Devaluation [€/yr]
Hydrolysis Facility	6,000
Rendering Unit	1,280,000
Biodiesel Production Unit	896,000
Fermentation Facility	382,535
Downstream Processing facility	64,750
WWTP	417,650
Biogas Generation Facility	597,459
Infrastructure	44,800
Total	**3,689,194**

TABLE 13.4
Operational Cost Estimation

Categories	Range [%]	Cost [€]
Labor Cost	2	196,834
Operation Cost	0.50	49,209
Maintenance Cost (civil, mechanical)	1.50	147,626
Insurance	0.20	19,683
Electrical energy	0.10	509,051
Heat	0.04	114,313
Fermentation Process Operational Cost		**1,036,715**

fermentation processes. The calculated operational cost for the fermentation process is 1,036,715 €/year, while the consumption of chemicals during this process accounts to 1,159,12 €/year. The cumulative operational cost for the fermentation process is 2,195,839 €/year. Likewise, the calculated operating and chemical cost for the hydrolysis process is 413,402 €/year, while for wastewater treatment and biogas units, the estimated values amount to 723,503 €/year and 96,483 €/year, respectively. The estimated cost for the waste material collection and transportation is 19,878,075 €/year, which is also taken into account in the overall process operational cost. Thus, the sum of overall operational costs for the PHA production process is 27,142,706 €/year. The breakdown of the overall operating cost is given in Table 13.5.

13.2.1.7 PHA Production Cost

The overall production cost is the cumulative operational cost for all sub-processes, the prices of the raw materials and chemicals used for the production of 10,000 t/ year PHA as well as annual plant depreciation. Table 13.6 displays the thorough breakdown of throughputs and their related costs in the PHA production process. It counts for 5.07 €/kg of PHA production.

TABLE 13.5
The Breakdown of Operation Cost Estimation

Processing Facility	Operational Cost [€/yr]
Hydrolysis Facility	413,402
Rendering Unit	9,529,109
Biodiesel Production Unit	14,184,371
Fermentation Facility	2,195,839
WWT Plant	723,503
Biogas Generation Facility	96,483
Overall Operational Cost	**27,142,706**

TABLE 13.6
The Breakdown of PHA Production Cost

Input (t)	Cost [€]
Organic Nitrogen	413,402
Inorganic Nitrogen	383,264
Operational Cost	26,346,041
Plant Depreciation	3,689,194
Transportation	19,878,075
Total Cost	**50,709,975**

13.2.2 REVENUE

The revenue is calculated using selling prices for the main product PHA and the byproducts, namely high-quality biodiesel and MBM. The selling prices for the commodities are 4 €/kg for PHA, 0.95 €/L for biodiesel, and 300 €/t for MBM. Net cash flow from the byproducts is shown in Table 13.7.

Overall revenue obtained from the byproducts is about 34,166,978 €/year. It makes the overall process economically more feasible, decreasing the PHA manufacturing cost from 50,709,975 €/year to 16,542,997 €/year, which results in a cost decrease per kg PHA produced from 5.07 € to 1.65 €.

As PHA is a renewable polymer, it can get a higher selling price than fossil-based (PE-LD) polymer as its conventional counterpart. A retail price of 4 €/kg of PHA was decided after discussion with M. Koller, an expert in PHA production utilizing alternate resources, and based on the current market situation of bioplastics. According to this estimation, 40,000,000 €/year of PHA sales revenue can be generated.

$$\text{Net Profit by selling PHA} = \text{PHA selling revenue} - \text{PHA production cost}$$

$$= 40,000,000 - 16,542,997$$

$$= 23,457,003 \ €/\text{year}$$

The estimated breakeven point for investment retrieval was calculated using the net profit value as shown in Figure 13.2. The estimated retrieval period for the complete investment was approximately 4.25 years.

13.2.2.1 Sensitivity Analysis

The overall economic results are affected by the change of different parameters like biodiesel prices, MBM prices, and source of organic nitrogen.

i) Impact of provision of organic nitrogen from alternative resources

Initially it was assumed that offal material is a waste and had no market price; hence, no price was assigned to the offal material used for hydrolysis in the economic analysis. With the change in the market situation, offal material is not a waste material anymore; rather, it is processed to

TABLE 13.7

Net Revenue Flow by Products

Commodity	Revenue [€]
Biodiesel HQ	13,864,203
MBM Selling	18,791,948
Saving From Heat	635,676
Savings From Electricity	875,152
Total	**34,166,978**

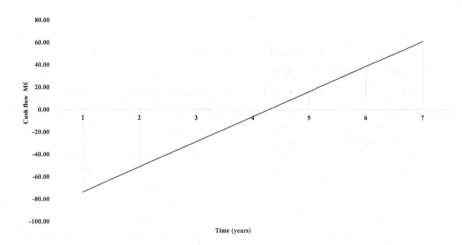

FIGURE 13.2   Cash flow representing investment recovery time.

produce pet food having a significant market value. Bearing in mind this new competitive circumstance, a comprehensive economic analysis for various resources, which can be used for the provision of organic nitrogen, was conducted. Such alternative resources for organic nitrogen provision include offal as waste, which is the base scenario, offal with market price, silage juice (SJ) as well as casamino acid (CA). In the second scenario, offal hydrolysate is the organic nitrogen source with offal material having a market value of 1.3 €/kg [29]. In the cases of CA being sources of organic nitrogen, its web price 114,646 €/t [30] was used, while for SJ 655 €/t [22, 31] was used for computation.

In the third and fourth scenario, SJ or CA act as sources of organic nitrogen, and their web prices are used for calculations. Table 13.8 represents the economic calculations for 10,000 t/year PHA production using different organic nitrogen sources.

As revealed in Table 13.8, organic nitrogen provision cost directly influences the overall PHA production cost. The organic nitrogen provision cost using offal material as waste, offal material market value, SJ or CA, accounts for 1, 5, 5.2, and 16.2% respectively of the overall annual production cost. The unit cost (€/kg) for PHA production gradually increased from 1.65 €/kg for offal as waste, 1.88 €/kg for offal having a market price, 1.90 €/kg for SJ as organic nitrogen source and 2.59 €/kg for utilizing CA as a source for organic nitrogen provision. Figure 13.3 shows the fiscal breakdown for PHA production using diverse organic nitrogen resource.

ii) Impact of biodiesel price fluctuations

Biodiesel price fluctuation has been calculated considering biodiesel prices 0.55, 0.95, and 1.35 €/L. The impact of price fluctuation on the overall project economics was considered for two basic scenarios, which are offal as waste material (scenario I) and using offal market price

TABLE 13.8

Monetary Investigation for PHA Production Consuming Various Organic Nitrogen Resources

Inputs	Waste [M€]	Offal [M€]	SJ [M€]	CA [M€]
Organic Nitrogen	0.413	2.658	2.863	9.788
Inorganic Nitrogen	0.383	0.383	0.383	0.383
Operational Cost	26.35	26.35	26.35	26.35
Plant Depreciation	3.689	3.689	3.689	3.689
Transportation	19.878	19.88	19.88	19.88
Total Cost	50.710	52.95	53.16	60.08
Revenue	34.17	34.17	34.17	34.17
PHA net price €/kg	**1.65**	**1.88**	**1.90**	**2.59**

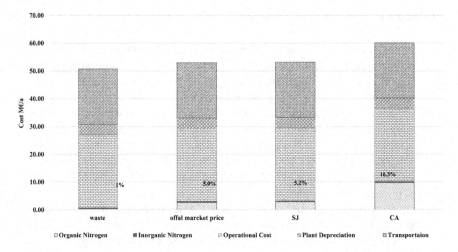

FIGURE 13.3 Fiscal analysis for PHA production utilizing various alternative resources for organic nitrogen.

(scenario II), considering its use as pet food while keeping MBM price 300 €/t (Table 13.9).

While Table 13.9 reflects the impact of biodiesel price variation and relevant cash flow, Figure 13.4 shows that the breakeven point for capital retrieval varies from 3 to 5.5 years.

Table 13.10 shows the effect of biodiesel selling price fluctuation on the overall production economics. The PHA production cost fluctuates between 2.25 and 1.09 €/kg considering offal material as waste, while these values fluctuate between 2.48 and 1.31 €/kg PHA utilizing offal market price for the analysis.

iii) Effect of change of MBM prices

The analysis carried out using a biodiesel price of 0.95 €/L is kept constant, while MBM price fluctuates between 300 and 500 €/t. The two

TABLE 13.9

Effect of Change of Biodiesel Prices Maintaining MBM at 300 €/t and Using Offal as Waste Material

Time [years]	Offal as Waste			Offal Market Price 1.3 €/kg		
	Revenue [M€] biodiesel 0.55	Revenue [M€] biodiesel 0.95	Revenue [M€] biodiesel 1.35	Revenue [M€] biodiesel 0.55	Revenue [M€] biodiesel 0.95	Revenue [M€] biodiesel 1.35
1	−74	−74	−74	−74	−74	−74
2	−56	−50	−45	−59	−53	−47
3	−39	−27	−16	−43	−32	−20
4	−21	−4	14	−28	−11	7
5	−4	19	43	−13	10	34
6	13	43	72	2	31	61
7	31	66	101	17	52	88

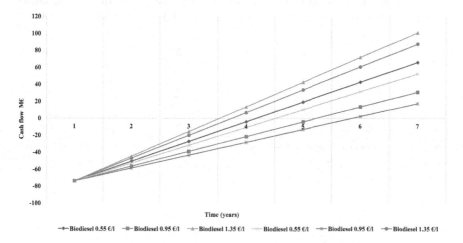

FIGURE 13.4 Impact of biodiesel selling price fluctuation using offal as waste material as well as offal market price.

scenarios simulated for analysis were based considering offal material as waste with no market price and offal material as raw material for pet food having a market price of 1.3 €/kg. The impact of MBM price variation is given in Table 13.11.

Figure 13.5 shows that MBM price variation has a substantial impact on the monetary analysis of the process compared to the biodiesel price impact. The breakeven point for investment retrieval for the first scenario lies between 2.5 to 3.75 years, while for the second scenario these values fall between 2.75 and 4 years, as shown in Figure 13.5.

TABLE 13.10

Biodiesel Price Effect on PHA Production Cost [€/kg]

Scenarios	Biodiesel 0.55 [€/L]	Biodiesel 0.95 [€/L]	Biodiesel 1.35 [€/L]
Waste	2.25	1.67	1.09
Offal	2.48	1.90	1.31

TABLE 13.11

The Impact of MBM Price Variation on the Monetary Analysis Using Biodiesel Price of 0.95 €/L, Offal as Waste, and Offal with a Market Value

	Offal as Waste			Offal Market Price 1.3 €/kg		
Time (years)	Revenue [M€] MBM € 300/t	Revenue [M€] MBM € 400/t	Revenue [M€] MBM € 500/t	Revenue [M€] MBM € 300/t	Revenue [M€] MBM € 400/t	Revenue [M€] MBM € 500/t
1	−74	−74	−74	−74	−74	−74
2	−50	−44	−38	−53	−46	−40
3	−27	−15	−2	−32	−19	−7
4	−4	15	34	−11	8	27
5	19	44	69	13	35	61
6	43	74	105	34	63	94
7	66	104	141	55	90	128

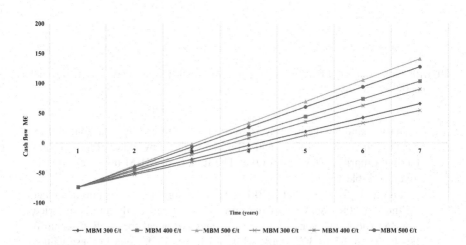

FIGURE 13.5   Effect of MBM prices using offal as waste.

TABLE 13.12

MBM Selling Price Variation Impact on PHA Production Cost €/kg

Scenarios	MBM 300 [€/t]	MBM 400 [€/t]	MBM 500 [€/t]
Waste	1.67	1.04	0.42
Offal	1.90	1.27	0.64

The impact of MBM price fluctuation on PHA production cost for both scenarios is shown in Table 13.12. For the first scenario it varies from 1.67 to 0.42 €/kg, while in the second scenario, it changes from 1.90 to 0.64 €/kg.

It has also been expected that the market price of MBM will rise further due to ever increasing demand for suitable animal feed. This is due to the worldwide increasin g trend of meat consumption [32, 33]. Similarly, it has also been reported to be used as an alternative source of phosphate (one of the scarcer resources in the future) and Ca^{2+} from the green manure obtained after incineration to harvest energy [34]. Similarly, combustion residues of MBM incineration have also been reported to be used for bioremediation of aqueous effluents by removing heavy metals from it [34]. So, the direct selling of MBM as animal feed as well as byproducts of its incineration hold potential for generating extra revenue in future.

13.3   SUSTAINABILITY OF PHA PRODUCTION

Sustainability is a well-recognized paradigm influencing future developmental policies and regulations at government as well as at intergovernmental level. It is considered as a model consisting of the three main pillars of a society, namely economic, environmental, and social aspects. It is normally represented by three intersecting circles, representing each aspect of sustainability. In other words, sustainability means development of human wellbeing, in accordance with the recognition of the existence of one diverse, yet ultimately finite planet. For decision makers it is becoming challenging how to fulfill human demands, while at the same time operating within the fixed limits of nature, and still to attain sustainability. This requires both the effective management of human demands as well as natural capital, while living within its ability to renew itself. In order to achieve this task, reliable measurement tools comparing the supply of natural income with human demand are crucial. They help decision makers to track progress, set targets and make policies for sustainability [35].

Sustainability assessment methods range from single issue measures, like Carbon Footprint [36], Water Footprint [37], etc. (measuring exchange of a single substance with the environment along the whole life cycle chain of product or service), to direct quantity and quality of energy measures (thermodynamic measures), like Exergy [38] and Energy Accounting [39], to more complex aggregated measures like Ecological Footprint [40], Sustainable Process Index – SPI [41], the Well Being Index [42], the Environmental Sustainability Index (ESI) [43] and material input efficiency based Material Input Per Service unit (MIPS) [44], just to name some

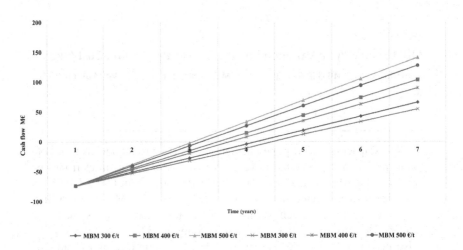

FIGURE 13.6   LCA study parts. [45]

common measurement categories. The Life Cycle Assessment (LCA)-based aggregate measure SPI was considered to evaluate the sustainability of cradle-to-grave PHA production process.

13.3.1   Life Cycle Assessment

LCA is a systemic approach to look at a specific product and its production. LCA is standardized and defined in the ISO 14040 norm [45]. The approach to get an LCA study is illustrated in Figure 13.6.

13.3.2   Life Cycle Impact Assessment with the Sustainable Process Index (SPI)

The SPI methodology was selected for Life Cycle Impact Assessment (LCIA) of the PHA production process. It measures the area required to embed the whole life cycle of processes, products, or provision of services sustainably in the ecosphere [2]. This methodology was developed based on considering solar energy as the only natural energy income of this planet. All natural processes and universal quantifiable material cycles (e.g. nitrogen, carbon, and water cycles) are driven by the energy provision through solar energy. The key resource required to harvest and transform this form of energy and to transform it into various usable materials, e.g. biomass and energy, is area. The direct applications of this resource include electricity or heat production through photovoltaic or solar thermal energy, or indirect utilization through photosynthesis to produce biomass [46]. Air, water, and productive soil are the key components of a sustainable economy. In order to maintain sustainable production, these components need to be retained in a well-maintained healthy condition. So, the emissions to these compartments have also been considered for calculating the ecological footprint in relation to the principles that global quantifiable material cycles must not be altered while making sure that the indigenous local quality of these compartments should not alter either. Thus, the SPI value represents the cumulation of seven different sub-areas. These sub-areas are indicated by different colors.

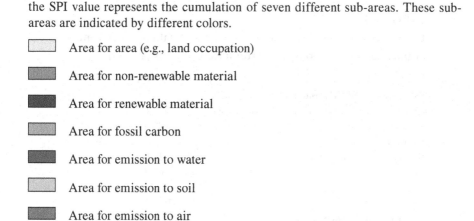

Area for area (e.g., land occupation)

Area for non-renewable material

Area for renewable material

Area for fossil carbon

Area for emission to water

Area for emission to soil

Area for emission to air

Summing up all the sub-areas represents the sustainable area required to fulfill material and energy provision as well as absorption of emissions over the life cycle of product production or service provision.

The SPI can be used making comparative analysis of diverse technologies [9], design and development of ecofriendly products, or measuring the environmental performance of the companies [47]. The latter is of special importance in the case of utilizing alternate energy from bagasse for PHA production using sugarcane as raw material [48]. The SPI realized as open access web-based software tool SPIonWeb (available at: http://spionweb.tugraz.at/) can design the whole life cycle product-service chain of PHA production. It also delivers genuine and comprehensive information about the ecological effects of the developments in question.

13.3.2.1 SPI Evaluation for Various Processes

The cumulated ecological footprint, i.e. SPI value (m^2 area) per unit (1 tonne) PHA production is obtained by evaluating each process and key step in the process design. Therefore, the production of 1 t of PHA is the base of evaluation for every key part which can be seen in the following Table 13.13 in the inventory column.

I. **Animal residue collection**

It is assumed that waste cost is equivalent to transportation cost for material collection from different slaughtering facilities. Similarly, residue material collected from slaughterhouses is considered to have an SPI footprint value of 0 m^2/t. This allocation is made because the footprint is already included in the main product of slaughtering, which is meat. Nevertheless, the transport of the residue material to the rendering facility was considered for SPI evaluation. It was estimated that for 1 t of PHA production 13.64 t of animal residue will be transported. A transportation radius of 75 km was considered for material collection causing about 150 km distance per trip. The amount of cargo transport per t of PHA produced amounts to 2046 tkm. The SPI value calculated for material transfer using 28 t transport truck is 84.95 m^2/tkm. Thus, the calculated SPI value for waste is 173,855 m^2.

TABLE 13.13

Life Cycle Ecological Assessment Inventory Data for Different Subprocesses of Biopolymer Production Process [49, 50]

Input	Amount	SI Units
Subprocess I: Offal Hydrolysis		
Offal Hydrolysis: 1 t Organic Nitrogen Equivalent		
Transportation	5951.657	tkm
Electric Energy EU27	0.957	MWh
Heating Energy, Natural Gas	7.092	MWh
HCl	46.798	t
Water	31.956	m^3
NaOH	16.902	t
Subprocess II: Rendering		
Rendering I: 1 MWh Heat Production		
Transportation	43.558	tkm
Electric Energy EU27	0.018	MWh
Heating Energy, Natural Gas	0.267	MWh
WWT	0.173	m^3
Water	0.076	m^3
Rendering II: 1 t Fat/Tallow Production		
Transportation	625	tkm
Electric Energy EU27	0.25	MWh
Heating Energy, Natural Gas	3.24	MWh
WWT	2.48	m^3
Water	1.08	m^3
Heat From Rendering I	0.30	MWh
Subprocess III: Trans-Esterification Process		
Trans-Esterification: 1 t Biodiesel Production		
Tallow	1.02	t
KOH	0.02	t
H_2SO_4	0.01	t
CH_3OH	0.11	t
Heating Energy, Natural Gas	0.05	MWh
Electric Energy EU27	0.07	MWh
WWT	0.10	m^3
Subprocess IV: Fermentation Process and Downstream Processing		
Fermentation: 1 t PHA Production		
Hydrolysate	0.004	t
Ammonium Hydroxide	0.077	t
Glycerol	0.237	t
Biodiesel	1.859	t
Inorganic Chemicals	0.078	t
Net Electricity EU27	0.321	MWh
WWT	8.118	m^3
Process Water	8.118	m^3
Process Energy, Natural Gas	0.292	MWh

II. **Hydrolysis**

The SPI value for offal hydrolysis includes the sum of estimated SPI values for offal transport, electricity provision for offal slicing and acid recovery, heat required for heating, and amount of acid consumed. The ecological assessment result for hydrolysis was carried out using SPI methodology based on inventory inputs data for unit hydrolysate production (equivalent to 1 t of organic nitrogen) by offal hydrolysis. The inventory data is given in Table 13.13, while Figure 13.7 shows the graphical footprint share produced by each input material. It shows that inorganic acid and base consumption are the main footprint contributors with 83% share of the overall footprint caused by the subprocess. Similarly, material transfer and provisions of heat and electricity also have noteworthy dividends. The higher footprint share corresponding to the inorganic acid and base is due to the fossil fuel–based, extremely energy-demanding production processes of these chemicals.

III. **Rendering process**

SPI footprint evaluation for the rendering subprocess is further separated into two portions based on the type of the material to be handled. Contaminated animal residue is exclusively utilized to produce heat for the rendering II subprocess. While on the other hand rendering II handles the main portion of the animal residue streams to harvest tallow as well as meat and bone meal (MBM) as main products.

Rendering I

The rendering I material constitutes contaminated material which is not allowed to be processed to obtain edible tallow. The SPI value for rendering I was simulated by utilizing inventory data for 1 MWh heat

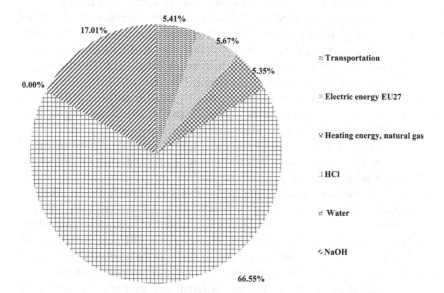

FIGURE 13.7 Ecological footprint contribution from various material, energy, and emission flow for hydrolysis.

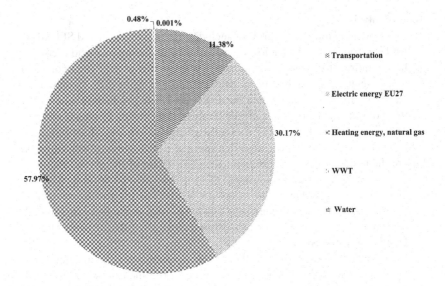

0.48% 0.001%

11.38%

30.17%

57.97%

Transportation

Electric energy EU27

Heating energy, natural gas

WWT

Water

FIGURE 13.8 Ecological footprint contribution from various inventory inputs for rendering I for heat production [49].

production as shown in Table 13.13. The ecological footprint share contributed by each input is shown in Figure 13.8. Process energy for heating, electricity, and transportation are the main contributors.

Rendering II

This is the key rendering sub-process, which processes non-risk material having extra fat, bones, and animal viscera as the main constituents to harvest tallow and MBM. Tallow is the main component used for biodiesel production through the transesterification process, while MBM will be traded on the market. The inventory data for harvesting 1 t of tallow in the rendering process is shown in Table 13.13. The chart given in Figure 13.9 represents the percentage footprint caused by each process input and discloses that energy provision (electric and heat) and material transfer are the key donors with 87.25 and 12.23% shares, respectively. This highlights that energy provision and transportation are the ecological hotspots and reveals the potential to minimize the overall footprint of the subprocess by utilizing energy from cleaner and renewable resources.

IV. **Biodiesel production**

As explained in the biodiesel production, this is accomplished by the alkaline-catalyzed transesterification of tallow with methanol. The inventory information for 1 t of biodiesel synthesis is given Table 13.13 and shows that tallow obtained from rendering process, KOH, H_2SO_4, CH_3OH, heat, electricity, and wastewater are the main energy and mass flows for the process. The graphical presentation of footprint analysis for the biodiesel production sub-process is provided in Figure 13.10. The input materials including tallow coming from the rendering II subprocess, methanol, and electricity consumption are the key donors to the all-inclusive footprint of the subprocess. As illustrated in

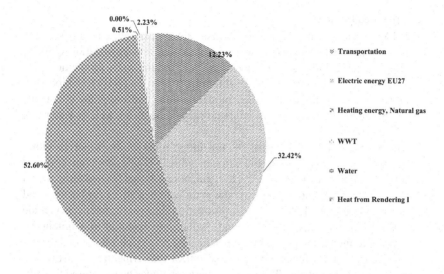

FIGURE 13.9 Ecological footprint contribution from various inventory inputs for rendering II process for tallow production [49].

Figure 13.9, fat extraction is an extremely energy exhaustive process. Also, it is the fundamental input material in the synthesis of biodiesel accounting up to 1.02 t/t biodiesel synthesis. It shares around 69% of the overall footprint of the subprocess accompanied by 17% coming from methanol and 11% induced by electricity consumption. Heating requirements have a very low input share of 1%, because heating requirements are fulfilled by burning the glycerol fraction produced as a byproduct in the biodiesel synthesis process and has a calorific value comparable to heavy fuel oil.

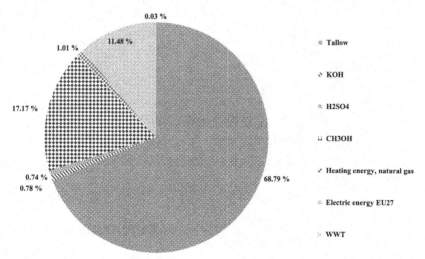

FIGURE 13.10 Ecological footprint contribution from various inventory inputs for bio-diesel production [49].

V. PHA production

PHA production, including the downstream processing, encompasses the fermentation sub-process as well as separation and purification of PHA from bacterial biomass. The inventory data for the PHA production process includes offal hydrolysate, ammonium hydroxide (NH_4OH), biodiesel, and glycerol as the main carbon input fractions and inorganic chemicals in agreement with fermentation media requirements, as well as water and energy inputs.

The hydrolysate liquid consists of vital amino acids for the bacterial culture and serves as a source of organic nitrogen, while NH_4OH acts as an inorganic nitrogen source, which also helps to regulate nutritional fermentation media conditions. TME and glycerol are the main carbon sources for bacteria and act as building blocks for the PHA production. Inorganic chemicals include several biochemicals and compounds, which are essential to maintain the desired fermentation process parameters. The energy consumption consists of electricity used for pumping in and out of the fermentation media into and from the bioreactor. The heating energy includes heat required for sterilization of the bioreactor and maintaining fermentation media temperature at about 37°C. Similarly, water is the main component of the fermentation media, and is also used in the downstream processing for separation and purification of PHA. The inventory input data for the fermentation process is given in Table 13.13.

The graph shown in Figure 13.11 illustrates the footprint distribution of 1 t PHA production according to the given inventory data. The main contributors to the overall footprint of this sub-process and downstream processing are biodiesel and glycerol along with electricity, which contribute around 68 and 19%, respectively.

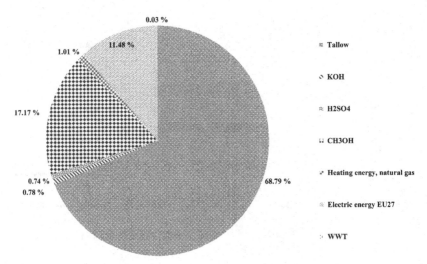

FIGURE 13.11 Graphical diagram of ecological footprint contributions caused by various inventory inputs of PHA production [49].

13.3.2.2 Comparative SPI Analysis for Biodiesel and PHA Production

The footprint investigation of the subprocesses discloses that transformation of the starting left-over material into a value-added usable material is an extremely energy-intensive procedure. In order to optimize the process based on the information obtained by the process results, an assessment of biodiesel and PHA synthesis considering electricity mixes [47] from various countries in Europe (Austria (AT), Poland (PL), Germany (DE), Denmark (DK), Norway (NO), Italy (IT) and France (FR)) as well as China and USA was carried out.

Figure 13.12 compares the ecological footprint per tonne of biodiesel production based on the ANIMPOL process (an EU funded FP7 project entitled " Biotechnological conversion of carbon containing wastes for eco-efficient production of high added value products"), considering rape seed oil (Bio-RSO) instead of tallow, and the effect of change of energy resources, i.e. renewable energy (renewable electricity and heat from biomass) and renewable electricity and natural gas as heating source (E+NG) with a conventional diesel production process [51, 52].

Figure 13.13 relates the ecological footprint per tonne of PHA production in accordance with the ANIMPOL process design, rape seed oil-based biodiesel as carbon source (PHA-RSO), using renewable energy and renewable E+NG with a conventional production process of poly(ethylene) LD (low-density PE; LD-PE).

13.3.2.3 Effect of Change of Energy Provision Resources

The footprint analysis of sub-processes shows that the conversion of slaughtering waste residues to a value-added product is a highly energy-intensive process. It indicates the potential of ecological optimization by using more ecofriendly energy from renewable energy resources. In order to assess the impact of change of energy from

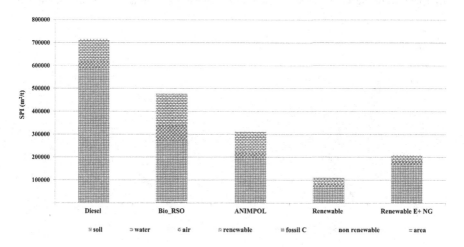

FIGURE 13.12 Comparative ecological footprint analysis of biodiesel production from different resources [49].

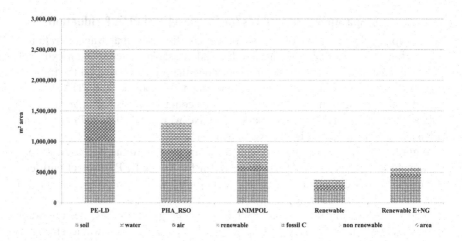

FIGURE 13.13 Production of polymer from different raw materials and energy resources [50].

business as usual (electricity from EU mix or national grid and heating with natural gas) to other energy resources, an analysis of biodiesel and PHA production utilizing electricity from coal, hydro-power, wind power, biomass, and replacement of both electricity and heat from biomass and biogas has been carried out.

13.3.2.3.1 Comparative Analysis of Biodiesel Production

The effect of a change of energy source on the ecological footprint of biodiesel production using SPI is shown in Figure 13.14. It shows that biodiesel production utilizing electricity form coal has the highest footprint, which, however, is still 55% lower than the footprint of diesel available at a regional store. Biodiesel production using electricity from the EU27 mix has a 61% lower footprint, while biodiesel production using electricity from hydropower, wind power, and biomass has a 70% lower footprint than diesel. The replacement of electricity as well as heat from biogas has

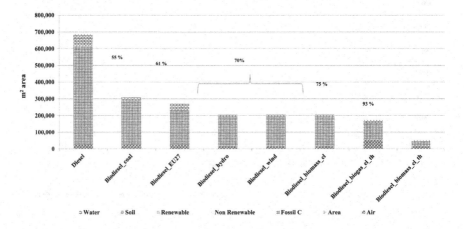

FIGURE 13.14 Comparison of biodiesel production utilizing different energy resources and diesel at ex regional store.

an even better impact, having a 75% lower footprint. The best scenario for biodiesel production is the use of electricity as well as heat from biomass, which has a 93% lower footprint than diesel. These simulation results along the lifecycle of biodiesel production show the effectiveness of energy use from different resources.

13.3.2.3.2 Comparative Analysis of PHA Production

The effect of change of energy resources on the ecological impact of PHA production along the whole lifecycle chain is shown in Figure 13.15. Among PHA production results, PHA production using electricity from coal has the highest SPI footprint. Its footprint value is still 59% less than the poly(ethylene) low density (PE-LD) footprint, a fossil-based polymer considered as a competitor to PHA. The footprint caused by PHA production using electricity from EU27 mix is 65% lower, while PHA production utilizing electricity from hydropower, wind power, and biomass has a 77% lower footprint than PE-LD. The replacement of electricity as well as heat from biogas has a slightly lower footprint than electricity from hydro power, wind power, and biomass. The ecologically best-optimized scenario for PHA production is use of electricity and heat from biomass. It has a 90% lower footprint value than PE-LD.

13.3.2.4 Effect of Topographical Context and Energy Provision Resources

The comparison of ecological footprint for biodiesel production utilizing electricity mixes from various countries, with renewable energy mix for EU, using rape seed oil for biodiesel production (bio-RSO), and diesel production is shown in Figure 13.16. Diesel production possesses the highest footprint value, i.e. 715,469 m^2/t among all these processes, whereas biodiesel production using the renewable energy mix has the lowest footprint value of 310,771 m^2/t.

The variation in the footprint value for various countries is dependent on the energy mix configuration; the higher the share of renewable energy generation, the lower the footprint value for the product or service will be. The categorical comparison of footprint reveals that "area for fossil C" accounts for more than 75% of the overall footprint area

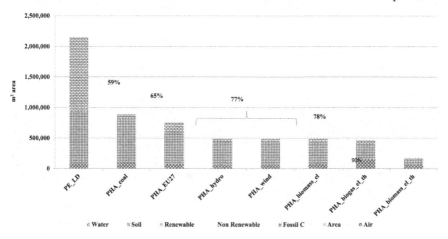

FIGURE 13.15 Comparison of PHA production utilizing different energy resources and Poly(ethylene) low density (PE-LD).

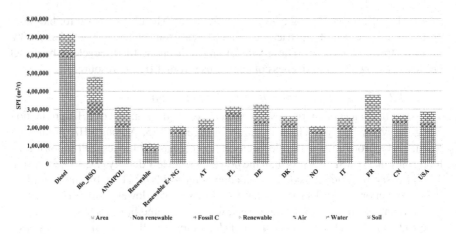

FIGURE 13.16 Footprint of biodiesel production using energy mix from different countries (modified from [50]).

excluding France, where emissions to water compartments are also very high. This high emission to water is representative of the higher share of power generation from nuclear technology which accounts for about 72% of the net electricity mix for France. Similarly, the highest input for "area for fossil carbon" is occupied in the case of the energy mix in Poland, which is representative of the higher share of electricity generation from coal incineration, amounting to almost 84% of the total energy production in Poland.

The conventional energy production technologies usually rely on fossil resources like crude oil, coal, and natural gas. Due to this reason, these technologies exert the largest pressure during operation by emitting enormous amounts of CO_2 into the atmosphere. It is a common perception that energy technologies based on renewable resources such as solar radiation, biomass, and wind power are environmentally friendly and can cope with the global warming problem. Figure 13.17 reveals the

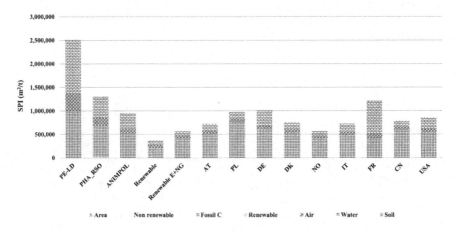

FIGURE 13.17 Evaluation of footprint for PHA production using energy mix from various countries.

footprint evaluation per t PHA production considering the energy mix from various countries in Europe with the EU mix, the renewable energy mix, rape seed oil as carbon source (PHA_RSO), and fossil-based equivalent low-density poly(ethylene) (PL-LD). PE-LD has the maximum footprint value of 2,508,409 m^2/t, while PHA produced using the renewable energy mix displays the lowest footprint value of 372,950 m^2/t.

Similarly, for Norway, PHA production has a significantly lower footprint of 567,392.86 m^2/t, which is almost equal to "renewable E+NG" 750,684.674 m^2/t (representing renewable electricity mix and natural gas for heating). This is because more than 91% of electricity generation takes place through hydropower production systems.

For PHA production emissions to different compartments, the emissions to the water category has increased significantly for both PE-LD and France as compared to biodiesel production results shown in Figure 13.6. PHA production from utilizing the EU energy mix shows a 62% lower footprint value compared to PE-LD, whereas PHA production using energy provision from renewable resources has 85 and 62% lower footprint value compared to PE-LD and EU energy mix scenarios, respectively. Likewise, the renewable E+NG scenario of PHA production exhibits a footprint value 70% lower than PE-LD and 21% lower than for the EU energy mix.

The life cycle CO_2 emissions or carbon footprint has also been calculated for diverse scenarios based on the subcategory "area for fossil C". The evaluation of CO_2 emissions per 1 t of PHA production is shown in Figure 13.18.

PE-LD has the maximum CO_2 emission of 7.4 t per t of PHA if compared to renewable energy resources with only 1.5 t CO_2 emissions per 1 t of PHA to be produced. Similarly, in the case of Poland, significantly higher CO_2 emissions (5.6 t per t PHA) are calculated if compared to other countries, particularly Norway, with only 2.97 t CO_2 emissions. Renewable E+NG exhibited high CO_2 emissions, which is due to natural gas consumption along with biomass burning and the usage of fossil fuel in the photovoltaic manufacturing plants. Germany, along with PR China, also has

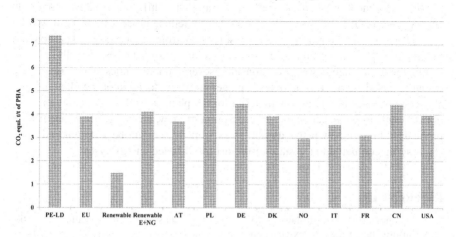

FIGURE 13.18 CO_2 Emissions for PHA production using energy mix from various countries.

higher CO_2 emissions ranging from 4.41 to 4.45 t, which is caused by the rather high share of coal-fired power generation systems in their energy mixes.

13.4   CONCLUSIONS AND OUTLOOK

The comprehensive financial investigation discloses the consequence of organic nitrogen provision from offal material as waste material (no market value) or considering it a primary resource for pet food with a defined market price along with MBM and biodiesel price fluctuations. The sensitivity analysis reveals that these parameters have a considerable economic effect on the overall process. The PHA manufacturing cost fluctuates between 1.65 and 1.88 €/kg assuming offal material as waste, while considering biodiesel 0.95 €/L and MBM prices of 300 €/t for the analysis. These prices considered for the analysis represent the current market prices for organic nitrogen provision source, biodiesel, and MBM, respectively. These values directly impact the annual cash flow and profit margin which eventually determines the investment breakeven time. The expected return of investment time fluctuates between 3.5 and 4 years.

An environmental footprint assessment of a renewable biopolymer shows that the environmental burden of a production process or merchandise is highly reliant on the accessible energy system. The environmental footprint of energy mixes has a ranking with Norway being at the top of the list, generating over 90% from hydropower, and France as the worst case because of generating the majority of its energy share from nuclear power plants. Traditional fossil-based power generation differs from alternative renewable resource-based power generation technologies by factors. It has been further elaborated by flow inspection of cradle to grave life cycle assessment of PHA production using electricity from the grid, PHA-RSO, PE-LD, and PHA production consuming energy from alternative renewable sources. The impact of energy provision from fossil-based business as usual energy systems is further exaggerated when this energy is utilized in extremely energy-intensive procedures, e.g. extraction of tallow and biodiesel production in this case study. As per the assessment, it would not be false to say that a shift to alternative renewable energy systems has the potential to dramatically reduce process impacts. It will be due to the shifting of environmental pressure away from the fossil-based carbon and making processes and products greener and more environmentally friendly.

On the other hand, we must consider several uncertainties about renewable resource-based energy technologies, e.g., geographical location of the plant, renewable resource availability as raw material for the plant as well as for energy production systems. From a chemical engineer's point of view, heat and energy integration in the designing of new plants as well as existing ones might be different and more challenging compared to the state-of-the-art fossil-based energy systems.

REFERENCES

1. Sandholzer D, Narodoslawsky M. SPIonExcel—Fast and easy calculation of the Sustainable Process Index via computer. *Resour Conserv Recy* 2005; 50(2): 130–142.
2. Narodoslawsky M, Krotschek C. The sustainable process index (SPI): Evaluating processes according to environmental compatibility. *J Haz Mat* 1995; 41(2–3): 383–397.

3. Kourmentza C, Plácido J, Venetsaneas N, Burniol-Figols A, Varrone C, Gavala HN, Reis MAM. Recent advances and challenges towards sustainable polyhydroxyalkanoate (PHA) production. *Bioengineering* 2017; 4(2): 55.
4. Koller M. Switching from petro-plastics to microbial polyhydroxyalkanoates (PHA): The biotechnological escape route of choice out of the plastic predicament? *The EuroBiotech J* 2019; 3(1): 32–44.
5. Koller M. Poly(hydroxyalkanoates) for food packaging: Application and attempts towards implementation. *Appl Food Biotechnol* 2014; 1(1): 3–15.
6. Zinn M, Witholt B, Egli T. Occurrence, synthesis and medical application of bacterial polyhydroxyalkanoate. *Adv Drug Deliver Rev* 2001; 53(1): 5–21.
7. Luef KP, Stelzer F, Wiesbrock F. Poly(hydroxyalkanoate)s in medical applications. *Chem Biochem Eng Q* 2015; 29(2): 287–297.
8. Khosravi-Darani K, Bucci DZ. Application of poly (hydroxyalkanoate) in food packaging: Improvements by nanotechnology. *Chem Biochem Eng Q* 2015; 29(2): 275–285.
9. Kettl KH, Titz M, Koller M, Shahzad K, Schnitzer H, Narodoslawsky M. Process design and evaluation of biobased polyhydroxyalkanoates (PHA) production. *Chem Eng Trans* 2011; 25: 983–988.
10. Riedl C. Stoffstromanalyse in Schlachthöfen. MSc Thesis. Graz University of Technology, Graz, Austria, 2003.
11. European Union. Regulation (EC) no 1774/2002 of the European Parliament and of the council of 2002 laying down health rules concerning animal by-products not intended for human consumption. *Off J Eur Commun* 2002; L273: 1–95.
12. Titz M, Kettl KH, Shahzad K, Koller M, Schnitzer H, Narodoslawsky M. Process optimization for efficient biomediated PHA production from animal-based waste streams. *Clean Technol Environ Pol* 2012; 14(3): 495–503.
13. Meher LC, Sagar DV, Naik SN. Technical aspects of biodiesel production by transesterification—A review. *Renew Sustain Energy Rev* 2006; 10(3): 248–268.
14. Shahzad K, Nizami AS, Sagir M, Rehan M, *et al.* Biodiesel production potential from fat fraction of municipal waste in Makkah. *PLoS One* 2017; 12(2): e0171297.
15. Koller M, Salerno A, Miranda de Sousa DM, Reiterer A, Braunegg G. Modern biotechnological polymer synthesis: A review. *Food Technol Biotechnol* 2010; 48(3): 255–269.
16. Jacobsen R, Buysse J, Gellynck X. Cost comparison between private and public collection of residual household waste: Multiple case studies in the Flemish region of Belgium. *Waste Manag* 2013; 33(1): 3–11.
17. Aremu AS. In-town tour optimization of conventional mode for municipal solid waste collection. *Niger J Technol* 2013; 32(3): 443–449.
18. OECD. 2002. Glossary of statistical terms [Online]. Available at: stats.oecd.org/glossary/detail.asp?ID=4097 (last accessed 08 Nov 2016).
19. Kromus S. Die Grüne Bioraffinerie Österreich – Entwicklung eines integrierten Systems zur Nutzung von Grünlandbiomasse. PhD Thesis. Graz University of Technology, Graz, 2002.
20. Shahzad K, Narodoslawsky M, Sagir M, Ali N, *et al.* Techno-economic feasibility of waste biorefinery: Using slaughtering waste streams as starting material for biopolyester production. *Waste Manag* 2017; 67: 73–85.
21. Pereira CG, Rosa PT, Meireles MAA. Extraction and isolation of indole alkaloids from Tabernaemontana catharinensis A. DC: Technical and economical analysis. *J Supercrit Fluids* 2007; 40(2): 232–238.
22. Online resource 1: MESPS steel price outlook. Available at: http://www.meps.co.uk/EU%20price.htm (last accessed 07 Nov 2016).
23. Inyang UE, Basey EN, Inyan JD. Characterization of brewery effluent fluid. *J Eng Appl Sci* 2012; 4: 67–77.

24. Von Sperling M, Chernicharo CAL. Biological wastewater treatment in warm climate regions, Freely downloadable at IWA Publishing, 2005, p. 1496. Available at: http://www.iwapublishing.com/open-access-ebooks/3567.
25. Henze M, Harremoës P, la Cour Jansen J, Arvin E. *Wastewater Treatment: Biological and Chemical Processes.* Springer-Verlag, Berlin, 2002.
26. Van Haandel A, Van der Lubbe J. *Handbook Wastewater Treatment Design and Optimization of Activated Sludge Systems, Model of Biological Phosphorus Removal 5.1.3,* 2007, pp. 197–198. IWA publishing, London.
27. Online resource 2: BIOGAS Netzspeisung, 2013. Modelanlage [online]. Available at: http://www.biogas-netzeinspeisung.at/wirtschaftlicheplanung/modellanlage.html (last accessed 07 Nov 2016).
28. Langdom D, Cost Model Wilkes M. Small industrial units. *Building.co.uk* 2006; 46: 64–68.
29. Personal communication Ulrike Reistenhofer GmbH. Available at: www.firmeninfo.at/firma/ulrike-reistenhofer-gesellschaft-mbh/1343994.
30. Online resource:3. Available at: www.mybiosource.com (last accessed 07 Nov 2016).
31. Narodoslawsky M, Kromus S, Wachter B, Koschuh W, Mandl M, Krotscheck C. The Green Biorefinery Austria - Development of an integrated system for green biomass utilization. *Chem Biochem Eng Q* 2004; 18(1): 7–12.
32. Dagevos H, Voordouw J. Sustainability and meat consumption: Is reduction realistic? *Sustain Sci* 2013; 9(2): 60–69.
33. Delgado CL. Rising consumption of meat and milk in developing countries has created a new food revolution. *J Nutr* 2003; 133(11): 3907S–3910S.
34. Deydier E, Guilet R, Sarda S, Sharrock P. Physical and chemical characterisation of crude meat and bone meal combustion residue: "waste or raw material?" *J Hazard Mater* 2005; 121(1): 141–148.
35. Keiner M. (ed.) *The Future of Sustainability.* Springer, Dordrecht, 2006. ISBN 978-1-4020-4734-3.
36. Wright L, Kemp S, Williams I. Williams I. 'carbon footprinting': Towards a universally accepted definition. *Carbon Manag* 2011; 2(1): 61–72.
37. Hoekstra AY, Chapagain AK *Globalization of Water: Sharing the Planet's Freshwater Resources.* Blackwell Publishing, Oxford, UK, 2008. doi:10.1002/9780470696224.ch2.
38. Wall G. Exergy flows in industrial processes. *Energy* 1988; 13(2): 197–208.
39. Odum HT. *Environmental Accounting: EMERGY and Environmental Decision Making.* John Wiley & Sons, Toronto, 1996.
40. Wackernagel M, Rees WE. The ecological footprint and appropriated carrying capacity: A tool for planning toward sustainability. Unpublished PhD Thesis. University of British Columbia School of Community and Regional Planning, Vancouver, 1994.
41. Narodoslawsky M, Krotscheck C, Sage J. The sustainable process index (SPI) – A measure for process industries. In: *Proceedings AFCET, International Symposium: Models of Sustainable Development. Exclusive or Complementary Approaches of Sustainability? Volume II,* March 16–18, Paris, 1994.
42. Prescott-Allen R. *The Wellbeing of Nations. A Country-By Country Index of Quality of Life and the Environment.* Island Press, Washington. Center for International Earth Science Information Network, 2001.
43. Center for International Earth Science Information Network. *Environmental Sustainability Index (ESI),* 2002. Available at: http://sedac.ciesin.columbia.edu/es/esi/ESI2002_21MAR02tot.pdf. (last accessed 15 Jun 2014).
44. Schmidt-Bleek F. *Wieviel Umwelt braucht der Mensch – Mips, das Ökologische Mass zum Wirtschaften.* Birkhäuser, Basel, Boston, Berlin, 1993; appeared in Japanese (4th edition, Pringer Tokyo), Chinese and Finnish. English version in www.factor10-institute.org under the title "The Fossil Makers.

45. ISO 2006. Environmental management - life cycle assessment - principles and framework, ISO 14040ISO 14067/TS.
46. Schnitzer H, Brunner C, Gwehenberger G. Minimizing greenhouse gas emissions through the application of solar thermal energy in industrial processes. *J Cleaner Pro* 2007; 15: 1271–1286.
47. Gwehenberger G, Narodoslawsky M, Liebmann B, Friedl A. Ecology of scale versus economy of scale for bioethanol production. *Biofuel Bioprod Bior* 2007; 1(4): 264–269.
48. Harding KG, Dennis JS, Blottnitz H, Harrison STL. Environmental analysis of plastic production processes: Comparing petroleum-based polypropylene and polyethylene with biologically based poly-β-hydroxybutyric acid using life cycle analysis. *J Biotechnol* 2007; 130(1): 57–66.
49. Kettl K-H, Shahzad K, Narodoslawsky M. Ecological footprint comparison of biobased PHA production from animal residues. *Chem Eng Trans* 2012; 29: 439–444.
50. Shahzad K, Kettl K-H, Titz M, Koller M, Schnitzer H, Narodoslawsky M. Comparison of the ecological footprint for biobased PHA production from animal residues utilizing different energy resources. *Clean Techn Environ Policy* 2013; 15(3): 525–536.
51. International Energy Agency (IEA). World energy outlook, 2009. Organization for Economic Cooperation and Development. Available at: www.iea.org/stats/index.asp (last accessed 15 Nov 2012.
52. Eurostat. Panorama of energy: Energy statistics to support EU policies and solutions, 2009. ISBN 978-92-79-11151-8, ISSN 1831-3256, doi:10.2785/26846, Cat. No. KS-GH-09-001-EN-C.

Index

Printed in the United States
by Baker & Taylor Publisher Services